Machine Intelligence for Materials Science

Series Editors

N. M. Anoop Krishnan⑩, Department of Civil Engineering, Indian Institute of Technology Delhi, New Delhi, India

Miao Liu, Songshan Lake Materials Laboratory, Dongguan, Guangdong, China

This book series is dedicated to showcasing the latest research and developments at the intersection of materials science and engineering, computational intelligence, and data sciences. The series covers a wide range of topics that explore the application of artificial intelligence (AI), machine learning (ML), deep learning (DL), reinforcement learning (RL), and data science approaches to solve complex problems across the materials research domain.

Topical areas covered in the series include but are not limited to:

- AI and ML for accelerated materials discovery, design, and optimization
- Materials informatics
- Materials genomics
- Data-driven multi-scale materials modeling and simulation
- Physics-informed machine learning for materials
- High-throughput materials synthesis and characterization
- Cognitive computing for materials research

The series also welcomes manuscript submissions exploring the application of AI, ML, and data science techniques to following areas:

- Materials processing optimization
- Materials degradation and failure
- Additive manufacturing and 3D printing
- Image analysis and signal processing

Each book in the series is written by experts in the field and provides a valuable resource for understanding the current state of the field and the direction in which it is headed. Books in this series are aimed at researchers, engineers, and academics in the field of materials science and engineering, as well as anyone interested in the impact of AI on the field.

Michael te Vrugt

Editor

Artificial Intelligence and Intelligent Matter

Nanoscience, Soft Matter, Philosophy

 Springer

Editor
Michael te Vrugt
Johannes Gutenberg University Mainz
Mainz, Germany

ISSN 2948-1813   ISSN 2948-1821 (electronic)
Machine Intelligence for Materials Science
ISBN 978-3-032-04128-9   ISBN 978-3-032-04129-6 (eBook)
https://doi.org/10.1007/978-3-032-04129-6

This work was supported by SFB1551 (CRC 1551 "Polymer concepts in cellular function"), SFB 1552 (CRC 1552 "Defects and Defect Engineering in Soft Matter") and Eberhard Karls Universität Tübingen.

This Springer imprint is published by the registered company Springer Nature Switzerland AG
The registered company address is: Gewerbestrasse 11, 6330 Cham, Switzerland

If disposing of this product, please recycle the paper.

Preface

Certainly, here is a possible introduction for your book:[1]

The nucleation point for this book was a workshop entitled *Artificial Intelligence and Intelligent Matter* that took place at the University of Münster in July 2023. Eight months before, in November 2022, the release of the chatbot ChatGPT by OpenAI had demonstrated to the wide public the enormous potential of artificial intelligence (AI) and the things that it is nowadays capable of. ChatGPT and similar software released by other companies had in particular led to many discussions at universities about the impact AI will have on university research and teaching. Will now all research papers be written by AI? And, God forbid, will now all student homework exercises be written by AI?

Owing to these discussions, the Center for Philosophy of Science (Zentrum für Wissenschaftstheorie, ZfW) in Münster initiated a workshop on this topic. This workshop featured, in addition to an interdisciplinary collection of research talks, a plenary discussion on the impact of large language models on teaching. Münster was at this time (and still is today) home to two further AI-related research initiatives, which then also participated in the workshop organization. These are on the one hand InterKIWWU, which is a program focused on AI teaching at universities, and the collaborative research center (Sonderforschungsbereich, SFB) 1459 "Intelligent Matter", which aims at the design of novel class of matter, so-called "intelligent materials". These are capable of processing external inputs and of adapting to their environment. They do, thereby, constitute a physical implementation of some operating principles of artificial intelligence.

The book then further developed its shape in the context of several coordinated research activities at the Johannes Gutenberg University in Mainz. First, there is here the SFB 1551 "Polymer concept in cellular function", which uses polymer science to understand the inner workings of a cell and to model things such as the formation

---

[1] A phrase of this form was found in the introduction of a peer-reviewed article published in the journal *Surfaces and Interfaces* published in March 2024 (and later retracted), which showed the article's text to be partially the result of a prompt and which thereby prompted a lively discussion about the use of AI in research.

of membraneless cell compartments and neurodegenerative diseases. It is thereby operating in a field that has been massively reshaped by AI tools such as AlphaFold (that allows to study protein folding), and such approaches are of key importance for the SFB. Then, there is the SFB 1552 "Defects and defect engineering in soft matter" that aims to make the idea of engineering material properties by using defects—an approach that has for a long time been successfully employed in hard condensed matter systems in the context of, e.g., semiconductors—fruitful also in soft matter. This links to the material science aspect of the intelligent matter problem, which requires one to develop materials that can adapt their state. Finally, there is the Mainz Institute of Multiscale Modelling (M^3ODEL), a top-level research initiative funded by the state of Rhineland-Palatinate that focuses on computational and datadriven modeling of complex systems connecting multiple scales. Also this is an area where much progress has been made using AI tools. The mere fact that so many collaborative research centers have been listed at this point—and that more working on similar topics could be easily added or will probably come to life in the future—already shows the enormous importance.

Also, the above "origin story" shows that this publication, probably unlike many others that basic research produces, has originally arisen not so much from purely curiosity-driven work, but from a specific societal need, namely that of addressing how large language models might reshape universities. Besides that, another and certainly much more severe societal need is to be mentioned at this point: Human technological activities require an enormous use of natural resources and directly damage the natural environment, leading (most prominently) to climate change. AI and similar technologies are far from innocent here due to the enormous amount of energy that their training requires. A key motivation behind unconventional computing approaches is to develop more energy-efficient approaches by bypassing restrictions imposed by the operating principles ("von Neumann architecture") of traditional computers.

And finally: Besides addressing societal needs, there is of course also a lot to be gained here for curiosity-driven basic research. The design of intelligent materials posesmany theoretical and experimental challenges for physics, chemistry, computer science, and engineering alike, and studying physical implementations of artificial intelligence will likely also lead to novel insights into natural intelligence (the kind of thing that we have). A good example for this is reservoir computing (a core topic of this volume), which is used to perform learning tasks with magnetic or optical systems, but which has also been suggested as an operating principle in biological systems. Another example is research on active matter systems, which are permanently kept out of thermodynamic equilibrium by local energy influxes. Biological systems generally fall into this category, but also many candidates for intelligent matter do. Research on active matter is therefore of practical relevance, but it has also raised many fundamental issues for statistical physics.

During the course of its development, this book rather quickly evolved away from the workshop it was originally based on (which had as a core topic also philosophical discussions of the notion of "intelligence" used in AI) and towards a volume providing a general overview of scientific research in the field. There is, due to the importance of this field, a clear need for such a book: On the one hand it is very active and rapidly growing, such that many young (and older) researchers start working on it and require an introduction. On the other hand, it has reached at least such a level of maturity that it is possible to identify some key research directions and present them on an introductory level, which is what books are especially suitable for. Researchers at any career stage seeking to get an overview over the field are the target audience of this volume.

The book covers the topic in a cross-sectional way to do justice to the crosssectional nature of this field of research. First, the study of intelligent matter involves studies of a quite diverse range of physical systems, ranging from electronics, spintronics, and photonics to soft and biological matter. A (certainly non-ideal) term sometimes used to unify all this is "nanosystem". Second, the interaction of AI with such physical systems can occur in two directions: The AI can be applied to the physical system (as one does, e.g., when developing a machine-learned density functional theory model for a fluid) or the physical system can be applied to the AI (as one does, e.g., when magnetic skyrmions are used for reservoir computing). In particular the latter distinction motivates the structure of this book, which has four parts:

1. Fundamental Concepts in Artificial Intelligence
2. Applications of Artificial Intelligence to Nanosystems
3. Implementations of Artificial Intelligence in Nanosystems
4. Philosophical Aspects

It is so far not clear which kind of nanosystem is most suitable for intelligent matter, and in practice it will probably turn out that different systems have different advantages and disadvantages in this regard. Therefore, this book does not focus on a specific kind of systems (such as lasers or soft matter), but gives broader overview.

Mainz, Germany   Michael te Vrugt

Contents

Introduction: Artificial Intelligence and Intelligent Matter

Michael te Vrugt

Abstract The investigation and application of artificial intelligence (AI) is a major research trend in virtually every field of science. The study of micro- and nanosystems, both in hard and soft condensed matter science, is no exception here. This involves both the use of artificial intelligence to understand static and dynamic properties of materials and the development of nanosystems that themselves possess basic features of (articial) intelligence in the framework of "intelligent matter". This book provides an overview over this rapidly developing field. It starts with introductory chapters on AI (part I) and continues with discussions of applications of AI *to* nanosystems (part II) and implementations of AI *in* nanosystems (part III). It concludes with some philosophical discussions (part IV).

The enormous progress in the field of artificial intelligence (AI) [28] and the importance it has gained for our everyday life has certainly not escaped anyone's attention. Similarly, it will not have escaped the attention of any researcher, be it in the natural sciences or the humanities, how central AI now is in all fields of research. The physical sciences are no exception here [6, 7, 19, 32]. A good example is the "derivation" of (free) energy functionals, which are crucial for understanding the structure of materials, via AI, which is possible in both the quantum [26] and the classical [29] case. It can surely be expected that AI will continue to make significant contributions in solving research problems. (That being said, one should of course avoid overhyping this field–AI is certainly not going to solve *all* of our scientific problems, and traditional analytical or computational approaches to physical problems will continue to remain valuable.)

At present, AI systems usually run on traditional computers. This is problematic insofar as AI applications generally require processing very large amounts of data that need to be moved between memory and processor, which requires significant energy consumption [17] and thus makes current approaches somewhat unsustainable. The human brain, in contrast, can make substantial achievements with a power consumption of just about 20 W [2]. The operating principles of the human brain are

M. te Vrugt (✉)
Institut für Physik, Johannes Gutenberg-Universität Mainz, Mainz, Germany
e-mail: tevrugtm@uni-mainz.de

© The Author(s) 2026   1
M. te Vrugt (ed.), *Artificial Intelligence and Intelligent Matter*, Machine Intelligence
for Materials Science, https://doi.org/10.1007/978-3-032-04129-6_1

quite diffent from those of a computer, and consequently, a major direction of current research is the development of hardware approaches that are closer to the way the brain operates [11]. See Refs. [10, 30] for broad reviews of computing with physical systems. One of the forerunners of this approach is the work by Hopfield [12].

As a part of this development, condensed matter system become not just the modeling target for AI systems, but also the hardware on which AI runs. One prominent manifestation of this approach is the idea of reservoir computing [24], where an input signal is fed into a large dynamical system, the output of which is then passed on to a readout layer. This readout layer is the only part that is trained. The dynamical system can also be a physical system, and consequently the idea of reservoir computing has been frequently used to perform computations with physical systems. This idea will be discussed in many chapters of this book. Another instance of AI being implemented in a physical system rather than just being applied to it are "smart" microswimmers that are able to navigate in noisy environments [25].

The eventual aim here is to realize materials capable of information processing, which thereby constitute "intelligent matter". To understand what this is supposed to mean, it is helpful to consider a classification suggested by Kaspar et al. [17]:

- **Structural matter**: Here, the material has a certain structure (maybe even a fairly complex one), but cannot change its properties.
- **Responsive matter**: The material can change its properties in response to an external input.
- **Adaptive matter**: The material can regulate its properties in response to its environment using internal feedback.
- **Intelligent matter**: The material can regulate its behavior and learn from received inputs.

These differences may require some further explanation: A responsive material always reacts to the same input in the same way. In contrast, an adaptive material may behave differently if a certain stimulus arrives once or a hundred times in a row– in the latter case, it has *adapted* to the fact that it is in an environment where this stimulus occurs frequently. Physically, this is manifest in the fact that a responsive system switches between different local minima in a fixed energy landscape when a stimulus is applied, whereas in an adaptive material the underlying energy landscape can be changed [31]. Fully intelligent matter requires four key functional elements [17]:

1. A **sensor** that records information about the environment, such as the presence of a stimulus.
2. An **actuator** by which the material changes its properties in reaction to a stimulus.
3. A **memory** by which the material stores information (e.g., about input or feedback signals).
4. A **network** connecting the first three elements.

For a deeper discussion of these classifications see the reviews in Refs. [17, 22, 31]. It is often argued [17, 31] that the systems best suited for designing adaptive materials are those consisting of "active matter" [3, 21], i.e., matter that is driven out of equilibrium by a constant local influx and dissipation of energy. Active matter has indeed been found to be quite promising for realizing programmable and adaptive materials [4, 5, 13], and will be one of the central topics of this book. An interesting direction for future work in this context is the application of defect engineering [15], which has already found several successful applications in computing with soft matter [18, 33]. Moreover, the statistical physics of active matter also allows for a theoretical understanding of the learning process itself [16, 20].

Finally, artificial intelligence has also been of great interest among philosophers [8], raising a significant number of conceptual and ethical challenges. What does it mean for a system to be intelligent, and can we attribute intelligence to an AI? How can we ensure that AI systems make ethical decisions, and what do they consist in (e.g., who should a self-driving car kill if in doubt)? As a consequence of such conceptual problems, explicit appeals to philosophy of science can be found also in the physical literature [19], something that is generally characteristic for a field that is in a state of paradigm shift [9]. Significant ethical issues also arise in the field of nanoscience [1]. The overlap between AI and nanoscience, which can consequently be expected to be of quite some interest, has received very little coverage in philosophy so far. This is a gap that the present book aims to fill.

The overall topic of this volume is the use of AI in micro- and nanoscience. Here, both "use" and "micro- and nanoscience" are understood rather broadly. "Use" means both the application of "traditional" AI to understand nanosystems and the implementation of AI in nanosystems that thereby themselves become intelligent. With the term "micro- and nanoscience", I aim to generally cover systems with small (micro- or nanometer) constituents. This is a broad field, ranging from artificial microswimmers whose size is in the micrometer range [3] to magnetic skyrmions which are only a few nanometers large [23]. It turns out that the physics of micro- and nanometer-sized objects is often quite similar, and that a lot is to be gained by studying them in conjunction [27].

The rather broad scope of this book has two reasons. First, it aims to give an overview over the field, which can of course only be done selectively, but which nevertheless should give an idea of its diversity. As of today, we do not know what the best approach for realizing intelligent matter is, and it is likely that this will depend on the context of application—consequently, it is important to be aware of different directions of work. This book can therefore also be used as teaching material for graduate students. (See Ref. [14] for teaching material on AI in general.)

Second, the book aims at highlighting possible links between different fields. For example, the idea of reservoir computing is now studied in quite a number of different fields. This books presents (in different chapters) reservoir computing approaches in quantum systems, active matter, DNA, photonics, and spintronics, and can therefore help researchers from one of these areas to also get an impression of what is going on in others (and researchers who are not from any of these areas to get an impression of which areas exist here).

Structure of this Book

This book consists of four parts:

1. **Introduction to artificial intelligence**: This part contains introductory chapters presenting some concepts from the field of AI that are relevant to the content of this book. This starts with a general introduction to AI (Tobias Wand), followed by discussions of reinforcement learning (Malte Schilling) and reservoir computing (Michael te Vrugt), which are AI concepts with a specific relevance for intelligent matter.
2. **Applications of artificial intelligence to nanosystems**: This part presents various ways in which AI can be used for the physical investigation of nanosystems. It start with a general introduction to data-driven equation learning methods (Oliver Kamps/Tim W. Kroll/Oliver Mai), followed by a discussion of two important cases where equations—namely density functionals–are obtained via machine learning in the classical (Alessandro Simon/Martin Oettel) and quantum (Thorsten Deilmann) case. Related to the latter is the AI-based design of materials (Teng Long/Yixuan Zhang/Hongbin Zhang). The final three chapters present applications to specific physical systems, namely topological lasers (Stephan Wong/Doris E. Reiter/Sang Soon Oh), semiconductors (Md Ashiqur Rahman Laskar/Srijan Chakrabarti/Umberto Celano), and active matter (Giovanni Volpe).
3. **Implementations of artificial intelligence in nanosystems**: This part addresses the case that the nanosystems themselves constitute the AI by discussing implementations of AI *in* nanosystems. This part is opened with a general introduction to neuromorphic hardware (Akhilesh Jaiswal/Md Abdullah-Al-Kaiser/Maryam Parsa). The later chapters again cover the cases of both soft and hard or quantum matter. On the soft side, active and biological matter is in the focus, specifically intelligent active particles (Hartmut Löwen/Benno Liebchen), active-matter based computing (Julian Jeggle/Raphael Wittkowski), DNA-based neural networks (Michael te Vrugt), and intelligence in slime molds (Jannes Freiberg/Roshani Madurawala). The next two chapter address computing based on optics from both an experimental (Lennart Meyer/Rongyang Xu/Wolfram Pernice) and a theoretical (Kathy Lüdge/Lina Jaurigue) perspective. Finally, approaches based on solid-state and quantum systems, specifically neuromorphic spintronics (Atreya Majumdar/Karin Everschor-Sitte) and quantum machine learning (Ivana Nikoloska) are presented.
4. **Philosophical aspects**: Intelligent matter research raises also a number of conceptual and ethical issues, which are the topic of the final part of this book. This starts with a philosophical analysis of the notion of "intelligent matter" (Barnaby Crook), followed by a discussion of its implications for debates about consciousness and intentions (Christian Kaernbach) and a contribution addressing applications in biology (Luis Lopez). The final chapter discusses ethical problems at the intersection of AI and nanoscience (John Weckert).

Acknowledgements A first big thank you goes to all authors who contributed chapters to this book–many thanks for your effort in writing! The nucleation point for this book was a workshop on "Artificial Intelligence and Intelligent Matter" that took place at the University of Münster in July 2023 and was jointly hosted by the Zentrum für Wissenschaftstheorie (Center for Philosophy of Science), the SFB 1459 "Intelligent Matter" (funded by the Deutsche Forschungsgemeinschaft), and the teaching program "InterKI WWU" (funded by the Bundesministerium für Bildung und Forschung, Germany). I am very grateful everyone who was involved in organizing and funding this workshop. Moreover, I would like to thank the Springer Nature book publishing team for their assistance in preparing this volume. Finally, I am very grateful to the anonymous reviewers who spent their time looking at the chapters of this book and making valuable suggestions for improving them.
The open access publication was possible due to funding by the Deutsche Forschungsgemeinschaft (DFG, German Research Foundation) in the framework of SFB 1551; Project No. 464588647 and SFB 1552; Project No. 465145163 and by the Mainz Institute of Multiscale Modeling, M^3ODEL.

References

1. Allhoff F, Lin P, Moor JH, Weckert J (eds) (2007) Nanoethics: the ethical and social implications of nanotechnology. John Wiley & Sons, Hoboken
2. Balasubramanian V (2021) Brain power. Proc Natl Acad Sci USA 118(32):e2107022118
3. Bechinger C, Leonardo R, Löwen H, Reichhardt C, Volpe G, Volpe G (2016) Active particles in complex and crowded environments. Rev Mod Phys 88(4):045006
4. Bickmann J, Bröker S, te Vrugt M, Wittkowski R (2023) Active Brownian particles in external force fields: field-theoretical models, generalized barometric law, and programmable density patterns. Phys Rev E 108(4):044601
5. Bröker S, Bickmann J, te Vrugt M, Cates ME, Wittkowski R (2023) Orientation-dependent propulsion of active Brownian spheres: from self-advection to programmable cluster shapes. Phys Rev Lett 131(16):168203
6. Carleo G, Cirac I, Cranmer K, Daudet L, Schuld M, Tishby N, Vogt-Maranto L, Zdeborová L (2019) Machine learning and the physical sciences. Rev Mod Phys 91(4):045002
7. Cichos F, Gustavsson K, Mehlig B, Volpe G (2020) Machine learning for active matter. Nat Mach Intell 2(2):94–103
8. Copeland J (1993) Artificial intelligence: a philosophical introduction. Blackwell, Oxford
9. De Haro S (2020) Science and philosophy: a love-hate relationship. Found Sci 25(2):297–314
10. Finocchio G, Incorvia JAC, Friedman JS, Yang Q, Giordano A, Grollier J, Yang H, Ciubotaru F, Chumak AV, Naeemi AJ, Cotofana SD, Tomasello R, Panagopoulos C, Carpentieri M, Lin P, Pan G, Yang JJ, Todri-Sanial A, Boschetto G, Makasheva K, Sangwan VK, Trivedi AR, Hersam MC, Camsari KY, McMahon PL, Datta S, Koiller B, Aguilar GH, Temporao GP, Rodrigues DR, Sunada S, Everschor-Sitte K, Tatsumura K, Goto H, Puliafito V, Åkerman J, Takesu H, Ventra M, Pershin YV, Mukhopadhyay S, Roy K, Wang IT, Kang W, Zhu Y, Kaushik BK, Hasler J, Ganguly S, Ghosh AW, Levy W, Roychowdhury V, Bandyopadhyay S (2024) Roadmap for unconventional computing with nanotechnology. Nano Futur 8:012001
11. Grollier J, Querlioz D, Camsari KY, Everschor-Sitte K, Fukami S, Stiles MD (2020) Neuromorphic spintronics. Nat Electron 3(7):360–370
12. Hopfield JJ (1982) Neural networks and physical systems with emergent collective computational abilities. Proc Natl Acad Sci USA 79(8):2554–2558
13. Huang ZF, te Vrugt M, Mayer Martins J, Wittkowski R, Löwen H (2024) Active pattern formation emergent from single-species nonreciprocity. arXiv:2305.16131
14. Huwer J, Becker-Genschow S, Thyssen C, Thoms LJ, Finger A, von Kotzebue L, Kremser E, Meier M, Bruckermann T (eds) (2024) Kompetenzen für den Unterricht mit und über Künstliche

Intelligenz. Perspektiven, Orientierungshilfen und Praxisbeispiele für die Lehramtsausbildung in den Naturwissenschaften. Waxmann, Münster

15. Jangizehi A, Schmid F, Besenius P, Kremer K, Seiffert S (2020) Defects and defect engineering in soft matter. Soft Matter 16(48):10809–10859

16. Jung G, Ozawa M, Bertin E (2025) Kinetic theory of decentralized learning for smart active matter. Phys Rev Lett 134(24):248302

17. Kaspar C, Ravoo BJ, van der Wiel WG, Wegner SV, Pernice WHP (2021) The rise of intelligent matter. Nature 594(7863):345–355

18. Kos Ž, Dunkel J (2022) Nematic bits and universal logic gates. Sci Adv 8(33):eabp8371

19. Krenn M, Pollice R, Guo SY, Aldeghi M, Cervera-Lierta A, Friederich P, Passos GG, Häse F, Jinich A, Nigam A, Yao Z, Aspuru-Guzik A (2022) On scientific understanding with artificial intelligence. Nat Rev Phys 4(12):761–769

20. Mandal R, Huang R, Fruchart M, Moerman PG, Vaikuntanathan S, Murugan A, Vitelli V (2024) Learning dynamical behaviors in physical systems. arXiv:2406.07856

21. Marchetti MC, Joanny JF, Ramaswamy S, Liverpool TB, Prost J, Rao M, Simha RA (2013) Hydrodynamics of soft active matter. Rev Mod Phys 85(3):1143–1189

22. Merindol R, Walther A (2017) Materials learning from life: concepts for active, adaptive and autonomous molecular systems. Chem Soc Rev 46(18):5588–5619

23. Nagaosa N, Tokura Y (2013) Topological properties and dynamics of magnetic skyrmions. Nat Nanotechnol 8(12):899–911

24. Nakajima K, Fischer I (eds) (2021) Reservoir computing. Springer, Singapore

25. Nasiri M, Löwen H, Liebchen B (2023) Optimal active particle navigation meets machine learning. EPL 142(1):17001

26. Pederson R, Kalita B, Burke K (2022) Machine learning and density functional theory. Nat Rev Phys 4(6):357–358

27. Poon W (2004) Colloids as big atoms. Science 304(5672):830–831

28. Russell SJ, Norvig P (2022) Artificial intelligence: a modern approach, 4th edn. Pearson, Harlow

29. Simon A, Weimar J, Martius G, Oettel M (2024) Machine learning of a density functional for anisotropic patchy particles. J Chem Theory Comput 20(3):1062–1077

30. Stern M, Murugan A (2023) Learning without neurons in physical systems. Annu Rev Condens Matter Phys 14:417–441

31. Walther A (2020) From responsive to adaptive and interactive materials and materials systems: a roadmap. Adv Mater 32(20):1905111

32. Zhang K, Gong X, Jiang Y (2024) Machine learning in soft matter: from simulations to experiments. Adv Funct Mater 34(24):2315177

33. Zhang R, Mozaffari A, de Pablo JJ (2022) Logic operations with active topological defects. Sci Adv 8(8):eabg9060

Fundamental Concepts in Artificial Intelligence

Introduction to Artificial Intelligence

Tobias Wand

Abstract Artificial Intelligence has recently achieved big successes in various fields of application such as natural language processing or image recognition. Despite having enormous amounts of parameters, the basic concepts behind those algorithms are fairly simple to understand. This chapter starts with an introduction to the machine learning methodology and its terminology as well as some general metrics to evaluate the quality of an algorithm. It then discusses decision and regression trees as well as their extensions via bagging and boosting as an example to illustrate basic machine learning concepts. The chapter concludes with an introduction to neural networks and explains their basic structure and how to train them.

1 Artificial Intelligence and the Machine Learning Methodology

People tend to hold wildly different associations when hearing "machine learning" (ML) and the grandeur of the word "artificial *intelligence*" (AI) makes things even more complicated. How to exactly define artificial intelligence is by no means an easy question and even controversial among experts [18]. Instead, it might make sense to focus on the major milestone achievements and surmounted challenges. In 1966, the first major natural language processing algorithm ELIZA was used to mimic basic human conversations [32]. ELIZA searched the input text for keywords and then responded accordingly via predefined rules. Just a few years later, the similarly logic-based algorithm SHRDLU managed to combine natural language with interactions with a (virtual) world; i.e. SHRDLU had access to an environment of various geometrical objects and connected its natural language output to the position of these objects in the environment. Thereby, it built a bridge between language comprehension and interactions with an environment, instead of being a self-contained language model [33]. In 1997, IBM's Deep Blue defeated world champion Garri

T. Wand (✉)
University of Münster, Institute of Theoretical Physics, Münster, Germany
e-mail: t_wand01@uni-muenster.de

© The Author(s) 2026

M. te Vrugt (ed.), *Artificial Intelligence and Intelligent Matter*, Machine Intelligence for Materials Science, https://doi.org/10.1007/978-3-032-04129-6_2

Kasparov in a match of chess [12]. Deep Blue used a combination of a tree search of possible moves on the chessboard and explicitly programmed expert knowledge to achieve this feat.

However, when we think of the word "intelligence", we have something different in mind than simply learning rules or optimal strategies by heart and then blindly following them like in the famous Chinese room thought experiment, where a person essentially uses a highly detailed manual-like dictionary to translate Chinese sentences into English without having any real understanding of the Chinese language. Intelligence, however, evokes the idea that some key insights have been figured out that go beyond the mere observations that the intelligent being has made such that general rules have been derived by it. And this is actually quite close to what is happening under the hood of modern AI algorithms under the label of *machine learning*: while they need a lot of data to learn, they generalise far beyond what has been presented to them in the data. A famous example is the game Go whose complexity and enormous amounts of different board states made it a tough challenge for any computer algorithm because simulating and evaluating each possible future board state is computationally infeasible [22]. Nevertheless, Google's AlphaGo was the first machine to win against human Go champions [17, 27]. Learning from data instead of learning predefined rules also allowed language models like the GPT-family to achieve almost human-like performance when confronted with completely new sentences and dialogues [3]. The key to understand how these algorithms managed to excel at such highly complex tasks with almost uncountably many different possible board states or word combinations is to regard machine learning as a methodology rather than as a collection of methods.

A rough sketch of the machine learning methodology is given in Fig. 1. Two key aspects characterise this methodology: first, there is not one data set that is used for the model training and evaluation, but two mutually exclusive sets of data. Second,

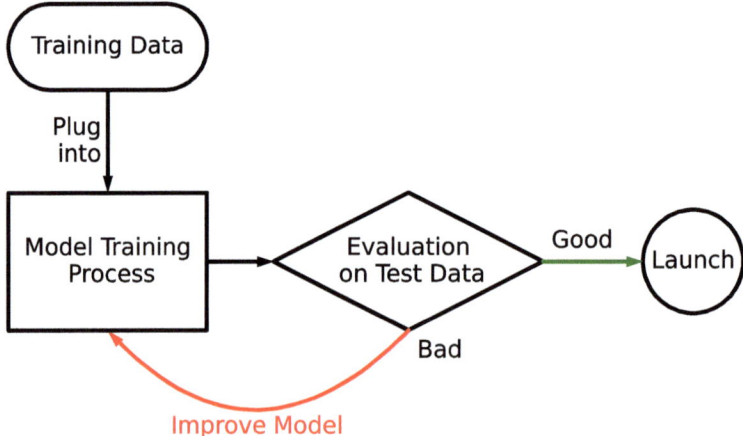

Fig. 1 Visualisation of the machine learning methodology as a flowchart sketched via [5]

there is a loop of evaluating the model and improving it until it is good enough to be launched. Similar to falsifiability being the cornerstone of the scientific method, the evaluation on unknown test data is the centre of the machine learning methodology. But let us go through this workflow step by step.

Usually, the data is first split into training and test data, so that no information of the test data can be used to inform the modelling decisions. Then, after some exploratory data analysis on the training data, the data scientist develops a model that might be a suitable choice for the given problem. The model is trained on the training data, i.e. its parameters are optimised so that its predictions are as close to the data as possible. Afterwards, the model is for the very first time confronted with the test data and the scientist checks, if the model still performs well enough on this previously unknown data. If it does not, then the model has to be improved: this can either happen automatically (i.e. some observations are given more weight than others or some parameters are modified) or manually, if e.g. the scientist realises that the model is not complex enough and needs some additional terms. Often, the scientist has no access to the test data, but only gets to see how well the model managed to predict it. The steps of evaluation and improvement are repeated until the model finally achieves a satisfying level of accuracy on the unknown data.

This methodology is oversimplified, but it already shows us how algorithms like AlphaGo managed to "learn" a highly complex game like Go: Specifically because they are tested on unknown data, these models cannot just stick to what they have seen in the training data, but have to find general patterns that still hold for completely new data. Note that there is another reason why it is prudent to use unknown test data. Otherwise, the model is prone to *overfitting* which means that it almost exactly reproduces the known data, but its accuracy is too good to be true and it fails spectacularly on new data. We will focus on this in Sect. 4.1. However, there is still an important question within this methodology: How do you evaluate the quality of your model? Before we answer this question (which is not as straightforward as one might hope it to be), we first focus on the machine learning terminology and some broad categories of algorithms in Sect. 2. Many readers will know these concepts by different terms and it is prudent to make everyone know the standard machine learning vocabulary for them. Then, we will focus on different performance metrics in Sect. 3 that evaluate the quality of our algorithms. We will then explore the many tree-based machine learning models in Sect. 4 as they are widely used, easy to interpret and understand and can be used to illustrate some key concepts. Finally, there will be a brief introduction to neural networks which are the basis of many current state-of-the-art machine learning tools in Sect. 5.

2 Terminology

Machine Learning follows its own vocabulary which might be confusing for readers who are rather used to the terminology from a different discipline. Generally, all machine learning models have the shape

$$f(\mathbf{x}) = \hat{y} \tag{1}$$

where f represents the model's algorithm, $\mathbf{x}$ the observed data that is used for the prediction and $\hat{y}$ the predicted *target* value that is supposed to be as close to the true target y as possible. Of course, we usually do not just have one observation $\mathbf{x}$ but many observations $\left(\mathbf{x}^{(i)}\right)_{1 \leq i \leq m}$ and targets $y^{(i)}$. Imagine that we want to predict the height of a human based on their weight, sex and nationality. Then for the ith human in our sample, our observation may tell us $\mathbf{x}^{(i)} = (80 \text{ kg, male, French})$ and our function predicts $f(\mathbf{x}^{(i)}) = \hat{y}^{(i)} = 185$ cm as the target while the true height is $y^{(i)} = 182$ cm. The entire vector of observations for the ith person $\mathbf{x}^{(i)}$ is called an *instance* and its entries $\mathbf{x}_j^{(i)}$ are called *features*. Here, the first feature is the weight, the second is the sex and the third is the nationality of the person.

Often, when you have m instances and n features, a matrix notation $\mathbf{X}$ for all features is used where $(X_{i,j})_{1 \leq j \leq n}^{1 \leq i \leq m}$ with $\mathbf{X}_{i,j} = \mathbf{x}_j^{(i)}$ is the jth feature of the ith instance. It is then convenient to combine all targets into one vector $\mathbf{y} = (y^{(i)})^{1 \leq i \leq m}$. If all entries of $\mathbf{X}$ are numerical, then some data science methods can make use of the matrix shape to do several calculations efficiently for the whole data set.

The above terms are useful to talk about your data, but there is also some standard vocabulary that helps to describe the model f: A very important distinction is made between **categorical models or classifiers** on the one hand and **regression** models on the other hand. Classifiers are used when the targets can be split into k different groups, e.g. if the algorithm is supposed to sort images into the categories "Dog", "Cat" and "Mouse". A sensible way of evaluating such an algorithm is to find out how often your prediction is correct.

On the other hand, a regression task predicts a continuous number, e.g. in the previous example the height of a human. It does not necessarily make sense to categorise the heights into different groups, but instead we would like to predict the numerical value. Your algorithm will almost never give an exactly correct prediction (and if it does, you should ask yourself if something went wrong in your machine learning pipeline as perfect predictions never occur in reality!), because the differences between prediction and true target may be arbitrarily small. The basic idea to evaluate the quality of a regression algorithm is to ask how large the difference between your prediction and the true target typically is.

Another important distinction between algorithms is whether we know the true targets y or not. In the above example, we knew the targets and can compare our prediction to them. This is called a **supervised** algorithm. But often, data science problems involve finding such labels in an unstructured data set. Take a look at the data in Fig. 2, where the two values of two features are plotted against each other. Even without knowing anything about the problem or data behind them, one can clearly identify two clusters centred around $(0, 0)$ and $(5, 4)$. When the existence of such clusters is not as obvious as in this example, the role of machine learning is to identify such clusters. Because we do not a priori know the instances' target (or even if the clusters actually exist), such a task is an **unsupervised** machine learning problem; e.g. identifying different groups of customers within a company's customer

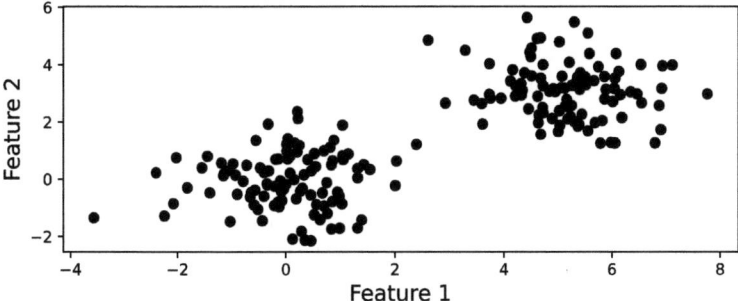

Fig. 2 A prototypical example of an unsupervised clustering problem

base. The distinction between unsupervised and supervised is not exhaustive, though. E.g. if labelling the data requires expensive expert judgement or a lot of time effort, then sometimes only parts of the data have been assigned a target value whereas the remaining data do not have a target. This is a **semi-supervised** scenario.

3   Performance Metrics

It is often not self-evident which metric you should use to evaluate your machine learning algorithm. While the distinction in classifiers and regressions already requires different metrics for these two groups, there are still a lot of questions that have to be answered: E.g. if you want to train a model that predicts the occurrence of cancer based on CT scans, do you want to make sure that no tumor is overlooked? Or do you want to minimise the amount of false alarms because you do not want to scare healthy people with an erroneous cancer diagnosis? Or a trade-off between both? There is no definite answer to questions like these as they strongly depend on the overall problem and context, but having a good overview of the different available metrics can help you to make an informed decision. Note that sometimes people focus on the model's error and wish to minimise the it which is equivalent to maximising the model's performance. However, before we focus on the quality of our model's *output*, let us briefly discuss the quality of its *input*.

3.1   Preface: Data Preprocessing–Garbage In, Garbage Out

The quality of any AI algorithm highly depends on the quality of its training and test data. If the data is not representative of the real-life situations in which the algorithm is supposed to be used, you will usually run into problems. But there are more subtle problems than that and this subsection highlights a few of them.

Many algorithms assume implicitly that all features in your data are distributed on a similar scale. Whenever an algorithm calculates the nearest neighbours of one instance, it assumes that distances in all features are equally valid. Imagine data on human body sizes with features $(x_1, x_2) = $ (height in meter, leg length in mm). If we now have two instances $\left(x_1^{(1)}, x_2^{(1)}\right) = (1.80, 831)$ and $\left(x_1^{(2)}, x_2^{(2)}\right) = (1.7, 805)$, then the difference between them is $(0.1, 26)$. The algorithm will assume that the difference is dominated by the second feature, but that is only because the 26 is given in mm, while the 0.1 is given in meter! Here, it is easy enough to convert both measurements into one unit, but what if we add a third feature "body mass in kg"? This cannot be converted into meters, so we need a different strategy. There are various **scaling** methods that can be used, but two are particularly easy to understand: The min/max normalisation takes all values $\left(x_j^{(i)}\right)^{1 \le i \le m}$ of feature j and transforms them into

$$
\tilde{x}_j^{(i)} = \frac{x_j^{(i)} - \min\limits_{1 \le i \le m} x_j^{(i)}}{\max\limits_{1 \le i \le m} x_j^{(i)} - \min\limits_{1 \le i \le m} x_j^{(i)}} \tag{2}
$$

resulting in all $\tilde{x}_j^{(i)}$ being in the interval $[0, 1]$. Another normalisation is derived from the well-known standardisation of random variables in statistics: First calculate the mean μ and standard deviation σ of the feature values $\left(x_j^{(i)}\right)^{1 \le i \le m}$ and then normalise them according to

$$
\tilde{x}_j^{(i)} = \frac{x_j^{(i)} - \mu}{\sigma}. \tag{3}
$$

This ensures that the rescaled $\tilde{x}_j^{(i)}$ have mean 0 and standard deviation (and variance) 1. The two methods greatly differ in the presence of outliers and some more advanced techniques exist to deal with them, too. Finally, one can also consider to rescale the target values $y^{(i)}$: if the target values are strictly positive (e.g. the length or weight of an object), but your algorithm might also produce negative values, then maybe using a dummy variable $z^{(i)} = \log y^{(i)}$ as the target can make sense. The algorithm will then predict z which you have to transform into strictly positive values for y via $y = \exp z$. Or if you wish to predict probabilities between 0 and 1, a logistic function can ensure that you values actually are within this interval.

For categorical targets, you often face the problem that one category is much less frequent in the data, i.e. the **data is imbalanced**. One can easily imagine a scenario in which you want to predict the presence or absence of a rare disease in a patient. If the disease is particularly rare, you might not have enough instances to train your model. There are methods that alleviate this problem by deliberately oversampling the minority class [16], but one has to take into account that this creates artificial samples that have not actually been recorded in the data.

Sometimes, you have **highly correlated features** in the data. If your data contains e.g. the left leg length and the right leg length of humans, then those features will essentially carry the same information. This can result in misleading model parameters and interpretation, but can easily be fixed by discarding one of the correlated features. Often, exploratory data analysis helps to identify correlated features, but model selection criteria such as the AIC can also help [1].

However, it is not always easy to see if your input data is "garbage" that you put into the model. There might be subtle issues about the sampling protocol or the different real-world interactions between different features that might not be spotted easily. Nobel laureate Herbert Simon has suggested that maybe "the right way to move is to **understand before one predicts**". Thoroughly analysing a problem might also reveal that spending time and effort on state-of-the-art AI methods might not even be necessary to solve it, but a much more simple solution may reveal itself.

3.2 Classifiers

A classifier predicts one of k classes as the target value, e.g. presence or absence of a disease or what kind of animal is shown in a picture. Therefore, at least in supervised machine learning scenarios, it is straightforward to tell if the algorithm was correct for one particular instance $\mathbf{x}^{(i)}$. An obvious method to aggregate this into an overall goodness-of-fit for the whole data is to calculate how often the algorithm was correct as a ratio of the full data. This quantity is called **accuracy** and always lies between 0 and 1. However, it can be a misleading criterion: Imagine again a scenario in which you want to predict the presence or absence of a rare disease that is only found in one in every 1000 patients. If our machine learning algorithm now decides to always predict "absence", it achieves a seemingly amazing accuracy of 0.999—but it has not learnt anything relevant from the data!

While the accuracy can be computed for classifiers with arbitrarily many classes, there are a few performance metrics specifically designed for binary classifiers which can only predict two classes (usually presence and absence or 0 and 1). However, every multiclass problem can be transformed into several binary problems. Imagine you want to predict if an image shows "Dog", "Cat" or "Mouse" in a multiclass scenario with three classes. You can turn this into three binary classifiers by predicting "Dog" vs. "No Dog", "Cat" vs. "No Cat" and "Mouse" vs. "No Mouse". For such binary classifiers, you often see the terminology of "positives" and "negatives": Positives refer to predictions of a 1 or presence of the object you are interested in, whereas negatives are a predicted 0 or the absence of the object. If the prediction is correct, it is called a true positive (predict 1 if 1 is true) or true negative respectively. If the prediction is wrong, you have false positives (predict 1, but 0 is true) and false negatives. Let now *TP*, *TN*, *FP* and *FN* denote the numbers of true positives, true negatives and false positives/negatives. The **precision** of an algorithm is then defined as

$$precision = \frac{TP}{TP + FP} = \mathbb{P}(\text{true value is 1} \mid \text{algorithm predicts 1}) \qquad (4)$$

where $\mathbb{P}(A|B)$ is the conditional probability of A given that B has been observed. Hence, the *precision* calculates the ratio between true positives and all positive predictions and gives you the probability $\mathbb{P}$ that your predicted 1 actually is a 1, the true positive rate (*TPR*). The false positive rate (*FPR*) is defined similarly as

$$FPR = \frac{FP}{FP + TN} = \mathbb{P}(\text{predict } 1 \mid \text{true value is } 0) \tag{5}$$

as the probability of predicting a 1 conditional on the true value being 0. Finally, the **recall** is defined as the probability of correctly identifying a 1 as a 1:

$$recall = \frac{TP}{TP + FN} = \mathbb{P}(\text{predict } 1 \mid \text{true value is } 1). \tag{6}$$

The difference might be subtle, but can be illustrated with an example: If the algorithm only predicts a 1 in extremely clear-cut cases, then its precision is very high because it almost never erroneously labelled a true 0 as a 1. However, its recall will be low because many not-so-obvious true 1s have been mislabelled as a 0 and are therefore false negatives. In fact, under the hood of many classifiers, you see that the classifier usually does not directly predict a 1 or 0, but rather computes a confidence score that the given instance is a 1. For $\mathbf{x}^{(1)}$, the confidence score for predicting a 1 may be 0.42, whereas it is 0.666 for $\mathbf{x}^{(2)}$. A threshold τ has to be chosen such that all scores below τ result in a prediction of 0 and everything above τ gets predicted as 1. You can see that by varying τ, it is possible to predict both $\mathbf{x}^{(1)}$ and $\mathbf{x}^{(2)}$ as a 0 ($\tau > 0.666$) or as 1 ($\tau < 0.42$) and an intermediate region in which only $\mathbf{x}^{(2)}$ is predicted as a 1. Often, the algorithms automatically choose a default value for the threshold, but you can usually manually change the threshold to fine-tune the predictions for the given problem (e.g. a high τ for a restrictive algorithm such that you have very few false positives). This decision depends on whether a false positive or false negative prediction is more harmful in the given problem, which may not have an objective answer. When screening for a dangerous disease, which of these outcomes is worse: Missing some infected people with false positives and not giving them early medication? Or scaring many people with false negatives and causing unnecessary stress and spending much time and health care budget on extensive additional tests, thereby not having enough time to treat other patients? The answer to this is highly subjective.

The considerations about a trade-off between precision and recall lead to another quantity that estimates the goodness-of-fit, the "area under curve" or *AUC*. The idea is to slide through a continuum of possible threshold τ and to then plot the *FPR* and *precision* of the resulting prediction. A perfect classifier has $FPR = 0$ and $precision = 1$ and one can use the distance to this point to estimate a good threshold τ for a given classifier. As the previous paragraph has already alluded to, tuning the threshold to extreme values will lead to a classifier with $FPR = 0$ and $precision = 0$ (nothing is predicted as 1) and $FPR = 1$ and $precision = 1$ (everything is predicted as 1). A random classifier that does not use any information from the training data

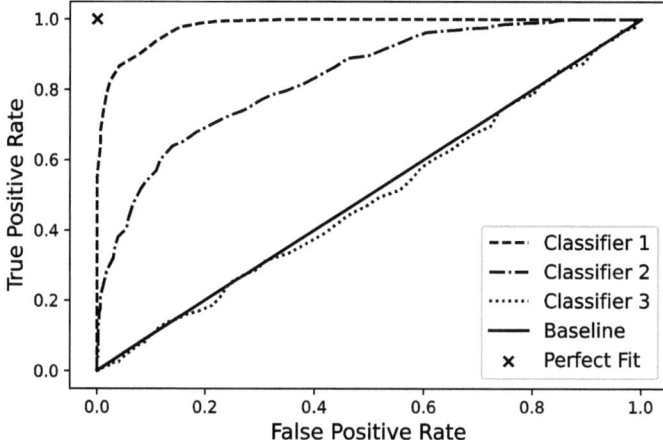

Fig. 3 ROC curves of different classifiers whose threshold is varied and hence, their *FPR* and *TPR* are also varied. Classifier 1 has the largest *AUC* and is therefore the best classifier. Classifier 2 is notably worse and Classifier 3 performs even slightly worse than the theoretical baseline for a random (i.e. uninformed) classifier

(e.g. a random number generator or an elaborate coin flip) will for all other thresholds τ lie on a straight line between those two points. If you plot the various (*FPR/TPR*) pairs that varying thresholds produce for each classifier, you produce the receiver operating characteristic curves (**ROC curve**) shown in Fig. 3. Here, the performance of three classifiers is shown and compared to the baseline of a theoretical random model. The *AUC* is the area under the ROC curve and large *AUC* values indicate a better performance for the respective classifier.

3.3 Regression

For a regression task, all targets $y^{(i)} \in \mathbb{R}$ and all predictions $\hat{y}^{(i)} = f\left(\mathbf{x}^{(i)}\right)$ are real numbers. The algorithm will almost never exactly hit the true value $y^{(i)}$, but one can use the difference between prediction and reality (the error or residual) $r^{(i)} = y^{(i)} - \hat{y}^{(i)}$ as an indicator of the algorithm's quality: the smaller the residuals' absolute values, the better the algorithm. But again, there are different ways to aggregate the prediction quality of all m instances to one performance value.

The gold standard for aggregating the residuals is the Root Mean Square Error **RMSE** whose name already gives away its mathematical definition as

$$RMSE = \sqrt{\frac{1}{m} \sum_{i=1}^{m} \left(y^{(i)} - \hat{y}^{(i)}\right)^2}. \tag{7}$$

Squaring the residuals ensures that positive and negative errors do not cancel each other out and the mean values aggregates the error across the sample. Finally, the root transforms the error measurement back to the same dimension of the targets (i.e. if the targets $y^{(i)}$ are given in m, it makes sense to have an error measurement in m instead of m^2).

Tightly connected to the *RMSE* is the coefficient of determination R^2. If μ is the mean value of the targets $y^{(i)}$, SS_{res} the sum of squared residuals and SS_{tot} the total variation in the data, the R^2 is defined as

$$R^2 = 1 - \frac{\sum_{i=1}^{m} \left(y^{(i)} - \hat{y}^{(i)}\right)^2}{\sum_{i=1}^{m} \left(y^{(i)} - \mu\right)^2} = 1 - \frac{SS_{res}}{SS_{tot}}. \tag{8}$$

It is interpreted as a measure how large the sum of squared residuals SS_{res} is compared to the total variation in the data quantified by the total sum of squares SS_{tot} (which is essentially the same as the variance of the data). Good models reach a R^2 close to 1 and there is a convenient interpretation for $R^2 = 0$: if you constantly use the mean value μ for prediction (and therefore do not use any information about the features $\mathbf{x}^{(i)}$), the resulting score will be $R^2 = 0$. Note that despite its exponent 2, it is possible to get negative R^2 values if the model is worse than the mean value of the targets. This may sound like a niche case because why would anyone even come up with such a bad model if they could use μ instead? But this can actually happen in practise because if you calculate the R^2 on the *test* data, you do not know μ_{test} beforehand, but only μ_{train}! Maybe your model had a $R^2 > 0$ on the training data, but failed spectacularly on the test data. The connection between R^2 and *RMSE* is that they both use the sum of squared residuals $SS_{res} = \left(y^{(i)} - \hat{y}^{(i)}\right)^2$.

However, the *RMSE* and R^2 can give enormous weight to outliers in the data because squaring the errors means that few outliers can dominate these metrics for the whole sample. The Mean Absolute Error *MAE* given by

$$MAE = \frac{1}{m} \sum_{i=1}^{m} \left| y^{(i)} - \hat{y}^{(i)} \right| \tag{9}$$

alleviates this problem.

3.4   Cross Validation

We have already stressed the importance of splitting your data into training and test data and the **cross validation** of your model takes this strategy even a step further and can be used to fine-tune your model. In a k-fold cross validation, the training data is split into k equally sized batches. The model is then trained on all but one of the batches and the remaining batch is used as a validation data set (think about this like test data that you have access to) to evaluate its performance. This is repeated k times where each time a different batch is used for testing. Hence, you do not end up with one R^2 or AUC value for your model, but k of these values. This can be especially useful to compare different models with each other: With one train-test-split, model one may have an $R_1^2 = 0.78$ and model two $R_2^2 = 0.74$—model one is better, but only by a slight margin. Is this margin significant or just down to random chance? You can now use a 10-fold cross validation to make a more thorough comparison of the models. If model one always has a better R^2 for each of the 10 folds, it probably is better than model two. But if both models are better than the other during 5 of the folds, you might conclude that they are essentially equally good candidates.

4 Trees and Forests

Many AI methods are based on trees because on the one hand, trees can be combined with ensemble methods to become powerful models, and on the other hand, their basic structure is very easy to understand and interpret even for laymen (an exemplary decision tree is given in Fig. 4). Especially the latter property means that tree-based methods are still widely used whenever the machine learning model not only has to be accurate, but also transparent and understandable for decision-makers or because of regulatory standards.

Although trees can be used both as a classifier and as a regression model, it is most intuitive to start with decision tree classifiers. For each node on the tree, the algorithm selects a decision criterion (e.g. "feature $x_j > 0.5$") and splits the elements of the parent node based on whether they fulfil the criterion or not. This produces two child nodes that are again split according to new decision criteria and so on. The decision criterion is selected to optimise the purity of the children nodes compared

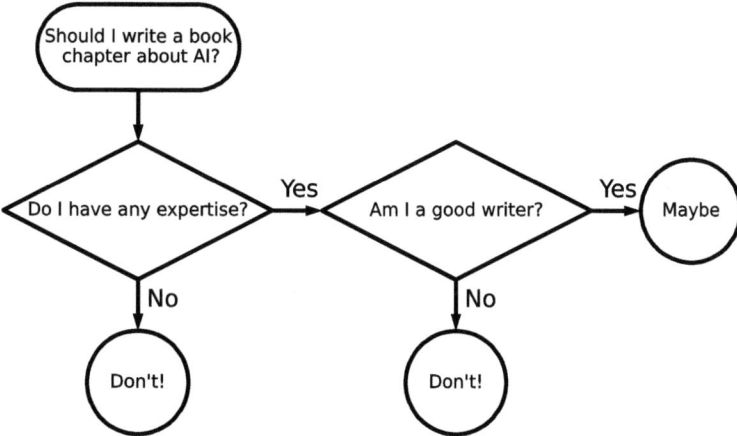

Fig. 4 Example of a decision tree. This tree may or may not have been used by the author

to the parent node. Purity means that ideally, a node should only contain members of one class. If a node is already completely pure, it is not split anymore; otherwise the tree continues to be split into subtrees until a maximum amount of splits has been performed (the maximum depth of the tree has to specified beforehand). The nodes which are no longer split into smaller nodes are called "terminal nodes". If we present the tree with a new instance, it will move along the flow chart of the different decision criteria based on the instance's feature values until it reaches a terminal node. The majority class of the training data which ended up in that terminal node is then given as the prediction for the new instance.

The purity of a tree's node is usually measured via one of two quantities, the **entropy** H or the **Gini index** G. For both of them, consider k different classes and the proportion of instances that are part of class l is given by $(p_l)_{1 \leq l \leq k}$ with $\sum_{l=1}^{k} p_l = 1$. These two quantities are then defined as

$$H = - \sum_{l=1}^{k} p_l \log_2 p_l \text{ and } G = 1 - \sum_{l=1}^{k} p_l^2. \tag{10}$$

Both are minimised with values $H = G = 0$ for perfectly pure nodes if for one class L, all instances fall into that class with probability $p_L = 1$ and all other classes $l \neq L$ do not occur at all and their probability is $p_l = 0$. They are maximised if $p_l = \frac{1}{k}$ is equally likely for all classes l (then $H_{\max} = \log_2 k$ and $G_{\max} = 1 - 1/k$). Usually, the Gini index is evaluated faster than the entropy because it does not require the computation of logarithms, but the entropy has a convenient interpretation as the information content measured in bits (hence the use of base 2 in the logarithm). While Gini index and Entropy often produce similar trees, there is a tendency for G to isolate the most frequent class whereas using H tends to result in more well-mixed trees [10].

For the well-known Iris data on the morphologic variation of Iris flower species [7], a visualisation of a decision tree (with maximum depth $= 2$) is shown in Fig. 5. Here, three different species of the Iris flower (setosa, versicolor and virginica) are used as the classes and different morphological features of the plants are used to classify them via the entropy H. The first decision criterion is based on the petal length of the plants which already results in a pure node for the setosa species and a mixed node which contains all versicolor and virginica instances. This node is then split according to the petal width and we receive two almost pure nodes (the left has 49 versicolor and 5 virginica samples, the right has 1 versicolor and 45 viriginica samples). Because the maximum depth of 2 has been reached, the algorithm terminates. The structure of the tree is easy to understand even if you do not know anything of the inner workings of the algorithm behind it.

Regression trees work very similar to their classifier counterparts, but instead of predicting a class, they predict a numerical target $y \in \mathbb{R}$. Given a new instance $\mathbf{x}$, they also follow a similar flowchart like in Fig. 5 until $\mathbf{x}$ ends up in a terminal node. But now, the prediction $f(\mathbf{x}) = \hat{y}$ is the mean value of all $y^{(i)}$ in that terminal node. Essentially this results in a piecewise constant function $f(\mathbf{x})$.

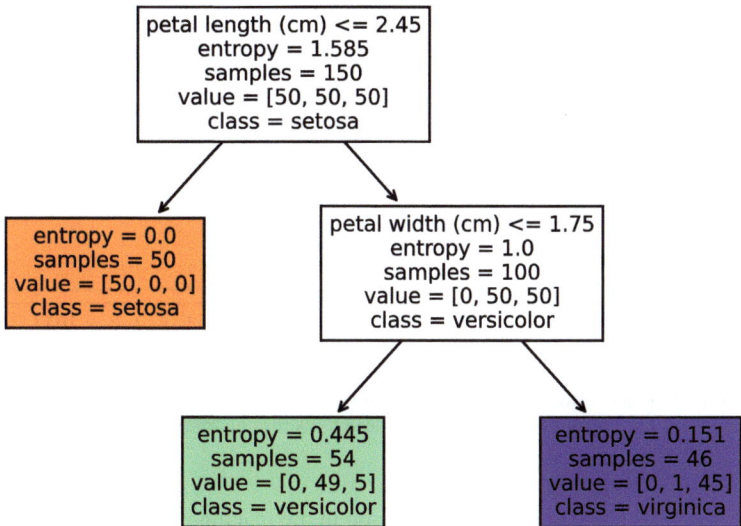

Fig. 5 Tree classifier for the famous Iris dataset learnt and visualised via [23]

4.1 Deep Trees and Overfitting

Setting the maximum depth to 2 made the tree classifier in Fig. 5 terminate although the terminal nodes are not yet pure. If the maximum depth had been larger, the tree would have grown further and produced purer terminal nodes. Especially for such a small data set, one can realistically hope to have only perfectly pure terminal nodes, i.e. that all terminal nodes of the tree only contain one class of instances. The tree might need to be very deep (and too large to be visualised on this page), but this is certainly possible—but not recommended.

Adding more and more depth to the tree means that some terminal nodes will only contain very few instances. E.g. take a look at the rightmost node in Fig. 5 which contains one versicolor specimen and 45 virginica specimens. If we want to purify this node even further, we have to split off the single versicolor instance from the 45 other instances. There surely will be some combination of criteria that achieves this, but it might be rather complicated and heavily tailored to that single versicolor specimen. Such extremely specific criteria often do not generalise well on unknown data. Think about the example with only one versicolor instance: Why should a terminal node that was only constructed based on a single instance be a good generalisation for unknown data? Why should the complex chain of decision criteria leading to this terminal node hold any real knowledge instead of just desperately attempting to model the noise that is inherent in every data set? It really encapsulates not seeing the wood for the trees, as the perfectly grown tree model distracts from the fact that you do not actually care about the training data, but need a stable model for new and unknown data. This phenomenon is called **overfitting** and is common in many machine learning

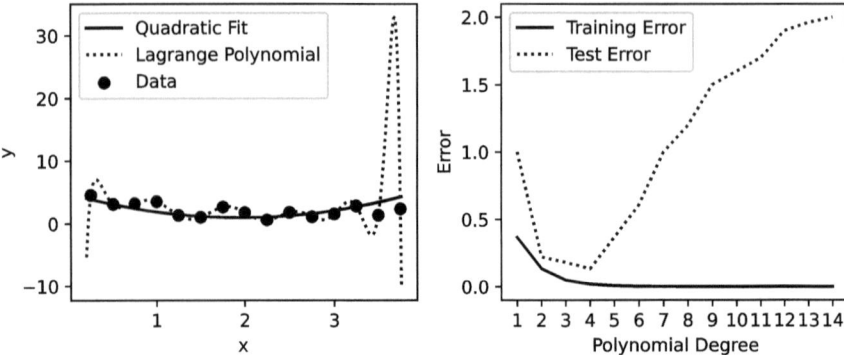

Fig. 6 Illustration of overfitting: The Lagrange polynomial with high degree manages to perfectly fit the data (left), but higher polynomial degrees lead to increasing training errors (right)

algorithms. It usually can be spotted early on in the machine learning pipeline when you notice that making your algorithm more complex (i.e. adding more parameters) improves your performance on the training data, but decreases its performance on the unknown test data. This marks the beginning of the overfitting regime.

This problem is also illustrated in Fig. 6, where we try to fit polynomials of degree d to N data points. A quadratic fit captures a lot of the data, but a Legendre polynomial whose degree is given by $d = N - 1$ can always fit the data exactly. However, it has a lot more parameters than the simple quadratic polynomial and you can see in Fig. 6 (left) that the Legendre polynomials can have a poor performance between the observed data. Hence, the test error of the polynomial fit starts to increase at high degrees, whereas their training error still decreases (right).

Sometimes, **regularisation** methods can prevent overfitting during the training process. A naive training algorithm varies the model parameters $\boldsymbol{\theta}$ until it minimises the model error. But with regularisation, you add a penalty term to the model error that punishes overly complicated models. As a simple example, the Lasso method simply adds the absolute values of all parameters θ_v to the model error [30], i.e. the optimiser now needs to minimise

$$\min_{\boldsymbol{\theta}} \left(\text{Error}(\boldsymbol{\theta}) + \lambda \sum_{v=1}^{N_\theta} |\theta_v| \right). \tag{11}$$

The regularisation hyperparameter $\lambda > 0$ determines how much the regularisation punishes complex models. For small $\lambda \gtrsim 0$, the regularisation has almost no effect, but if $\lambda \gg 0$, it will ensure that only very few parameters θ_v are significantly different from zero. Looking at the polynomials in Fig. 6, you can probably see that the Lagrange polynomial needs many more nonzero parameters than the quadratic fit and would be punished heavily by a Lasso regularisation.

4.2 Boosting

If a tree is simply not accurate enough for the given problem, but you do not want to risk overfitting, you can turn to ensemble methods that combine many weak models to one good prediction. Think about this as a "wisdom of crowds"-scenario. One type of ensemble methods is **boosting** the weak model by training new versions of it that improve their predecessor's mistakes. The two most frequently used boosting methods are Adaptive Boosting or **AdaBoost** [2] and **Gradient Boosting** [8].

AdaBoost first trains an initial weak model M_1 (such as a tree with a low maximum depth) and trains it to the data set. It then identifies the instances for which the model's prediction were wrong and gives them more weight in the next training process. A new weak model M_2 is then trained with the updated weights and the process is repeated until the ensemble of trained models $\{M_e\}_{e=1,\ldots,E}$ is large enough. For a new instance $\mathbf{x}$, AdaBoost then uses each model in its ensemble to make an ensemble of predictions. These predictions are combined to one prediction by assigning weights to the model's prediction $M_e(\mathbf{x})$ based on how well the model performed on the training data. For classifiers, a weighted majority decision can be used to aggregate the ensemble's predictions to one final prediction.

Gradient Boosted Regression Trees (GBRT) have a similar idea, but each subsequent model M_{e+1} is trained on the residuals of its predecessor M_e. GBRT also starts with a base model M_1 whose prediction errors $\mathbf{r}_1 = \mathbf{y} - M_1(X)$ are used to train M_2. If M_2 manages to achieve perfect accuracy, then the sum of the two predictions will exactly result in the target values $M_1(X) + M_2(X) = \mathbf{y}$. As this is usually not possible you train another tree M_3 for the residuals $\mathbf{r}_2$ of M_2 and so on until you have enough trees. For any new instance $\mathbf{x}$, the prediction of the ensemble is then given by the sum of every individual tree's prediction

$$f(\mathbf{x}) = \sum_{e=1}^{E} M_e(\mathbf{x}). \tag{12}$$

4.3 Bagging and Random Forests

Another group of ensemble methods uses the Bootstrapping technique [6], whose name derives from the saying "to pull oneself up by one's bootstraps" because when you initially read about this method, you tend to believe that it is just about as non-sensical as pulling yourself up by your bootstraps. The core idea behind statistical bootstrapping is taking your sample $Z = (z_1, \ldots, z_n)$ and resampling with replacement. I.e. you construct a new sample $\tilde{Z} = (z_{i_1}, \ldots, z_{i_n})$, where some z_i from the original sample are used more than once, whereas others have been dropped completely (mathematically, there are $l \neq m$ with $i_l = i_m$). This is repeated n_B times such that n_B new samples $\{\tilde{Z}_1, \ldots, \tilde{Z}_{n_B}\}$ are created. Statisticians can now calculate any statistic of interest—e.g. the median—on each of the n_B new samples and get an ensemble of median values from which they can calculate the standard deviation of the median. This is especially useful if no analytically closed form exists to calculate uncertainty intervals for the statistic you are interested in (imagine you want to evaluate a complicated function across the sample Z but also need uncertainty intervals for it).

As machine learning engineers, we can now use a weak classifier such as a tree and train it on all n_B bootstrapped samples to get an ensemble $\{M_1, \ldots, M_{n_B}\}$ of trees. Their predictions can be aggregated via e.g. majority decision for classifiers or with the mean value for regression trees and will result in a powerful and stable model. The bootstrap aggregation is abbreviated as **bagging**. Because bootstrapping results in such a variety of new samples $\tilde{Z}_b$, the bagging model will rarely be tempted to overfit the data.

An extension of the bagging method is the **random forest**. Just like bagging, the random forest also uses bootstrapped samples $\tilde{Z}$ to train many trees. The key difference between random forests and bagging is that random forests have an additional layer of randomness in the selection of features: Trees usually choose the most optimal combination of feature and threshold value for their splitting criteria. In random forests, they can only choose the optimal splitting criterion from a subset of features, not from all features in the data. This is another safety precaution against overfitting. Of course, both methods can be combined by bagging boosted trees.

5 Neural Networks

In recent years, artificial neural networks have pushed the field of machine learning to new limits and allowed artificial intelligence to overcome more and more spectacular challenges. AlphaGo's famous victory against the Go master Lee Sedol and the impressive performance of the various GPT models in natural language processing are just two examples of algorithms which use neural networks as their key architecture and which manage to come close to or even surpass human performance. Often, engineering solutions have been inspired by biology (the lotus effect is a famous example for this) and as the name suggests, artificial neural networks are no exception to this. The fact that its basic structure is loosely modelled after our brain probably helps to evoke strong associations with human-like intelligence whenever artificial intelligence based on neural networks is discussed. However, a look at the mathematics behind neural networks helps to demystify them.

The fundamental unit of a neural network, a neuron, is directly inspired by the neurons in our brain. A biological neuron receives several signals and if their sum is strong enough, it "fires" and sends out another signal as an output. Artificial neurons or **perceptrons** work very much the same [19]: they receive input signals $i_1, \ldots, i_n$ and use a weighted sum to combine them to one value $\sum_{j=1}^{n} w_j i_j$. If this sum is large, the neuron produces a strong output according to an activation function $h(\cdot)$, otherwise its output is close to zero or exactly zero. Often, a bias term b is also added to the sum. The neuron's output is therefore given as

$$h\left(\sum_{j=1}^{n} w_j i_j + b\right) \tag{13}$$

where i_j are the inputs and the tuneable parameters are the weights w_j and bias b. Initially, h was usually given by a step function with $h(x) = \mathbb{1}_{[0,\infty)}(x)$, but for reasons discussed below in Sect. 5.1, other functions have started to become popular. Some frequent choices are depicted in Fig. 7.

However, one neuron alone is not a neural network. Instead, there are usually several hidden layers of neurons and each layer contains many neurons. The input values (i.e. the features of an instance) are plugged into each neuron of the first hidden layer and these neurons compute an output according to their activation function like (13). Their output values are then passed into the next deepest layer where they are used as inputs for the next neurons and so on. Finally, all outputs from the deepest layer are plugged into one or more output nodes which then give the prediction of the network. For classification tasks with k classes, you usually have k output nodes (one for each class) whose outputs are then normalised to form a probability distribution. Output O_κ will then give you the probability that the instance is in class κ. For a regression task, one output node is usually enough as it will give the numerical value that the algorithm predicts. Such networks are called **multilayer perceptrons** or **MLPs** and a sketch of a network with one output is shown in Fig. 8.

Here, the network receives two features as inputs and passes them into the first hidden layer with three neurons. All neurons receive both input values, i.e. this is a fully connected layer. Their outputs are then passed to the second hidden layer with four neurons. However, this layer is not fully connected (e.g. neuron $N_{2,1}$ only receives inputs from two of the three neurons in the previous layer). Finally, their output is aggregated into an output layer. The network depicted in Fig. 8 has 19

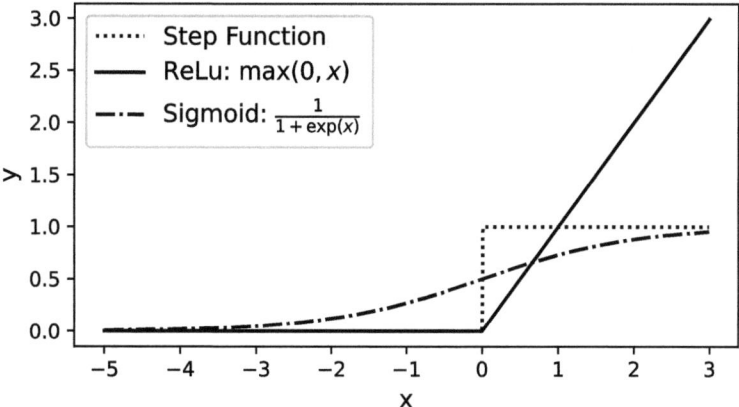

Fig. 7 Popular activation functions for neural networks include the ReLu function and the smooth sigmoid function

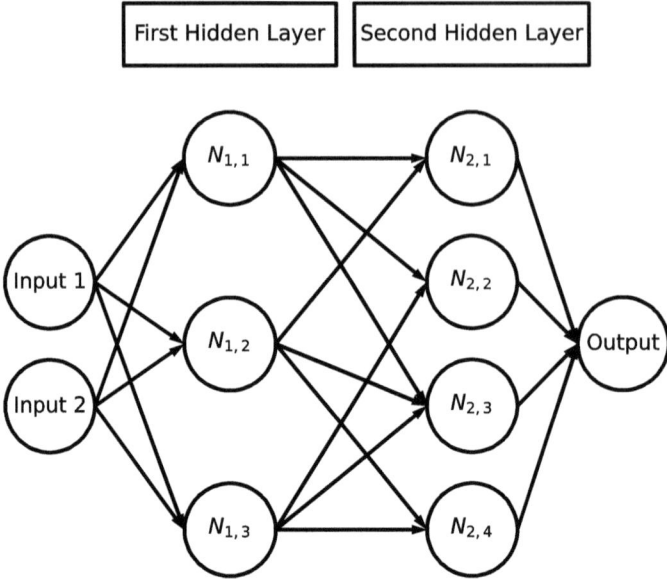

Fig. 8 Sketch of a neural network's architecture made via [5]. Each $N_{i,j}$ is a neuron with an activation function according to (13) and the arrows show input/output relationships

connections between inputs, neurons and output node and (including the output) 8 neurons, i.e. it has 27 tuneable parameters (19 weights and 8 biases). And we are dealing here with a very small network! It is not uncommon to see layers of 2^6 or 2^7 neurons and deep neural networks have many more layers than just two. Thus, it is easy to see that their number of parameters essentially explodes.

You can interpret the neural network as an ensemble method similar to the boosting and bagging techniques from Sect. 4: Like gradient boosting, many simple neurons are essentially chained after each other to correct their predecessors' mistakes. And similar to bagging, the neurons in the first hidden layer are many versions of the same base model working on slightly modified input data, but while bagging resamples the data via bootstrapping for each base model, each neuron has different weights w that change how it views the input data.

5.1 Gradient Descent and Backpropagation

The parameters of neural networks are estimated such that the network's predictions are as good as possible according to a given evaluation metric (see Sect. 3). However, simultaneously tuning so many parameters $\theta = (\mathbf{w}, \mathbf{b})$ is not an easy task and elaborate algorithms have been developed without which the practical applicability of neural networks would be extremely limited. The foundation of these optimisation

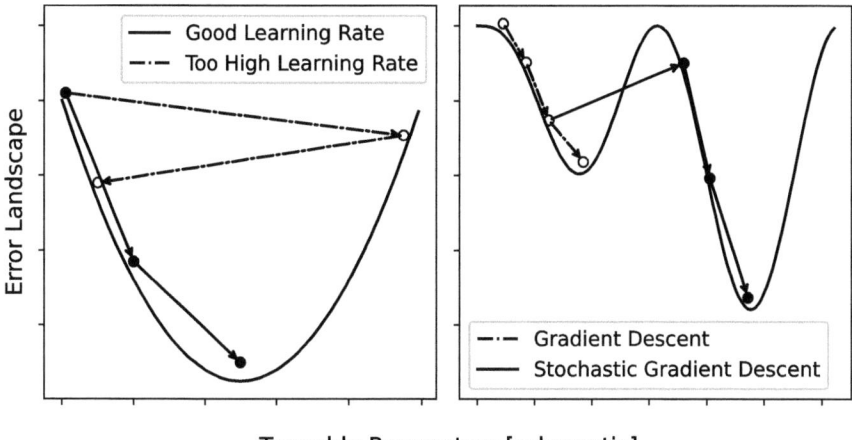

Fig. 9 Schematic representations of error landscapes during a training process. Left: gradient descent with a high learning rate can overshoot and miss the optimum. Right: regular gradient descent can get stuck in local minima, whereas stochastic gradient descent can escape from them

algorithms is the **gradient descent** on the landscape of an error function which can be regarded as the opposite of a performance metric (the larger the error is, the worse the algorithm performs). The optimiser starts with randomly initialised parameters θ_0 and then calculates the local gradient of the error function for the given parameters, i.e. the error function's derivative or slope. It then moves a step into the direction of the decreasing gradient and arrives at a new parameter vector θ_1 which (hopefully) has a lower error than its predecessor (see the arrows in Fig. 9, left); it *descends* along the error landscape. This process is then repeated until a minimum of the error function is found. The size of the steps is a crucial parameter called the **learning rate**. If the steps are small, then the optimiser needs a lot of steps to find the optimum, whereas a large step size might overshoot and miss the optimum (illustrated for the high learning rate in Fig. 9, left). Because gradient descent can get stuck in local minima and fail to find the global optimum, **stochastic gradient descent** has been invented as an extension. It does not deterministically follow the direction of the gradient, but instead has a stochastic component that allows jumps across the error landscape to escape local minima (shown in Fig. 9, right).

Moreover, the gradient descent technique is supplemented with another important trick, the **backpropagation** algorithm, to achieve a fast training process for neural networks [26]. In each epoch of the training process, the neural network first makes a prediction on the training set (called a "forward pass") and each neuron's intermediate result is saved for later calculations. The prediction error is then measured and propagated backwards through the neural network. The algorithm evaluates the contribution to the overall error of each connection between neurons. This is done via the chain rule to efficiently calculate the gradients for gradient descent. As shown

in introductory calculus classes, the derivative of a chain of functions is the product
of their derivatives

$$\frac{d}{dx}h(f(x)) = \frac{df}{dx}\frac{dh}{df}.$$ (14)

Thus, even long chains of neuron outputs like (13) can be evaluated with relative
ease as long as the derivatives of each step exist. This is why the step function is
no longer used for neural networks: at their flat steps, their derivative is zero, and at
the position of the step, the function is not differentiable. Hence, functions like the
ReLu or Sigmoid (shown in Fig. 7) are used instead so that the full gradient can be
calculated rather easily to finally apply (stochastic) gradient descent.

The backpropagation algorithm can explain why neural networks have become
the premier machine learning model after having been neglected by researchers for
decades. While neural network training iterates over the m training instances, the
previous gold standard of artificial intelligence, support vector machines, inverts an
$m \times m$ matrix. Thus, for large m, the computational complexity of neural networks
scales linearly as $\mathcal{O}(m)$, but the complexity of support vector machines as $\mathcal{O}(m^3)$[1].
For twice as much training data, the required effort to train a neural network is
doubled, whereas it increases eightfold for support vector machines. Because an
abundance of training data is necessary for good performance and the era of "big
data" provides machine learning engineers with plenty of data, scaling efficiently
with large data is a key advantage of neural networks.

5.2   Advanced Architectures

Multilayer perceptrons can already solve a lot of machine learning problems with
great success. However, research did not stop with MLPs and numerous specialised
architectures have been developed to tackle specific machine learning problems. This
section will give a brief introduction to some of them, while being fully aware that
each of these could easily warrant a full book chapter just for themselves.

Just like MLPs and the concept of a neuron have been inspired by our biological
neural networks, a specialised neural network was loosely modelled after our brain's
visual cortex to excel at image classification. **Convolutional neural networks** (CNN)
use small filters that scan over a given image and analyse if a given pattern is found
in any location of the image's (x/y)-plane [9]. One filter might be used to identify
vertical stripes, another one for horizontal stripes and other filters for curved stripes
and so on. This means that neurons on the first layer of the CNN are not connected
to every pixel of the image, but only to the small subset of their so-called receptive
field as depicted in Fig. 10. Often, several convolutional layers are stacked on top of
each other until a regular MLP uses the CNN's output to classify the image as e.g. a

---

[1] The author would like to thank Prof. Stefan Harmeling (TU Dortmund) for pointing that out.

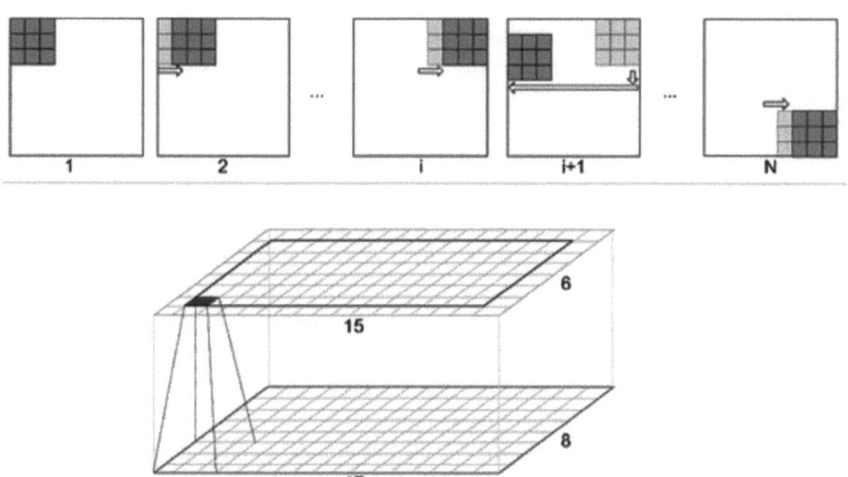

Fig. 10 Illustration of a convolutional layer in a CNN, taken from [28] and reproduced with permission from Springe Nature (SNCSC). A 17×8 pixel image is analysed by running a 3×3 pixel filter across it (top line). The filter can e.g. detect vertical or horizontal lines and its detection results are then recorded in a 15×6 pixel map (bottom). This map now essentially encodes if the geometrical structure of the filter appears in a given part of the image or not

house (with many straight lines vertically and horizontally) or a ball (many curves). Note that all of these filters are learnt during the training process and not given a priori.

One could theoretically build an MLP that uses all pixels of the image as input features, but CNNs have two major advantages over such architectures: First, even low-resolution images have at least 10^5 pixels, meaning that an MLP needs an enormous amount of parameters to process them. Second, and maybe most importantly, because the CNN scans its filter across the entire image, it can detect the same visual elements at different positions in the image. For example, images of a dog usually depict two eyes and two ears. CNNs can recognise that there are two of these elements present in the image, whereas MLPs struggle much more to do this.

Another important task for machine learning algorithms is to predict time series, like e.g. forecasting the weather or stock market prices. CNNs can actually be used for this purpose, too, if they have one-dimensional filters that can learn to detect ups and downs of a time series. A more tailored solution for this kind of task is a **recurrent neural network** (RNN). For each time step t, the recurrent neurons not only receive the input for that time step (i.e. the features $\mathbf{x}(t)$), but also is output from the previous time step $h(\mathbf{x}(t-1))$. It therefore has a memory of the previous time step and can use this to more accurately predict the next step.

However, RNNs have a rather short-lived memory, which can be detrimental for some applications. **LSTM** cells (whose abbreviation stands for the paradoxical long short-term memory) can remedy this issue [11]. The LSTM cell has an explicit long-term state and learns which information has to be stored in this state to improve its performance.

Especially for natural language processing, the **transformer** architecture has become an irreplaceable building block [31]. Transformers consist of two elements, an encoder and a decoder. First, the given input sentence is encoded into an abstract vector representation that e.g. captures the sentences general sentiment (hopeful, sad etc.). This vector is then decoded back into human language and the decoder can be used to predict the next word in the sentence (i.e. to write new texts) or to decode the original sentence in another language. Such translator greatly profit from the transformer's abstract representation and work much better than e.g. word-by-word translators.

Sometimes, machine learning is used to forecast models which we have essentially already understood on a microscopic level, but are too computationally expensive to explicitly simulate from the bottom up. While it is certainly possible to let the machine learning algorithm simply "do its magic", it is unwise to disregard the knowledge that we already have about the system. **Physics-informed neural networks** (PINNs) have been developed to incorporate such knowledge into the machine learning framework [24]. If we are e.g. faced with a problem from fluid dynamics, we might know which partial differential equations (PDEs) the model should follow. But explicitly simulating them might be too difficult with the given initial conditions. Then, we can train the model as a regular neural network, but impose an additional condition that the model should not only accurately capture the data, but also the underlying PDEs. Essentially, this corresponds to regularisation methods like (11), but using a highly informed penalty function.

5.3 Implementation

After spending many pages on discussing the theory of neural networks, practically minded readers might now demand some advice on how to *use* neural networks and train them themselves. In the past years, the programming language Python has established itself as the premier language to implement machine learning models because its rather simple syntax makes it accessible for people without coding experience and its multitude of packages include many popular machine learning algorithms. The package *scikit-learn* (usually abbreviated as sklearn) provides a good general machine learning toolkit alongside some data wrangling tools [23] and also an implementation of multilayer perceptron networks. However, for large-scale neural networks, the packages *tensorflow.keras* and *Pytorch* are much more frequently used [4, 29]. These packages also include straightforward implementations of the advanced architectures from the previous section.

6 Closing Words

While the first major successes of artificial intelligence were achieved with logical algorithms that used explicitly given rules, there has been a paradigm shift over the past decades towards machine learning, i.e. algorithms that are capable of deriving rules themselves from data. Of course, innovation does not simply stop here. Another extension to the machine learning methodology is to make the model interact with an environment and receive a reward based on how well its algorithm performs. This strategy, **reinforcement learning**, aims to more closely mimic how intelligent beings learn and will be discussed in detail in the next chapter.

Within the latest AI-craze, it is prudent to remember that interest in artificial intelligence has already experienced two "AI winters" during which funding for researchers decreased and innovation was slowing down. This is at least partially because of human nature, as new achievements in AI can easily lead to a hype of optimism and overconfidence which simply cannot live up to reality and have to lead to disappointment. However, this also means that some technologies had been forgotten for decades of hibernation. After the concept of perceptrons had been sharply criticised by one of the leading AI experts who argued that it was a pointless endeavour to study them [20], research on multilayer perceptrons struggled to keep on during the first AI winter and was dormant for decades. As a result, neural networks had to basically be rediscovered afterwards by a new generation of researchers. Maybe the next big thing has already been invented years ago and waits to be rediscovered?

Of course, this short introductory chapter is lacking both in scope and depth. The interested reader is recommended to consult [10], which is one of the most widely used introductory textbooks on artificial intelligence. Equipped with a good general grasp on machine learning, one can then find tutorials and blog posts on almost any AI topic on the internet. Although statistics has been touched only briefly throughout this section, probability theory and statistics are nevertheless an important foundation of machine learning. Sometimes, it might even be necessary to refresh one's knowledge about these topics in order to fully understand how to apply an algorithm. Especially for readers with a background in physics, [14] is a great introduction to probability theory, whereas the classic [15] is a more in-depth mathematical textbook.

An interesting new development in the field of AI is the need for greater transparency, often driven by regulatory agencies, but also by the desire to better understand what the model is actually doing. As a great example of the "garbage in, garbage out"-paradigm, one image classifier learnt to recognise pictures of horses only because they all had the same photographer's tag in one corner of the image [13]. This situation is quite amusing, but what if something like this occurs if the AI is supposed to make high stakes decisions?

Explainable AI (XAI) has been suggested to increase trust in AI by making the "AI black box" more transparent [21]. XAI methods try to explain in simple terms why the AI made its decisions the way it did, e.g. by producing heat maps of influential regions for image classifiers. An alternative methodology is to design inherently interpretable algorithms as suggested by Cynthia Rudin in [25]. Rudin argues in favour of engineering meaningful features from the data by using experts and domain knowledge such that even a simple and easily understandable algorithm can make accurate predictions based on those features. This is an interesting throwback to the older generations of AI algorithms such as Deep Blue, whose chess strategies were devised by domain experts, but Rudin argues to supplement the feature engineering with the machine learning methodology from recent AI models. Only time will tell which of these approaches will be more successful, but greater transparency of AI and better understanding of its inner workings is certainly a valuable goal for new researchers.

Acknowledgements The author is funded by the Studienstiftung des deutschen Volkes and would like to thank Jonathan Wolf, Fabian Zelesinski and the anonymous reviewer for valuable feedback.

References

1. Akaike H (1998) Information theory and an extension of the maximum likelihood principle. In: Parzen E, Tanabe K, Kitagawa G (eds) Selected papers of hirotugu akaike. Springer, New York, NY, pp 199–213. ISBN: 978-1-4612-1694-0. https://doi.org/10.1007/978-1-4612-1694-0_15
2. Breiman L (1997) Arcing the edge. Tech. rep, Citeseer
3. Brown TB et al (2020) Language models are few-shot learners. arXiv: 2005.14165 [cs.CL]
4. Chollet F et al (2015) Keras. https://keras.io
5. Delker CJ, Schemdraw. https://schemdraw.readthedocs.io/en/stable/index.html
6. Efron B (1979) Bootstrap methods: another look at the jackknife. Ann Stat 7(1):1–26. https://doi.org/10.1214/aos/1176344552
7. Fisher R (1936) Iris [Dataset]. UCI machine learning repository. https://doi.org/10.24432/C56C76
8. Freund Y, Schapire RE (1997) A decision-theoretic generalization of on-line learning and an application to boosting. J Comput Syst Sci 55(1):119–139. ISSN: 0022-0000. https://doi.org/10.1006/jcss.1997.1504. https://www.sciencedirect.com/science/article/pii/S002200009791504X
9. Fukushima K (1980) Neocognitron: a self-organizing neural network model for a mechanism of pattern recognition unaffected by shift in position. Biol Cybern 36(4):193–202. https://doi.org/10.1007/bf00344251
10. Geron A (2022) Hands-on machine learning with Scikit-Learn, Keras, and TensorFlow. Inc, O'Reilly Media. ISSN: 978-1-098-12247-8
11. Hochreiter S, Schmidhuber J (1997) Long short-term memory. Neural Comput 9(8):1735–1780
12. Hsu F-H (2022) Behind deep blue–building the computer that defeated the world chess champion. Princeton University Press, Kassel. ISBN: 978-0-691-23513-4
13. Lapuschkin S et al (2019) Unmasking clever Hans predictors and assessing what machines really learn. Nat Commun 10. https://doi.org/10.1038/s41467-019-08987-4
14. Lawrence A (2019) Probability in physics. Springer International Publishing. https://doi.org/10.1007/978-3-030-04544-9
15. Lehmann EL, Romano JP (2022) Testing statistical hypotheses. Springer International Publishing. https://doi.org/10.1007/978-3-030-70578-7
16. Lemaître G, Nogueira F, Aridas CK (2017) Imbalanced- learn: a Python toolbox to tackle the curse of imbalanced datasets in machine learning. J Mach Learn Res 18(17):1–5. http://jmlr.org/papers/v18/16-365.html

17. Mackenzie D (2021) Update: why this week's man-versus-machine go match doesn't matter (and what does). https://doi.org/10.1126/science.aaf4152.
18. McCarthy J (2007) What is artificial intelligence?
19. McCulloch WS, Pitts W (1943) A logical calculus of the ideas immanent in nervous activity. Bull Math Biophys 5(4):115–133. https://doi.org/10.1007/bf02478259
20. Minsky M, Papert S (1969) Perceptrons; an introduction to computational geometry. MIT Press, Cambridge. ISBN: 978-0-262-13043-1
21. Molnar C (2022) Interpretable machine learning. In: A guide for making black box models explainable, 2nd edn. https://christophm.github.io/interpretable-ml-book/
22. Müller M (2002) Computer go. Artif Intell 134(1):145–179. ISSN: 0004-3702. https://doi.org/10.1016/S0004-3702(01)00121-7. https://www.sciencedirect.com/science/article/pii/S0004370201001217
23. Pedregosa F et al (2011) Scikit-learn: machine learning in Python. J Mach Learn Res 12:2825–2830
24. Raissi M, Perdikaris P, Karniadakis GE (2019) Physics-informed neural networks: a deep learning framework for solving forward and inverse problems involving nonlinear partial differential equations. J Comput Phys 378:686–707. ISSN: 0021-9991. https://doi.org/10.1016/j.jcp.2018.10.045. https://www.sciencedirect.com/science/article/pii/S0021999118307125
25. Rudin C (2019) Stop explaining black box machine learning models for high stakes decisions and use interpretable models instead. Nat Mach Intell 1(5):206–215. https://doi.org/10.1038/s42256-019-0048-x
26. Rumelhart DE, Hinton GE, Williams RJ et al (1985) Learning internal representations by error propagation
27. Silver D et al (2016) Mastering the game of go with deep neural networks and tree search. Nature 529(7587):484–489. https://doi.org/10.1038/nature16961
28. Skansi S (2018) Convolutional neural networks. In: Introduction to deep learning: from logical calculus to artificial intelligence. Springer International Publishing, Cham, pp 121–133. ISBN: 978-3-319-73004-2. https://doi.org/10.1007/978-3-319-73004-2_6.
29. Stevens E, Antiga L, Viehmann T (2020) Deep learning with PyTorch. Simon and Schuster, New York. ISBN: 978-1-617-29526-3
30. Tibshirani R (1996) Regression shrinkage and selection via the lasso. J R Stat Soc Ser B (Methodol) 58(1):267–288. ISSN: 00359246. http://www.jstor.org/stable/2346178. Accessed 17 Jul 2023
31. Vaswani A et al (2017) Attention is all you need. In: Proceedings of the 31st international conference on neural information processing systems. NIPS'17. Long Beach, California. Curran Associates Inc., USA, pp 6000–6010. ISBN: 9781510860964
32. Weizenbaum J (1966) ELIZA-a computer program for the study of natural language communication between man and machine. Commun ACM 9:36–45
33. Winograd T (1971) Procedures as a representation for data in a computer program for understanding natural language. Massachusetts Institute of Technology Cambridge, p 462

An Introduction to Reinforcement Learning–in Artificial and Biological Control Systems

Malte Schilling

Abstract Reinforcement Learning takes an active perspective on the learning of internal models, focusing on an agent acting in an environment. The agent's control system is tasked with selecting actions for the agent based on the current sensed state. In Reinforcement Learning, the control system is optimized to gain high rewards over time. While the optimization is driven by a usually external reward signal, the process is explorative, as the agent tries to find areas within the environment that offer high rewards. In Deep Reinforcement Learning (DRL), neural networks are employed as function approximators to deal with continuous sensory input spaces. Considering actions as the output of the control system, the exploration of large action spaces quickly requires a high number of exploratory interactions or, even worse, becomes intractable. In this chapter, we will introduce Reinforcement Learning, its conceptual formulation as a Markov Decision Process, and highlight current problems of DRL as well as biologically-inspired approaches that address these issues.

1 Introduction

Central to the approach of Machine Learning is the idea of learning structure from data [7, 37]. The goal is to train internal models that, after training, can be applied to novel inputs and provide reasonable outputs for these, which is considered generalization. These models can be represented as functions, and learning is viewed as training such models in a way that, over time, they better fulfill a given task, measured by specified metrics. In general, there is one major distinction in Machine Learning concerned with how such models learn: First, in supervised learning, there is labeled example data. In this case, data points are described by possibly a large number of features, and the model has the task of assigning a label (or target) to such data points. For the training dataset, these labels are already given and serve as ground truth. The two most common cases inside supervised learning are classification and

M. Schilling (✉)
University of Münster, Münster, Germany
e-mail: malte.schilling@uni-muenster.de

© The Author(s) 2026
M. te Vrugt (ed.), *Artificial Intelligence and Intelligent Matter*, Machine Intelligence for Materials Science, https://doi.org/10.1007/978-3-032-04129-6_3

regression tasks. For example, in classification, a model receives an image as input, and the model should produce as output a label which indicates the class depicted in the image, e.g., a cat, a dog, or which digit in the classic MNIST dataset is shown. The availability of large data sets has led to the widespread success of supervised learning approaches, and today, when provided with sufficient data (ranging in the thousands to millions), supervised learning solves classification problems in visual domains when there is sufficient representative data available. Secondly, in unsupervised learning, the data is not labeled, and there are no given targets. In such cases, the goal is to uncover underlying structures from the data and represent the distribution of the data. Typical tasks include the detection of outliers—does a novel data point fit into the data provided during training—or clustering approaches—do the different data points form specific clusters?

A third approach to Machine Learning is given by Reinforcement Learning, which takes a quite different and more enactive perspective. In Reinforcement Learning, the learning model is considered as making (or supporting) decisions [2, 53]. A current state of an environment is used as an input to the model. The task for the model is to produce an action signal as output. This view is inspired by biology, in which Reinforcement Learning is used as a framework to describe exactly this type of sequential decision making [38]. An agent acts in an environment, and the actions depend on the current sensed state of the environment. In Reinforcement Learning, the agent learns how to act, driven by an explicitly given reward signal which is directly inspired from biology, where in such reward-driven learning, an animal is rewarded after performing a certain action [52]. In this conceptual learning framework, an agent not only observes the environment (these observations represent the input to the model and the current state), but in addition, an explicit reward signal is required that indicates to the agent how well it currently chooses actions (Fig. 1). The goal of Reinforcement Learning is to develop a control model that selects actions that—in the long run—maximize the accumulated reward, i.e., select what is considered optimal behavior.

Reinforcement Learning is an active area of research (see as well the chapter by Hartmut Löwen/Benno Liebchen in this book) and, over the last years, has shown success in multiple application areas, particularly in combination with deep neural networks. For example, it has been demonstrated that better than human performance could be learned in playing computer games in a complete end-to-end fashion [35]. More recently, this has been extended to areas such as robotics [9, 22, 27, 29]. Reinforcement Learning is particularly attractive as it provides a simple framework for learning that does not rely on expensively labeled data, but instead is guided by a simple reward signal. Recently, this conceptual simplicity has been applied and became essential in the success of large language models like ChatGPT. Tuning of such large language models was known to be effective but still required huge amounts of human training data. As a consequence, such models could not be sufficiently trained towards dialogue models. The current series of chat models became successful when Reinforcement Learning was applied in training, and instead of supervised data, the training was simplified and purely driven by a simple—and easily trainable—reward function [41]. The intuition that it is much simpler to assign a simple reward value

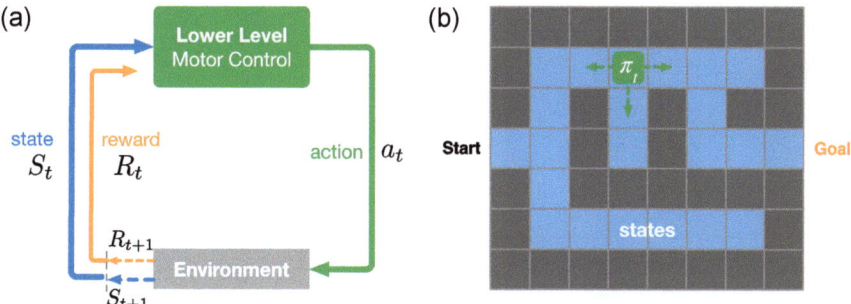

Fig. 1 **a** Conceptual view of Reinforcement Learning: Based on a current state S_t, a controller selects an action a_t. During Reinforcement Learning, this controller is optimized to maximize a returned reward signal R_t that is provided by the environment together with a new observation of the subsequent state S_{t+1}. **b** Illustration of a maze environment as used as a typical textbook example. In such map based scenarios states are often considered as discrete areas on a map (shown here as grey squares; the blue background represents part of the hallway of the maze and the dark color represents walls). The agent (shown in green) is controlled by a policy π_t that has to pick in which direction to move (out of a set of a discrete options as cardinal directions)

compared to generating large amounts of high-quality text proved right and could actually be delegated to a different (and much simpler reward model) neural network. Training such a reward model that later on was able to guide large numbers of interactions of a large language model turned out to be an essential step towards the current state of dialogue models.

Reinforcement Learning has proven as an ideal framework for learning in scenarios that can easily be related to a specified reward function [2]. The learning task is usually considered from the perspective of an agent that is trying to maximize the accumulated long-term reward. As a consequence, the agent is tasked with making optimal decisions in the long-run. So, the agent must learn which experience leads to which rewards (and possibly how this affects future rewards when taking a decision). In the end, the agent should be able to estimate long term consequences for each of the possible actions and should select the appropriate action. But this introduces a problem for the agent: Has he already collected enough experience to make an optimal decision? Or could he be missing out on some high reward as he has not sufficiently explored the long term consequences of one particular and underexplored action? In Reinforcement Learning an agent is always dealing with this trade-off during training: An agent aims at exploiting the acquired knowledge on the environment, but at the same time the agent has to explore and learn more about the environment as well as the underlying reward structure [53]. This is theoretically well researched for single decisions in multi-armed bandit problems in which there is no distinction of different states, but for all possible actions an estimate can be learned over time. But this particular problem becomes more renounced in sequential decision making tasks in which the agent is dealing with a series of decisions that can influence each other. In many such sequential tasks rewards are sparse and only experienced quiet

(a)
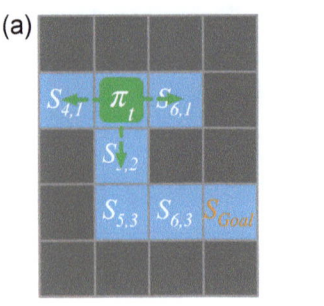

(b)
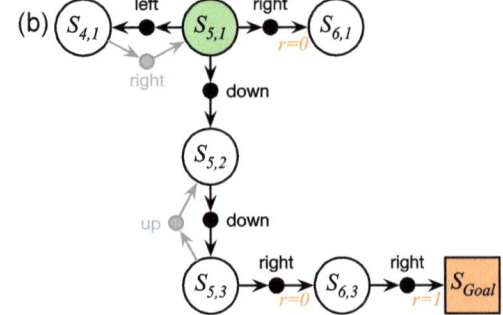

Fig. 2 a State representation for a maze environment. Illustration of a map consisting of discrete square areas, each represented as a single state $S_{x,y}$ of the MDP (shown here as grey squares; the blue background represents part of the hallway of the maze, and the dark color represents walls). The agent (shown in green) is controlled by a policy π_t that has to pick in which direction to move (out of a set of a discrete options such as cardinal directions). **b** Markov Decision Process visualization. States are depicted as circles (the current state has a green shading). The graph can be traversed by picking an action shown as a black dot (or gray one). A terminating state is depicted as a square, rewards (r) are shown in orange attached to transitions after selecting an action. In the maze case only the final state is rewarded ($r > 0$)

late in the sequence or even at the end of a sequence, e.g., considering an animal running through a maze that has to find some rewarding food (Fig. 2a). The early decisions are equally important for finding the goal, but this only becomes apparent after experiencing the reward [14].

In this chapter, we will introduce Reinforcement Learning (RL). First, we provide the formal framework given by a Markov Decision Process (MDP). Afterwards, we introduce a taxonomy for the field of Reinforcement Learning and describe more in detail the main difficulties in (Deep) Reinforcement Learning. Last, we will turn back towards biological inspiration for robotic control and how the structure of biological control architectures can positively influence and alleviate some of the current problems in Deep Reinforcement Learning.

2 Reinforcement Learning Formalization: Markov Process

In Reinforcement Learning (RL), an agent observes the state of an environment. The task is to choose an action based on the current observation. As a result, the agent receives a reward for the chosen action and an updated state of the environment. Over time, the agent aims to maximize this reward. This optimization problem can be formalized as a Markov Decision Process (MDP), which provides a model for sequential decision-making (for a detailed introduction, see [53]). A Markov Decision Process is defined as a tuple of:

- observable states $\mathcal{S}$,
- possible actions $\mathcal{A}$,
- a reward signal $\mathcal{R}$ providing an immediate reward for choosing an action depending on the current state,
- transition probabilities $\mathcal{P}$ that describe the probability distribution over states when taking an action when currently being in that specific state, and
- a discount factor γ that describes how to decrease the weight of future rewards (usually taken from [0, 1]).

In the following, we will briefly explain and discuss each of these components. Both, the state space and the action space can be either discrete or continuous. The goal in RL is to find a policy $\pi(s)$ that chooses optimal actions depending on the current state s for the Markov Decision Process. Discrete spaces provide a conceptually simpler case. When both the observation as the state space and the action space are discrete, a policy can be represented as a table and the MDP can be nicely visualized as a graph of states and actions leading to transitions for traversing the graph. As a simple example, consider a maze with a discrete set of locations in a fixed grid and the actions correspond to cardinal direction for navigating the maze (Fig. 2).

As one example for continuous spaces, in robot motor control, both state space and action space are usually continuous. The state space as the input space is given by sensory signals, and the continuous action space corresponds directly to motor signals, as in the control of a robot when moving the joint of a robot into a specified position. In the continuous case, learning operates over high-dimensional continuous spaces. Therefore, the agent can't learn an exhaustive list matching all possible states to optimal actions. Instead, we want to learn a high-dimensional continuous function over the state space. In Deep Reinforcement Learning (DRL), deep neural networks are used as non-linear function approximators for learning such a function [2].

In a Markov Decision Process, an agent decides which action to take depending on the current sensed state. When performing the selected action the agent is provided with a new, updated state and, in addition, a reward signal. Both can be probabilistic as well, for example, in the well-known frozen lake environment an agent chooses actions to navigate a frozen lake with holes in the ice [53]. The task is to navigate the maze-like frozen lake without falling into the holes (Fig. 3). Importantly, as the surface of the lake is considered slippery, the agent can't be certain to end up exactly in the intended direction. Whenever he chooses an action, in a fraction of the cases he might end up in a different location; for example, when he wants to move to the right, only in a third of the cases does he actually move right and might instead move up or down in the grid world with equal probability (these are the transition probabilities $\mathcal{P} : \mathcal{S} \times \mathcal{A} \rightarrow \mathcal{S}$).

Reinforcement Learning is driven by rewards. After an action has been taken, a new state and reward are returned. A positive reward should reinforce the sequence of actions that led to this reward, e.g., in the maze case not only the last action should be reinforced, but the whole path. Therefore, rewards are usually distributed along the whole sequence of decisions. But the further away in time a decision

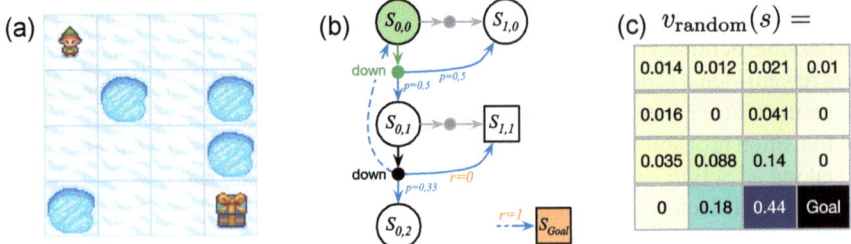

Fig. 3 a Illustration showing the frozen lake environment realized in the OpenAI gym [8]. **b** Section of the Markov Decision Process. Shown are the states for the upper left part, and from the upper left corner, the action 'down' has been selected. This only moves the agent downwards to the state $S_{0,1}$ with a 50% chance. Otherwise, the agent might end up in the state $S_{1,0}$ to the right. Only when reaching a goal is a reward (of $r = 1$) earned. **c** Value function for the states of the frozen lake environment depicted as a heatmap. Shown are the approximated expected values for the agent for the given place and when following a random policy. As an episode is terminated when falling into a hole, the expected reward always stays 0 in these places. In the cases of the other states, states close to the goal have a higher value, as randomly selecting an action towards the goal is more likely

has been from finally receiving a reward, the less such an early decision might have contributed towards receiving this reward, or this contribution becomes usually much more uncertain. Therefore, we want to decrease the contribution of early decisions (in our maze example in Fig. 1b, the agent might initially have turned downwards to the dead end at the bottom and only much later turned towards the upper part of the maze that leads to the goal; after receiving a reward, only the last steps that caused reaching the goal should be strongly reinforced). The discount factor—typically less than one—distributes a reward across a sequence of decisions, while progressively reducing the importance of future decisions, as their contribution to the final reward becomes more uncertain.

The goal of the optimization is to find a policy $\pi(s)$ for which the accumulated reward over time is maximized. This accumulated reward is the return $G_t = R_{t+1} + \gamma R_{t+2} + \cdots = \sum_{k=0}^{\infty} \gamma^k R_{t+k+1}$. A value function can be used to estimate this return for each state (or state-action combination):

$$v_\pi(s) = \mathbb{E}\left[R_t \mid s\right] = \sum_a \pi(a|s) \sum_r \sum_{s'} P\left(s', r|s, a\right)\left[r + \gamma v_\pi(s')\right] \qquad (1)$$

In this general formulation the individual parts are all considered stochastic: First, action selection itself is modeled as a probability distribution, and the next action would be sampled from this probability distribution, e.g., randomly during exploration. It is important to note that a value function always depends on the given policy π as the estimated value directly depends on how actions are selected which is defined by the policy the agent is following. The second and third sums in the definition of the value function deal with the joint probability distribution that describes the reaction of the environment. Both, the returned reward r as well as the next state s' could be non-deterministic. Therefore it has to be integrated over all possible rewards and possible next states. Further note, that only the transition towards the next state is estimated while the following trajectory is approximated using the value function itself recursively (this is called the Bellman equation [4]). Finding the value function for a given policy is termed policy evaluation. In addition, one can formulate an action-value function that estimates the expected return when starting from a current state and choosing a specific action followed by adhering to the current policy:

$$q_\pi(s, a) = \mathbb{E}_\pi[G_t|S_t = s, A_t = a] \tag{2}$$
$$= \sum_r \sum_{s' \in S} p(r, s'|s, a)\left(r + \gamma \sum_{a' \in \mathcal{A}} \pi(a'|s')q_\pi(s', a')\right)$$

Reinforcement Learning algorithms can either directly use and optimize a policy for selecting the next action when in a given state, or value functions can be used to estimate the competency of a current policy, and these can be iteratively exploited [2]. In that case, for a given policy—initially, this may be a random policy or later on, the agent is acting mostly greedily on a currently considered optimal policy—an agent subsequently evaluates the value function for the given policy. Afterwards, the policy gets updated: It should start to exploit the found value function—the policy selects those actions for which a higher return is expected according to the value (or action-value) function. In many current approaches, such as actor-critic learning [24, 34], both approaches are combined. While, on one hand, a policy is directly learned that selects from current state information the next action, there is, on the other hand, a critic involved which provides information on how valuable the currently considered state is. This kind of information is particularly important for ongoing learning as it allows bootstraping if a gained reward is more due to being in a high rewarding state as such or the selection of the action was responsible for yielding a reward.

3   Characterization of RL Approaches

As there are a number of differentiations among Reinforcement Learning approaches (see [2]), this section aims to provide an overview of the main characterizations of Reinforcement Learning (RL). We have already discussed how action selection is driven as one important distinction—either by exploiting the estimates of a value function or by directly employing and learning a policy.

A further distinction in Reinforcement Learning approaches deals with our knowledge of the environment: Do we possess a complete model representing the transition probabilities of the assumed Markov Decision Process (MDP)? Having complete knowledge of these joint probabilities allows us to plan ahead and apply model-based methods [30]. This means, from a given state, all possible alternatives can be considered, and we can calculate for each possible successor state its probability and the associated reward (or a distribution over rewards). Consequently, in model-based

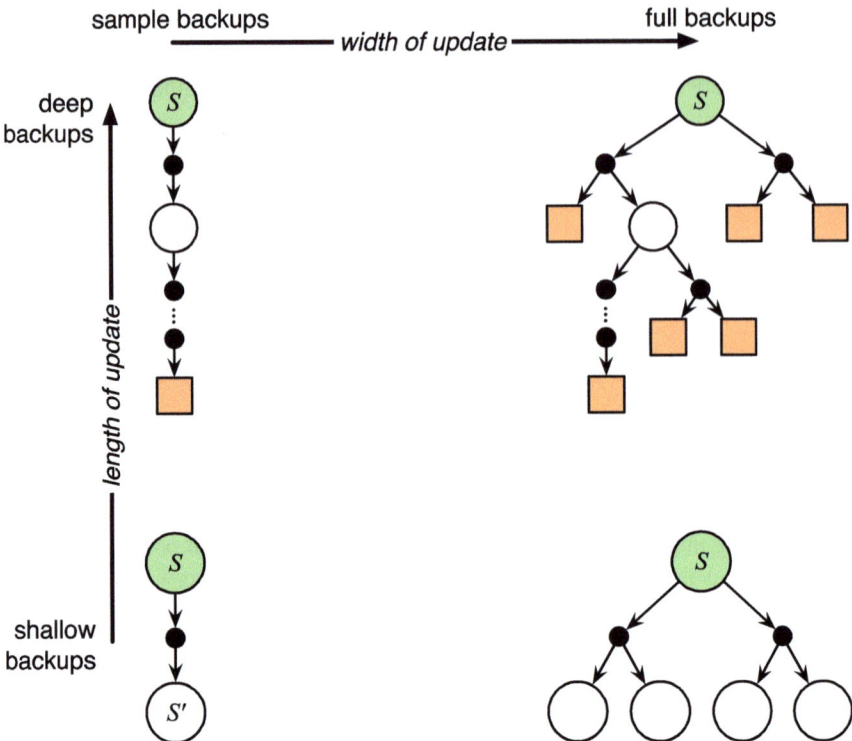

Fig. 4 Two main dimensions characterize Reinforcement Learning algorithms. The distinction is based on how backups over possible decisions are integrated during learning or constructing a policy. Vertically, the length of sequences is distinguished, i.e., either full traces (top) are considered, and the agent only learns after finishing an episode. In contrast (bottom), an agent can focus only on learning from the reward of a single step as it relies on a prediction for the rest of the episode. The horizontal organization deals with addressing a single or multiple options at each stage of the decision. On the one hand (left), only a single action is picked, and a new state plus reward are experienced. On the other hand (right), when the environment response is known and a model is given, we can exhaustively estimate all possible actions and their long-term consequences. Image adapted from [53]

methods, all possible actions can be considered and evaluated one after another (assumed for a discrete set of actions; otherwise, we deal with probability distributions over actions). Examples of such model-based approaches can be found in robotics [44, 51, 57], presupposing a form of internal model [39, 43]. The different possibilities for a decision are often envisioned as a tree spanning all these decisions in backup diagrams (see Fig. 4, top right). In a backup diagram [53], the succession of states and actions is represented: from a state (shown as an open circle as in the MDP visualization), all selectable actions are shown below (as filled black circles). Each action is further connected downwards with all possible resulting states when performing that action (these transitions could be stochastic, as in the frozen lake example, and selecting an action does not necessarily result in a given state, but one could end up in different states and receiving different rewards, as indicated in the example backup diagrams). Model-based approaches utilize the given model and evaluate over the whole width of the spanned backup diagrams. Such approaches are only applicable if a model is known or has been learned, e.g., in game-like scenarios with a clear defined progression of states following an action. In contrast, in model-free methods, the transitions themselves are not known or not needed. Such model-free approaches are considered as sample-based methods, as in each state a single decision is made, carried out, and the agent continues from the resulting state without considering alternative decisions and their consequences. Model-free methods are particularly useful when it is not possible to evaluate alternative actions, as in most real-world scenarios, e.g., when considering a robot interacting with an environment, it can be very hard to model following states faithfully (the only possibility would be to actually test a selected action on the real system without having the possibility to go back and evaluate other alternative actions).

As a second differentiation, it is often distinguished if learning is based on full traces (or a complete depth backup when considering planning) of decisions or, in contrast, only narrowed down focusing on a single decision individually with estimated consequences. When using full traces until ending up in a final state, the return over this episode can be distributed onto each of the sequential decisions. As a disadvantage, this is only possible after finishing a sequence. The results often appear quite noisy, depending on the taken path. When dealing with non-deterministic settings, the same sequence of decisions might lead to different results. When using the full trace to assign the received return from one single episode to all decisions, all these single-step decisions are considered as contributing equally (only weighted by the discount factor). In contrast, an agent could exploit already gained knowledge from previous interactions for in-between steps along a sequence of decisions. This knowledge can be integrated into the evaluation. In such a case, not the return gathered over the whole episode is used in calculating backward how much a decision has contributed to receiving this return. Instead, only the immediate reward for a particular decision is taken, and for the whole following sequence, an estimate of the expected return can be used and exploited, which is much less afflicted by noise. We have used this recursive approximation of estimating a value in the formulation of the value function in equation (2) (which is based on the Bellman equation [4]): the immediately received reward is taken and for the rest of the episode the return is estimated by the recursive application of the value function.

For the frozen lake example, when using a value function: Whenever randomly sliding into the ice—even though the agent selected the correct direction towards the goal—the whole trace would have zero return, and the agent couldn't learn anything from this experience. In contrast, for the example scenario in which the agent falls into the hole directly above the goal position, there are still parts of that trace that appear positive and should be reinforced. When considering a single decision—in the one-step approach—only the immediate reward for taking that decision would be considered, and all the consequences for the rest of the sequence would be approximated as given by the value function. In that case, all steps that lead towards states close to the goal should become more and more valuable, providing an incentive during learning and reinforcing over time a path towards the goal.

Lastly, a distinction is made between on-policy methods and off-policy methods. Central to Reinforcement Learning is the trade-off between exploration and exploitation. In on-policy control, a single policy is used to, on one hand, gather experience and, on the other hand, is directly improved based on the reward received from that experience. This is known as on-policy learning. However, when applied to many real-world tasks, it can become problematic to sufficiently explore while following and improving a single policy. It is therefore often considered beneficial to use a different policy aimed at exploration for making decisions, compared to the policy that is being improved over time. A particular example is when one gathers a lot of experience in parallel that should be integrated into one optimal policy. In off-policy learning, this optimal policy is subsequently updated in larger time intervals based on the gathered experience. The behavioral policy that is used to make decisions is usually derived from the optimal policy in a way that still guarantees, to a certain degree, some form of exploration, e.g., by occasionally selecting an action randomly while otherwise acting greedily.

4 Current State of Deep Reinforcement Learning

In recent years, Deep Reinforcement Learning (DRL) has achieved success in many areas. Initially, the groundbreaking work of [35] applied neural networks to learn an action-value function in Deep-Q-Learning in an end-to-end fashion in the area of computer games. In this approach, a control system was trained over time, collecting experience as sequences of images from Atari computer games coupled with rewards. The task of the system was to select the next discrete action, which corresponded to a specific key press in these games. The system learned an approximation of a q-function as a deep neural network that consisted of convolutional layers for processing the image input and some dense layers. Over time, the system continuously improved and was able to solve many of the tested games or to play many of these at a level better than human. Two crucial ingredients of the Deep Q-Network (DQN) approach were, first, the collection of experience in a buffer from which data was randomly sampled for training the neural networks. Secondly, in the estimation of the value for the recursive part—the estimate for the subsequent trajectory after deciding on the next action—a second q-network was used that was less frequently updated. This stabilized these estimates and proved crucial for the convergence of the q-function approximation. As a result, the DQN approach was able to solve many of the Atari computer games and perform in many of these on a better than human level.

In robotics control, policy-based approaches also led to great results, and successes could be demonstrated in areas such as manipulation [28] or locomotion [18, 21, 22, 33, 58]. However, application in robotic scenarios highlights one particular drawback of DRL approaches. Deep Reinforcement Learning requires vast amounts of training data and a large number of iterations. This is problematic

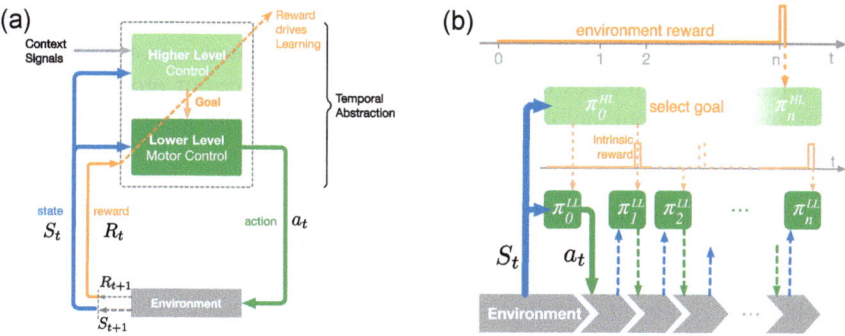

Fig. 5 **a** Extending the standard view of interaction with the environment in RL to a hierarchical approach [26, 32]. It consists of two interleaved loops of reinforcement learners that are usually assumed to learn at different rates. For higher-level control (shown in light green), this is in agreement with what we know about the structure of motor control in animals [13]. **b** Learning over time is shown for the two different levels of the hierarchy. This demonstrates a form of temporal abstraction in which the different levels operate on different temporal scales and rates. The higher level (light green) only operates at a slow timescale, being activated every couple of time steps. The higher level might aim for a sparse environmental reward. The lower level operates on a more fine-grained timescale and provides detailed action during each control step. It is conditioned on the higher level output, which provides a form of goal context and, during learning, can set intrinsic rewards that guide learning towards a diverse set of lower-level actions

for application on real robots, and in many cases, systems were initially trained in simulation and only afterward were the trained controllers transferred onto a real robot [54]. But such a transfer appeared problematic in many cases as well, pointing out a second general problem of Deep Reinforcement Learning: DRL can be prone to forms of overfitting [59]. When trained in a specific setting, a learned controller narrows down the mapping from observations to the action space. While neural networks provide very good function approximators for interpolation, they tend to perform worse on extrapolation [17, 19]. As a consequence, when a neural network is trained in simulation on a specific setting fulfilling one particular task, it becomes highly specialized and will mostly focus on a small part of the whole input space that is common to the system's experiences [36]. If the real robotic system behaves differently or operates in a slightly different part of the observation space, the control system might produce quite different and unexpected behavior.

A third general problem concerns the specification of rewards: Rewards guide the learning process in Deep Reinforcement Learning. Ideally, describing a high-level goal, such as reaching a certain goal location in a locomotion task, would be sufficient. Unfortunately, a simple sparse reward that is received after accomplishing a goal usually does not provide enough guidance [20]. In many cases, reaching such a rewarded state would require a series of optimal decisions that could only be discovered through random exploration, which quickly becomes intractable. Such sparse rewards are therefore problematic because, on the one hand, it is hard to encounter them at all. On the other hand, it is often unclear which of all the previous decisions contributed to receiving the reward (a credit assignment problem).

Current research focuses on how to deal with these types of problems and is taking inspiration from biology and the organization of motor control in animals. We will briefly mention and discuss two main characteristics that can mitigate these problems [47]. First, we will discuss the hierarchical organization of control systems. Secondly, the decentralization of the control system and concurrent processing. Key to hierarchical systems is the idea of an abstraction of tasks [6, 31]. Instead of solving a problem through a detailed sequence of decisions, the general idea is to deconstruct a problem into different subtasks that can be solved individually [32]. Hierarchical Reinforcement Learning [16, 21] deals with different levels of typical reinforcement learning loops (Fig. 5): On a lower level, individual subtasks have to be solved by finding sequences of decisions that solve certain subgoals. A higher level selects subgoals subsequently and is tasked with finding a sequence of subgoals. Actions correspond on the higher level to selecting certain subgoals. The choice of subgoals is crucial, but in many tasks, these can be specified quite naturally [55]. As one example from the original DQN study [35]: The simple DQN approach failed in scenarios with sparse and delayed rewards, e.g., in the game Montezuma's revenge, the player has to first find and pick up a key found in the given map before the player could, secondly, receive a reward for opening a door and finishing a level. Stumbling on such a long sequence by chance appears intractable. In a hierarchical approach, Kulkarni and colleagues [26] broke the problem down into two different levels of a hierarchy. On the higher level, a simple visual system identified all visible objects as subgoals. Reinforcement Learning on this level explored random sequences of visiting these objects one after another. On the lower level, for each selected visible object, a policy was trained towards reaching that target object unharmed. This approach was successful in solving the game as it learned quite quickly how to reach all objects. And afterward, could rely on these lower-level policies when selecting a lower-level policy on the higher level by selecting a new subgoal in a novel sequence. Similar approaches have already been applied in robotic tasks as well [21, 22, 49], but specifying and identifying sensible subgoals appears as a more difficult question. For a more detailed overview of hierarchical reinforcement learning and the biological inspiration, see [32].

As a second organizational principle, we will focus on the decentralization [10, 15] of motor control structures. Biological control systems operate in such a decentralized fashion [1]. Local concurrently operating control modules enable fast (re-)actions, as control is distributed into local control circuits that handle the detailed control of movement. This enables adaptive behavior of animals and dealing with difficult control tasks that require immediate responses. Importantly, decentralized control is not restricted to the lowest level responsible for reflex-movements of single joints. But decentralization is a central part of motor control that fundamentally contributes to coordinated behavior. One example is the control of locomotion and the contribution of—sensory driven—oscillations. Detailed work in invertebrates, as for example in insects, allows for a closer look at underlying neuronal control circuits that drive this behavior. On lower levels, there is a decentralized organization into concurrent,

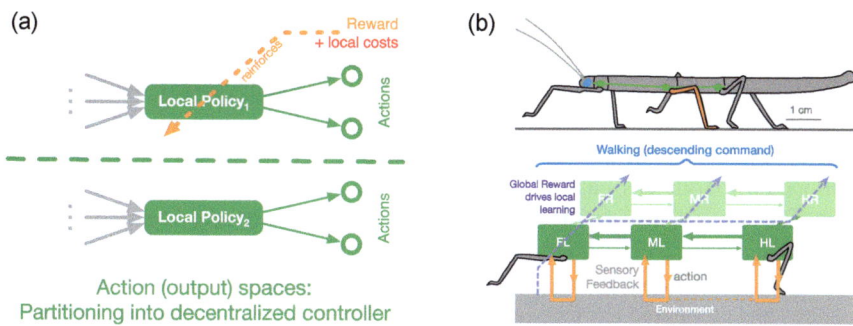

Fig. 6 Conceptual view of decentralization in Deep Reinforcement Learning. Panel (a) illustrates the notion of partitioning the action space. In a decentralized case, multiple control circuits (shown in green) operate on smaller and as independent assumed action spaces (shown on the right) based on given observations (left). As RL is based on the exploration of possible action combinations, this decrease in the number of degrees of freedom can dramatically reduce the space of possible mappings from states to actions that need to be explored. **b** As an example, a decentralized motor control structure as assumed in insects is shown. On top, a schematic of a stick insect is shown. At the bottom, a schematic of the decentralized organization of the motor control system in which each leg is controlled by an individual controller (shown in green; FL–front left leg, ML–middle left leg, HL–hind left leg, FR–front right leg, MR–middle right leg, HR–hind right leg). Behavior emerges as a result of decentralized and locally interacting concurrent control structures [48]. In decentralized DRL, each of the leg controllers acts independently and is trained individually. RL is driven by a (partially) shared reward signal [50]

small control circuits [5, 23, 46]. Control is constituted by multiple, partially independent decentralized and concurrently operating controllers. As explained above, these operate on different levels of abstraction [13], e.g., it is assumed that there is a control circuit for each individual leg, but even further, control is distributed on a lower level onto the single joint level [45]. This form of modularization into decentralized local control clusters reflects the natural organization of the body itself. Overall, behavior emerges from the interactions of all these local control circuits that are coupled through local connections [5, 11] and operate in an environment. This decentralization is assumed to be central in facilitating the adaptivity of the overall system.

Decentralization, as a form of partitioning the action space of a system into smaller subgroups of local actuators, has mostly been neglected in DRL research (Fig. 6 provides a conceptual view). Current DRL approaches focus on optimization and learning in a single central control unit. DRL is constantly tasked with handling the trade-off between exploration and exploitation over the action space. Increasingly larger action spaces lead to exponentially more possibilities for exploration and, expectedly, therefore require much longer training times (as observed in current DRL approaches and discussed above as a major problem for current DRL approaches). Decentralization offers an approach to break control down into multiple control modules, which reduces the dimensionality of action spaces for the individual controller [50]. In legged robots, the structure of the system naturally suggests distributing control into a single controller per each leg [12, 25, 48].

In an experiment [50], we tested such a decentralized control approach for a four-legged simulated robotic agent from the OpenAI gym environment [8]. The agent is equipped with four legs, each with two joints overall, leading to eight degrees of freedom. The action space is quite limited, which allows even standard centralized DRL approaches to succeed in a reasonable time (training in the ranges of hours, which still equates to millions of simulation steps). We set up a series of decentralized architectures and compared learning in these decentralized control architectures with the standard DRL approach [50]. In the decentralized case, there were four independent controllers that were trained individually. The individual leg controller was restricted to control of the two degrees of freedom of that particular leg, and the action space was therefore reduced to two dimensions. As a first observation, we found that the decentralized systems performed after training on the exact same level as a standard approach (Fig. 7). Secondly, with respect to the training phase, the decentralized controller learned much quicker and never got stuck in local minima [50]. It appears that in decentralized control, the controller learns a simpler mapping from sensory inputs towards outputs. As a third result, decentralized controllers were tested in novel and more difficult terrains, e.g., climbing through small hilly areas. In that case, different types of decentralized controllers performed differently, and it showed as important for such a more difficult case to offer enough sensory information from other legs as well to the decentralized controllers. Offering additional information to individual decentralized control circuits didn't affect learning quality or speed, but for climbing through difficult terrain, it was required information [50]. Decentralized control approaches that include additional information from other legs performed more robustly in such more demanding and novel terrains. These approaches appear to better generalize and transfer to untrained scenarios. In contrast, quite detailed information in centralized approaches might have induced a form of overfitting, as relations between sensory inputs were constructed during learning that actually did not help for transfer to a slightly changed task [50]. Decentralization appears as a promising direction for DRL, leading to more robust control systems and accelerated training that has now been adapted in a couple of approaches [3, 12, 56].

5 Conclusion

Deep Reinforcement Learning (DRL) has become quite popular over the last few years, as it has seen more and more applications in many diverse areas, ranging from robotics to Large Language Models. There still remain quite severe conceptual obstacles, but this has made DRL an interesting and very active field for research. In this chapter, we briefly introduced the general concept of Reinforcement Learning as well as Deep Reinforcement Learning and pointed out a couple of current problems or difficulties. This should provide the reader with a good starting point

(a)

(b) Individual **Learning Curves** of individual seeds

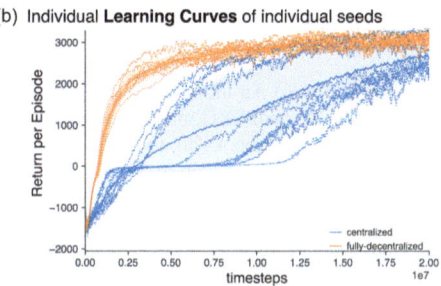

Fig. 7 In (**a**), the quadruped walking agent is shown on a slightly uneven terrain (based on an OpenAI standard DRL task). **b** Comparison of two different control architectures for learning to walk for this four-legged simulated walker. Shown are learning curves—return per episode given on the vertical axis shown over time (horizontal axis)—for a centralized controller (blue) and a decentralized controller (orange, each leg is controlled by an independent controller which partitions the action space). Given are ten individual traces (dashed lines) and the average trace (solid line) each representing a single training run [50]

for understanding what makes a problem a Reinforcement Learning problem or how to reformulate a given task within the conceptual framework of Reinforcement Learning—framing it as an active exploration task that can be connected towards a reward that should drive learning. Furthermore, we highlighted in two examples how Deep Reinforcement Learning can benefit from taking closer inspiration from biological control systems and taking into account considerations on the underlying architectures. Overall, DRL is a quickly evolving field as it is applied in different areas and there is much more inspiration as well as current developments on DRL for a growing set of different problems. As one example, we want to point out the case of multi-agent scenarios. These are intrinsically difficult for any learning system, as while an agent is learning from interactions with an environment that includes other agents, the response of the environment can change by itself as the other agents' response characteristics are not fixed but could also be subject to change [42]. There have been first successes in such multi-agent settings [40] that are based on Deep Reinforcement Learning as well. This is another fascinating area of research that will show interesting developments in the next few years and will probably contribute massively towards interactive robots and agents that should support us at work or home.

Declaration of AI and AI-Assisted Technologies in the Writing Process

During the preparation of this work the author(s) used ChatGPT (OpenAI (2024), ChatGPT, Version January 10, 2024) in order to do copy editing. After using this tool/service, the author(s) reviewed and edited the content as needed and take(s) full responsibility for the content of the publication.

References

1. Alon U (2006) An introduction to systems biology: design principles of biological circuits. Chapman & Hall/CRC mathematical and computational biology. Taylor & Francis. https://books.google.de/books?id=tcxCkIxzCO4C
2. Arulkumaran K, Deisenroth MP, Brundage M, Bharath AA (2017) A brief survey of deep reinforcement learning. IEEE Signal Process Mag 34(6):26–38. https://doi.org/10.1109/MSP.2017.2743240. arXiv:1708.05866
3. Bellegarda G, Ijspeert A (2022) CPG-RL: learning central pattern generators for quadruped locomotion. IEEE Robot Autom Lett 7(4):12547–12554
4. Bellman R (1952) On the theory of dynamic programming. Proc Natl Acad Sci 38(8):716–719
5. Bidaye SS, Bockemühl T, Büschges A (2018) Six-legged walking in insects: how CPGs, peripheral feedback, and descending signals generate coordinated and adaptive motor rhythms. J Neurophysiol 119(2):459–475. https://doi.org/10.1152/jn.00658.2017
6. Binder MD, Hirokawa N, Windhorst U (2009) Motor control hierarchy. In: Encyclopedia of neuroscience. Springer, Berlin, Heidelberg, pp 2428–2428. https://doi.org/10.1007/978-3-540-29678-2_3583
7. Bishop CM, Nasrabadi NM (2006) Pattern recognition and machine learning, vol 4. Springer
8. Brockman G, Cheung V, Pettersson L, Schneider J, Schulman J, Tang J, Zaremba W (2016) Openai gym. arxiv:1606.01540
9. Chatzilygeroudis KI, Vassiliades V, Stulp F, Calinon S, Mouret J (2020) A survey on policy search algorithms for learning robot controllers in a handful of trials. IEEE Trans Robotics 36(2):328–347. https://doi.org/10.1109/TRO.2019.2958211
10. Clune J, Mouret JB, Lipson H (2013) The evolutionary origins of modularity. Proc R Soc B Biol Sci 280(1755):20122863. https://doi.org/10.1098/rspb.2012.2863. https://royalsocietypublishing.org/doi/abs/10.1098/rspb.2012.2863
11. DeAngelis BD, Zavatone-Veth JA, Clark DA (2019) The manifold structure of limb coordination in walking Drosophila. Elife 8:e46409. https://doi.org/10.7554/eLife.46409. https://elifesciences.org/articles/46409
12. Deshpande AM, Hurd E, Minai AA, Kumar M (2023) Deepcpg policies for robot locomotion. IEEE Trans Cogn Dev Syst
13. Dickinson MH, Farley CT, Full RJ, Koehl MAR, Kram R, Lehman S (2000) How animals move: an integrative view. Science 288(5463):100–106. https://doi.org/10.1126/science.288.5463.100. http://science.sciencemag.org/content/288/5463/100
14. Ecoffet A, Huizinga J, Lehman J, Stanley KO, Clune J (2021) First return, then explore. Nature 590(7847):580–586
15. Ellefsen KO, Huizinga J, Torresen J (2020) Guiding neuroevolution with structural objectives. Evol Comput 28(1):115–140
16. Eppe M, Gumbsch C, Kerzel M, Nguyen PD, Butz MV, Wermter S (2022) Intelligent problem-solving as integrated hierarchical reinforcement learning. Nat Mach Intell 4(1):11–20
17. Fujimoto S, Conti E, Ghavamzadeh M, Pineau J Benchmarking batch deep reinforcement learning algorithms. arXiv:1910.01708 [cs, stat]
18. Haarnoja T, Ha S, Zhou A, Tan J, Tucker G, Levine S (2019) Learning to walk via deep reinforcement learning. arXiv:1812.11103 [cs, stat]
19. Haley PJ, Soloway D (1992) Extrapolation limitations of multilayer feedforward neural networks. In: [Proceedings 1992] IJCNN international joint conference on neural networks, vol 4. IEEE, pp 25–30
20. Hare J (2019) Dealing with sparse rewards in reinforcement learning. arXiv:1910.09281
21. Heess N, Wayne G, Tassa Y, Lillicrap T, Riedmiller M, Silver D (2016) Learning and transfer of modulated locomotor controllers. arXiv:1610.05182
22. Hwangbo J, Lee J, Dosovitskiy A, Bellicoso D, Tsounis V, Koltun V, Hutter M (2019) Learning agile and dynamic motor skills for legged robots. Sci Robot 4(26):eaau5872. https://doi.org/10.1126/scirobotics.aau5872. http://robotics.sciencemag.org/lookup/doi/10.1126/scirobotics.aau5872

23. Kano T, Kanauchi D, Ono T, Aonuma H, Ishiguro A (2019) Flexible coordination of flexible limbs: decentralized control scheme for inter- and intra-limb coordination in brittle stars' locomotion. Front Neurorobot 13:104. https://doi.org/10.3389/fnbot.2019.00104 https://www.frontiersin.org/article/10.3389/fnbot.2019.00104
24. Konda V, Tsitsiklis J (1999) Actor-critic algorithms. Adv Neural Inf Process Syst 12
25. Konen K, Korthals T, Melnik A, Schilling M (2019) Biologically-inspired deep reinforcement learning of modular control for a six-legged robot. In: 2019 IEEE international conference on robotics and automation workshop on learning legged locomotion workshop, (ICRA) 2019, Montreal, CA, May 20–25
26. Kulkarni T, Narasimhan K, Saeedi A, Tenenbaum JB (2016) Hierarchical deep reinforcement learning: integrating temporal abstraction and intrinsic motivation. Adv Neural Inf Process Syst 3675–3683
27. Levine S, Finn C, Darrell T, Abbeel P (2016) End-to-end training of deep visuomotor policies. J Mach Learn Res 17(39):1–40
28. Levine S, Wagener N, Abbeel P (2015) Learning contact-rich manipulation skills with guided policy search. arXiv:1501.05611
29. Lillicrap TP, Hunt JJ, Pritzel A, Heess N, Erez T, Tassa Y, Silver D, Wierstra D (2015) Continuous control with deep reinforcement learning. arXiv:1509.02971
30. Luo FM, Xu T, Lai H, Chen XH, Zhang W, Yu Y (2024) A survey on model-based reinforcement learning. Sci China Inf Sci 67(2):121101
31. Mengistu H, Huizinga J, Mouret JB, Clune J (2016) The evolutionary origins of hierarchy. PLOS Comput Biol 12(6):e1004829
32. Merel J, Botvinick M, Wayne G (2019) Hierarchical motor control in mammals and machines. Nat Commun 10(1):1–12. https://doi.org/10.1038/s41467-019-13239-6. https://www.nature.com/articles/s41467-019-13239-6
33. Miki T, Lee J, Hwangbo J, Wellhausen L, Koltun V, Hutter M (2022) Learning robust perceptive locomotion for quadrupedal robots in the wild. Sci Robot 7(62):eabk2822
34. Mnih V, Badia AP, Mirza M, Graves A, Lillicrap TP, Harley T, Silver D, Kavukcuoglu K (2016) Asynchronous methods for deep reinforcement learning. CoRR arXiv:1602.01783
35. Mnih V, Kavukcuoglu K, Silver D, Rusu AA, Veness J, Bellemare MG, Graves A, Riedmiller M, Fidjeland AK, Ostrovski G, Petersen S, Beattie C, Sadik A, Antonoglou I, King H, Kumaran D, Wierstra D, Legg S, Hassabis D (2015) Human-level control through deep reinforcement learning. Nature 518(7540):529–533. https://doi.org/10.1038/nature14236
36. Mozifian M, Higuera JCG, Meger D, Dudek G (2019) Learning domain randomization distributions for training robust locomotion policies. arXiv:1906.00410 [cs, stat]
37. Murphy KP (2012) Machine learning: a probabilistic perspective. MIT Press
38. Neftci EO, Averbeck BB (2019) Reinforcement learning in artificial and biological systems. Nat Mach Intell 1:133–143. https://doi.org/10.1038/s42256-019-0025-4 https://www.nature.com/articles/s42256-019-0025-4
39. Nguyen-Tuong D, Peters J (2011) Model learning for robot control: a survey. Cogn Process 12:319–340
40. Oroojlooy J, Hajinezhad D (2019) A review of cooperative multi-agent deep reinforcement learning. arXiv:1908.03963 [cs, math, stat]
41. Ouyang L, Wu J, Jiang X, Almeida D, Wainwright C, Mishkin P, Zhang C, Agarwal S, Slama K, Ray A et al (2022) Training language models to follow instructions with human feedback. Adv Neural Inf Process Syst 35:27730–27744
42. Papoudakis G, Christianos F, Rahman A, Albrecht SV (2019) Dealing with non-stationarity in multi-agent deep reinforcement learning. arXiv:1906.04737 [cs, stat]
43. Schilling M, Cruse H (2007) Hierarchical MMC networks as a manipulable body model. In: Proceedings, pp 2141–2146
44. Schilling M, Cruse H (2017) ReaCog, a minimal cognitive controller based on recruitment of reactive systems. Front Neurorobotics 11. https://doi.org/10.3389/fnbot.2017.00003
45. Schilling M, Cruse H (2020) Decentralized control of insect walking: a simple neural network explains a wide range of behavioral and neurophysiological results. PLoS Comput Biol 16(4):e1007804. https://doi.org/10.1371/journal.pcbi.1007804

46. Schilling M, Cruse H (2023) NeuroWalknet, a controller for hexapod walking allowing for context dependent behavior. PLoS Comput Biol 19(1):e1010136
47. Schilling M, Hammer B, Ohl FW, Ritter HJ, Wiskott L (2023) Modularity in nervous systems—a key to efficient adaptivity for deep reinforcement learning. Cogn Comput 1–16
48. Schilling M, Konen K, Ohl FW, Korthals T (2020) Decentralized deep reinforcement learning for a distributed and adaptive locomotion controller of a hexapod robot. In: 2020 IEEE/RSJ international conference on intelligent robots and systems (IROS). IEEE, pp 5335–5342
49. Schilling M, Melnik A (2018) An approach to hierarchical deep reinforcement learning for a decentralized walking control architecture. In: Biologically inspired cognitive architectures 2018. Proceedings of the ninth annual meeting of the BICA society, vol 848. https://pub.uni-bielefeld.de/record/2934190
50. Schilling M, Melnik A, Ohl FW, Ritter HJ, Hammer B (2021) Decentralized control and local information for robust and adaptive decentralized deep reinforcement learning. Neural Netw 144:699–725
51. Schilling M, Paskarbeit J, Ritter H, Schneider A, Cruse H (2021) From adaptive locomotion to predictive action selection–cognitive control for a six-legged walker. IEEE Trans Robot 1–17. https://doi.org/10.1109/TRO.2021.3106832
52. Schultz W, Dayan P, Montague PR (1997) A neural substrate of prediction and reward. Science 275(5306):1593–1599
53. Sutton RS, Barto AG (2018) Reinforcement learning: an introduction, 2nd edn. The MIT Press. http://incompleteideas.net/book/the-book-2nd.html
54. Tan J, Zhang T, Coumans E, Iscen A, Bai Y, Hafner D, Bohez S, Vanhoucke V (2018) Sim-to-real: learning agile locomotion for quadruped robots. arXiv:1804.10332 [cs]
55. Vezhnevets AS, Osindero S, Schaul T, Heess N, Jaderberg M, Silver D, Kavukcuoglu K (2017) Feudal networks for hierarchical reinforcement learning. In: Proceedings of the 34th international conference on machine learning, vol 70, pp 3540–3549
56. Wang T, Liao R, Ba J, Fidler S (2018) Nervenet: learning structured policy with graph neural networks. In: International conference on learning representations
57. Zhang B, Liu P (2021) Model-based and model-free robot control: a review. In: RiTA 2020: proceedings of the 8th international conference on robot intelligence technology and applications. Springer, pp 45–55
58. Zhang C, Rudin N, Hoeller D, Hutter M (2023) Learning agile locomotion on risky terrains. arXiv:2311.10484
59. Zhang C, Vinyals O, Munos R, Bengio S (2018) A study on overfitting in deep reinforcement learning. arXiv:1804.06893 [cs, stat]

An Introduction to Reservoir Computing

Michael te Vrugt

Abstract There is a growing interest in the development of artificial neural networks that are implemented in a physical system. A major challenge in this context is that these networks are difficult to train since training here would require a change of physical parameters rather than simply of coefficients in a computer program. For this reason, *reservoir computing*, where one employs high-dimensional recurrent networks and trains only the final layer, is widely used in this context. In this chapter, I introduce the basic concepts of reservoir computing. Moreover, I present some important physical implementations coming from electronics, photonics, spintronics, mechanics, and biology. Finally, I provide a brief discussion of quantum reservoir computing.

1 Introduction

You know that it is possible to perform machine learning tasks on a computer, but did you know that it is also possible to do so with a bucket of water? Precisely that was demonstrated in Ref. [12]. Input data was mechanically fed into a bucket, recordings of the water surface then could be used for classification tasks. This is a form of *reservoir computing* (RC), which this chapter will provide an introduction to.

The core idea of RC is that a significant portion of the computing task is performed not by a trained network, but by a very high-dimensional system (reservoir) that is essentially treated as a black box and whose output is fed into a single readout layer, which is the only component of the network that is trained. This setup is illustrated in Fig. 1. Since physical systems are generally much more difficult to train than neural networks on a computer, RC is a very promising approach for implementing artificial intelligence in a physical system (for example a bucket of water or–more relevant in

M. te Vrugt (✉)
Institut für Physik, Johannes Gutenberg-Universität Mainz, Mainz, Germany
e-mail: tevrugtm@uni-mainz.de

© The Author(s) 2026
M. te Vrugt (ed.), *Artificial Intelligence and Intelligent Matter*, Machine Intelligence for Materials Science, https://doi.org/10.1007/978-3-032-04129-6_4

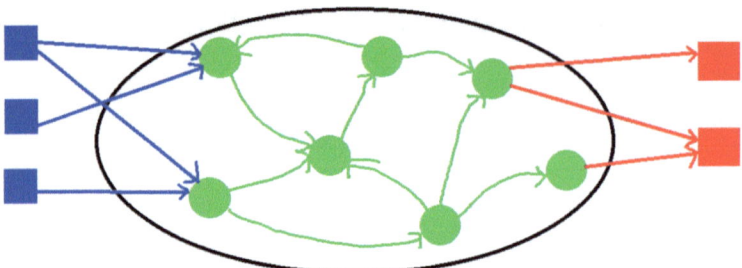

Fig. 1 Illustration of the basic framework of reservoir computing. The input (blue) is fed into a recurrent neural network (green) and subsequently into an output layer (red). (A similar figure can be found in Ref. [42])

practice–optical or magnetic systems). Consequently, RC has attracted considerable interest among physicists in recent years. From a computer science point of view, on the other hand, RC is useful for the otherwise difficult task of training recurrent neural networks.

The foundations of what is nowadays referred to as reservoir computing were independently developed by Jaeger [23], who referred to it as *echo state network*, and by Maass et al. [34], who called it *liquid state machine*. (There were some earlier ideas in this direction–see Refs. [27, 28, 44]–that are largely unknown today [38].) The name "reservoir computing" was coined in Refs. [52, 53] to unify these concepts. RCs are particularly useful for tasks involving some temporal dynamics without needing external memory, since what they do depends also on the value of the input signal at previous times. Reviews of RC can be found in Refs. [10, 11, 32, 33, 42], a book on the topic was edited by Nakajima and Fischer [38].

2 Basic Concepts of Reservoir Computing

See Fig. 1.

2.1 How Reservoir Computing Works

The discussion follows Refs. [10, 36] (and also adapts the notation used there).

Suppose that we have an n-dimensional training input signal $\mathbf{u}_{\text{train}}(t)$ (depending on time t) that is supposed to lead to a certain m-dimensional output signal $\mathbf{y}_{\text{train}}(t)$. For instance, if the system is supposed to predict time series data, $\mathbf{u}_{\text{train}}(t)$ would be the first part of a time series and $\mathbf{y}_{\text{train}}(t)$ the second part (that we want to predict from the first part) [10]. The input signal is used to drive a high-dimensional nonlinear dynamical system that is referred to as the *reservoir*.

Specifically, the state $\mathbf{X}$ of the reservoir at time t_i takes the form [36]

$$\mathbf{X}(t_i) = \mathbf{f}(\mathbf{X}(t_{i-1}), \mathbf{u}(t_i)) \tag{1}$$

with the input signal $\mathbf{u}$ and a function $\mathbf{f}$. A sufficiently large subset of the state variables, summarized in a vector $\mathbf{x}(t)$, must be accessible. Finally, a readout function $\mathbf{F}$ maps the vector $\mathbf{x}(t)$ to the output $\mathbf{y}(t)$, i.e.,

$$\mathbf{F}(\mathbf{x}(t)) = \mathbf{y}(t). \tag{2}$$

This readout function is the only thing that is touched during the training process. It is designed to minimize a loss function that typically that depends on the difference between the actual output $\mathbf{y}(t)$ and the desired output $\mathbf{y}_{\text{train}}(t)$. In most cases, $\mathbf{F}$ is obtained via linear regression. Once $\mathbf{F}$ is known, one can apply the system to new input signals $\mathbf{u}(t)$.

This strategy has a number of advantages:

- The training usually only consists in linear regression, which is simple, computationally inexpensive, and easy to implement.
- One only needs to train the readout function $\mathbf{F}$ and not the entire system. This is helpful if we are dealing with a physical system where the precise interactions between the parts are difficult to modify and perhaps not even fully known.
- One can use the same reservoir for different computing tasks by simply using different readout functions $\mathbf{F}$.
- It is a good approach for dealing with time series data.

Almost every physical system can be described by an equation of the form (1), and therefore a large variety of physical system can be used as reservoirs (although in practice there are restrictions, see Sect. 3). A notable feature of RCs is their temporal dynamics. Iterating Eq. (1) gives [36]

$$\mathbf{X}(t_i) = \mathbf{f}(\mathbf{f}(\mathbf{f}(\mathbf{X}(t_{i-3}), \mathbf{u}(t_{i-2})), \mathbf{u}(t_{i-1})), \mathbf{u}(t_i)), \tag{3}$$

showing that the system's state at a certain time depends on the system states and inputs at previous times. This allows the system to react to temporal input signals, such as spoken language.

2.2 Recurrent Neural Networks

An essential distinction in this context (see also the chapter by Tobias Wand in this volume) is that between a *feed-forward neural network* (FNN) and a *recurrent neural network* (RNN). In a FNN, signals propagate in one direction—the first layer activates the second layer, the second layer the third layer and so on. In a RNN, on the other hand, information can also "move backwards". This allows the system to have memory: If a system is supposed to process a signal $\mathbf{u}(t)$, then reacting to $\mathbf{u}(t_2)$ might require also information about $\mathbf{u}(t_1)$ with $t_1 < t_2$. This can be achieved by feeding back information about previous states of nodes. This distinguishes RC from the related concept of an *extreme learning machine* [21, 54], which employs FNNs and does therefore not have memory of this type [7].

The existence of closed loops implies in particular that RNNs can have a temporal dynamics even in the absence of external inputs. Therefore, while FNNs are (mathematically speaking) *functions*–they map a certain input to a certain output–RNNs are (mathematically speaking) *dynamical systems* [32]. This already indicates how they might be related to RC, which, as discussed in Sect. 2.1, is based on dynamical systems. Frequently, in computer science applications, the dynamical system constituting the reservoir is a RNN.

In general, RNNs are difficult to train. FNNs are typically trained via gradient descent methods, where the parameters giving the connection weights are gradually changed to move the network's output closer to the target output. Gradient descent methods are also frequently, and in many different forms, applied to RNNs [2]. However, such methods are considerably more difficult to apply in the context of RNNs. Reasons for this include that a gradual parameter change might lead to a bifurcation that spoils convergence and that training times are very long since a single parameter change requires running the temporal dynamics for some time. RC was proposed as a new method for training RNNs that allows to avoid these problems by changing the weights only in a non-recurrent readout layer. The RNN itself can be created randomly and is unchanged during the training [32].

2.3 Why Does This Work?

At first sight, the idea of RC is somewhat counterintuitive. While it is of course easier to train only the final layer of a neural network rather than the entire one, there is certainly a reason why one usually trains all layers. Usually one would not get away with only changing the final layer, so why is it possible here?

What is exploited here is the high dimensionality of the reservoir. The information that we are interested in is in principle encoded in the input signal, but it is mixed up, nonlinearly, with a huge amount of other stuff that we are not interested in. The projection onto a higher-dimensional space allows for separability. This idea is illustrated in Fig. 2. Suppose we want to classify some inputs into stars and circles. Unfortunately, due to the way the data is distributed, it does not exhibit linear separability, i.e., we cannot find a hyperplane (in this case a line) that gives us the border between a region with stars and a region with circles (cf. Fig. 2a). If, however, we project the data onto a three-dimensional space (cf. Fig. 2b), we *can* find a hyperplane (in this case a plane) that separates stars and circles. The separation problem has thereby become much simpler, and can be solved already by a single

Fig. 2 Projection onto a high-dimensional space allows for separability. (A similar figure can be found in Ref. [1])

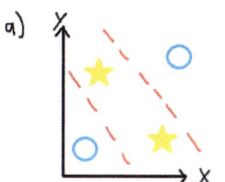 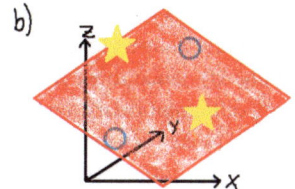

layer that is trained via linear regression. In RC, feeding the data into the reservoir corresponds to the projection from the two- to the three-dimensional plane (usually more dimensions will be involved) and the training of the readout layer corresponds to finding the plane [46].

3   What Is a Good Reservoir?

While the procedure discussed above can in principle be applied very generally, not every dynamical system does in practice make a good reservoir. A number of criteria have been developed that a reservoir should satisfy to be useful in this context–although the broad range of systems that have been used for this purpose (consider, for example, the water bucket mentioned above) suggests that in practice these criteria are quite flexible [46].

First, *reproducibility* is of course important. If two input signals are very similar, the output signals should be similar as well. This is related to the ability of the system to generalize from the data it has been trained on to other input data [46]. Second, if the input signals are sufficiently different, the output signals should also be different (*separation*). This determines the reservoir's ability to distinguish different sorts of inputs. A reservoir computer that always produces the same output regardless of the input obviously satisfies the reproducibility property, but would not be really useful.

Moreover, *fading memory* is an important property of reservoir computers [10]. The computer is supposed to process a time series $\mathbf{u}(t)$, not just the instantaneous value of $\mathbf{u}$ (the latter would not require a RNN). Therefore, it needs to have memory. On the other hand, the current value of $\mathbf{u}$ should still be what is most important, values at earlier times become gradually less important. In other words, the memory should gradually fade. The computer should relax to a quiescent state if there is no external input [42], and its behavior should not depend on the initial condition $\mathbf{X}(0)$ of the reservoir. This is referred to as the *echo state property*—the influence of initial conditions should gradually vanish [57]. The timescale of the fading memory should be comparable to the timescales of the input signal [26]. Since it depends on the application what the timescale of the input signal is, the adequate reservoir system may differ depending on the application.

Also, it is often considered advantageous to operate the reservoir computer in a parameter range close to an instability (for example, in the vicinity of a transition to chaos), since there its behavior is particularly complex and therefore has a very high computational complexity. Intuitively, the reason is that such regimes present a compromise between ordered phases (where the behavior is stable and thus reliable, but where initial conditions are also quickly erased by the approach to an attractor making the system less able to react to an input) and chaotic phases (where the system's behavior is strongly affected by small differences in the initial conditions, harming separability) [46]. There are, however, also counterexamples, i.e., systems where moving close to an instability actually decreases the performance of a reservoir computer [8]. In fact, Herbert Jaeger (in his foreword for Ref. [38]) argues that the idea that RC should operate close to chaos or criticality is a "myth", neither mathematically well-defined nor empirically confirmed.

4 Physical Reservoir Computing

The discussions so far were concerned with general concepts in artificial intelligence. This book, however, has a specific focus on the relation of AI to physics, and there is a reason that RC features so prominently in the introductory part of this book. This reason is *physical reservoir computing*.

In physical RC, the reservoir is a physical system. Thereby, physical RC allows to perform machine learning tasks using physical systems. There has, in recent years, been a considerable amount of work on the development of experimental setups that can be used for this purpose. In this section, I will briefly discuss some examples. Later chapters will address specific such as in optics (in the chapter by Kathy Lüdge/Lina Jaurigue), spintronics (in the chapter by Atreya Majumdar/Karin Everschor-Sitte), and soft matter (in the chapter by Julian Jeggle/Raphael Wittkowski). The classification in different types of reservoirs (electronic, photonic, spintronic, mechanical, biological, chemical) is adapted from Ref. [49]. Note that this list is not exhaustive.

4.1 Electronic Reservoir Computing

Standard computers are based on electronics, and therefore it is a very natural idea to use electronic systems. This has been achieved in a broad number of ways (see Refs. [31, 49] for an overview), of which I discuss here just one, namely *memristors*. A memristor is a resistor that possesses memory, i.e., whose resistance changes based on the current that has passed through it. Memristors are therefore useful devices in applications like RC where memory is important. They can be used to mimic the plasticity of biological neurons, and allow for nonlinear transformations of input signals [49]. A memristor-based RC was first proposed by Kulkarni and Teuscher [30].

4.2   Photonic Reservoir Computing

In *photonics*, information processing is based not (solely) on electric currents, but on light (photons). RC has developed into a widely used approach in photonics, see Ref. [42] for a review (I am loosely following this reference here). An important approach is the implementation of RC in on-chip-photonics, where photonic systems are integrated into chips. This allows the systems to be produced and sold on industrial scales and to therefore use them for high-speed low-power-consumption computing. This approach was theoretically suggested in Ref. [50] and later realized in hardware [51].

A further interesting approach is *delay reservoir computing* [22]. Delay systems have been of considerable interest for optics in the past years [29, 45]. They are described mathematically by delay differential equations, which differ considerably from ordinary differential equations since they possess an infinite-dimensional phase space–for solving it, one needs to specify not only the state of the system at a single initial time, but on an entire time interval. Using delay systems, one can therefore achieve a high-dimensional phase space (which is advantageous in the context of RC) even with a very simple setup [42].

Photonic approaches to neuromorphic computing are discussed further in the chapters by Kathy Lüdge/Lina Jaurigue and by Lennart Meyer/Rongyang Xu/Wolfram Pernice in this book.

4.3   Spintronic Reservoir Computing

Spintronics is a field of technology where information processing is based not only on the electric charges (as in electronics), but also on the spins (elementary magnetic moments) of electrons. Spintronics has become increasingly popular in neuromorphic computing in general and RC in particular, see Refs. [11, 13, 18, 31, 56, 58] for reviews. Spintronic systems can be used to build artificial synapses, thereby mimicking the structure and functionality of biological brains [18]. An introduction is provided in the chapter by Atreya Majumdar/Karin Everschor-Sitte in this book.

An interesting recent proposal in this context is *Brownian reservoir computing* based on skyrmions [3–5, 41]. Brownian motion [6] is the random thermal motion of particles, which is a central phenomenon in soft matter physics, but also arises in magnetic systems. An example are magnetic skyrmions, which are whirl-like topological magnetic nanostructures that have particle-like diffusion behavior reminiscent of neurotransmitters and can be used as information carriers in spintronics [18]. In Brownian computing, one employs thermal fluctuations–which in most systems are present anyway–for computing purposes to achieve a high energy efficiency. It is of course helpful if the employed Brownian system can be easily integrated into a computer, which is why magnetic nanosystems such as skyrmions are useful here [4]. Brownian RC based on skyrmions was realized experimentally by Raab et al. [41], who demonstrated that this approach is very promising for energy-efficient computing.

4.4   Mechanical Reservoir Computing

Mechanical systems can make for useful reservoirs. Robotic systems, in particular from soft robotics (where the bodies of the robots are flexible), have been repeatedly used in this context. Hauser et al. [20] have modeled this using the example of a nonlinear mass-spring-damper system connected to a mechanical network, which was intended to represent in a simple way the body of a soft robot (or biological system) and which exhibits the complex nonlinear dynamics required for successful RC. Nakajima et al. [39] employed a silicon-based robot arm inspired by the arm of an octopus, with the input being the rotation of the arm and the output being measured strain. While the noisy and nonlinear dynamics of soft robots is often perceived as disadvantageous, it can be very useful in the context of RC. (This paragraph follows Ref. [19].)

4.5   Biological and Chemical Reservoir Computing

Reservoir computing has always had a close connection to neurobiology. In particular, work on RC has been motivated by attempts to understand information processing in mammalian brains [47]. For instance, it has been proposed that the cerebellum might work like a liquid state machine [55], and experiments on mice [9] suggest that the mouse brain exploits principles of RC [9, 31]. Moreover, RC–which is frequently based on random neural networks and noisy systems–might explain why the brain works so accurately despite being a rather noisy system [32]. It is therefore a promising direction of work to use biological neural networks for RC tasks [47]. Biological neurons are in a sense the most obvious, but not the only approach to biological RC. For example, it has been proposed to realize RC based on Escherichia coli bacteria [24]. Another variant, namely DNA reservoir computing [17], will be discussed in the chapter on DNA neural networks in this book. This approach is based on employing chemical systems for RC, an idea that is also used in non-biological contexts (for example based on electrolyte solutions [25]). See the chapter by Julian Jeggle/Raphael Wittkowski in this volume for a discussion of the related concept of active matter RC.

5   Quantum Reservoir Computing

In the wake of the currently growing interest in quantum computing in general and quantum-mechanical approaches to machine learning in particular, *quantum reservoir computing* [15, 16, 35, 36, 48] has attracted some interest. Here, one employs quantum-mechanical reservoirs and thereby aims to exploit the advantages of quantum computers for RC. In this section, I will introduce the elementary ideas of how this works. The discussion follows Ref. [15], which was one of the first articles on this topic. A more general introduction to quantum machine learning can be found in the chapter by Ivana Nikoloska in this volume.

Quantum states are represented by vectors in complex Hilbert states. In the context of quantum computing, the minimal information unit is a *qubit*, corresponding to a two-dimensional complex vector in a vector space spanned by the vectors $|0\rangle$ and $|1\rangle$. In general, the state of a system of N qubits is described by a $2^N x 2^N$ Hermitian matrix ρ, the *density matrix*. The quantum system is said to be in a *pure state* if ρ can be written as $\rho = |\psi\rangle\langle\psi|$, where $|\psi\rangle$ is a 2^N-dimensional vector and $\langle\psi|$ is a covector to $|\psi\rangle$. (For instance, if $|\psi\rangle = (1, 2, 3)^T$, then $\langle\psi| = (1, 2, 3)$.) If the density matrix at time t is $\rho(t)$, then the density matrix at time $t + \tau$ is

$$\rho(t + \tau) = e^{-iH\tau}\rho(t)e^{iH\tau} \tag{4}$$

with the Hamiltonian H (a $2^N x 2^N$ Hermitian matrix that determines the dynamics and whose eigenvalues correspond to the energy levels of the quantum system). For an arbitrary observable A_i, which is also represented by a $2^N x 2^N$ Hermitian matrix, the expectation value is given by

$$a_i(t) = \text{Tr}(\rho(t)A_i) \tag{5}$$

with the trace Tr.

What is now required is a way to feed an input signal $\mathbf{u}(t)$ into the system and to get an output signal $\mathbf{x}(t)$ that can then be fed into the readout function $\mathbf{F}$. Let us consider for simplicity a one-dimensional input signal $u(t)$, which we sample in M discrete time intervals of length τ to get a sequence $\{u_k\}$ with $u_k = u(k\tau)$ and $k = 0, 1, ...M$. At each time $k\tau$, the state of the first qubit is changed to $\rho_{u_k} = |\psi_{u_k}\rangle\langle\psi_{u_k}|$ with

$$|\psi_{u_k}\rangle = \sqrt{1 - u_k}|0\rangle + \sqrt{u_k}|1\rangle. \tag{6}$$

The density matrix ρ is thereby replaced by

$$\rho_{u_k} \otimes \text{Tr}_1(\rho), \tag{7}$$

where $\otimes$ is a tensor product and Tr_1 is a trace over the degrees of freedom of the first qubit. Afterwards, the density matrix is time evolved via Eq. (4) for a time τ. This time has to be optimized in order to optimize the performance of the computer (see Ref. [15]). For the output $\mathbf{x}(t)$, we then pick some observables A_i and assemble them in a vector $\mathbf{A}$. Then, we can obtain the output vector from their expectation values as

$$\mathbf{x}(t) = \text{Tr}(\rho(t)\mathbf{A}). \tag{8}$$

Specifically, Fujii and Nakajima [15] choose A_i as the Pauli operator acting on the ith qubit.

The dimension of the quantum-mechanical Hilbert space increases exponentially with the number of qubits N, giving rise to an exponentially increasing number of nodes in the reservoir. For readout purposes, these are split into *true nodes* (the observed ones) and *hidden nodes* (the rest). The signals are sampled not only at the time $k\tau$, but also at several times in between. Dividing the time interval into V parts gives rise to V *virtual nodes* and allows to increase the number of nodes from N to NV via temporal multiplexing. Thereby, the exponentially large Hilbert space is monitored via a polynomial number of signals. This is the distinguishing feature of quantum RC compared to other RC approaches [15]. Changing τ corresponds to a change of the dynamics of the reservoir, whereas changing V corresponds to a change of the way it is observed [40].

An important feature of quantum systems is also the way in which they interact with the environment. Such interactions lead to dissipation and decoherence [43], where quantum states are destroyed by interactions with the environment. Moreover, performing a measurement of a quantum state generally changes it, a phenomenon giving rise to the famous quantum measurement problem [14]. Usually, interactions with the environment are not beneficial for the performance of quantum computers. One can, however, also try and exploit such effects in quantum reservoir computing, as has recently been demonstrated for both measurements [37] and dissipation [43].

6 Outlook: Relation to Intelligent Matter

If we loosely understand "intelligent matter" as "physical materials perform tasks similar to those expected from computer systems that we would refer to as (artificially) intelligent", then RC seems to be, if not an instance of it, then at least an important step towards it. We have here physical systems that can be employed in computational tasks of the form that appear in machine learning.

Nevertheless, according to Kaspar et al. [26], reservoir computing systems in the form described here do not constitute "intelligent matter" in the technical sense:

- The systems possess fading memory, whereas intelligent matter needs to have long-term memory.
- The readout function **F** still needs to be trained manually, the system does not adapt on its own.

Regarding the first point, it should be noted, however, that the fading memory can be tuned to fade rather slowly if this is desired in a certain context.

RC does nevertheless have significant potential for the development of "true" intelligent matter, in particular when considering its relation to evolutionary dynamics (a topic reviewed in Ref. [46]). After all, RC is a possible working principle of biological brains. It is conceivable that RC emerges in evolutionary contexts, as it has certain advantages (such as the low cost of learning and the fact that external sytems can be used to carry out computations) that could give biological systems exploiting this paradigm a fitness advantage. An evolutionary evolving RC system would be a system that evolves its computing capabilities in adaptation to the environment, bringing it closer to actual intelligence. A possible disadvantage of RC in evolutionary contexts, Seoane [46] suggests, is that (since the reservoir needs to be high-dimensional), it requires systems to perform a lot of activity that is not really used for computing, making it energetically costly (which leads to a fitness disadvantage). A fine-tuned neural network can have a smaller number of nodes.

7 Summary

In this chapter, I have introduced the basic ideas of reservoir computing. Here, one uses a very high-dimensional recurrent neural network and trains only the final layer. This makes it possible to use for the rest of the network a physical system whose properties might be difficult to tune or not fully known. A variety of systems have been used here, ranging from buckets of water to optical and magnetic setups. Reservoir computing is a very promising tool for implementing artificial intelligence in nanosystems, and will continue to be a thriving field of research in the coming years.

Acknowledgements I thank Raphael Wittkowski for very helpful discussions on this topic. This work was funded by the Deutsche Forschungsgemeinschaft (DFG, German Research Foundation) in the framework of SFB 1551; Project No. 464588647 and SFB 1552; Project No. 465145163. The author also acknowledges funding from the Mainz Institute of Multiscale Modeling, M^3ODEL.

References

1. Appeltant L, Soriano MC, Van der Sande G, Danckaert J, Massar S, Dambre J, Schrauwen B, Mirasso CR, Fischer I (2011) Information processing using a single dynamical node as complex system. Nat Commun 2(1):468
2. Atiya AF, Parlos AG (2000) New results on recurrent network training: unifying the algorithms and accelerating convergence. IEEE Trans Neural Netw 11(3):697–709
3. Brems MA, Kläui M, Virnau P (2021) Circuits and excitations to enable Brownian token-based computing with skyrmions. Appl Phys Lett 119(13)
4. Brems MA, Raab K, Virnau P, Kläui M (2023) Brownscher reservoir-computer mit Skyrmionen. Phys Unserer Zeit 54(2):60–61
5. Brems MA, Raab K, Zázvorka J, Beneke G, Winkle T, Rothörl J, Kammerbauer F, Virnau P, Mentink JH, Kläui M (2023b) Non-conventional computing using thermal and driven Skyrmion dynamics. In: 2023 IEEE international magnetic conference—short papers (INTERMAG short papers), pp 1–2. https://doi.org/10.1109/INTERMAGShortPapers58606.2023.10228647
6. Brown R (1828) A brief account of microscopical observations made in the months of June, July and August 1827, on the particles contained in the pollen of plants; and on the general existence of active molecules in organic and inorganic bodies. Phil Mag 4(21):161–173
7. Butcher JB, Verstraeten D, Schrauwen B, Day CR, Haycock PW (2013) Reservoir computing and extreme learning machines for non-linear time-series data analysis. Neural Netw 38:76–89
8. Carroll TL (2020) Do reservoir computers work best at the edge of chaos? Chaos Interdiscip J Nonlinear Sci 30(12):121109

9. Cazettes F, Mazzucato L, Murakami M, Morais JP, Augusto E, Renart A, Mainen ZF (2023) A reservoir of foraging decision variables in the mouse brain. Nat Neurosci 26(5):840–849

10. Cucchi M, Abreu S, Ciccone G, Brunner D, Kleemann H (2022) Hands-on reservoir computing: a tutorial for practical implementation. Neuromorphic Comput Eng 2:032002

11. Everschor-Sitte K, Majumdar A, Wolk K, Meier D (2024) Topological magnetic and ferroelectric systems for reservoir computing. Nat Rev Phys 6:455–462

12. Fernando C, Sojakka S (2003) Pattern recognition in a bucket. In: Banzhaf W, Ziegler J, Christaller T, Dittrich P, Kim J (eds) Advances in artificial life. ECAL 2003, Berlin, Heidelberg, pp 588–597

13. Finocchio G, Di Ventra M, Camsari KY, Everschor-Sitte K, Amiri PK, Zeng Z (2021) The promise of spintronics for unconventional computing. J Magn Magn Mater 521:167506

14. Friebe C, Kuhlmann M, Lyre H, Näger PM, Passon O, Stöckler M (2018) The philosophy of quantum physics. Springer, Wiesbaden

15. Fujii K, Nakajima K (2017) Harnessing disordered-ensemble quantum dynamics for machine learning. Phys Rev Appl 8(2):024030

16. Fujii K, Nakajima K (2021) Quantum reservoir computing: a reservoir approach toward quantum machine learning on near-term quantum devices. In: Nakajima K, Fischer I (eds) Reservoir computing. Springer, Singapore, pp 423–450

17. Goudarzi A, Lakin MR, Stefanovic D (2013) DNA reservoir computing: a novel molecular computing approach. In: Soloveichik D, Yurke B (eds) DNA Computing and molecular programming. DNA 2013, Springer, Cham, pp 76–89

18. Grollier J, Querlioz D, Camsari KY, Everschor-Sitte K, Fukami S, Stiles MD (2020) Neuromorphic spintronics. Nat Electron 3(7):360–370

19. Hauser H (2021) Physical reservoir computing in robotics. In: Nakajima K, Fischer I (eds) Reservoir computing. Springer, Singapore, pp 169–190

20. Hauser H, Ijspeert AJ, Füchslin RM, Pfeifer R, Maass W (2011) Towards a theoretical foundation for morphological computation with compliant bodies. Biol Cybern 105:355–370

21. Huang G, Huang GB, Song S, You K (2015) Trends in extreme learning machines: a review. Neural Netw 61:32–48

22. Hülser T, Köster F, Jaurigue L, Lüdge K (2022) Role of delay-times in delay-based photonic reservoir computing. Opt Mater Express 12(3):1214–1231

23. Jaeger H (2001) The "echo state" approach to analysing and training recurrent neural networks. Bonn, Germany: German National Research Center for Information Technology GMD Technical Report 148

24. Jones B, Stekel D, Rowe J, Fernando C (2007) Is there a liquid state machine in the bacterium *Escherichia Coli*? In: 2007 IEEE symposium on artificial life. IEEE, pp 187–191

25. Kan S, Nakajima K, Asai T, Akai-Kasaya M (2021) Physical implementation of reservoir computing through electrochemical reaction. Adv Sci 9:2104076

26. Kaspar C, Ravoo BJ, van der Wiel WG, Wegner SV, Pernice WHP (2021) The rise of intelligent matter. Nature 594(7863):345–355

27. Kirby K (1991) Context dynamics in neural sequential learning. In: Proceedings Florida AI research symposium (FLAIRS), pp 66–70

28. Kirby KG, Day N (1990) The neurodynamics of context reverberation learning. In: Proceedings of the twelfth annual international conference of the IEEE engineering in medicine and biology society. IEEE, pp 1781–1782

29. Koch ER, Seidel TG, Javaloyes J, Gurevich SV (2023) Temporal localized states and square-waves in semiconductor micro-resonators with strong time-delayed feedback. Chaos 33(4):043142

30. Kulkarni MS, Teuscher C (2012) Memristor-based reservoir computing. In: Proceedings of the 2012 IEEE/ACM international symposium on nanoscale architectures, pp 226–232

31. Liang X, Tang J, Zhong Y, Gao B, Qian H, Wu H (2024) Physical reservoir computing with emerging electronics. Nat Electron 7(3):193–206

32. Lukoševičius M, Jaeger H (2009) Reservoir computing approaches to recurrent neural network training. Comput Sci Rev 3(3):127–149

33. Lukoševičius M, Jaeger H, Schrauwen B (2012) Reservoir computing trends. KI-Künstliche Intelligenz 26:365–371
34. Maass W, Natschläger T, Markram H (2002) Real-time computing without stable states: a new framework for neural computation based on perturbations. Neural Comput 14(11):2531–2560
35. Martínez-Peña R, Giorgi GL, Nokkala J, Soriano MC, Zambrini R (2021) Dynamical phase transitions in quantum reservoir computing. Phys Rev Lett 127(10):100502
36. Mujal P, Martínez-Peña R, Nokkala J, García-Beni J, Giorgi GL, Soriano MC, Zambrini R (2021) Opportunities in quantum reservoir computing and extreme learning machines. Adv Quantum Technol 4(8):2100027
37. Mujal P, Martínez-Peña R, Giorgi GL, Soriano MC, Zambrini R (2023) Time-series quantum reservoir computing with weak and projective measurements. NPJ Quantum Inf 9(1):16
38. Nakajima K, Fischer I (eds) (2021) Reservoir computing. Springer, Singapore
39. Nakajima K, Li T, Hauser H, Pfeifer R (2014) Exploiting short-term memory in soft body dynamics as a computational resource. J R Soc Interface 11(100):20140437
40. Nakajima K, Fujii K, Negoro M, Mitarai K, Kitagawa M (2019) Boosting computational power through spatial multiplexing in quantum reservoir computing. Phys Rev Appl 11(3):034021
41. Raab K, Brems MA, Beneke G, Dohi T, Rothörl J, Kammerbauer F, Mentink JH, Kläui M (2022) Brownian reservoir computing realized using geometrically confined skyrmion dynamics. Nat Commun 13(1):6982
42. Van der Sande G, Brunner D, Soriano MC (2017) Advances in photonic reservoir computing. Nanophotonics 6(3):561–576
43. Sannia A, Martínez-Peña R, Soriano MC, Giorgi GL, Zambrini R (2024) Dissipation as a resource for quantum reservoir computing. Quantum 8:1291
44. Schomaker LRB (1991) Simulation and recognition of handwriting movements: a vertical approach to modeling human motor behavior. PhD thesis, Nijmeegs Instituut voor Cognitie-onderzoek en Informatietechnologie, Nijmegen. https://repository.ubn.ru.nl/handle/2066/113914
45. Seidel TG, Gurevich SV, Javaloyes J (2022) Conservative solitons and reversibility in time delayed systems. Phys Rev Lett 128:083901
46. Seoane LF (2019) Evolutionary aspects of reservoir computing. Philos Trans R Soc B 374:20180377
47. Sumi T, Yamamoto H, Katori Y, Ito K, Moriya S, Konno T, Sato S, Hirano-Iwata A (2023) Biological neurons act as generalization filters in reservoir computing. Proc Natl Acad Sci USA 120(25):e2217008120
48. Suzuki Y, Gao Q, Pradel KC, Yasuoka K, Yamamoto N (2022) Natural quantum reservoir computing for temporal information processing. Sci Rep 12(1):1353
49. Tanaka G, Yamane T, Héroux JB, Nakane R, Kanazawa N, Takeda S, Numata H, Nakano D, Hirose A (2019) Recent advances in physical reservoir computing: a review. Neural Netw 115:100–123
50. Vandoorne K, Dierckx W, Schrauwen B, Verstraeten D, Baets R, Bienstman P, Van Campenhout J (2008) Toward optical signal processing using photonic reservoir computing. Opt Express 16(15):11182–11192
51. Vandoorne K, Mechet P, Van Vaerenbergh T, Fiers M, Morthier G, Verstraeten D, Schrauwen B, Dambre J, Bienstman P (2014) Experimental demonstration of reservoir computing on a silicon photonics chip. Nat Commun 5(1):3541
52. Verstraeten D, Schrauwen B, Stroobandt D (2005) Reservoir computing with stochastic bit-stream neurons. In: Proceedings of the 16th annual ProRISC workshop, pp 454–459
53. Verstraeten D, Schrauwen B, d'Haene M, Stroobandt D (2007) An experimental unification of reservoir computing methods. Neural Netw 20(3):391–403
54. Wang J, Lu S, Wang SH, Zhang YD (2022) A review on extreme learning machine. Multimed Tools Appl 81(29):41611–41660
55. Yamazaki T, Tanaka S (2007) The cerebellum as a liquid state machine. Neural Netw 20(3):290–297

56. Yan M, Huang C, Bienstman P, Tino P, Lin W, Sun J (2024) Emerging opportunities and challenges for the future of reservoir computing. Nat Commun 15(1):2056
57. Yildiz IB, Jaeger H, Kiebel SJ (2012) Re-visiting the echo state property. Neural Netw 35:1–9
58. Zhou J, Chen J (2021) Prospect of spintronics in neuromorphic computing. Adv Electron Mater 7(9):2100465

Applications of Artificial Intelligence
to Nanosystems

Learning Dynamical Systems from Data

Oliver Kamps, Tim W. Kroll, and Oliver Mai

Abstract Learning dynamic equations from data has shown great promise in various fields of research, such as physics, engineering, and biology. This short review provides a comprehensive overview of the methods, challenges, and applications involved in learning governing equations from time-series data. We begin by highlighting the importance of dynamic equations in modeling complex systems. Subsequently, we present different approaches used for learning dynamic equations, including symbolic regression and neural networks. These methods are contextualized within different modeling scopes, which here are largely defined by the properties of the underlying differential equations. We explore the advantages and limitations of these methods, with a particular focus on symbolic regression. Through various studies and examples, we demonstrate the utility of learning dynamic equations from data and showcase their applicability in the domain of density functional theory. Finally, we identify promising directions for future research and discuss potential applications of this methodology in advancing scientific understanding and solving real-world problems.

1 Introduction

One might wonder why there is a demand for discovering dynamic equations in the first place. Although alternative methods exist for predicting outcomes or even modeling complex interactions in the real world, the language of mathematics and the theory of complex dynamical systems are uniquely suited for understanding and categorizing the spatio-temporal evolution of virtually anything. Beyond their broad applicability, once a set of differential equations is established, it can help simplify problems or identify similar patterns across different disciplines.

O. Kamps (✉) · T. W. Kroll · O. Mai
Center for Nonlinear Science, University of Münster, Münster, Germany
e-mail: okamp@uni-muenster.de

© The Author(s) 2026

M. te Vrugt (ed.), *Artificial Intelligence and Intelligent Matter*, Machine Intelligence for Materials Science, https://doi.org/10.1007/978-3-032-04129-6_5

Being able to fully capture complex dynamics opens up a wealth of possibilities, including the use of other analytical tools like stability analysis. It also enables the design of control strategies or the optimization of system performance.

However, real-world systems often exhibit intricate and nonlinear behavior that is challenging to capture with traditional analytical methods alone. Commonly used methods either derive dynamic equations from physical principals or more general equations, which may not always be known, or project known equations into a different basis, e.g. Galerkin models. First there are systems, where we do not know the true governing equations, as one could argue in the case of brain dynamics. Secondly even with known equations we can run into troubles. One famous example is the cylinder wake in classical fluid mechanics. Here one tries to project the Navier-Stokes equations onto a set of basis functions, obtained through PCA. Even though the governing equations are known, the reduced order model does not correspond to the correct dynamics [1, 2]. With the proliferation of sensors and computational resources, vast amounts of high-dimensional time-series data are becoming increasingly available across various domains. This availability makes the estimation of dynamic equations a viable method to complement traditional theories and experiments. It provides an additional avenue for hypothesis testing, model validation, and the discovery of novel phenomena. Existing theoretical models may be adapted to accommodate new scenarios or domains, thereby enhancing their generalization capabilities.

In general, modeling approaches can be categorized into several types: analytical, numerical, data-driven, and hybrid approaches [3]. Specifically, data-driven approaches, the main focus of this work, can be subdivided into parametric and non-parametric methods. These categories differ in their underlying assumptions and estimation techniques. Parametric models, like those employed in spatial point pattern analysis [4], are based on specific assumptions about the data and utilize methods such as least squares and maximum likelihood estimation. In contrast, non-parametric models impose fewer assumptions and employ techniques like kernel and spline smoothing [5] or artificial neural networks [6]. These methodological differences can affect efficiency and robustness: parametric models are more efficient when their assumptions are met but can be misleading otherwise, while non-parametric models offer greater flexibility and robustness to model mis-specifications. The selection between these approaches hinges on the data's characteristics and the required flexibility and robustness.

Within the realm of complex dynamical systems, a distinction is often made based on system dimensionality. Systems can be classified as either finite-dimensional, where ordinary differential equations (ODEs) suffice for modeling, or spatially extended, necessitating partial differential equations (PDEs) for accurate description. While many data-driven modeling techniques are applicable to both scenarios, the choice of method may depend on its computational scalability in relation to the system's dimensionality. Consequently, there is a slight variation in common practices between employing ODEs and PDEs, and this work differentiates between these two use cases

In the sections that follow, we will first provide an overview of the most commonly used methods in both parametric and non-parametric data-driven modeling. Many of these methods are applicable to both ordinary differential equations (ODEs) and partial differential equations (PDEs). Subsequently, we will illustrate these methods with examples for both types of equations.

2   Methods

The necessity or advantage of modeling with ordinary differential equations (ODEs) arises from at least two distinct perspectives. Firstly, in some disciplines, there are only a few variables that need modeling. Examples include certain areas of biology or models for drag in fluid mechanics [7]. Take for example the Navier-Stokes equations, which by themselves are hard to solve analytically and computationally expensive to simulate. But for specific problems they can be reduced to be handled more easily, such as in the case of the Jeffery-Hamel flow, where the flow between two planes that meet at an angle can be reduced to an ODE problem under suitable assumptions. Secondly, in many fields, reduced-order models are employed to derive lower-dimensional representations of more complex ODE or PDE systems. These are particularly useful when the large-scale dynamics are unknown or when modeling the system with a few order parameters or structures is advantageous. From a scientific standpoint, reduced-order models are valuable for understanding how a system evolves in terms of macroscopic structures, offering insights more readily than by modeling a wider or more detailed scope, such as using the full extend of the Navier-Stokes equations. In such models, modal decompositions like principal component analysis (PCA) are often utilized to model the dynamics of PCA modes. Typically, Galerkin models are applied in this context, but their use is limited to systems where the general dynamics are understood, and they cannot be extended to areas like brain dynamics, where no equivalent to the Navier-Stokes equations exists yet.

In both scenarios, there is significant interest in developing models that accurately capture the dynamics. Historically, various approaches have been adopted for this purpose. Notably, there is a distinction between 'black box' models, which are less interpretable, and more transparent models that utilize symbolic ordinary differential equations.

For completeness, we will mention a few attempts to use 'black box' models for ordinary differential equations, focusing on how to model the behavior of an ODE with neural networks. In recent years, several attempts have been made to model the dynamics of ordinary differential equations using neural networks. Initial attempts utilized Long-Short-Term-Memory networks (LSTM) to model chaotic systems [8]. This approach was also combined with reduced-order modeling in [9]. Other attempts involved recurrent neural networks, for example, in psychology/neuroscience [10, 11]. Progressing from recurrent neural networks, the concept of neural ODEs was introduced [12], which, although initially presented from a neural network perspective without dynamical systems in mind, can be applied to the problems we are considering here. Stemming from recurrent neural networks, echo state networks [13], a type of reservoir computing, have also been explored and can be augmented with physical information [14–16]. Another significant development is the use of physically informed neural networks (PINNs) [17], where neural networks are trained to model dynamics while being augmented by the knowledge of physical or general laws of the underlying system.

This last approach is directly linked to the modeling of partial differential equations, where PINNs are commonly used. For certain complex systems, instead of relying on reduced-order models, we aim to model the underlying partial differential equations of the system. With PINNs, for example, we may constrain our neural network to satisfy the Navier-Stokes equations or the Maxwell equations [18]. PINNs, which incorporate known physical principles into their architecture, have proven effective in modeling complex systems [19, 20]. These networks are part of non-parametric data-driven modeling approaches, which also include non-parametric neural networks [21] and other data-driven models, such as those based on multi-objective evolutionary algorithms [22]. These approaches are particularly useful for identifying and controlling nonlinear systems, offering a promising way to enhance model accuracy and reduce uncertainty [17, 23–27].

In this article, we will focus on the symbolic modeling of differential equations. One of the initial attempts in this area involved the development of the multiple shooting method [28]. Unfortunately, this method had several flaws that were challenging to overcome and numerically quite costly. The field experienced a significant transformation with the introduction of the SINDy method in 2016 [29]. This method relies on two fundamental ideas: Firstly, it assumes that a set of time-series data can be modeled by an ordinary differential equation, which is a sum of various candidate functions that we define. Along with knowledge of the system's time derivative, obtained through any suitable numerical differentiation technique, we can frame the problem of estimating an ordinary differential equation as a least-squares problem. This problem can be solved quickly and reliably, thus estimating the coefficients of our test functions. Secondly, based on the observation that most dynamical systems are underpinned by simple dynamics (e.g., the Lorenz system, which exhibits a wide array of dynamical behaviors with just seven terms up to quadratic order in its three variables), we assume our system is sparse in the space of possible test functions. We can enhance the least-squares problem by introducing an L1-prior for the coefficients of the test functions. Although this concept had been explored previously, Brunton et al. applied a novel method called sequential least squares thresholding to efficiently find a sparse ODE model. Brunton's approach has since been further developed for various applications. For example, it is now possible to integrate physical knowledge to constrain the system a priori [30], or to apply this method to identify right-hand sides with rational expressions, such as $\frac{x}{x+y}$, commonly found in mathematical biology [31].

A different approach to SINDy in this domain is symbolic regression using genetic algorithms [32, 33], where the right-hand side of the ODEs is encoded by a tree-like structure with variables and mathematical expressions serving as leaves and nodes.

This method has the advantage of utilizing a richer space of possible right-hand sides. A notable Python library in this area is PySR [34], which is currently limited to estimating symbolic expressions for time series and is not directly applicable for differential equations estimation out of the box. However, it has attracted interest for postprocessing models that were estimated using neural networks.

Recently, the symbolic regression approach was augmented by a Bayesian scheme to sample the space of possible tree structures more efficiently, as demonstrated in [35]. This method has been successfully applied to model various data sets, including synthetic data from a noisy Rössler attractor, salmon data from the Fraser River in British Columbia, Canada, and friction in turbulent rough pipes.

A significant advancement for this field is the extension of the SINDy method to the estimation of partial differential equations [36]. While parameter estimation methods for PDEs date back to the late 1990s [37–40], the adaptation of SINDy rejuvenated the field, leading to various methodological enhancements, such as the introduction of weak formulation [41] and the incorporation of information criteria [42]. The general workflow for these methods remains consistent with that of SINDy, albeit with increased computational demand, as spatial derivatives typically need to be calculated. This, in turn, opens up opportunities for approximations or other methods to accelerate computations. For another relatively recent overview, see [43].

3 Examples

In the following sections, we will demonstrate the application of sparse symbolic regression on two example cases: one from the field of ordinary differential equations and another from the area of partial differential equations.

3.1 Ordinary Differential Equations

In this part we use the SINDy-method from [29] to estimate a model from data. The Sparse Identification of Nonlinear Dynamical Systems (SINDy) is a method for discovering sparse governing equations of a dynamical system directly from data. SINDy identifies the underlying dynamical system by constructing a model for the time derivatives of the system's states based on a library of candidate functions.

First we assume we have a set of measurements $\mathbf{X} = [\mathbf{x}(t_1), \mathbf{x}(t_2), \ldots, \mathbf{x}(t_m)]^T$, where $\mathbf{x}(t_i) = [x_1(t_i), x_2(t_i), \ldots, x_n(t_i)]^T$ represents the state of the system at time t_i. The biggest assumption is that our system can be described by a set of differential equations:

$$\frac{d\mathbf{x}}{dt} = \mathbf{f}(\mathbf{x}), \tag{1}$$

where $\mathbf{f}(\mathbf{x})$ is an unknown function we aim to identify. In general this function could be as complicated as one could imagine, but in real systems we often assume that only a few simple terms are sufficient to describe the dynamics. Since we do not know a priori which terms those are, we construct a library $\boldsymbol{\Theta}(\mathbf{x})$ of candidate functions. This library can include various functional forms, such as polynomials, trigonometric functions, and other basis functions, but most of the time polynomials up to a certain degree in the states variables are used. For example, if $\mathbf{x} = [x_1, x_2]$, a polynomial library up to second order could be: $\boldsymbol{\Theta}(\mathbf{x}) = [1, x_1, x_2, x_1^2, x_1 x_2, x_2^2]$. The dynamics can then be expressed as a linear combination of these candidate functions: $\frac{d\mathbf{x}}{dt} \approx \boldsymbol{\Theta}(\mathbf{x})\Xi$, where we want to find the entries of Ξ. To solve this problem we calculate time-differences of $\mathbf{x}$ via a suitable numerical scheme like finite differences. This leads to the regression problem:

$$\dot{\mathbf{X}} = \boldsymbol{\Theta}(\mathbf{X})\Xi \tag{2}$$

Solving this regression problem leads to spurios terms, which are not present in the correct underlying dynamical system and therefore we need to regularise the problem and instead solve:

$$\Xi = \arg\min_{\Xi} \|\boldsymbol{\Theta}(\mathbf{X})\Xi - \dot{\mathbf{X}}\|_2^2 + \tilde{\lambda}\|\Xi\|_1, \tag{3}$$

where we promote sparsity by the regularisation parameter $\tilde{\lambda}$. To solve this problem the SINDy method now uses sequential least squares thressholding. Instead of directly solving the regularised problem, we solve the non-regularised problem, but discard every coefficient of Ξ, that is below a certain threshhold λ. After doing this the non-regularised problem is solved again, but only with terms, that were not discarded before. This procedure is repeated for some iteration number n. This converges to a sparse solution. To demonstrate the capabilities of current methods, we will employ the SINDy method to estimate the Lorenz equations from data within the chaotic regime. The governing equations we aim to estimate are:

$$\frac{dx}{dt} = \sigma(y - x)$$
$$\frac{dy}{dt} = x(\rho - z) - y$$
$$\frac{dz}{dt} = xy - \beta z$$

As mentioned above, the SINDy method utilizes a hyperparameter λ to determine the number of terms that remain in the final equations To illustrate the impact of this parameter, we present the integrated time series of models estimated by SINDy for different values of λ. To obtain these results we implemented an own version of the SINDy-method. It is possible to recover dynamics close to the original data from the Lorenz model; however, the success strongly depends on the choice of λ. To date, no rigorous method has been proposed to efficiently determine λ. One approach,

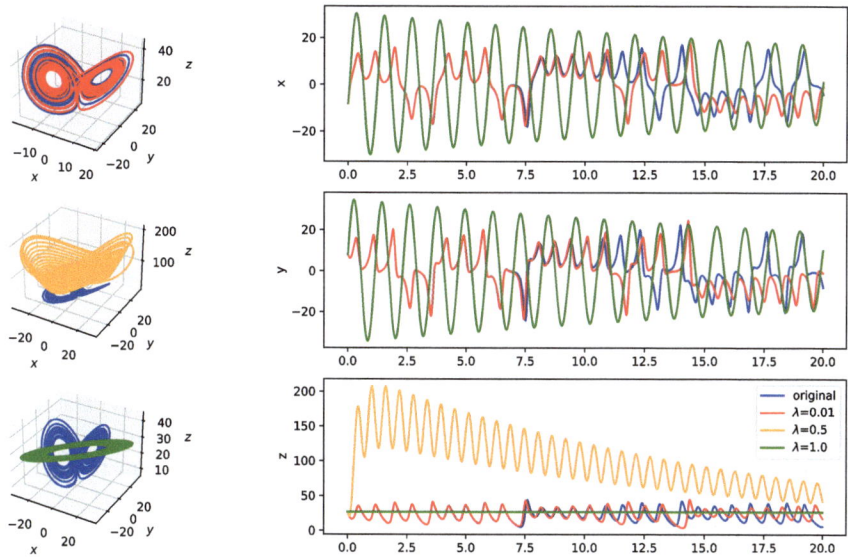

Fig. 1 Simulated data of the Lorenz-attractor (blue) in comparison with different estimated attractors (left) and comparison of the corresponding timeseries (right)

augmented by an information criterion, was suggested in [44]. Unfortunately, this method was only tested on data with exactly known numerical derivatives, meaning it does not account for errors from the derivatives. It merely demonstrates that removing too many or too few terms fails to recover the original system. For instance, selecting λ to eliminate all coefficients or to retain them all results in either a model that represents the data's mean or a model with the maximum number of non-zero coefficients. While the latter may yield a low error concerning the derivatives, it does not produce a reliable model of the integrated time series.

As depicted in Fig. 1, the accuracy of the estimated model heavily depends on the choice of the hyperparameter λ. Depending on λ, we can recover a model that closely resembles the original Lorenz attractor ($\lambda = 0.01$), a model that vaguely captures the overall structure of two wings but is otherwise inaccurate ($\lambda = 0.5$), or a model that merely produces an oscillation on a two-dimensional manifold ($\lambda = 1$).

3.2 Partial Differential Equations

As in [36] we now consider a nonlinear PDE of the general form

$$\partial_t u = \mathcal{N}(u, \partial_x u, \partial_x^2 u, \cdots, x),$$ (4)

where $\partial_\bullet$ is shorthand for the differential operator $\frac{\partial}{\partial \bullet}$ in either time t or space x and $\mathcal{N}(\cdot)$ is our unknown right-hand side, which we assume to be some nonlinear function of our observed quantity $u(x, t) \in \mathbb{R}^n$ and its spatial derivatives. The goal now is to find $\mathcal{N}(\cdot)$, which typically only consists of a few terms, thus leading to sparse regression methods, which take as many (nonlinear) candidate functions as are permissible and try to find those that contribute to the dynamics the most. We now write the corresponding PDE similar to Sect. 3.1:

$$\partial_t u = \Theta(u)\Xi, \tag{5}$$

but with $\Theta \in \mathbb{R}^{n \times k}$ as a matrix of k (possibly nonlinear, spatially extended) functions that make up our ansatz and we combine them linearly with parameters Ξ_i. The task of any sparse-regression algorithm in this case is to find a sparse set of these parameters (and their values), that best describe the observed time series data of u.

To illustrate this methodology, we look at the Korteweg-De Vries equation, which is used to model waves on shallow water surfaces. The equation reads

$$\partial_t u = -\partial_x^3 u - 6u\partial_x u, \tag{6}$$

where u describes the height displacement of the water surface from its equilibrium height. As we can see our right-hand side consists of only two terms, a dispersion term $\partial_x^3 u$ and an advection term $u\partial_x u$. We first generate time series data using numerical time integration methods of our choosing[1] and add a slight amount of white noise to it. For the initial conditions, we use two soliton peaks of different heights. The results can be seen in Fig. 2.

Most regression-based discovery methods adhere to a similar methodology: denoising the data, followed by differentiation in time (and space in case for PDEs), constructing a library of candidate terms, and ultimately employing a sparse regression technique to identify significant terms. For use on PDEs the key difference lies in the data structure, i.e. the dimensions of the library, although discretization allows the procedures to be virtually identical to those for ODEs. While there may be additional procedures related to model selection or hyperparameter optimization, this general framework is applicable to a majority of methods. Now we again solve the problem 3 using a sequential threshold least-squares (STLS) algorithm. For more details see [36] and its supplementary material. This method can be adapted to accommodate coefficients that vary in space or time [45], or optimization can be improved using auxiliary variables in the penalty term [46].

---

[1] Here the initial value problem solver 'solve_ivp' of the python package Scipy has been used, after discretizing the problem.

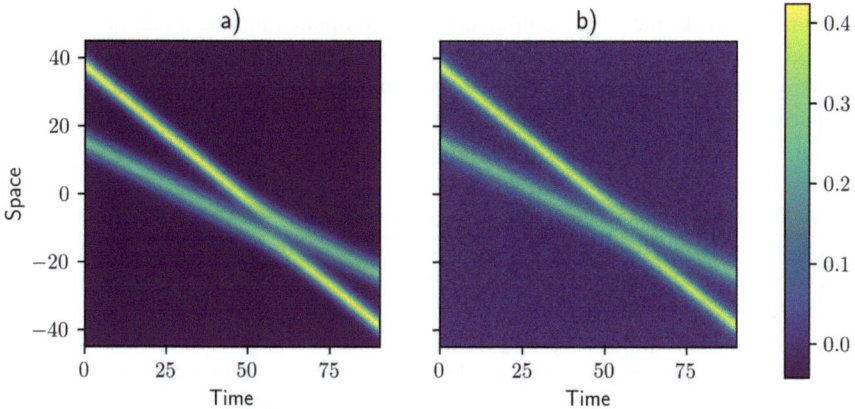

Fig. 2 Data simulated from Korteweg-De Vries equation (6). Left (a) shows the true data and right (b) shows noisy data with $\gamma = 0.01$

As in the example that can be found at https://github.com/snagcliffs/PDE-FIND we use the following candidate terms in Θ for re-discovering the Korteweg-De Vries equation (6):

$$\left[1, u, u^2, u^3, \partial_x u, u\partial_x u, u^2\partial_x u, u^3\partial_x u, \partial_x^2 u, u^2\partial_x^2 u, u^3\partial_x^2 u, \partial_x^3, u\partial_x^3 u, u^2\partial_x^3 u, u^3\partial_x^3 u\right]$$

Using the noisy simulation data from before as input, the data is de-noised within the optimization procedure and derivatives are being calculated. After the procedure is complete the discovered equation reads:

$$\partial_t u = -1.04\,\partial_x^3 u - 6.07\,u\partial_x u, \tag{7}$$

which correctly identifies the two relevant terms with no superfluous terms and gives coefficients with errors within single-digit per cent.

4 Learning Functionals in DFT

One application of data-driven modeling is the estimation of functionals in density functional theory (DFT). Most DFT calculations are conducted within the Kohn-Sham framework, where complex many-body terms are expressed by an exchange-correlation (XC) functional, while maintaining a one-body Kohn-Sham equation. Although this formulation with the XC-functional can be theoretically exact, the explicit form of the XC functional remains unknown. Consequently, approximations of the XC-functional are necessary and have been approached in various ways in the literature. Techniques ranging from Bayesian methods and genetic algorithms to neural networks have been employed to approximate the functional, given neural networks' ability to approximate any continuous function.

While neural network approaches have achieved high accuracy in DFT calculations, their interpretability is limited, as they typically function as black-box approximators. In light of this, we turn our attention to a specific publication [47], which employs the previously mentioned symbolic regression, yielding an interpretable description of the functional. This method encodes a symbolic representation of a functional in a machine-readable format, which is then optimized through a regularized evolutionary search algorithm known as symbolic functional evolutionary search (SyFES). SyFES iteratively refines functionals by manipulating those with optimal fitness within a subset of potential functionals, where fitness is determined by a weighted root mean square error between known and computed energies from the functional. This methodology was evaluated using the MGCDB84 dataset [48]. After demonstrating the capability of SyFES by recovering the well-known B97 XC-functional, the technique was applied to evolve the known ωB97M-V functional into a novel, more accurate functional named GAS. While this example highlights the use of symbolic methods in DFT, numerous other approaches, including those employing neural networks in quantum and fluid realms of DFT, exist. Notable contributions in this area include the research by Martin Oettel, as reported in [49–51]. Additional relevant research can be found in publications such as [46, 52, 53].

5   Conclusion

This review has explored the diverse landscape of data-driven methods for learning dynamic equations from time-series data, highlighting their significance across various domains like physics, engineering, and biology. We have discussed the transition from traditional modeling approaches to advanced data-driven techniques, emphasizing the shift towards harnessing high-dimensional data for uncovering underlying dynamic equations.

Through this exploration, we have underscored the critical role of symbolic regression and neural networks in modeling complex systems. While neural networks offer a powerful tool for approximation, their black-box nature poses interpretability challenges. In contrast, symbolic regression, particularly within frameworks like SINDy, provides a more interpretable model by identifying sparse representations of dynamic systems.

The application of these methods in specific domains, such as the estimation of functionals in density functional theory, illustrates their practical value and potential for advancing scientific understanding. The ability to recover known functionals and discover novel ones underscores the transformative power of data-driven approaches in refining existing theories and facilitating new discoveries.

The continuous advancement of data-driven modeling methods promises further insights into complex systems, offering new possibilities for research and application.

References

1. Noack BR, Afanasiev K, Morzynski M, Tadmor G, Thiele F (2003) J Fluid Mech 497:335
2. Ali H, Walter L (1998) JSME Int J Ser C 41(3):510–531. https://doi.org/10.1299/jsmec.41.510
3. Habib MK, Ayankoso SA, Nagata F (2021) 2021 IEEE international conference on mechatronics and automation (ICMA). IEEE. https://doi.org/10.1109/icma52036.2021.9512658
4. Diggle PJ, Mateu J, Clough HE (2000) Adv Appl Probab 32(2):331–343. https://doi.org/10.1239/aap/1013540166
5. Mahmoud HFF (2021) Int J Stat Probab 10(2):90. https://doi.org/10.5539/ijsp.v10n2p90
6. Ruano AE, Ferreira PM, Fonseca CM (2005) An overview of nonlinear identification and control with neural networks. Institution of Engineering and Technology, pp 37–88. https://doi.org/10.1049/pbce070e_ch2
7. Loiseau JC, Brunton SL (2018) J Fluid Mech 838:42–67. https://doi.org/10.1017/jfm.2017.823
8. Vlachas PR, Byeon W, Wan ZY, Sapsis TP, Koumoutsakos P (2018) Proceedings of the royal society a: mathematical. Phys Eng Sci 474(2213):20170844. https://doi.org/10.1098/rspa.2017.0844
9. Wan ZY, Vlachas P, Koumoutsakos P, Sapsis T (2018) PLoS One 13(5):1
10. Koppe G, Toutounji H, Kirsch P, Lis S, Durstewitz D (2019) PLoS Comput Biol 15(8):e1007263. https://doi.org/10.1371/journal.pcbi.1007263
11. Durstewitz D, Koppe G, Thurm MI (2023) Nat Rev Neurosci 24(11):693. https://doi.org/10.1038/s41583-023-00740-7
12. Chen RT, Rubanova Y, Bettencourt J, Duvenaud DK (2018) Adv Neural Inf Process Syst 31
13. Jaeger H (2001) The "echo state" approach to analysing and training recurrent neural networks. GMD Report 148, GMD—German National Research Institute for Computer Science. http://www.faculty.jacobs-university.de/hjaeger/pubs/EchoStatesTechRep.pdf
14. Doan NAK, Polifke W, Magri L (2019) Physics-informed echo state networks for chaotic systems forecasting. Springer International Publishing, pp 192–198. https://doi.org/10.1007/978-3-030-22747-0_15
15. Doan NAK, Polifke W, Magri L (2020) Learning hidden states in a chaotic system: a physics-informed echo state network approach. Springer International Publishing, pp 117–123. https://doi.org/10.1007/978-3-030-50433-5_9
16. Doan N, Polifke W, Magri L (2020) J Comput Sci 47:101237. https://doi.org/10.1016/j.jocs.2020.101237
17. Raissi M, Perdikaris P, Karniadakis GE (2019) J Comput Phys 378:686
18. Lim J, Psaltis D (2022) APL Photonics 7(1). https://doi.org/10.1063/5.0071616
19. Ouala S, Nguyen D, Drumetz L, Chapron B, Pascual A, Collard F, Gaultier L, Fablet R (2020) Chaos Interdiscip J Nonlinear Sci 30(10) (2020). https://doi.org/10.1063/5.0019309
20. Robinson H, Pawar S, Rasheed A, San O (2022) Neural Netw 154:333–345. https://doi.org/10.1016/j.neunet.2022.07.023
21. Lehrmann A, Sigal L (2017) Adv Neural Inf Process Syst 30
22. Ferreira PM, Ruano AE (2011) Evolutionary multiobjective neural network models identification: evolving task-optimised models. Springer, Berlin, Heidelberg, pp 21–53. https://doi.org/10.1007/978-3-642-11739-8_2
23. Long Z, Lu Y, Dong B (2019) J Comput Phys 399:108925. https://doi.org/10.1016/j.jcp.2019.108925
24. Zubov K, McCarthy Z, Ma Y, Calisto F, Pagliarino V, Azeglio S, Bottero L, Luján E, Sulzer V, Bharambe A et al (2021) arXiv:2107.09443
25. Raissi M, Perdikaris P, Karniadakis GE (2017) arXiv:1711.10561
26. Raissi M, Perdikaris P, Karniadakis GE (2017) arXiv:1711.10566
27. Dwivedi V, Srinivasan B (2019) Physics informed extreme learning machine (pielm)–a rapid method for the numerical solution of partial differential equations. https://doi.org/10.48550/ARXIV.1907.03507. arXiv:1907.03507

28. Bock H, Plitt K (1984) IFAC Proc Vol 17(2):1603–1608. https://doi.org/10.1016/s1474-6670(17)61205-9

29. Brunton SL, Proctor JL, Kutz JN (2016) Proc Natl Acad Sci 113(15):3932. https://doi.org/10.1073/pnas.1517384113

30. Loiseau JC, Noack BR, Brunton SL (2018) J Fluid Mech 844:459–490. https://doi.org/10.1017/jfm.2018.147

31. Mangan NM, Brunton SL, Proctor JL, Kutz JN (2016) IEEE Trans Mol Biol Multi Scale Commun 2(1):52. https://doi.org/10.1109/TMBMC.2016.2633265

32. Bongard J, Lipson H (2007) Proc Natl Acad Sci 104(24):9943. https://doi.org/10.1073/pnas.0609476104

33. Schmidt M, Lipson H (2009) Science 324(5923):81. https://doi.org/10.1126/science.1165893

34. Cranmer M (2020) Milescranmer/pysr v0.2. https://doi.org/10.5281/ZENODO.4041459. https://zenodo.org/record/4041459

35. Guimerà R, Reichardt I, Aguilar-Mogas A, Massucci FA, Miranda M, Pallarès J, Sales-Pardo M (2020) Sci Adv 6(5). https://doi.org/10.1126/sciadv.aav6971

36. Rudy SH, Brunton SL, Proctor JL, Kutz JN (2017) Sci Adv 3(4):e1602614. https://doi.org/10.1126/sciadv.1602614

37. Bär M, Hegger R, Kantz H (1999) Phys Rev E 59(1):337. https://doi.org/10.1103/physreve.59.337

38. Ljung L (1998) System identification: theory for the user. Prentice Hall information and system sciences series, 2nd edn. Prentice Hall, Philadelphia, PA

39. Voss HU, Kolodner P, Abel M, Kurths J (1999) Phys Rev Lett 83(17):3422. https://doi.org/10.1103/physrevlett.83.3422

40. Voss H, Bünner M, Abel M (1998) Phys Rev E 57(3):2820. https://doi.org/10.1103/physreve.57.2820

41. Messenger DA, Bortz DM (2021) J Comput Phys 443:110525. https://doi.org/10.1016/j.jcp.2021.110525

42. Dong X, Bai YL, Lu Y, Fan M (2022) Nonlinear Dyn 111(2):1485

43. North JS, Wikle CK, Schliep EM (2023) Int Stat Rev 91(3):464–492. https://doi.org/10.1111/insr.12554

44. Mangan NM, Kutz JN, Brunton SL, Proctor JL (2017) Proc R Soc A Math Phys Eng Sci 473(2204):20170009. https://doi.org/10.1098/rspa.2017.0009

45. Rudy S, Alla A, Brunton SL, Kutz JN (2019) SIAM J Appl Dyn Syst 18(2):643. https://doi.org/10.1137/18m1191944

46. Zheng P, Askham T, Brunton SL, Kutz JN, Aravkin AY (2018) A unified framework for sparse relaxed regularized regression: Sr3. https://doi.org/10.48550/ARXIV.1807.05411. arXiv:1807.05411

47. Ma H, Narayanaswamy A, Riley P, Li L (2022) Sci Adv 8(36). https://doi.org/10.1126/sciadv.abq0279

48. Mardirossian N, Head-Gordon M (2016) J Chem Phys 144(21). https://doi.org/10.1063/1.4952647

49. Lin SC, Martius G, Oettel M (2020) J Chem Phys 152(2). https://doi.org/10.1063/1.5135919

50. Shang-Chun L, Oettel M (2019) SciPost Phys 6(2). https://doi.org/10.21468/scipostphys.6.2.025

51. Simon A, Weimar J, Martius G, Oettel M (2024) J Chem Theory Comput 20(3):1062–1077. https://doi.org/10.1021/acs.jctc.3c01238

52. Colen J, Han M, Zhang R, Redford SA, Lemma LM, Morgan L, Ruijgrok PV, Adkins R, Bryant Z, Dogic Z, Gardel ML, de Pablo JJ, Vitelli V (2021) Proc Natl Acad Sci 118(10). https://doi.org/10.1073/pnas.2016708118

53. Sammüller F, Hermann S, de las Heras D, Schmidt M (2023) Proc Natl Acad Sci 120(50). https://doi.org/10.1073/pnas.2312484120

Machine Learning Approaches to Classical Density Functional Theory

Alessandro Simon and Martin Oettel

Abstract In this chapter, we discuss recent advances and new opportunities through methods of machine learning for the field of classical density functional theory, dealing with the equilibrium properties of thermal nano– and microparticle systems having classical interactions. Machine learning methods offer the great potential to construct and/or improve the free energy functional (the central object of density functional theory) from simulation data and thus they complement traditional physics– or intuition–based approaches to the free energy construction. We also give an outlook to machine learning efforts in related fields, such as liquid state theory, electron density functional theory and power functional theory as a functionally formulated approach to classical nonequilibrium systems.

1 Introduction

Density functional theory (DFT) is a powerful reductionist scheme for classical and quantum many-body systems in equilibrium. The reduction comes about by the existence of a unique (free) energy functional, depending only on the one–body density of classical or quantum particles. From this functional, all other properties of interest, most notably higher order correlations can be derived. For quantum systems at zero temperature, $T = 0$, the unique mapping between an external potential $V^{\mathrm{ext}}(\mathbf{r})$ and the particle density $\rho(\mathbf{r})$ entails the existence of a unique energy functional $E[\rho]$, not depending on V^{ext} [1]. For finite T, the argument can be generalized to show the existence of a unique free energy functional $\mathcal{F}[\rho]$ both in the quantum case [2] and in the classical case [3]. However, in general the functional $\mathcal{F}[\rho]$ is different for differing internal Hamiltonians of the system (i.e. differing particle–particle interaction potentials) and it is not known except for a few exceptional cases

A. Simon · M. Oettel (✉)
Institute of Applied Physics, University of Tübingen, Tübingen, Germany
e-mail: martin.oettel@uni-tuebingen.de

A. Simon
e-mail: alessandro-rodolfo.simon@uni-tuebingen.de

M. te Vrugt (ed.), *Artificial Intelligence and Intelligent Matter*, Machine Intelligence for Materials Science, https://doi.org/10.1007/978-3-032-04129-6_6

83

(like the ideal gas and a system of one–dimensional (1D) hard rods in the classical case). Moreover, in the classical case there are a number of Hamiltonians of interest, ranging from those of atomic and molecular systems to those of polymeric and colloidal systems where the basic particles are macromolecular in nature and their particle–particle interactions are already coarse–grained.

Constructing classical free energy functionals is occasionally more of an art than a systematic procedure and entails the use of specific physical and mathematical insight into the system of interest. For hard–body systems, e.g., one can use concepts from integral geometry to derive fundamental measure theory (FMT) [4]. These FMT functionals are perhaps examples for being most advanced and accurate compared to simulation data (however, they are not exact except for the 1D case). For systems with other interactions one has not reached yet such a level of insight and precision. Here, efforts have gone into defining simplified model systems for which functionals are constructed with differing success. These simplified model systems include simple fluids with repulsive cores and attractive tails (as embodied by the Lennard-Jones (LJ) potential) [5], particles with (screened) electrostatic interactions [6], polymeric fluids with simple connectivity assumptions between monomers [7], patchy particles as examples for associating fluids [8], ... An exception to these simplified model systems is water (due to its overwhelming importance) whose functional building can be viewed as archetypical for liquids with anisotropic molecules [9].

The "cheap" alternative to the difficulties of classical DFT (cDFT) are classical simulations, either Monte Carlo (MC) or Molecular Dynamics (MD). One simply needs to specify the underlying potential energies and forces and acquires the desired properties as statistical averages over snapshots of the system. Such simulations, however, can be costly if higher–order correlations are needed, or if free energies need to be computed via thermodynamic integrations. Additionally, physical insights into the system (such as schematic behavior of certain correlations) necessitate running simulations for a large number of parameters (thermodynamic ones such as density and temperature, or specific parameters in the interaction Hamiltonian). Nevertheless simulation data constitute "ground truth" for a classical model which, as said, can be generated in a relatively cheap way, and this situation appears to be highly suitable for big data techniques as exemplified by machine learning (ML). Thus, there is hope to combine the precision of simulation data with the conceptual power of the cDFT formulation, which in the end also would allow for very resource–efficient computations. A vision for the description of a classical many–body system would consequently be the systematic use of simulation data to construct or "learn" an interpretable and manipulable (functionally differentiable) free energy functional. The systematic use would include the possibility to refine and improve the "learned" functional if needed.

This small review intends to cover the efforts of the past years to apply ML techniques to cDFT. In Sect. 2, we briefly summarize basic relations of density functional theory. Work in the past years has concentrated on the "simplest" of the simplified model systems mentioned above, and these are introduced in Sect. 3. Specific approaches are reviewed in Sect. 4. At the moment, there appears to be no preference for or clear advantage of a specific ML technique, so we attempt to describe the gist of the used techniques in a tractable manner in this section. Finally, in Sect. 5 we describe the relation to the integral equation method of liquid state theory, to some ML approaches to the problem of electron (quantum) DFT and give an outlook to the general nonequilibrium, time–dependent problem which allows a formulation akin to cDFT in terms of a unique power functional [10].

2 Classical DFT: Basic Theory

There are excellent books on classical liquid state theory and more specifically excellent reviews on cDFT, for a selection we refer to [3, 4, 11, 12].

We consider rigid particles with positional and orientational degrees of freedom, thus particles can be anisotropic to allow for the description of molecular fluids and nonspherical colloidal systems. We follow standard classical statistical mechanics in the grand canonical ensemble. For the Hamiltonian we assume the following form

$$H = K + u(\mathbf{r}^N, \omega^N) + \sum_i V^{\text{ext}}(\mathbf{r}_i, \omega_i) \tag{1}$$

where K is the kinetic energy of translational and rotational motion. Furthermore, $\mathbf{r}_i$ is the position, and ω_i the orientation (in general specified by three Euler angles) of particle i. The position–dependent part of the internal energy $u(\mathbf{r}^N, \omega^N)$ is often taken to be a sum of 2–body pair potentials but this is not necessary. The external potential $V^{\text{ext}}(\mathbf{r}_i, \omega_i)$ is a one–body term acting in general on both position and orientation of the individual particle. We introduce the collective variable $x_i = [\mathbf{r}_i, \omega_i]$ combining both position and orientation for brevity. The one-body density is then defined as the statistical average of all particles' positions and orientations

$$\rho(x) = \left\langle \sum_i \delta(x - x_i) \right\rangle \tag{2}$$

where δ is the (Dirac) delta function. Classical density functional theory is based on the existence of a functional for the grand potential $\Omega[\rho(x)]$ whose minimization gives the equilibrium density $\rho_{\text{eq}}(x)$. The functional $\Omega[\rho(x)]$ reads

$$\Omega[\rho] = \mathcal{F}_{\text{id}}[\rho] + \mathcal{F}_{\text{ex}}[\rho] + \int dx \left(V^{\text{ext}}(x) - \mu \right) \tag{3}$$

Here, the one–body piece containing V^{ext} and the chemical potential μ is separated out, and $\mathcal{F}[\rho] = \mathcal{F}_{\text{id}}[\rho] + \mathcal{F}_{\text{ex}}[\rho]$ is the unique free energy functional only depending on the density (and not on the external potential). It consists of the ideal (non-interacting) part $\mathcal{F}_{\text{id}}[\rho]$ and the excess (over ideal) part $\mathcal{F}_{\text{ex}}[\rho]$. The ideal gas part is given by

$$\beta \mathcal{F}_{id} = \int dx \, \rho(x) \left[\ln(\rho(x)\lambda^3) - 1\right], \tag{4}$$

and is the exact free energy functional for noninteracting particles ($u = 0$). Here, $\beta = 1/(k_B T)$ is the inverse temperature and λ^3 is a volume factor containing the de–Broglie thermal wavelength and a normalization factor of the orientational integral. Minimization of Ω w.r.t. the density results in the Euler–Lagrange equation

$$\rho_{eq}(x) = \exp\left(-\beta V^{ext}(x) + \beta\mu + c_1[\rho_{eq}(x)]\right) \tag{5}$$

Here, $c_1[\rho]$ is the first member of the hierarchy of direct correlation functions (DCF), defined by functional derivatives of $\beta \mathcal{F}_{ex}[\rho]$ w.r.t. the density. Specifically,

$$c_1(x)[\rho] = -\beta \frac{\delta \mathcal{F}_{ex}[\rho]}{\delta\rho(x)}, \tag{6}$$

and (owing to its importance)

$$c_2(x, x')[\rho] = -\beta \frac{\delta^2 \mathcal{F}_{ex}[\rho]}{\delta\rho(x)\delta\rho(x')}, \tag{7}$$

is the second–order direct correlation function (often "the" DCF in the literature).

The excess free energy functional $\mathcal{F}_{ex}[\rho]$ is in general not known. The most famous exception is the system of 1D hard rods which therefore has played an important role in the past years to test ML methods (see below). The full functional $\mathcal{F}_{ex}[\rho]$ is not directly accessible in simulations (as "ground truth"), easily computable is only the equilibrium density profile $\rho_{eq}(x)$ for a chosen $V^{ext}(x)$. The Euler–Lagrange equation (5) entails a map

$$\rho_{eq}(x) \quad \leftrightarrow \quad c_1(x)[\rho_{eq}(x)], \tag{8}$$

and thus simulations can provide us with individual points $\{\rho_{eq}, c_1\}$ of this map. Thus the reconstruction of the functional $c_1(x)[\rho_{eq}(x)]$ should be suitable for ML methods given enough points of the map. Having learned the functional $c_1(x)[\rho_{eq}(x)]$ gives access to higher–order correlation functions and specific physics contained in those (e.g. sum rules) as long as the ML methods allow for functional differentiation.

Note that the original DFT proof [2, 3] rests on the unique map

$$\rho_{eq}(x) \quad \leftrightarrow \quad V^{ext}(x). \tag{9}$$

If the functional of the excess free energy is not known, the arrow to the right from ρ_{eq} to V^{ext} is actually a typical difficult inverse problem for simulations. Given that simulations allow the computation of $\rho_{eq}(x)$ for given $V^{ext}(x)$ with comparable ease (arrow to the left), ML methods should be in principle suited to learn the functional $\rho_{eq}(x)[V^{ext}(x)]$ which from the functional perspective complements $c_1(x)[\rho_{eq}(x)]$ as follows. We define a "local chemical potential" by $\psi(x) = \mu - V^{ext}(x)$. The grand potential functional Ω of Eq. (3) can be viewed as a functional of $\psi(x)$ whose functional derivatives generate the density profile and the higher–order density fluctuation functions [3]. Specifically

$$\frac{\delta\Omega[\psi]}{\delta\psi(x)} = -\rho_{\text{eq}}(x) \tag{10}$$

and

$$\frac{1}{\beta}\frac{\delta^2\Omega[\psi]}{\delta\psi(x)\delta\psi(x')} = -H_2(x, x') \tag{11}$$

where $H_2(x, x')$ is the density–density correlation function defined by

$$H_2(x, x') = \langle [\rho(x) - \langle\rho(x)\rangle] [\rho(x') - \langle\rho(x')\rangle] \rangle \tag{12}$$

One can write $H_2(x, x') = \rho_{\text{eq}}(x)\delta(x - x') + \rho_{\text{eq}}(x)\rho_{\text{eq}}(x')h(x, x')$ where $h(x, x')$ is linked to the standard pair correlation function $g(x, x')$ by $h = g - 1$. The functions $h(x, x')$ and $c_2(x, x')$ are linked by the famous Ornstein–Zernike relation

$$h(x, x') - c_2(x, x') = \int dx'' h(x, x'') \rho_{\text{eq}}(x'') c_2(x', x''), \tag{13}$$

an integral equation of formidable difficulty in the case of anisotropic fluids and a general $V^{\text{ext}}(x)$.

The link between the structural functions h and c_2 is the problem of the integral equation approach to liquid state theory and can be seen as a specific subtopic of the general cDFT problem. We will comment upon recent ML advances in integral equation theory briefly in Sect. 5.1.

3 Model Systems

The ML methods which are described more in detail below in Sect. 4 concentrate on different aspects of cDFT and also apply to different model systems. Here, for completeness, we briefly introduce the used model systems.

3.1 Hard Sphere System in 1D and 3D

An easy to handle, yet non-trivial model is the one dimensional hard-rod system. Here the particles of width σ are constrained to a line with coordinate z without the possibility of overlapping and no additional interaction between them. It is also one of the few models where the exact excess free energy functional is known.

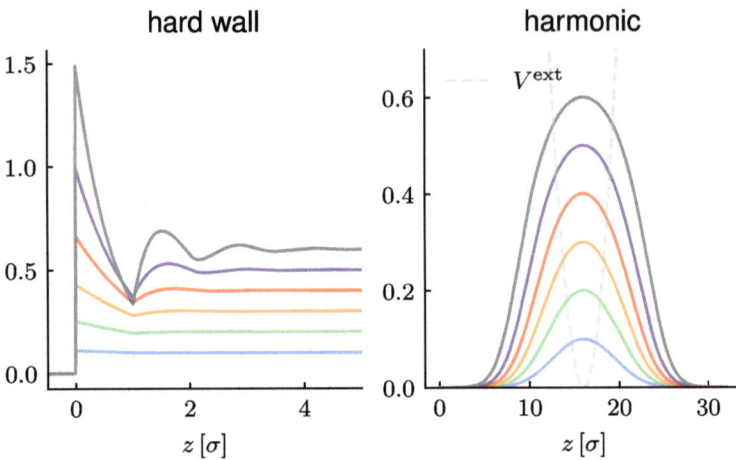

Fig. 1 Solutions for the density profiles $\rho(z)$ of the one dimensional hard-rod system for two kinds of external potentials. The bulk densities are linearly increasing $\rho_b \sigma = 0.1, 0.2, \ldots, 0.6$, and the rod length is σ

$$\beta \mathcal{F}_{\text{ex}}^{\text{hr}}[\rho(z)] = \int \Phi(n_0, n_1) dz = -\int n_0 \ln(1 - n_1) dz \qquad (14)$$

with the weighted densities

$$n_i(z) = \int dz' \rho(z') \omega_i(z - z') = \rho \otimes \omega_i \qquad (15)$$

and the two kernel functions $\omega_0(z) = \delta(\sigma/2 - |z|)/2$ and $\omega_1(z) = \Theta(\sigma/2 - |z|)$. This exact functional has a form characteristic for many of the approximate functionals. The local excess free energy density, $\beta f_{\text{ex}}(z) = \Phi(z)$, depends non-locally on the density profile $\rho(z)$ through weighted (or "smeared") densities with characteristic weight functions. In Fig. 1, we show characteristic density profiles resulting from this exact functional, namely adsorption at a hard wall (left panel, showing the characteristic layering effect) and confinement in a trapping potential (right panel). Such profiles are frequently used in the machine learning routines described later.

The hard rod system can also be extended to higher dimensions, although no exact functionals are known for the 2D or 3D case. There are however very accurate functionals based on fundamental measure theory (FMT), see Ref. [4] for a review.

3.2 Lennard-Jones

The Lennard-Jones (LJ) system is undoubtedly one of the most extensively studied models for a simple fluid, featuring typical short–range repulsion (Pauli exclusion of closed electron shells) and longer–ranged attraction (van der Waals interaction), giving rise to a gas, liquid and solid phase. It is realistic for noble gases. The LJ potential has the form

$$u_{\mathrm{LJ}}(r) = -4\varepsilon \left[(r/\sigma)^6 - (r/\sigma)^{12} \right], \tag{16}$$

where σ is a particle diameter. No exact free energy functional for the LJ system is known, owing mainly to the complications of the attractive part. Here, the random phase approximation [11, 12] (RPA) is semiquantitative for supercritical state points at high temperature but gives also insights and a qualitative account of the phase diagram for lower temperatures. The RPA functional is given by:

$$\mathcal{F}_{\mathrm{ex}}^{\mathrm{RPA}}[\rho] = \mathcal{F}_{\mathrm{ex}}^{\mathrm{ref}}[\rho] + \frac{1}{2} \int \mathbf{dr}\mathbf{dr}' \rho(\mathbf{r}) \rho(\mathbf{r}') u_{\mathrm{LJ}}^{\mathrm{att}}(\mathbf{r} - \mathbf{r}') \tag{17}$$

where $\mathcal{F}_{\mathrm{ex}}^{\mathrm{ref}}$ is a reference functional (usually the one of a hard sphere system with optimized σ). The attractive part is of typical mean–field form (with the assumption of uncorrelated densities), and through defining $n^{\mathrm{att}} = \rho \otimes u_{\mathrm{LJ}}^{\mathrm{att}}$ can be written in the weighted density form described above. $u_{\mathrm{LJ}}^{\mathrm{att}}$ is a suitably defined attractive part of the LJ potential, e.g. from the WCA prescription [11, 12].

3.3 Kern–Frenkel

The Kern–Frenkel (KF) potential is a popular model for anisotropic interactions. In addition to the repulsive hard-sphere interaction, every particle is equipped with N so called patches, which in the KF model are cones emanating from the particle center outwards up to a certain cut-off radius. If any two cones belonging to different particles overlap, they are considered to be bonded, with an associated decrease of the energy of the system by ε (see Fig. 2). The bonding energy, the cut-off radius δ, together with the cone angle θ and number of patches are tunable parameters which allow us to adapt the model to different situations. Popular are choices for which ensure that the single bond condition between two particles is fulfilled, which is the particular limit of an associating fluid where Wertheim theory [12] gives a semiquantitative account of the phase diagram and bonding statistics.

4 Machine Learning Approaches

The machine learning approaches in the literature can be roughly classified into the following categories:

- Parameterization of the excess functional [13–17] ($\rho_{\mathrm{eq}} \to \mathcal{F}$ map)

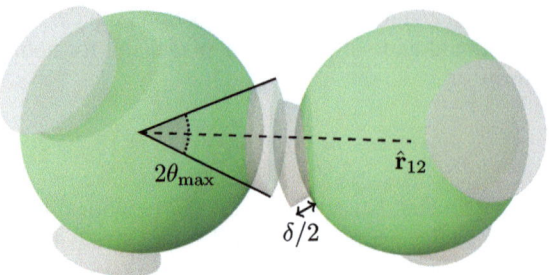

Fig. 2 Patch–patch interaction in the Kern–Frenkel model between two particles connected by the vector $\hat{\mathbf{r}}_{12}$. The angle $\theta^{\max}$ specifies the opening of the patch cone and the parameter δ the extension in the radial direction

Table 1 Different methods and their respective input/output pairs and loss terms, if applicable. The star superscript denotes the outputs of the neural network

Approach	Input	Output	Parameters	Loss
Direct parameterization [13–16]	$\mu, V^{\text{ext}}(\mathbf{r}), (\beta)$	$\mathcal{F}_{\text{ex}}[\rho]$	Interaction kernels, symbolic structure	MSE$(\rho^\star, \rho)$ + reg.
Neural DFT [18, 19]	$\rho(\mathbf{r})$	$c_1(\mathbf{r}, [\rho])$	Network weights	MSE$(c_1^\star, c_1)$
c_2-matching [17]	$\rho(\mathbf{z})$	$c_2(|z_i - z_j|)$	Networks weights	MSE$(c_2^\star, c_2)$ + reg.
Bayesian model [21]	$\rho(\mathbf{r})$	$V^{\text{ext}}(\mathbf{r})$	External potential parameterization	[Posterior sampling]
Gaussian process [22]	$\mu, V^{\text{ext}}(\mathbf{r})$	$\rho(\mathbf{r})$	Prior hyperparameters	[Posterior sampling]

- Parameterization of the one-body correlation function [18, 19] ($\rho_{\text{eq}} \to c_1$ map)
- (Parametric) Bayesian methods on the $\rho_{\text{eq}} \leftrightarrow V^{\text{ext}}$ map [20, 21]
- Gaussian Processes for the $\rho_{\text{eq}} \to V^{\text{ext}}$ map for reinforcement learning [22].

Below we try to categorize the approaches according to their main idea/ingredient, even though one should be aware of possible overlaps (Table 1).

4.1 Direct Parameterization of the Functional

In the historical development of cDFT for simple fluids, it was quickly noticed that the local density approximation

$$\mathcal{F}_{\text{ex}}[\rho] = \int d\mathbf{r}\, f_{\text{ex}}(\rho(\mathbf{r})),\tag{18}$$

or the (square) gradient approximation

$$\mathcal{F}_{ex}[\rho] = \int d\mathbf{r} \left[f_{ex}(\rho(\mathbf{r})) + \frac{a}{2} (\nabla \rho(\mathbf{r}))^2 \right] \tag{19}$$

have very limited accuracy and cannot capture especially the correlation effect due to repulsive cores ("layering"). Nevertheless, the square gradient approximation contributed a lot to our understanding of the physics of liquid–vapour interfaces. For the subsequent development, the insight was crucial that the functional has a non-local dependency on the density distribution $\rho(x)$ through convoluted, weighted densities $n_i = \rho \otimes \omega_i$ with weight functions ω_i which are usually of finite range, see also Sect. 3. The weighted-density form for the free energy entails that the minimizing equation for the equilibrium profile ρ_{eq}, Eq. (5), is a nonlinear integral equation. It is usually solved using the Picard iteration scheme, which starts with an initial guess $\rho^{i=0}(\mathbf{r})$, usually the uniform bulk density. Inserting ρ^i into the EL equation gives a new density distribution

$$\underline{\rho^i}(\mathbf{r}) = \exp\left(-\beta V^{ext}(\mathbf{r}) + \beta\mu + c_1[\rho^i(\mathbf{r})]\right). \tag{20}$$

Owing to stability reasons, the next iteration ρ^{i+1} is obtained by mixing:

$$\rho^{i+1}(\mathbf{r}) = (1-\alpha)\rho^i(\mathbf{r}) + \alpha\underline{\rho^i}(\mathbf{r}), \tag{21}$$

where α is a small parameter which can be dynamically adapted. Upon reaching a small norm of the difference between two iterations, $||\rho^{i+1}(\mathbf{r}), \rho^i(\mathbf{r})|| \leq \epsilon$, one can speak of a "self–consistent" solution, i.e. the free energy functional permits a minimization of the grand potential functional up to a certain numerical accuracy. For machine–learned functionals, this is a nontrivial condition.

In the context of machine learning, it appears natural to assume (parametrize) a new excess functional $\mathcal{F}_{ex}^{ML}[\rho; \theta]$, which depends on some unknown internal parameters θ. These parameters could either constitute the weights of a universal approximator (e.g. a multilayer perceptron (MLP)) or the variables in an *ansatz* built on existing knowledge of the system. Using a black-box model such as an MLP, prevents us from any direct interpretation of the learned internal representation. On the other hand, less general, parameterized models might suffer from limited generalizability. One may roughly differentiate between the two (idealized) camps of ML/DFT practitioners:

(i) The ones mainly interested in an accurate emulation of the physical system in question, taking advantage of the efficiency of the DFT formalism and

(ii) those interested in uncovering ("fitting") interpretable representations of the functional maps.

4.1.1 Mean-Field and Third Order Terms, Isotropic Case

The LJ system was investigated in 1D [13] and then later in 3D [15] using a "camp (ii)" approach aiming at learning mean-field and higher order correction terms for the attractive part of the interaction. Apart from one important detail, namely whether self-consistency of the functional was imposed during training, both approaches are similar and we will limit the exposition to the more recent Ref. [15]. The topic of self-consistency will be brought up again at the end of this section.

In the work by Cats et al. [15] the authors consider the LJ fluid in the reference scheme. The standard RPA functional of Eq. (17) is compared with the RPA functional plus ML corrections, where the corrections are parametrized as

$$\beta \Delta \mathcal{F}_{\text{ex}}^{\text{ML2}} = \frac{1}{2} \int d\mathbf{r} d\mathbf{r}' \rho(\mathbf{r}) \rho(\mathbf{r}') \Omega_2(|\mathbf{r} - \mathbf{r}'|) \tag{22}$$

$$\beta \Delta \mathcal{F}_{\text{ex}}^{\text{ML3}} = \frac{1}{3} \int d\mathbf{r} d\mathbf{r}' \rho(\mathbf{r})^2 \rho(\mathbf{r}') \Omega_3(|\mathbf{r} - \mathbf{r}'|) \tag{23}$$

with the unknown kernels Ω_i. Here, ML2 is second order in density and provides a correction to the RPA mean–field kernel, while ML3 adds a contribution of third order in density. Training was performed at one supercritical temperature in flat wall geometry with external potentials (varying in their steepness), here $\rho(z)$ only depends on the Cartesian coordinate z. In the flat wall geometry, the functionals (22), (23) retain their form with $\mathbf{r}, \mathbf{r}' \to z, z'$ and $\Omega_i \to \omega_i(z)$ and the new kernels $\omega_i(z)$ are differentiable parameters of the network. Owing to the isotropy of the LJ interaction, the kernels are related by

$$\omega_i(z) = 2\pi \int_z^\infty dr \, r \, \Omega_i(r), \tag{24}$$

$$\Omega_i(|\mathbf{r}|) = -\frac{1}{2\pi} \frac{1}{z} \frac{d\omega_i(z)}{dz} \bigg|_{z=|\mathbf{r}|}, \tag{25}$$

thus the training in the flat wall geometry is sufficient, and the resulting functional can be used in any other geometry.

Training

The loss consists of two parts. The principal term L_1 quantifies the difference between the profiles resulting from the parameterized ansatz and the Monte Carlo data. The regularizer term L_2 constrains the interaction kernels to be localized around the center regions and smoothly decaying. (Regularizers are usually needed to prevent overfitting in ML models.) In order to evaluate the derivative of the complete loss with respect to the network parameters it is necessary to solve the Euler–Lagrange equation to obtain the ML profile $\rho^\star$. All other necessary derivatives can be computed analytically once the equilibrium density distribution is known for a certain set of parameters.

Results

The density profiles generated with the learned correction terms show a large improvement compared to the RPA ansatz alone, even for the external potential being highly irregular and very different from potentials used during training. ML3 does not give substantial improvements over ML2 such that the obtained functional is a vindication of the RPA *ansatz* with an optimized kernel (at least for the particular temperature chosen). Here, an extension to a larger temperature region is clearly of interest. (Note that the 1D investigations of Ref. [13] included more temperatures and here a correction to the functional of third order in density showed more substantial improvements.) Results in radial geometry (not the training geometry!) were obtained for the direct correlation function c_2 and the pair correlation function and showed good agreement with simulations.

Self-consistency

As mentioned earlier, it is possible to either impose self-consistency already during training or to check for it after training is complete. While it is of course desirable to have this property included into the training already it is not straightforward to do so if the ML functional is somewhat "noisy". The self-consistency condition can only be ensured after solving the Euler–Lagrange equation Eq. (5) for ρ_{eq}, which may be time consuming. Further, the solution is usually found using iterative methods which makes the computational graph leading to the solution grow very fast, posing a problem to automatic differentiation approaches. In Ref. [13] this issue was circumvented by using only the generative output of the rhs of the Euler–Lagrange equation (5) as $\rho^{\star}$, using the ML functional but evaluated with the ground truth $\rho_{sim} \approx \rho_{eq}$ from simulations. Having $\rho^{\star} \approx \rho_{sim}$ up to a reasonable precision is a necessary condition for the existence of the fixed point but not a sufficient one. The approach used in Ref. [15] is based on manually calculating the necessary partial derivatives in order to do a back-propagation pass with respect to the parameters of the network. However, more complicated *ansätze* will make this procedure more laborious.

4.1.2   A Mean-Field Functional for the Kern–Frenkel Fluid

Similar in method (reference functional plus ML mean–field functional) is the investigation of an anisotropic fluid in Ref. [16]. A major complication results from the fact that the model fluid interacts through the angle-dependent Kern–Frenkel potential, making it necessary to include orientational degrees of freedom beyond the orientationally averaged density distribution alone. The training was performed on simulated density profiles between hard walls for a range of densities and supercritical temperatures not far from the critical point. In this geometry, the density profile $\rho(x) \rightarrow \rho(z)\alpha(z, \omega)$ is the product of an orientationally averaged profile ρ and a position– and orientation–dependent orientational profile α.

The approach identifies a set of reduced orientational profiles $\alpha_i(z)$ invariant under the symmetry group of the particle (tetrahedral symmetry) and include all linearly-

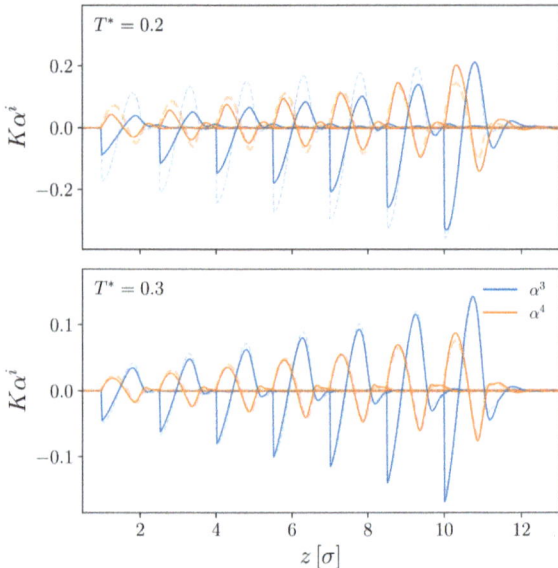

Fig. 3 Self-consistent, leading orientational moments from learned mean-field ansatz for two isotherms with reduced temperature $T^* = 0.2$ and 0.3. The y axis shows the first two (scaled) orientational moments at distance z to the wall, which starts at $z = 0.5$ in comparison to simulations (faint dashed lines). The result for each bulk density (in the range $\rho\sigma^3$ between 0.1 and 0.3, covering the critical region) was shifted to the right by $\Delta z = 1.5$, starting from the lowest density on the far left. The critical temperature is $T^* \approx 0.17$. Reprinted with permission from Ref. [16]

dependent combinations. The excess free energy $f_{\mathrm{ex}} = f_{\mathrm{ex}}^{\mathrm{ref}} + f_{\mathrm{ex}}^{\mathrm{ML}}$ (here per area) is then augmented by the ML mean-field term

$$\beta f_{\mathrm{ex}}^{\mathrm{ML}}[\rho, \{\alpha_i\}] = \frac{1}{2} \int \mathrm{d}z\mathrm{d}z' \rho(z)\rho(z') \sum_{ij} M^{ij}(z - z')\alpha^i(z)\alpha^j(z') \qquad (26)$$

where the kernels $M^{ij}(z)$, discretized on a grid, are the differentiable parameters of the network. The reference part comes from "functionalized" Wertheim theory for associating fluids and depends on the orientationally averaged profile ρ only.

Due to the larger state space, the training data (especially for the orientations) is more noisy than that of an isotropic systems. This makes training by evaluating on the fixed point alone difficult, as it introduces unphysical components to the learned parameters. It is therefore necessary to evaluate the derivatives of the loss with respect to a self-consistent solution of the Euler–Lagrange equations (fixed point), similar to Ref. [15]. While this approach is already sufficient to constrain the fixed point in a numerically stable way, it is sensible to assume that the gradient for points in the vicinity of the fixed point can help to stabilize the procedure further. This could be done, for example, by saving the computational graph of the whole iterative procedure up to the final fixed point, in order to later do a backpropagation on it.

Unfortunately this becomes too memory intensive even for relatively few iterations (say in the hundreds). There are some techniques to save memory on repeated function evaluations in loops (e.g. `jax.lax.scan`) and also more sophisticated solvers that can reach a solution after fewer iterations. Another way to reduce memory consumption is by using implicit differentiation. One trades, effectively, the memory savings for the need of solving a linear equation for every backpropagation. This can however be done, rather fast. For a fixed point Euler–Lagrange equation of the form $\alpha = g(\alpha, \theta)$ (both α, θ are vectors) one finds the gradient of the fixed point α^*

$$
\frac{\partial \alpha^*}{\partial \theta} = \left(\frac{\partial g}{\partial \theta} \right)_{\alpha^*} \left[\mathbb{I} - \left(\frac{\partial g}{\partial \alpha} \right)_{\alpha^*} \right]^{-1}
\tag{27}
$$

While this result is exact, solving the matrix inversion can become unstable. By approximating the inverse with just the identity matrix, one can save computing time and in many cases stabilize the procedure [23].

Results

The mean field kernels were trained on different isotherms since there is an implicit temperature dependence of the kernels, that is not explicitly included. As expected from a mean-field ansatz, the reproduction of the observed orientation works better for high-temperature state points. At lower temperatures, but still above the critical temperature, the mean-field ansatz is not able to accurately represent the orientations for all considered bulk densities (average reduced densities $\rho \sigma^3$ ranged from 0.10 to 0.30), see Fig. 3. In every case, the trained kernels were much stronger than those predicted by RPA, which were computed using Monte Carlo integration of the relevant orientational integrals. This is very different to the LJ case.

Due to the training method that was used, the functional could be minimized self-consistently and could be extrapolated to higher densities. As opposed to the case of an isotropic fluid, the ML functional with the learned kernel M^{ij} can not be used for other geometries since in the flat wall geometry the elements of a general mean-field kernel are projected and integrated, and can not be reconstructed.

4.1.3 Equation Learner Network

The approach used in Ref. [14] makes use of the Equation Learner Network (EQL) [24, 25], which is a neural network used for symbolic regression, i.e. the problem of finding symbolic expressions that describe the relationship between two datasets. The goal in this context was to discover the symbolic form the free energy functional and not just an opaque representation of it. When using the EQL it is necessary to specify beforehand the kind of building blocks (sin, cos, $\times$, $\div$ etc.) that can appear in the learned expression. By additionally specifying the maximum depth or number of layers, the amount of representable expression is fixed and finite. It is therefore not guaranteed that the EQL is able to find the exact equations with the selected basis functions, but the interpretability of the results might be able to steer further investigations.

The main advantage of the EQL consists in being fully differentiable during training, one can therefore not only use its output $f(x)$ to compute losses (and their gradients, w.r.t. θ) but also $f'(x)$, $f''(x)$ and so on. Indeed this is needed for the evaluation of $c_1 = -\beta \delta \mathcal{F}_{ex}/\delta \rho$ during training.

The network was applied to the 1D hard rod and 1D LJ–like system and functionally represents the free energy density $f_{ex}[\rho; \theta]$ where θ are the parameters of the network. An example for the network, aiming at the full functional for the LJ system, is shown in Fig. 4. To train the network on ground truth density distributions $\rho_{eq}(\mathbf{r})$, the predicted ML output density profile $\rho^* = \rho_{ML}$ is defined as in Ref. [13]:

$$\rho_{ML}(\mathbf{r}) = \rho_b \exp\left(-\beta V^{ext} + \beta \mu_{ex} + c_1^*[\rho_{eq}(\mathbf{r})]\right) \qquad (28)$$

where the chemical potential was split into the ideal part $\mu_{id} = k_B T \lambda^3/\rho$ and the excess part μ_{ex}. For the 1D hard rod fluid, the training data can be generated fast and exact using the functional in Eq. (14), and for the LJ fluid ground truth data were simulated using Monte Carlo methods.

Results

For the hard rod system, the network in Fig. 4 was used with just the upper three weighted densities. Since no other training constraints besides the density profiles

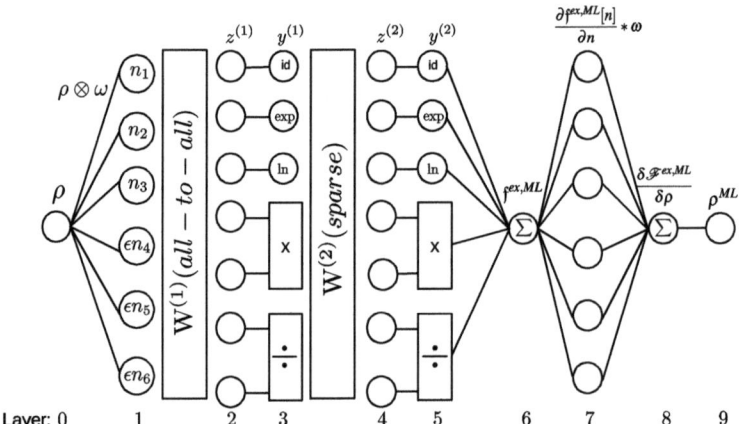

Fig. 4 Architecture of the functional equation learner network for learning $\mathcal{F}_{ex}$ for the 1D LJ system. The density is first fed into a convolution layer with learnable kernels, defining six weighted densities n_i. Three of the n_i are multiplied with the reduced LJ interaction strength $\epsilon = \beta \varepsilon$ in order to capture the temperature dependence. After that, two layers of non–linear symbolic transformations follow, producing finally the ML free energy density (here denoted as $f^{ex,ML}$). The discretized functional derivative is performed and the resulting ML density profile produced, see Eq. (28). Reprinted from [14], with the permission of AIP Publishing

were used, the network could not find the exact functional (see the SI of Ref. [14]) and also not the exact virial coefficients but the overall agreement for the equation of state and density profiles was very good, also outside the training region.

For the LJ system, two approaches were pursued. On the one hand, the reference *ansatz* was used in the form

$$\mathcal{F}_{ex}[\rho; \theta] = \mathcal{F}_{ex}^{hr}[\rho] + \epsilon F_{ex}^{ML}[\rho; \theta] \tag{29}$$

where $\epsilon = \beta\varepsilon$ is the reduced LJ strength and ML accounts for the attraction part of the functional. Secondly, the full excess functional was learned, using the network in Fig. 4, i.e. the temperature dependence (via ϵ) is completely transferred into the ML functional.

The calculated self-consistent density profiles for test and extrapolation showed a good performance. In addition to the density profiles, the equation of state $P(\rho)$ and direct correlation function $c_2(x)$ was evaluated and compared to the MC case. In general one sees that the ML version with splitting ("physics informed") performs better than the one without, especially for the direct correlation function, which agrees semiquantitatively with simulation.

The complexity of the produced symbolic expressions can be controlled by a regularization hyperparameter that penalizes large values of the expansion parameters (cf. $W^{(i)}$ in Fig. 4) and setting those below a certain threshold (e.g. 1×10^{-5}) to zero. While being simple, this regularization produces sparsity only indirectly by putting a pressure on the weights, but without accounting for the underlying algebraic structure of the expressions. The consequence is, that in order to achieve good results for the density profiles, a large regularizer value had to be chosen, producing rather complex expressions which are too complicated indeed to gain further insight about the mathematical structure of the functional (see SI of Ref. [14]).

4.2 Parameterization of $c_1[\rho]$

Here, the map between the first functional derivative, $c_1 = -\beta\delta\mathcal{F}_{ex}/\delta\rho$, and ρ is learned directly [18, 19] and was dubbed "neural functional theory" by the authors. Through Eq. (5), also c_1 is obtainable as ground truth from simulations and thus the map can be checked directly on simulation data. The novelty of this approach is two-fold. First, it avoids the use of the self-consistent iteration procedure of Eq. (5) in training by using the density $\rho(\mathbf{r})$ as the input and the one-body direct correlation function $c_1(\mathbf{r})$ as the target. An additional helpful observation is that $c_1(\mathbf{r})$ is short ranged and thus easier to learn. The second idea is to use only a part of the input array for inference, instead of the complete information over all of $\mathbf{r}$. This is physically sensible as the effect of the external potentials cannot extend arbitrarily far in c_1 and is mostly felt in the local neighborhood of the point in question, see Fig. 5. Corresponding to this is an imposed finite range of weight functions in the functional equation learner of Fig. 4.

Fig. 5 Relationship between input and output of the neural DFT network, parametrized with a standard MLP. The value of $c_1(z)$ depends only on density points within a certain finite interval around z. Reproduced with permission from Ref. [18]

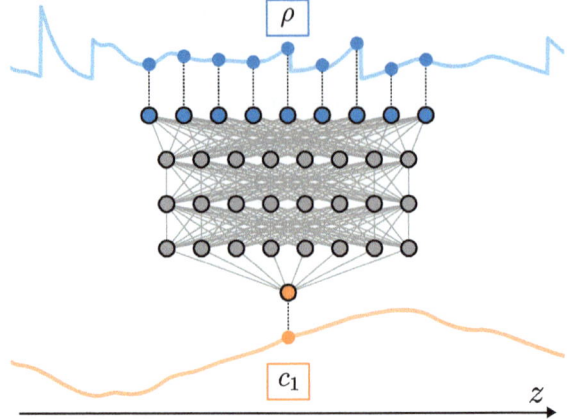

Examples for this neural functional theory were given in Ref. [18] for the 3D hard sphere fluid in flat wall geometry and in Ref. [19] for the 1D hard rod system. In both cases, density profiles $\rho(z)$ depend on one Cartesian coordinate. Here we discuss Ref. [18]. The neural network is structured as follows: The input layer consists of 513 nodes which are fed with the discretized values of the density profile $\rho(z_i)$ for $z_i \in [z - \Delta z, z + \Delta z]$ i.e. only in a window characterized by the cutoff Δz around the value z. Then follows a fully-connected multilayer perceptron of three layers and finally the scalar output $c_1(z)$. Since c_1 is short-ranged, the input window around the point of interest z has enough information content for an accurate prediction. Further, one simulation profile can be split into multiple such windows, which increases the size of the training set considerably. A further increase can be achieved by using mirrored profiles. By making use of automatic differentiation one is able to extract the "wall geometry" two-body direct correlation function

$$\bar{c}_2(z, z'; [\rho]) = \frac{\delta c_1(z; [\rho])}{\delta \rho(z')} \tag{30}$$

which is sufficiently smooth, due to the chosen activation function of the MLP.

Derived quantities

Starting from the one-body direction correlation function, which results directly from the network, it is possible to extract other quantities of interest such as the mentioned two-body correlation function. Note that the system is trained only in the planar wall geometry, but through automatic differentiation (evaluated at bulk densities) the bulk 3D direct correlation function $c_2(r)$ is obtainable as $c_2(r) = -\frac{1}{2\pi z} \bar{c}_2(z)' \big|_{z=r}$, equivalent to the relation between the mean–field kernels (25) of Sect. 4.1.1. From that, the structure factor $S(k)$, or the radial distribution function $g(r)$ can be computed using the Ornstein–Zernike relation. Note that the excess free energy is also available through functional line integration, which needs to be done numerically

$$\beta \mathcal{F}_{\text{ex}}[\rho] = -\int_0^1 d\alpha \int dz \rho(z) c_1(z; [\alpha \rho]) \tag{31}$$

The network representation for c_1 can be evaluated at any $\rho(z)$, i.e. also along the path. All these quantities were not included in the original training and can therefore be used to reason about the internal consistency of the learned model. An additional way to test accuracy and internal consistency is to see whether Noether sum rules (following from symmetry transformations on a thermal system) are fulfilled.

Similar to the case of the anisotropic particles of Sect. 4.1.2, the machine–learned c_1 is not applicable to problems in other than flat wall geometries. Extensions to genuine 3D situations are certainly desirable.

Self-consistency

Since the training process does not involve the self-consistent iteration of Eq. (20), it is not clear a priori, whether the neural $c_1^*(z)$ can be used to solve the Euler–Lagrange equation self-consistently and if so, whether the result $\rho^*(z)$ corresponds to the correct density distribution. This point was investigated by solving the Euler–Lagrange equation together with c_1^* using the Picard iteration scheme with mixing. In order to achieve numerical stability the mixing parameter needed to small in the beginning and could later be increased to a larger value, making convergence faster. The usual stringent convergence criterion in standard cDFT calculations however needed to be relaxed due to possible fluctuations in the MLP representation c_1^*.

As the training was done on windows around a point of interest, in applications one is not restricted to the box sizes that was used for generating the training data. The authors show that the network is able to accurately model the density in slits which are larger by an order of magnitude.

Results

The model was trained using data from the whole liquid regime with average densities ranging from $0.003\sigma^{-3}$ to $0.803\sigma^{-3}$, and is able reproduce the one-body correlations up to the accuracy of the input simulation data. This means that the architecture, with its choice of parameterization and size, is capable of capturing the relevant physics. Of greater interest is the accuracy of derived quantities like the self-consistent density profiles $\rho(z)$, the DCF c_2 and the (integrated) excess free energy. Here we observe that the network is exceeding current standard analytical descriptions, like the FMT excess free energy or the Carnahan-Starling equation of state, in accuracy and speed. This means that in spite of its rather simple design, the neural network is versatile enough to encode complex many-body information of the hard-sphere fluid beyond the one-body correlation function.

4.3 Learning $\mathcal{F}_{ex}$ by c_2 Matching

The approach described in Ref. [17] is again similar to those mentioned in 4.1, as it is based on finding a parameterized, approximative version of $\mathcal{F}_{ex}$, denoted by $F_\theta^{(2)}$. This approximative ML network is designed for planar geometry, i.e. $V^{ext} \equiv V^{ext}(z)$ and the parameters are not directly interpretable as they constitute the internal representation of the neural approximator for the free energy (this is similar to Refs. [18, 19]). As the novel element, instead of using the equilibrium density profile resulting from $F_\theta^{(2)}$ to train the set of parameters θ, the authors take advantage of the full differentiability of the network (here a convolutional neural network or CNN) and calculate the (planar) DCF by means of automatic differentiation, according to

$$\bar{c}_2(\theta, |z_i - z_j|) = -\frac{\partial^2 F_\theta^{(2)}}{\partial \rho(z_i), \partial \rho(z_j)} \frac{\beta}{A(\Delta z)^2} \tag{32}$$

and impose it to be similar to the (planar) DCF determined from simulations. The quantity Δz is the grid spacing of the discretized grid and A the area of the simulation box in the xy direction.

Initially this training procedure looks like a combination of previous approaches as it is i) learning a parametrized version of $\mathcal{F}_{ex}$ and ii) using a derivative of F in the loss computation (although a higher-order correlation function than in Ref. [18]). However, using c_2 in the loss has a major advantage. Training data of $\bar{c}_2$ can be extracted in a rather straightforward way using simulations of homogeneous bulk systems. This is in contrast to the other approaches, that need as inputs inhomogeneous simulation profiles to compute either ρ_{eq} or c_1, which depend on the imposed external potential.

Training

The necessary training data is collected by performing grand-canonical Monte Carlo simulations of a Lennard-Jones system above the critical temperature at different chemical potentials. From these configurations the radial distribution function $g(r)$ and $h(r) = g(r) - 1$ are extracted. For the simulation bulk DCF $c_2(r)$, one needs to solve the bulk Ornstein–Zernike equation

$$c_2(r) = \frac{1}{2\pi^2} \int \frac{\sin(kr)}{kr} \left(\frac{\hat{h}(k)}{1 + \rho_b \hat{h}(k)} \right) k^2 dk \tag{33}$$

where the hat symbol denotes Fourier transformed quantities. Finally, the conversion from radial to planar geometry needs to be performed on $c_2(r)$ as described earlier, see Eq. (24).

The (scalar) free energy functional is approximated by a convolutional neural network with periodic and dilated convolutions, where the latter means increasing the reach of the kernel by stretching and padding with zeros, without adding more parameters to the model. The input density profiles are discretized on 320 points, the convolution kernel size is 3, the dilation factor 2 and the number of layers 6, with [16, 16, 32, 32, 64, 64] channels for the individual layers.

Applying auto-differentiation to the scalar output twice produces a Hessian $\partial^2 F_\theta^{(2)} / \partial \rho_i, \partial \rho_j$ of rank two. In order to limit the computational cost, a randomly sampled subset of 10 rows was used for every loss evaluation, defined by

$$L(\theta) = \sum_{i,j} \left(\int_{|z_i - z_j|}^{\infty} dr \, 2\pi r c_2(r) - \bar{c}_2(\theta, |z_i - z_j|) \right)^2 \tag{34}$$

By fixing the second derivative of the free energy to a certain value one is still left with two "integration constants" in the parametrization of F_{ex}. The first one is determined in a consistent way by adding a regularization loss that constrains the first derivative of the free energy to be equal to the excess chemical potential μ_{ex} (which is the direct correlation function c_1 (times $k_B T$) for bulk densities), i.e.

$$L_{\mathrm{reg}}(\theta) = \frac{1}{n} \sum_i \left(\frac{1}{A \Delta z} \frac{\partial F_\theta^{(2)}}{\partial \rho_i} - \mu_{\mathrm{ex}}(\rho_b) \right)^2 \tag{35}$$

where $\beta \mu_{\mathrm{ex}}(\rho_b) = \beta \mu - \ln(\lambda^3 \rho_b)$ and ρ_b is the measured bulk density in the grand canonical simulations at chemical potential μ. The second integration constant can be determined by the fact that $F_{\mathrm{ex}}[0] = 0$, i.e. by subtracting the value $F_\theta^{(2)}[0] = C$ from the neural functional.

After the training is completed the resulting functional $F_\theta^{(2)}$ can be used to compute equilibrium profiles for arbitrary (planar) external and chemical potentials. Similar to Ref. [18], the network was never trained on self–consistent density profiles. Therefore it is a priori unclear whether self–consistent, converged inhomogeneous equilibrium profiles can be determined and are accurate, especially since during training the network encountered only bulk systems.

Results

The method is compared to two other functionals and Monte Carlo simulations, which serve as ground truth. The first functional F_{MF}, is an analytical mean-field approximation of the Lennard-Jones potential using fundamental measure theory for the hard-sphere reference system. The other is an identically parametrized neural network, trained on the output of the *first* functional derivative of $F_{\mathrm{ex}}(\theta)$, essentially a version of the method described in Ref. [18], leading to a free energy $F_\theta^{(1)}$.

A variety of quantities can be compared in order to assess the quality of the learned functional approximation. Of special interest are the resulting equilibrium density profiles for varying external and chemical potentials, the equation of state and the value of the free energy itself. The general observation is that $F_\theta^{(2)}$ performs comparatively or better than $F_\theta^{(1)}$, especially (far) outside the training region, where the other methods often fail to converge. Ensuring numerical stability and self-consistent fixed points outside the training region is notoriously difficult for parameterized free energy functionals, probably due to unwanted overfitting to the training data. Conditioning on a higher–order correlation function seems to be a beneficial strategy in this regard.

In summary, pair-correlation (c_2) matching is able to learn a very stable and accurate neural density functional without having "seen" during training inhomogeneous density profiles due to an explicit external potential (other than the inter-particle potential through the radial distribution function). Taking advantage of the convertability between radial and planar geometry the network could be trained on the computationally simpler to handle geometry without losing information about the radial components. Nevertheless, this approach rests on the fact that DCFs are available through the solution of the Ornstein-Zernike relation, which especially in the case of anisotropic pair potentials becomes a highly non–trivial task.

4.4 Learning the Map $\rho \leftrightarrow V^{\text{ext}}$

4.4.1 Bayesian Methods

The methods described in Refs. [20, 21] are based on Bayes' theorem, namely

$$P(Q|\mathcal{D}) \propto P(Q)P(\mathcal{D}|Q), \tag{36}$$

where $\mathcal{D}$ is short for the observed data, Q are the model parameters and $P(Q|\mathcal{D})$ is the posterior distribution of Q given $\mathcal{D}$. While both works are based on the same statistical method, they differ in their aims. The approach in Ref. [20] focuses on learning the free energy functional of a 1D hard rod fluid, while in Ref. [21] the mapping of $\rho(x) \rightarrow V^{\text{ext}}(x)$ is investigated. Here we limit ourselves to the later work as the methods used in both are rather similar.

The principal question formulates a typical inverse problem: *Given a density profile $\rho(\mathbf{r})$ (i.e. the data $\mathcal{D}$ in the Bayesian sense) resulting from an unknown external potential V^{ext} is it possible to infer the potential (i.e. its parameters Q along with their uncertainty) given the density profile alone?* The answer to this is yes and it is given by the distribution over Q, i.e. $P(Q|\mathcal{D})$. The interesting point is that $P(Q|\mathcal{D})$ contains the uncertainty in the final solution for the external potential; this is relevant if the input data would come from noisy simulations, say.

The authors start by parameterizing the external potentials using a set of parameters $\{Q_i, Q_i'\}$, with $i \in [1, \ldots, i_{\max}]$

$$V(z) = \sum_{i=1}^{i_{\max}} Q_i \exp\left[-(z - z_i)^2 / \exp Q_i'\right] \tag{37}$$

where z_i are points in the simulation box, corresponding to the center of each Gaussian. The choice of parameterization is essentially arbitrary but a good compromise between simplicity and expressiveness is helpful for training and generalization. In order to determine $P(Q|\mathcal{D})$, the two distributions on the right hand side of Eq. (36) need to be specified. The prior distribution $P(Q)$ is somewhat arbitrary and chosen to be Gaussian with zero mean and a diagonal covariance matrix, $P(Q) = \mathcal{N}(0, \Sigma_Q)$. It remains to specify the quantity $P(\mathcal{D}|Q)$, which tells us how probable a certain density distribution is given the parameters Q of the external potential. For this, the probabilistic interpretation of the particle density itself is used:

$$P(\mathcal{D}|Q) = \prod_i \rho(z_i|Q), \tag{38}$$

where one multiplies the probabilities over all space points. This mapping of the parameters Q to the density profile is not analytically known, but numerically through simulations (in general) or through cDFT using the exact functional in the specific case of hard rods considered here. In Markov chain Monte Carlo simulations, the Q are included in the sampling and a detailed balance criterion is formulated which generates a histogram according to the distribution of Eq. (36). From this histogram the final V^{ext} follows from the most probable values of Q, and a band of uncertainty in V^{ext} can be drawn from the quantiles.

Performance

In order to test the approach it was trained with sets of three different sizes (800, 2400 and 6000) with configurations coming from Monte Carlo simulations. With increasing dataset size the histograms for the individual parameters Q_i become narrower, signaling an increased confidence in the parameter. This also carries through to the prediction of the external potential, whose confidence intervals gets smaller with increased training set size. As for the maximum a posteriori (MAP) prediction of the potentials, one observes slight discrepancies between the ground truth and the prediction of the external potentials. When looking at the corresponding density profiles the difference between ground truth and predicted value becomes smaller, signaling a certain freedom in the choice of external potentials, see Fig. 6. (This appears to be similar to the problem of finding the exact excess free energy functional for hard rods in Ref. [14].) Nevertheless the exact V^{ext} is for the largest part contained in the 99% probability region, even for the smaller training set size of 800. Indeed it was observed that the minimal training dataset size at which the predictions fail to recover the results of the small training set is around 360.

4.4.2 Gaussian Process

Reference [22] addresses the map from the external potential to the density profile. It will be represented by a typical surrogate model of machine learning, namely Gaussian processes (GP), which replaces the cDFT calculation or the direct simulation. The authors exemplify the method for the 1D hard rod system, for the sake of laying out the principle. A particular focus lies in formulating a reinforcement learn-

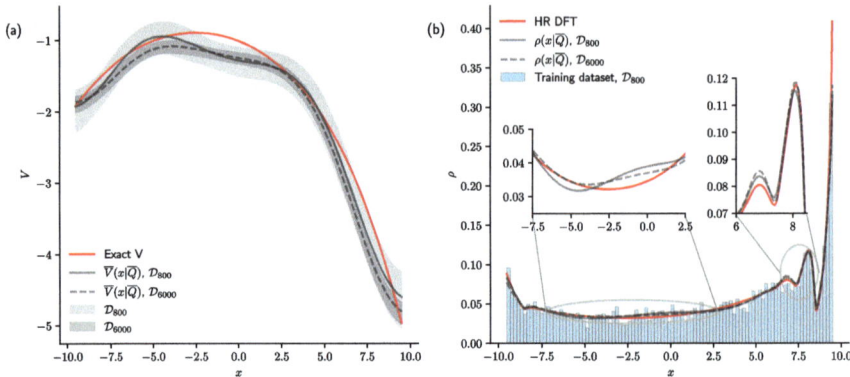

Fig. 6 Predicted external potentials (**a**) and corresponding density profiles (**b**) for two different dataset sizes of 800 and 6000 with 99% confidence intervals. The differences in the density profiles are smaller than those in the potential representation. Here the 1D coordinate is denoted by x. Figure taken from Ref. [21]

ing scheme, i.e. in the automated inclusion of new training data if the output of the surrogate model for a particular new external potential is of potential low accuracy.

The input for the map $V^{\text{ext}} \rightarrow \rho$ is discretized with p points on the Cartesian z axis, forming the vector

$$\mathbf{x} = \mu - V^{\text{ext}}(z) = \psi(z) \in \mathbb{R}^p, \tag{39}$$

and the resulting output is the discretized density vector $\boldsymbol{\rho} \in \mathbb{R}^p$ with $\rho_i = \rho(z_i)$. For multiple inputs $\mathbf{x}^{(1)}, \ldots, \mathbf{x}^{(n)}$ we have the density outputs $\boldsymbol{\rho}^{(1)}, \ldots, \boldsymbol{\rho}^{(n)}$, and the assumption of the GP model is that the vector of densities $\hat{\boldsymbol{\rho}}_j$ at *one space point* j follows a multivariate normal distribution

$$\hat{\boldsymbol{\rho}}_j = [\rho_j^{(1)}, \rho_j^{(2)}, \ldots, \rho_j^{(n)}]^\top \sim \mathcal{MN}(\boldsymbol{\mu}_j, \sigma_j^2 \mathbf{R}^{(n)}) \tag{40}$$

where $\mathbf{R}^{(n)}$ is the $n \times n$ correlation matrix between densities at a fixed grid point with elements $R_{ij}^{(n)} = K(||\mathbf{x}^{(i)}, \mathbf{x}^{(j)}||)$. It is through this kernel function that the density at one grid point depends on the external potential at different grid points. In Ref. [22], so-called Matérn kernel functions were used, which depend on a range parameter γ that needs to be estimated. The point is now to consider the n densities $\boldsymbol{\rho}^{(1)}(\mathbf{x}^{(1)}), \ldots, \boldsymbol{\rho}^{(n)}(\mathbf{x}^{(n)})$ as input data $\mathcal{D}$ from simulations, say. One adds a new external potential $\mathbf{x}^{(n+1)} = \mathbf{x}^*$ with unknown density output $\rho_j^{(n+1)} = \rho_j^*$ and is interested in the distribution $P(\rho_j^*|\mathcal{D})$ given the input data. Since the joint distribution $P(\rho_j^{(1)}, \ldots, \rho_j^{(n+1)})$ still follows the multivariate distribution (40), one has $P(\rho_j^*|\rho_j^{(1)}, \ldots, \rho_j^{(n)}) = P(\rho_j^{(1)}, \ldots, \rho_j^{(n+1)})/P(\rho_j^{(1)}, \ldots, \rho_j^{(n)})$ which is still Gaussian but depends on the means $\boldsymbol{\mu}_j$ and variance parameters σ_j^2. Marginalizing out these with certain assumptions on their prior distributions finally leads to a closed form for $P(\rho_j^*|\mathcal{D})$.

The remaining challenge is to find a suitable strategy for choosing an optimal set of inputs for $\hat{\boldsymbol{\rho}}_j$ and estimating the range parameter γ in the kernel function K. The proposed architecture is termed ALEC (active learning with error control) and functions as an adaptive emulator in the following way. After supplying the first batch of training profiles the internal state of the estimator is built, i.e. the correlation matrix and the corresponding guesses for θ and σ are computed. When a new test input $\mathbf{x}^*$ (external potential) is given to the system, it predicts the output $\rho_j(\mathbf{x}^*)$ by sampling from the appropriate probability distribution with mean $\hat{\theta}$ and variance $\hat{\sigma}$, both depending on the covariance between samples in the training set. The parameter γ is estimated such that the performance is maximized. Since the predictive distribution $P(\rho_j^*|\mathcal{D})$ is known analytically, the predictive error can be directly assessed. In case the variance exceeds a certain threshold, the systems generates new training data for $\mathbf{x}^*$ and includes it into its training set (augmented set). Since the training set is now increased by one, it is necessary to recompute the covariance matrix $\mathbf{R}^{(n+1)}$ and its inverse. Reference [22] discusses methods to speed up this process. For benchmarking, ALEC is compared to training setups with a random choice of training samples and a different adaptive scheme (D-optimality) and shown to perform better in all three cases.

Although interesting in terms of the adaptive learning strategy, the approach is perhaps the most remote of the discussed methods to address fundamental or applied problems in cDFT. The map $V^{\text{ext}} \rightarrow \rho$ is still the easiest, also for simulations to generate ground truth data. A surrogate model for simulations would be needed in perspective only for very costly simulations, such as complex biomolecules with many internal degrees of freedom.

5   Outlook to Related Problems

5.1   Liquid State Theory

A recent review article [26] gives an overview on recent advances in liquid state theory using ML methods which we recommend to the reader and therefore restrict ourselves to a few points only.

The integral equation (IE) approach to liquid state theory aims at determining the direct correlation function $c_2(x, x')$ and the total correlation function $h(x, x')$ which are linked by the Ornstein–Zernike (OZ) Eq. (13). A second equation is needed which is usually termed *closure equation* and which for a fluid with intermolecular pair potential $\phi(x, x')$ takes the form [11]

$$\ln(h(x, x') + 1) + \beta\phi(x, x') = h(x, x') - c_2(x, x') + b(x, x'). \tag{41}$$

The unknown function $b(x, x')$ is called the bridge function. In terms of the Mayer expansion it has a clear diagrammatic definition which, however, does not allow for analytic resummations, even approximate ones. For simple liquids with repulsive cores it has been noted that the approximation of b by a b^{ref} from a reference system system (usually hard spheres) gives good results [27] (bridge function universality). Empirical closures have expressed b pointwise as functions of h, c_2 and $\beta\phi$ [11]. Recent work [28] has investigated an ML closure for simple fluids where b is a pointwise function of h, c_2, the derivative of $h - c_2$ and the fluctuations of the pair correlation function $g = h + 1$. Indeed it has been found that a variant of bridge function universality holds approximately, i.e. that for "hard" potentials (steeply repulsive near the origin) one approximate ML closure is found while for "soft" potentials (weakly diverging or permitting core overlap) another one is found.

An explicit connection to cDFT arises if one considers the cDFT derivation of the closure Eq. (41) [29–31]. If the excess free energy functional is functionally expanded around a (homogeneous or [if $V^{ext} \neq 0$] inhomogeneous) reference density profile and the grand potential is minimized in the additional presence of a test particle ($V^{ext} \rightarrow V^{ext} + \phi$), then Eq. (41) arises and $b(x, x')$ is the functional derivative of the Taylor expanded excess functional from which all terms up to second order are subtracted. Thus ML representations of the excess free energy functional would also contain an approximate solution to the IE problem.

5.2 Electron DFT

A dedicated and more comprehensive review on the connection between ML and electron DFT is in the chapter on *Machine learning in quantum density functional theory* by Thorsten Deilmann in this book. Here we limit ourselves to discussing ML assisted functional building approaches in electronic DFT and relate these to the cDFT approaches, if possible. In contrast to cDFT, in quantum DFT (qDFT) only one system is of major interest: interacting electrons with their pairwise Coulomb repulsion in the presence of nuclei. The Hohenberg–Kohn proof [1] entails that there exists a unique *energy* functional $E[n(\mathbf{r})]$ where the electron density is denoted by $n(\mathbf{r})$. Conventionally (in the Kohn-Sham approach) it is split

$$E[n] = K_S[n] + E^{ext}[n] + E_H[n] + E_{XC}[n] \tag{42}$$

where $K_S[n] = -\hbar^2/2m \sum_i \int d\mathbf{r}\phi_i^*[n]\nabla^2\phi_i[n]$ is the kinetic energy of noninteracting electrons in occupied Kohn–Sham orbitals ϕ_i [33]. $E^{ext}[n] = \int d\mathbf{r}n(\mathbf{r})V^{ext}(\mathbf{r})$ is the external energy due to the external potential V^{ext} exerted by the nuclei. $E_H[n] = (e^2/8\pi\epsilon_0) \int d\mathbf{r} \int d\mathbf{r}'n(\mathbf{r})n(\mathbf{r}')/|\mathbf{r} - \mathbf{r}'|$ is the Hartree energy which is the classical interaction energy of an inhomogeneous electron distribution and which is of typical mean-field form. The quantum–mechanical exchange and correlation effects are buried in the *exchange–correlation* functional $E_{XC}[n]$ whose exact form is unknown; it is the equivalent of the excess free energy functional $\mathcal{F}^{ex}[\rho]$ minus the mean–field part in cDFT. Once a specific form for $E_{XC}[n]$ is assumed, a typical iteration scheme for finding the minimum of E proceeds via the solution of a one–electron Schrödinger equation in an external potential (Kohn–Sham potential $v_S[n_j]$) which depends on the electron density n_j in the previous step j,

$$\left(-\frac{\hbar^2}{2m}\nabla^2 + v_S[n_j]\right)\phi_i(\mathbf{r}) = \epsilon_i\phi_i(\mathbf{r}) \tag{43}$$

$$n_{j+1}(\mathbf{r}) = \sum_i |\phi_i(\mathbf{r})|^2 \quad \text{(occupied orbitals)}, \tag{44}$$

and the construction of the density n_{j+1} in the next step $j+1$ through the sum of the densities in occupied orbitals (fixed by the total number N of electrons). Hereby the Kohn–Sham potential is given by $v_S[n] = V^{ext} + \delta(E_H + E_{XC})/\delta n$ (thus $-\delta(E_H + E_{XC})/\delta n$ is the equivalent of c_1 in cDFT). The procedure is repeated until the density is converged. This iteration is the equivalent of the iterative solution (20), (21) of the fixed point Euler–Lagrange equation (5) in cDFT.

ML methods in qDFT can be used in similar ways as in cDFT. In difference to classical systems, ground truth data do not come exclusively from dedicated simulations (like quantum Monte–Carlo [34, 35]) owing to numerical challenges. For test problems in 1D and few electrons, the many–body Schrödinger equation can be solved exactly. For "real world" problems, data sets from experiment or advanced (but laborious) and more systematic electron structure calculations can be used (see e.g. the Main Group Chemistry DataBase (MGCDB84) [36]) which are inherently approximative but deemed more precise than the average qDFT result.

Calculations in 1D (mostly for electrons with an exponentially screened potential) have been a test bed for ML methods. One can introduce a local energy per particle $e(\mathbf{r})$ through $E = \int d\mathbf{r}n(\mathbf{r})e(\mathbf{r})$. Nonlocal mappings $n \leftrightarrow e[n]$ can be represented by standard convolutional networks or MLPs. Examples for this can be found in Refs. [37–39] which differ in the forms of their loss function (taking into account the energy density, its spatial derivative, and the density distribution). An obvious choice is to use the density $\rho^\star$ and energy density $e^\star$ from converged Kohn–Sham (KS) iterations (with an ML functional) in the loss function to guarantee that the ML functional delivers self–consistent solutions (this is similar to Refs. [15, 16] in cDFT). However, large training data sets are needed and together with the numerical costs of the KS iterations some doubts to the scalability to interesting 3D systems arise. An interesting proposal for a partial remedy is in Ref. [39] which suggests to use also intermediate energies and densities from the KS iterations in the loss function and training process to multiply the training data.

Nevertheless, ML functionals for realistic 3D systems (i.e. molecules) have been published recently by Google collaborations [40, 41]. The Deepmind21 (DM21) functional of Ref. [40] addresses in particular the problem of correct dissociation curves of molecules by approximately solving the fractional charge and spin problem of DFT. The price to be paid is the use of approximate density distributions in the training and training/evaluation of the ML functional in a non–selfconsistent, perturbative manner. As for the other work, the idea followed in Ref. [41] is to learn a symbolic (analytic) exchange–correlation functional based on a reduced set of analytical terms and operations which is motivated by the structure of the most reliable "human–made" exchange–correlation functionals. The analytic form allows for self–consistency in all functional evaluations. The authors claim a superior description of their GoogleAcceleratedScience22 (GAS22) functional across the MGCDB84 database. In methodology, there is a strong similarity to the Functional Equation Learner of Ref. [14] for cDFT (see Sect. 4.1.3).

The Kohn–Sham approach to qDFT via KS orbitals and the self–consistent iteration of Eqs. (43) and (44) is a quite special trick to approximate the kinetic energy $K[n]$ in qDFT. In cDFT, the kinetic energy of particles is buried in the ideal part $\mathcal{F}_{id}[\rho]$ of the free energy, and the corresponding free energy density is a simple local functional. This is not so in qDFT, but it would seem natural to search for an explicit kinetic energy functional $K[n]$ here as well ("orbital free DFT"). In fact this was the subject of early work on finding functionals for 1D electrons with ML [42, 43] where the kinetic energy functional $T[n]$ was represented using a rather simple kernel ridge regression. The obtained functionals still suffered from a comparatively poor representation of the functional derivative $\delta T/\delta n$. These problems had been solved including a proper training on the functional derivative [44] and substituting the numerically expensive kernel ridge regression by more effective convolutional networks representing the map to the energy per particle $n \to e[n]$. It remains to be seen whether these insights for 1D will translate to a workable 3D orbital–free ML functional.

5.3 Power Functional Theory

We have attempted to show the similarity of the ML problem in cDFT and qDFT (despite the differences in the specific treatments) which is simply a consequence of the existence of the functional map $\rho \leftrightarrow \mathcal{F}[\rho]$ or $n \leftrightarrow E[n]$. If the general (classical or quantum–mechanical) nonequilibrium problem could be formulated with a similar functional map, one would expect that ML techniques can be of similar use. For classical systems, a functional formulation, power functional theory (PFT), has been proposed about 10 years ago which shows a close resemblance to cDFT and suggests itself for extending the ML techniques. The current status of PFT is reviewed in Ref. [45]. Here we restrict ourselves to the case of Brownian (overdamped) dynamics which is of superior relevance in the colloidal domain and describes the thermal motion of macroobjects in a solvent. Specifically we consider isotropic particles for which the force on a single particle i is proportional to its velocity, $\mathbf{f}_i = \gamma \mathbf{v}_i$.

The central quantity in Brownian PFT is the ensemble-averaged one–particle current $\mathbf{J}(\mathbf{r}, t) = \langle \sum_i \delta(\mathbf{r} - \mathbf{r}_i)\mathbf{v}_i \rangle$ (where the average is over initial conditions with a prescribed n–particle distribution function in space). The space and time dependent density distribution is linked to the current via the continuity equation $\dot{\rho}(\mathbf{r}, t) = -\nabla \cdot \mathbf{J}(\mathbf{r}, t)$. It can be shown that there exists a functional $R[\mathbf{J}]$ such that the physical nonequilibrium single particle current $\mathbf{J}_{\text{neq}}(\mathbf{r}, t)$ is determined by functional minimization

$$\left. \frac{\delta R[\mathbf{J}]}{\delta \mathbf{J}(\mathbf{r}, t)} \right|_{\mathbf{J}=\mathbf{J}_{\text{neq}}} = 0. \tag{45}$$

For the original proof see Ref. [10], some mathematical issues have been clarified and corrected in Ref. [46]. Similar to the grand potential functional in DFT, $R[\mathbf{J}]$ is split into an intrinsic part and a part accounting for the interactions with a space and time dependent external potential $V^{\text{ext}}(\mathbf{r}, t)$:

$$R[\mathbf{J}] =: R^{\text{int}}[\mathbf{J}] + \int d\mathbf{r} \, \mathbf{J} \cdot \nabla V^{\text{ext}} + \int d\mathbf{r} \, \rho \dot{V}^{\text{ext}}. \tag{46}$$

From Eq. (45) then follows an Euler-Lagrange equation of the form $\frac{\delta R^{\text{int}}}{\delta \mathbf{J}} = -\nabla V^{\text{ext}}$. This internal part $R^{\text{int}}[\mathbf{J}]$ (like the free energy functional $\mathcal{F}[\rho]$ in cDFT) is a unique functional *independent of the external potential* and can be split further by the *ansatz*:

$$R^{\text{int}}[\mathbf{J}] = \int d\mathbf{r} \, \mathbf{J} \cdot \nabla \frac{\delta \mathcal{F}[\rho]}{\delta \rho} + P_{\text{id}}[\mathbf{J}] + P_{\text{ex}}[\mathbf{J}] . \tag{47}$$

$P_{\text{id}} = \gamma \int d\mathbf{r} \, \mathbf{J}^2 / (2\rho)$ is the ideal gas part of the functional for the dissipated power, and P_{ex} is the (generally unknown) excess part. Note that R^{int} is strictly a functional of the current $\mathbf{J}$ only, and densities $\rho(\mathbf{r}, t)$ are determined through the continuity equation and only depend on the current $\mathbf{J}(t' < t)$ at earlier times. With these definitions, Eq. (45) becomes an implicit equation for the exact non-equilibrium current $\mathbf{J}_{\text{neq}}$:

$$\gamma \mathbf{J}_{\text{neq}} = -k_B T \nabla \rho - \rho \nabla V^{\text{ext}} \underbrace{-\rho \nabla \frac{\delta \mathcal{F}_{\text{ex}}[\rho]}{\delta \rho} - \rho \left. \frac{\delta P_{\text{ex}}}{\delta \mathbf{J}} \right|_{\mathbf{J}=\mathbf{J}_{\text{neq}}}}_{-\rho \delta R_{\text{ex}}^{\text{int}} / \delta \mathbf{J}}. \tag{48}$$

The first two terms here give the exact nonequilibrium current of an ideal gas in an external potential. With the third term added, dynamic DFT [47] is recovered in which the dynamics solely depends on the *instantaneous* density profile $\rho(\mathbf{r}, t)$ through the equilibrium free energy functional. The fourth term with the functional derivative of the excess power corrects the quasi–equilibrium approximation of dynamic DFT to recover the full nonequilibrium current. The sum of the third and fourth term can be viewed as the functional derivative of the excess part of the internal part of the functional R^{int}.

In view of this, the full classical equilibrium and (Brownian dynamics) nonequilibrium many–particle problem is the problem of finding the two unknown excess functionals $\mathcal{F}_{\text{ex}}[\rho]$ and $P_{\text{ex}}[\mathbf{J}]$. As described earlier, over the past decades knowledge about $\mathcal{F}_{\text{ex}}[\rho]$ has been accumulated and has entered the diverse ML approaches described in Sect. 4. This is different in the case of $P_{\text{ex}}[\mathbf{J}]$: currently we do not know an exact excess power functional for a simple (even 1D!) model system, and knowledge on the general structure of $P_{ex}[\mathbf{J}]$ has only begun to be gathered [45].

The problem of a nonequilibrium steady state ($\mathbf{J}$ = const.) is an excellent example for which the relevance of $P_{ex}[\mathbf{J}]$ has been demonstrated and for which the afore-mentioned full problem has been tackled with ML methods [48]. The system is 3D Lennard–Jones in a box with periodic boundary conditions to which an external force has been applied which produces a steady flow in the z direction with constant current J_z. The ensemble–averaged density $\rho(z)$ and velocity $v_z(z)$ are inhomogeneous but linked through $J_z = \rho(z)v_z(z)$. With dedicated Brownian dynamics simulations, the averaged internal force density

$$\mathbf{F}_{int}(\mathbf{r}, t) = -\langle \sum_i \delta(\mathbf{r} - \mathbf{r}_i) \nabla_i u(\mathbf{r}^N) \rangle = -\rho \frac{\delta R_{ex}^{int}}{\delta \mathbf{J}} \qquad (49)$$

can be obtained as ground truth. This internal force density is the density profile times the negative of the functional derivative of the excess part of $R^{int}[J]$ (it contains both contributions by $\mathcal{F}_{ex}[\rho]$ and $P_{ex}[\mathbf{J}]$) and thus offers itself for an ML representation. For the steady state problem, a functional dependence on the time–independent ρ and $\mathbf{J}$ is needed but since simulations only realize constant currents it is better to include the velocity $\mathbf{v}$ as a functional variable. The authors of Ref. [48] proceeded to represent the force per particle $\mathbf{f}_{int} = \mathbf{F}_{int}/\rho$ as a convolutional network $f_{z,int}^\star(z)[\rho(z'), v_z(z')]$ where the density and velocity distribution around a point z were mapped to the force at z. From the results, it is gratifying to see that the ML functional representation of $\mathbf{f}_{int} = -\delta R_{ex}^{int}/\delta \mathbf{J}$ indeed worked *independent of the external force*. Furthermore, principal shortcomings of the dynamic DFT approximation could be highlighted. Conceptually and methodwise, there is a strong similarity to Refs. [18, 19] (from the same group) which tackled the ML representation $c_1^\star[\rho]$ of the excess free energy functional derivative $-\beta\delta\mathcal{F}_{ex}/\delta\rho$ in equilibrium cDFT, see the discussion in Sect. 4.2.

6 Summary and Conclusion

We have discussed in some detail the recent developments in classical density functional theory connected with the use of machine learning techniques. Here, we laid particular focus on work which addressed the maps between density distribution and free energy functional (or its first derivative) and density distribution and external potential. We have not treated the use of machine learning techniques to classify and interpret results of high through–put DFT calculations [49] which is a topic which appears to be particularly relevant in quantum DFT and materials science (see also the chapter on *Machine learning in quantum density functional theory* by Thorsten Deilmann in this book). Most of the tested techniques (explicit parametrizations of functionals, multi-layer perceptrons, analytical learning schemes, Bayes tech-niques) have been applied to the one-dimensional hard rod system where the exact functional is known. One can fairly say that these demonstrations were successful and the representability of functional maps using ML is possible with quantitative accuracy. However, to date there have been limited applications to more realistic, three–dimensional systems with truly novel results. This will certainly constitute a challenge for the future. Possible roads for the future are (i) the standard inclusion of a $\rho \rightarrow c_1[\rho]$ training in simulations as an efficient way to organize the simulation database [18], (ii) further elucidations of the analytic structure of functionals in 3D systems [15, 16] and (iii) accounting for uncertainties in machine–learned representations [21]. It will be interesting to see whether machine–learning supported cDFT can expand to systems where currently there are little analytical insights into functionals, such as particles with internal degrees freedom (which are not decoupled), particles of complex anisotropic shape or biomolecules. Regarding functional building and the use of machine learning, we have noticed a strong conceptual overlap to quantum DFT, in particular with regard to orbital–free approaches. The problem setting in classical DFT is however much more diverse: While in quantum DFT only the system of interacting electrons is of paramount importance, there are many Hamiltonians in the classical realm which deserve scrutiny and which need differing attention. In our outlook we also noted the great potential which lies in the functional formulation of classical many–body dynamics which then can be tackled similarly to equilibrium DFT problems using methods of machine learning.

Acknowledgements We gratefully acknowledge funding by the Deutsche Forschungsgemeinschaft (DFG, German Research Foundation) under Germany's Excellence Strategy EXC no. 2064/1, Project no. 390727645.

References

1. Hohenberg P, Kohn W (1964) Inhomogeneous electron gas. Phys Rev 136:B864
2. Mermin ND (1965) Thermal properties of the inhomogeneous electron gas. Phys Rev 137:A1441
3. Evans R (1979) The nature of the liquid-vapour interface and other topics in the statistical mechanics of non-uniform, classical fluids. Adv Phys 28:143–200
4. Roth R (2010) Fundamental measure theory for hard-sphere mixtures: a review. J Phys Condens Matter 22:063102
5. Lutsko JF (2008) Density functional theory of inhomogeneous liquids. II. A fundamental measure approach. J Chem Phys 128:184711
6. Bültmann M, Härtel A (2022) The primitive model in classical density functional theory: beyond the standard mean-field approximation. J Phys Condens Matter 34:235101
7. Forsman J, Woodward CE (2017) Variational methods in molecular modeling. In: Wu J (ed). Springer Singapore, Singapore, pp 101–136
8. Stopper D, Hirschmann F, Oettel M, Roth R (2018) Bulk structural information from density functionals for patchy particles. J Chem Phys 149:224503
9. Ding L, Levesque M, Borgis D, Belloni L (2017) Efficient molecular density functional theory using generalized spherical harmonics expansions. J Chem Phys 147:094107
10. Schmidt M, Brader JM (2013) Power functional theory for Brownian dynamics. J Chem Phys 138
11. Hansen J-P, McDonald IR (2013) Theory of simple liquids: with applications to soft matter. Academic Press

12. Solana JR (2013) Perturbation theories for the thermodynamic properties of fluids and solids. CRC Press
13. Lin S-C, Oettel M (2019) A classical density functional from machine learning and a convolutional neural network. SciPost Phys 6:025
14. Lin S-C, Martius G, Oettel M (2020) Analytical classical density functionals from an equation learning network. J Chem Phys 152:021102
15. Cats P et al (2021) Machine-learning free-energy functionals using density profiles from simulations. APL Mater 9:031109
16. Simon A, Weimar J, Martius G, Oettel M (2024) Machine learning of a density functional for anisotropic patchy particles. J Chem Theory Comput 20:1062–1077
17. Dijkman J et al (2024) Learning neural free-energy functionals with pair- correlation matching. arXiv:2403.15007
18. Sammüller F, Hermann S, de las Heras D, Schmidt M (2023) Neural functional theory for inhomogeneous fluids: fundamentals and applications. Proc Natl Acad Sci 120:e2312484120
19. Sammüller F, Hermann S, Schmidt M (2024) Why neural functionals suit statistical mechanics. J Phys Condens Matter 36:243002
20. Yatsyshin P, Kalliadasis S, Duncan AB (2022) Physics-constrained Bayesian inference of state functions in classical density-functional theory. J Chem Phys 156
21. Malpica-Morales A, Yatsyshin P, Durán-Olivencia MA, Kalliadasis S (2023) Physics-informed Bayesian inference of external potentials in classical density-functional theory. J Chem Phys 159
22. Fang X, Gu M, Wu J (2022) Reliable emulation of complex functionals by active learning with error control. J Chem Phys 157
23. Chang M, Griffiths T, Levine S (2022) Object representations as fixed points: training iterative refinement algorithms with implicit differentiation. Adv Neural Inf Process Syst 35:32694–32708
24. Martius G, Lampert CH (2016) Extrapolation and learning equations. arXiv:1610.02995
25. Sahoo S, Lampert C, Martius G (2018) Learning equations for extrapolation and control. In: International conference on machine learning, pp 4442–4450
26. Wu J, Gu M (2023) Perfecting liquid-state theories with machine intelligence. J Phys Chem Lett 14:10545–10552
27. Rosenfeld Y, Ashcroft N (1979) Theory of simple classical fluids: universality in the short-range structure. Phys Rev A 20:1208
28. Goodall RE, Lee AA (2021) Data-driven approximations to the bridge function yield improved closures for the Ornstein Zernike equation. Soft Matter 17:5393
29. Rosenfeld Y (1993) Free energy model for inhomogeneous fluid mixtures: Yukawa-charged hard spheres, general interactions, and plasmas. J Chem Phys 98:8126
30. Oettel M (2005) Integral equations for simple fluids in a general reference functional approach. J Phys Condens Matter 17:429
31. Borgis D, Luukkonen S, Belloni L, Jeanmairet G (2021) Accurate prediction of hydration free energies and solvation structures using molecular density functional theory with a simple bridge functional. J Chem Phys 155
32. Deilmann T (2024) Artificial intelligence and intelligent matter. In: te Vrugt M (ed). Springer, Cham
33. Kohn W, Sham LJ (1965) Self-consistent equations including exchange and correlation effects. Phys Rev 140:A1133
34. Reynolds PJ, Ceperley DM, Alder BJ, Lester WA Jr (1982) Fixed- node quantum Monte Carlo for molecules. J Chem Phys 77:5593
35. Anderson J (2007) Quantum Monte Carlo: origins, developments, applications. Oxford University Press, Oxford, England
36. Mardirossian N, Head-Gordon M (2017) Thirty years of density functional theory in computational chemistry: an overview and extensive assessment of 200 density functionals. Mol Phys 115:2315

37. Nagai R, Akashi R, Sasaki S, Tsuneyuki S (2018) Neural-network Kohn- Sham exchange-correlation potential and its out-of-training transferability. J Chem Phys 148:241737
38. Schmidt J, Benavides-Riveros C, Marques M (2019) Machine learning the physical nonlocal exchange-correlation functional of density functional theory. J Chem Phys Lett 10:6425
39. Li L et al (2021) Kohn-Sham equations as regularizer: Building prior knowledge into machine-learned physics. Phys Rev Lett 126(3):036401
40. Kirkpatrick J et al (2021) Pushing the frontiers of density functionals by solving the fractional electron problem. Science 374:1385
41. Ma H, Narayanaswamy A, Riley P, Li L (2022) Evolving symbolic density functionals. Sci Adv 8:eabq0279
42. Snyder JC, Rupp M, Hansen K, Müller K-R, Burke K (2012) Finding density functionals with machine learning. Phys Rev Lett 108:253002
43. Li L et al (2016) Understanding machine-learned density functionals. Int J Quantum Chem 116:819
44. Meyer R, Weichselbaum M, Hauser AW (2020) Machine learning approaches toward orbital-free density functional theory: simultaneous training on the kinetic energy density functional and its functional derivative. J Chem Theory Comput 16:5685
45. Schmidt M (2022) Power functional theory for many-body dynamics. Rev Mod Phys 94:015007
46. Lutsko JF, Oettel M (2021) Reconsidering power functional theory. J Chem Phys 155:094901
47. te Vrugt M, Löwen H, Wittkowski R (2020) Classical dynamical density functional theory: from fundamentals to applications. Adv Phys 69:121
48. de las Heras D, Zimmermann T, Sammüller F, Hermann S, Schmidt M (2023) Perspective: how to overcome dynamical density functional theory. J Phys Condens Matter 35:271501
49. Zhang T et al (2022) Machine learning prediction of photocatalytic lignin cleavage of CC bonds based on density functional theory. Mater Today Sustain 100256

Machine Learning in Quantum Density Functional Theory

Thorsten Deilmann

Abstract In this chapter we discuss applications of machine learning (ML) to the field of quantum-mechanical simulations. Since the discovery of the quantum mechanics more than hundred years ago, a large number of computational methods have been developed to enable the calculation the physical and chemical properties of molecules and solids. One of the most used *first-principles* methods is the *density functional theory* (DFT). Compared to similar methods, it provides a reasonable approximation while keeping manageable computational costs. Nowadays, it can even be used as a black box tool which allows for high-throughput studies. The abundance of data makes DFT appealing for machine learning. On the one hand, it enables the learning of general structure-property relations with or without the aid of human theories, and, on the other hand, these data can be used to develop ML models for the functionals employed in the quantum-mechanical calculations.

1 Introduction

One of the long-standing goals of physics, chemistry, and materials science is the understanding of the direct link between the *atomic structure* and the *properties* of the resulting molecules or solids [1–6]. The phrase atomic structure refers to the positions and chemical type of the atomic cores, as well as the electrons spread around and in between them. Roughly speaking, solid materials can be classified depending on the number of atoms and their periodicity. Atomic structures are typically called molecules if the number of atoms is small, crystalline solids if periodic patterns are present, or amorphous solids otherwise. This chapter will focus on the first two. The properties that may be of interest are numerous. For example, this could be mechanical properties, like the volumetric mass density or the bulk modulus, chemical, electrical, magnetic, optical, or thermal properties. While the dream to connect structure and property has been alchemy in older days, modern science has

T. Deilmann (✉)
Universität Münster, Institut für Festkörpertheorie, Münster, Germany
e-mail: thorsten.deilmann@uni-muenster.de

© The Author(s) 2026
M. te Vrugt (ed.), *Artificial Intelligence and Intelligent Matter*, Machine Intelligence for Materials Science, https://doi.org/10.1007/978-3-032-04129-6_7

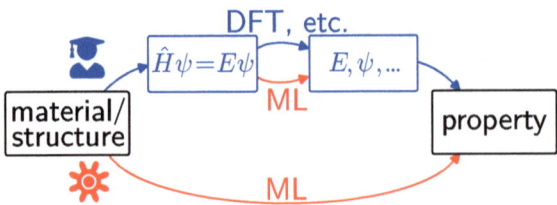

Fig. 1 The general structure-property problem. In blue the typical human approach is shown, in red possible starting points for machine learning. Note that the reverse direction is even more demanding

found several ways to describe and even to predict optimal atomic structures as well as their resulting properties [7–10]. The reverse direction is even more challenging, i.e. to ask which atomic structure is required (or the most efficient) to achieve a desired property (see Fig. 1).

From an engineer's point of view, the input to our algorithm are suitable positions and types of atoms and the output are all materials properties. To find a general algorithm seems practically hopeless. The quantity of potential combinations is just too high, already if the number of atoms is small. By studying several materials experimentally, humans have deduced the Schrödinger and, in more general, the Dirac equation. These have been very successful in describing systems ranging from atoms to infinitely crystal solids and are the foundation of quantum mechanics [7–10]. However, the possibility to write down the Schrödinger equation for a system does not necessarily mean that we can solve it. While this is possible analytically for the hydrogen atom only (and other ions with only one electron), all other systems have many-body character and require more sophisticated ways to handle them. In Sect. 2 we discuss the approach used in density functional theory in a simplified way. Several features are similar compared to the previous chapter by Alessandro Simon and Martin Oettel on classical density functional theory [11].

Machine learning approaches promise help at different levels, prominently demonstrated by AlphaFold [12]. On the one hand, one may try to learn the structure-property link directly. This mostly bridges quantum mechanics, or in other words, such approaches should include all quantum phenomena implicitly. Of course, a lot of high quality data is required already if we are only interested in materials similar to the training set. Due to its reasonable accuracy at moderate computing times, DFT is often used for the purpose of high-throughput studies which can serve these training data for neural networks but also for active learning [13, 14]. In Sect. 3 we discuss some of the large number of works in this direction and briefly motivate the construction of simpler methods like force fields [15, 16]. Another example is demonstrated in the next chapter discussing the inverse design by Teng Long, Yixuan Zhang, and Hongbin Zhang. On the other hand, machine learning can be employed to assist solving the Schrödinger equation. In particular, in DFT it is required to find a relationship (and in practical applications this means an approximation) between the density of electrons and their energy contribution, the so-called exchange-correlation functional. Here machine learning may be applied to challenge the previous human proposals in Sect. 4.

As this field is on the border between physics, chemistry and material science, there is rapidly growing literature from different perspectives. For example, with a focus on electronic structure methods the roadmap of Kulik et al. [1] is closely connected to physical and chemical techniques, while Himanen et al. [3] also discuss the infrastructural challenges, and Cai et al. and many more [18–23] tend towards new materials.

2   Introduction to Quantum Mechanics

To understand the challenges of quantum-mechanical calculations and possibilities to use machine learning, we need to introduce/recap some basics depending on the background of the reader. Experiments at the beginning of the 20th century have shown that classical physics was insufficient in the description at atomic scales, i.e. physical effects occurring on the order of nanometre.

To explain the measurements, Schrödinger proposed

$$\hat{H}\psi(\mathbf{r}, t) = i\hbar\frac{\partial \psi(\mathbf{r}, t)}{\partial t} \ . \tag{1}$$

Here $\hat{H}$ is the Hamilton operator which contains all interactions of the system, $\psi(\mathbf{r}, t)$ is the wave function, its absolute square can be understood as probability density, and on the right-hand side i, $\hbar$, and $\frac{\partial}{\partial t}$ are the imaginary unit, the reduced Planck constant, and the time derivative, respectively. The equation puts the wave function in the center of the quantum-mechanical description of a system and characterises its evolution. Later it turned out that this Schrödinger equation can be generalized to the Dirac equation to include the description of spin and relativistic effects. For the sake of simplicity we stick to Eq. (1) here.

The first step to solve the Schrödinger equation is typically to decouple the dynamics of electrons and cores, see e.g. [24]. The implied errors are often small due to the large difference of the masses of electrons and atomic cores. Furthermore, we are often only interested in static phenomena, i.e. after some steps Eq. (1) becomes the eigenvalue problem $\hat{H}_{el}\psi_{el}(\mathbf{r}) = E_{el}\psi_{el}(\mathbf{r})$, where E_{el} is the total energy (as usual, we suppress the index in the following). $\hat{H}$ includes the kinetic energy of each electron, the repulsive Coulomb interaction in between each pair of electrons, and the attractive Coulomb interaction to each atomic core. When we write the Hamiltonian for the electrons explicitly (the interaction with the nuclei is denoted as external potential V_{ext}),

$$\hat{H} = \sum_i \hat{T}_e(i) + \sum_i V_{ext}(i) + \frac{1}{2}\sum_{i \neq j} V_{ee}(i, j), \tag{2}$$

where the kinetic energy is denoted by $\hat{T}_e$, and the Coulomb repulsion between the electrons i and j as V_{ee}, this leads to a few up to 10^{23} terms (order of the Avogadro constant) for molecules to macroscopic systems, respectively. Such large numbers are only manageable if the system is periodic (i.e. the Bloch theorem [25] for crystals can be applied). The wave function becomes similarly complex as it needs to be antisymmetric under the exchange of each pair of electrons.

From a mathematical point of view, the Schrödinger equation is a high dimensional differential equation. Besides for the Hydrogen atom, no analytic solution has been found (for many-electron systems) yet. Therefore, further approximations are necessary to get closer to the solution. Several methods have been proposed [25], like Hartree-Fock or density functional theory. We will only briefly motivate the latter here. In 1964 Hohenberg and Kohn showed that unique functionals[1] connect all properties to the ground state density n_0. One year later, Kohn and Sham found the Kohn-Sham equation

$$\left(\hat{T}_e + V_{\text{ext}} + V_{\text{H}}(n) + \frac{\delta E_{\text{xc}}}{\delta n} \right) \varphi_i = \epsilon_i \varphi_i \tag{3}$$

by extremalising the Hohenberg-Kohn functional (we suppress the dependence on i on all four terms of the Hamiltonian). This is an effective one-particle equation for the i-th particle with wave function φ_i and energy ϵ_i. Besides the kinetic part and the external potential, the remaining interaction is given by the repulsion of the electron φ_i with the density n (the Hartree term V_{H}) and the so-called exchange-correlation (xc) potential. The latter is written as the functional derivative of the xc energy. E_{xc} is the central quantity in DFT. All non-trivial many-body effects resulting from the exchange (i.e. the interchange of two electrons) and the correlation (i.e. remaining effects between electrons) enter. By Eq. (3) we have reformulated the problem without introducing any further approximations. Consequently, the form of E_{xc} is unknown and the challenge of DFT is to find a good approximation. Until today several approximations have been proposed. In this short introduction we are only able to name a few of them. The most simple is the local density approximation (LDA). Here the idea is to replace the evaluation in the real system at the position $\mathbf{r}$ by using the result of a single particle in the homogeneous electron gas of the corresponding density $n(\mathbf{r})$, i.e. $E_{\text{xc}}^{\text{LDA}}[n] = \int n(\mathbf{r}) \epsilon_{\text{xc}}(n(\mathbf{r})) d\mathbf{r}$. For $\epsilon_{\text{xc}}(n(\mathbf{r})) = \epsilon_{\text{x}}(n(\mathbf{r})) + \epsilon_{\text{c}}(n(\mathbf{r}))$ the exchange part can be approximated analytically, while the remaining correlation is typically parametrized by the results of Monte Carlo simulations for the electron gas [26]. In addition to the local density $n(\mathbf{r})$, the generalized gradient approximation (GGA) includes the derivative $\nabla n(\mathbf{r})$, meta-GGA the second derivative $\nabla^2 n(\mathbf{r})$,

---

[1] A functional is the analog to a function (which maps the argument $x \mapsto f(x)$). In contrast, a functional maps an entire function f to the value, e.g., the area underneath the graph $f \mapsto \int_0^1 f(x)dx$ or just the value of function $f \mapsto f(a)$ (at the point/parameter a). As a simple example, we might be interested in minimizing the expression $I[f] = \int_a^b f(x)(\frac{d}{dx}f(x))dx$. For finding the minimal value, we have to vary the entire function f, or in other words perform a functional derivative $\frac{\delta I[f]}{\delta f}$. For more information on the mathematical details, we refer to the relevant literature [25].

etc. Furthermore, these functionals have been mixed with exact exchange to so-called hybrid functionals. Although more complex functionals typically improve the comparison to experiment for some materials or quantities, several fundamental problems remain which are further discussed in Sect. 4.

Besides the discussion which functional to choose in which case, many of them are able to predict many properties with a very reasonable accuracy. For this reason and because the required computational time is lower than for comparable or more accurate many-particle methods,[2] DFT is well suited for high-throughput screening discussed in next Sect. 3. We note in passing that several methods beyond DFT (e.g. GW for extended system [10, 27]) can also become highly relevant for machine learning, as they might help to learn and overcome the drawbacks in typical functionals.

3 Direct ML: DFT Results for Machine Learning

In the ideal case, sufficient experimental high quality data of various materials with various properties under various conditions is available for machine learning. But this is not the case today. Due to the favorable computational price-quality ratio, DFT is appealing for high-throughput studies and is able to offer large data sets with the possibility for further data where required. Several databases like AFLOWLIB, CMR, GDB, Material Project, NOMAD, or OQMD (see e.g. [3, 28] for more extended lists) have been collected mostly in the last decade. These databases store thousands of experimentally observed as well as hypothetical structures alongside various calculated physical and chemical properties.

A typical high throughput calculation is shown in Fig. 2. Depending on the database different strategies to select the initial structures have been used. Either they are inspired by experiments or hypothetical structures are generated. After these materials have been optimized, they are classified and checked for decomposition and duplicates. For reasonable candidates their stability and further properties are evaluated in DFT and stored in the database. In a next step the desired quantities are identified, e.g., if the material is a metal or a semiconductor, whether it is magnetic, etc. The final selected (few) candidates are then investigated in more detail.

This "classical" procedure opens several interesting applications for machine learning. The large set of data including the structures, basic and advanced mechanical, electronic, magnetic, and optical properties can serve as training and test data sets. ML can investigate the relationship between different properties or between the structure and a property, e.g. the band gap. We will discuss an example below. In addition, ML can be utilized to predict novel materials with promising properties.

---

[2] While typical approximations of the DFT exchange-correlation potential employ only the density (and its first/second derivatives), many-particle methods describe the correlations more accurately (e.g. by configuration interaction or many-body perturbation theory [10, 27]). .

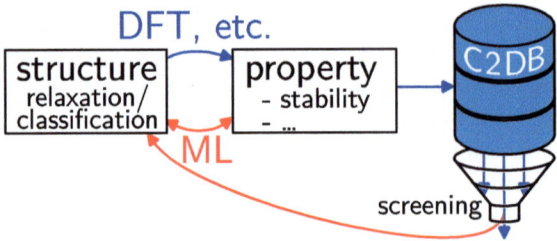

Fig. 2 Typical flow of a high throughput calculation. Multiple structures are relaxed, classified, and several properties are determined. Finally, the results are screened

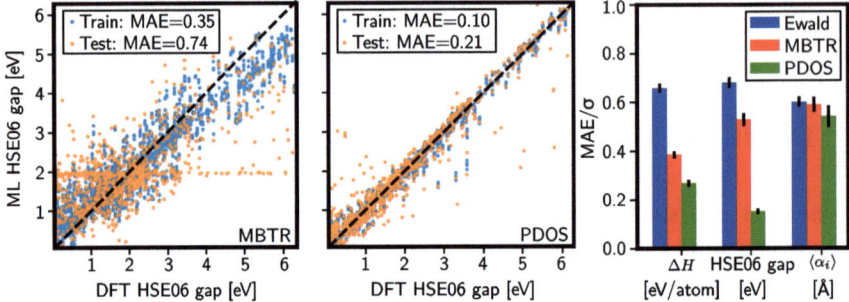

Fig. 3 Machine learning (Gaussian process regression) predicted gap vs. DFT gap for the many-body tensor representation (MBTR, left) and the PDOS fingerprint (middle). The right panel shows the prediction scores (mean absolute error by standard deviation) of the heat of formation (ΔH), the gap, and the static polarization ($\langle \alpha_i \rangle$), respectively, comparing different fingerprints. Modified from Ref. [29]

Figure 3 shows an example for two-dimensional materials in the CMR [30]. In a sufficiently complex ML model the materials should be characterised only by its atomic positions including the chemical elements and the periodicity. As such fundamental models are not possible yet, further information need to be included in the fingerprint describing the material. Three possibilities are evaluated: (i) The Ewald fingerprint which includes Coulomb matrix elements (not shown in detail). (ii) The many-body tensor representation (MBTR) uses atomic numbers, distances and angles. (iii) The PDOS fingerprint includes the projected density of states. The results of the latter two using Gaussian process regression (Fig. 3 left and middle panel) show a moderate success. MBTR shows a mean absolute error of 0.35 eV for the training set, and even much larger for the test set with 0.74 eV. Employing the PDOS fingerprint the accuracy can be improved (MAE of 0.1 and 0.21 eV, respectively), however several outliers remain and the required chemical accuracy is not achieved. These numbers are given for the band gap using the HSE06 hybrid functional, a property which is closely connected to the PDOS. For further properties like the heat of formation which is an important property describing the stability, the

MAE is larger which becomes even worse for the static polarization. The latter is problematic for all the three fingerprints.

Similar studies have been performed for many sets of materials and different properties [2, 18, 23, 28]. Examples include molecules for solar cells [20], materials for CO_2 reduction [21], or the class of MXenes [22]. Nowadays, basically all material classes and properties are covered [23]. In addition to these DFT results, machine learning has been employed to predict the wave function or densities [31, 32], which can then be employed to calculate further properties. On the one hand, more advanced ab initio methods[3] like the random phase or the GW approximation [33–35] have been used, while on the other hand, force fields (which are conceptional simpler compared to ab initio methods) are parametrized by ML [36, 37]. While traditional force fields use a static set of parameters, ML allows adapting and improve the parameters by fitting DFT results and thus drastically increase their accuracies.

In their current state, many ML models have appeared to described various properties individually. As discussed in Refs. [2, 23], for instance, it is currently unclear which algorithms will prevail. For larger data sets deep neural networks may be favorable. If the data sets are smaller, active learning (using Gaussian process regression) is an alternative in which DFT can be used to calculate the required new input. Even if much data is available publicly, this is much less true for the details of the ML models which makes their quantitative comparison very challenging.

4 Assisting ML: Learning the Density Functional

Besides the direct material-property link, ML can be utilized to improve the solution of the Schrödinger equation. In the case of the DFT this means the improvement of the chosen functional. Even if many researchers have previously worked on this topic (e.g. [38, 39] and many more), the topic has been highlighted when Kirkpatrick et al. published their work [40] in 2021. Here we will briefly discuss this work.

The previous discussion in Sect. 2 has been given in a simplified way. On the one hand, we have not discussed the spin, a purely quantum mechanical property similar to the angular momentum. In DFT, this requires the use of spin up and down densities $n(\mathbf{r}) = n_\uparrow(\mathbf{r}) + n_\downarrow(\mathbf{r})$ (and in general spin-orbit interactions coupling them to the structure). Especially for magnetic systems and molecules/radicals with free electron spin this make the state more complex. Besides the spin, the typical DFT functionals introduced above suffer from the general problem of *fractional electrons* [41]. Even if the concept of fractional electrons may seem counterintuitive, their quantum mechanical nature can lead to situations where they are part of two subsystems, i.e. only a fraction can be part of the system to be calculated. E.g. if

---

[3] As discussed in Sect. 2, the many-body Schrödinger equation (1) can be solved approximately. The Random Phase Approximation approximates the total correlation energy relying on the dielectric response function of the material. The GW approximation leads to reliable predictions of band gaps and positions employing the self energy given by $\Sigma = iGW$.

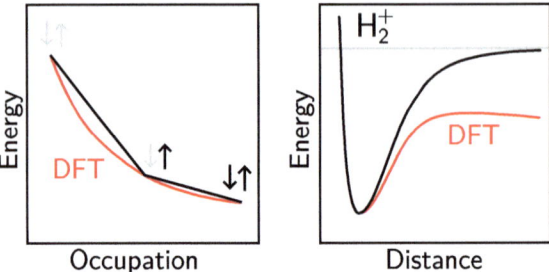

Fig. 4 Schematic dependence of the energy on the occupation (left). Grey (black) arrows indicate unoccupied (occupied) electron spins. The energy of H_2^+ depending on the distance is shown on the right. The corresponding DFT results with typical functionals are marked in red

we consider a hydrogen atom and vary the number of electrons between zero and two, the energy of this system should vary linearly from zero to one and one to two (Fig. 4). This derivative discontinuity is typically not well reproduced in functionals and is a result of their construction using the homogeneous electron gas. Another demonstrative example is the H_2^+ molecule. At the optimal bonding distance the situation can be described approximately by two positively charged cores with the electron in between them. However, if we separate the cores the electron has to decide and should be found close to only one of the cores. The allowed fractional electrons lead to a situation in which half an electron can be located close to each of the cores. Such unphysical situations are often not problematic if we take the entire system (or time averages) into account, but in common functionals they are imprinted separately at every point where the exchange-correlation is evaluated.

Other electronic structure methods like Hartree-Fock or similar take the exact exchange into account and are able to handle such systems more precisely. Also, density functionals have been proposed with efforts in this direction. E.g. the GLLBSC functional separates the Coulomb part of the exchange hole and the response part from the pair correlation function [42], and often leads to band gaps much closer to experimental findings.

As an alternative to the human designed functionals, ML has been employed to design new functionals for some time, e.g. Refs. [38, 39]. In the following we will focus on the recently developed version DM21 [40]. This functional aims to tackle the factional electron and spin problem and its general idea is shown in Fig. 5 including its results. On each mesh point the spin-dependent density, the norm of the gradient, the kinetic energy density $\tau_\pm$ and the Hartree Fock exchange energy densities e^{HF} are evaluated. By a neuronal network (multilayer perceptron, MLP) enhancement factors are evaluated and used to integrate the exchange-correlation energy. From a physical point of view, inputs similar to common functionals ($n_\uparrow$, $n_\downarrow$, etc.) are no longer connected by strict mathematical conditions, but their importance and their interdependencies are hidden in the neural network. As training data highly accurate calculations for several hundreds of atoms/molecules have been used. Depending on the exact constraints applied, Kirkpatrick et al. have fitted several versions of the

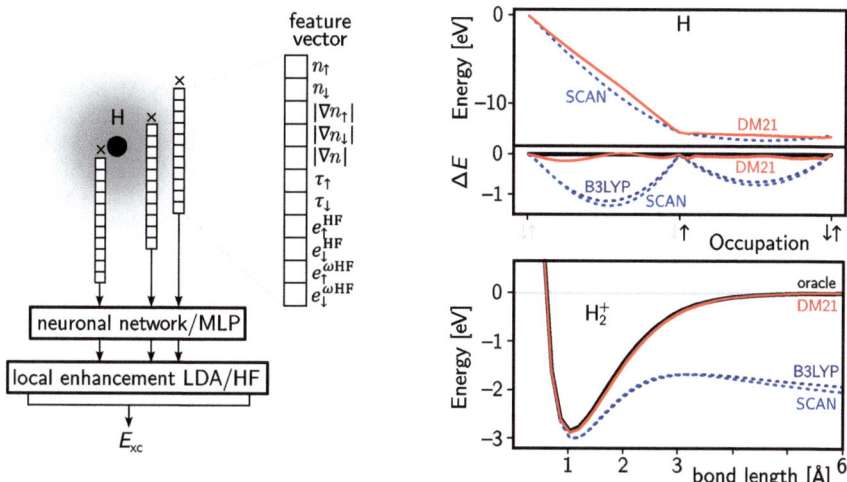

Fig. 5 At each mesh point the DM21 functional calculates the feature vector with different quantities like spin densities, Hartree Fock exchange energy densities, etc. which are weighted by a neuronal network (multilayer perceptron, MLP). The right panels show the fractional occupation of H comparing Monte Carlo (oracle) with three functionals B3LYP, SCAN, and DM21. The inset shows the deviation from the linear trend. Below the binding energies of H_2^+ is shown depending on the distance. Data taken from Ref. [40]

functional. Here we focus on DM21 which respects fractional charge and spin but violates the uniform electron gas limit.

The result for different occupations of a hydrogen atom are show in the right panel of Fig. 5. The most precise results can be expected from Monte Carlo calculations (labeled oracle). State-of-the-art results from hybrid and meta-GGA functionals B3LYP and SCAN deviate from the linear trend by up to 25 kcal/mol (~ 1 eV, see inset). Even if DM21 is not linear, its deviation is much smaller compared to the previously mentioned. For the dissociation of H_2^+ a similar improvement of DM21 compared to state-of-the-art functionals is found. Especially the limit of zero binding energy is fulfilled. Eventually further insights on the importance of the different contributions may also help to improve current standard functionals.

Further results on fractional spin systems or reactions suggest an overall improvement using DM21. However, DM21 only focuses on main group elements in non-periodic system. Nevertheless, it clearly shows the potential of ML to improve functionals more generally in the future.

Acknowledgements We acknowledge financial support from the Deutsche Forschungsgemeinschaft (DFG, German Research Foundation) through Project No. 426726249 (DE 2749/2-1 and DE 2749/2-2).

References

1. Kulik HJ, Hammerschmidt T, Schmidt J, Botti S, Marques MAL, Boley M, Scheffler M, Todorović M, Rinke P, Oses C, Smolyanyuk A, Curtarolo S, Tkatchenko A, Bartók AP, Manzhos S, Ihara M, Carrington T, Behler J, Isayev O, Veit M, Grisafi A, Nigam J, Ceriotti M, Schütt KT, Westermayr J, Gastegger M, Maurer RJ, Kalita B, Burke K, Nagai R, Akashi R, Sugino O, Hermann J, Noé F, Pilati S, Draxl C, Kuban M, Rigamonti S, Scheidgen M, Esters M, Hicks D, Toher C, Balachandran PV, Tamblyn I, Whitelam S, Bellinger C, Ghiringhelli LM (2022) Roadmap on machine learning in electronic structure. Electron Struct 4(2):023004. https://doi.org/10.1088/2516-1075/ac572f

2. Schmidt J, Marques MRG, Botti S, Marques MAL (2019) Recent advances and applications of machine learning in solid-state materials science. NPJ Comput Mater 5(1):83. https://doi.org/10.1038/s41524-019-0221-0

3. Himanen L, Geurts A, Foster AS, Rinke P (2019) Data-driven materials science: status, challenges, and perspectives. Adv Sci 6(21):1900808. https://doi.org/10.1002/advs.201900808

4. Bartók AP, De S, Poelking C, Bernstein N, Kermode JR, Csányi G, Ceriotti M (2017) Machine learning unifies the modeling of materials and molecules. Sci Adv 3(12):e1701816. https://doi.org/10.1126/sciadv.1701816

5. Behler J (2016) Perspective: Machine learning potentials for atomistic simulations. J Chem Phys 145(17):170901. https://doi.org/10.1063/1.4966192

6. Ramprasad R, Batra R, Pilania G, Mannodi-Kanakkithodi A, Kim C (2017) Machine learning in materials informatics: recent applications and prospects. NPJ Comput Mater 3(1):54. https://doi.org/10.1038/s41524-017-0056-5

7. Fock V (1930) Näherungsmethode zur Lösung des quantenmechanischen Mehrkörperproblems. Z Phys 61:126–148. https://doi.org/10.1007/BF01340294

8. Hohenberg P, Kohn W (1964) Inhomogeneous electron gas. Phys Rev 136:B864–B871. https://doi.org/10.1103/PhysRev.136.B864

9. Kohn W, Sham LJ (1965) Self-consistent equations including exchange and correlation effects. Phys Rev 140:A1133–A1138. https://doi.org/10.1103/PhysRev.140.A1133

10. Hedin L (1965) New method for calculating the one-particle green's function with application to the electron-gas problem. Phys Rev 139:A796–A823. https://doi.org/10.1103/PhysRev.139.A796

11. Simon A, Oettel M, Machine learning approaches to classical density functional theory. In: Artificial intelligence and intelligent matter

12. Jumper J, Evans R, Pritzel A, Green T, Figurnov M, Ronneberger O, Tunyasuvunakool K, Bates R, Žídek A, Potapenko A, Bridgland A, Meyer C, Kohl SAA, Ballard AJ, Cowie A, Romera-Paredes B, Nikolov S, Jain R, Adler J, Back T, Petersen S, Reiman D, Clancy E, Zielinski M, Steinegger M, Pacholska M, Berghammer T, Bodenstein S, Silver D, Vinyals O, Senior AW, Kavukcuoglu K, Kohli P, Hassabis D (2021) Highly accurate protein structure prediction with alphafold. Nature 596(7873):583. https://doi.org/10.1038/s41586-021-03819-2

13. Smith JS, Nebgen B, Lubbers N, Isayev O, Roitberg AE (2018) Less is more: sampling chemical space with active learning. J Chem Phys 148(24):241733. https://doi.org/10.1063/1.5023802

14. Wang AY-T, Murdock RJ, Kauwe SK, Oliynyk AO, Gurlo A, Brgoch J, Persson KA, Sparks TD (2020) Machine learning for materials scientists: an introductory guide toward best practices. Chem Mater 32(12):4954–4965. https://doi.org/10.1021/acs.chemmater.0c01907

15. Chmiela S, Sauceda HE, Müller K-R, Tkatchenko A (2018) Towards exact molecular dynamics simulations with machine-learned force fields. Nat Commun 9(1):3887. https://doi.org/10.1038/s41467-018-06169-2

16. Jinnouchi R, Karsai F, Kresse G (2019) On-the-fly machine learning force field generation: application to melting points. Phys Rev B 100(1):014105. https://doi.org/10.1103/PhysRevB.100.014105

17. Long T, Zhang Y, Zhang H, Generative deep learning for the inverse design of materials. In: Artificial intelligence and intelligent matter

18. Cai J, Chu X, Kun X, Li H, Wei J (2020) Machine learning-driven new material discovery. Nanoscale Adv 2(8):3115. https://doi.org/10.1039/d0na00388c
19. Wang H, Lei Z, Zhang X, Zhou B, Peng J (2019) A review of deep learning for renewable energy forecasting. Energy Convers Manag 198:111799. https://doi.org/10.1016/j.enconman. 2019.111799
20. Jørgensen PB, Mesta M, Shil S, García Lastra JM, Jacobsen KW, Thygesen KS, Schmidt MN (2018) Machine learning-based screening of complex molecules for polymer solar cells. J Chem Phys 148(24):241735. https://doi.org/10.1063/1.5023563
21. Ulissi ZW, Tang MT, Xiao J, Liu X, Torelli DA, Karamad M, Cummins K, Hahn C, Lewis NS, Jaramillo TF, Chan K, Nørskov JK (2017) Machine-learning methods enable exhaustive searches for active bimetallic facets and reveal active site motifs for co2 reduction. ACS Catal 7(10):6600. https://doi.org/10.1021/acscatal.7b01648
22. Frey NC, Wang J, Bellido GIV, Anasori B, Gogotsi Y, Shenoy VB (2019) Prediction of synthesis of 2D metal carbides and nitrides (MXenes) and their precursors with positive and unlabeled machine learning. ACS Nano 13(3):3031. https://doi.org/10.1021/acsnano.8b08014
23. Schleder GR, Padilha AC, Acosta CM, Costa M, Fazzio A (2019) From DFT to machine learning: recent approaches to materials science-a review. J Phys Mater 2(3):032001. https://doi.org/10.1088/2515-7639/ab084b
24. Born M, Oppenheimer R (1927) Zur Quantentheorie der Molekeln. Ann Phys 84(20):457–484. https://doi.org/10.1002/andp.19273892002
25. Martin RM (2020) Electronic structure. Cambridge University Press. https://doi.org/10.1017/9781108555586
26. Perdew JP, Zunger A (1981) Self-interaction correction to density-functional approximations for many-electron systems. Phys Rev B 23(10):5048. https://doi.org/10.1103/PhysRevB.23.5048
27. Onida G, Reining L, Rubio A (2002) Electronic excitations: density-functional versus many-body Green's-function approaches. Rev Mod Phys 74(2):601. https://doi.org/10.1103/RevModPhys.74.601
28. Butler KT, Davies DW, Cartwright H, Isayev O, Walsh A (2018) Machine learning for molecular and materials science. Nature 559(7715):547. https://doi.org/10.1038/s41586-018-0337-2
29. Gjerding MN, Taghizadeh A, Rasmussen A, Ali S, Bertoldo F, Deilmann T, Knøsgaard NR, Kruse M, Larsen AH, Manti S, Pedersen TG, Petralanda U, Skovhus T, Svendsen MK, Mortensen JJ, Olsen T, Thygesen KS (2021) Recent progress of the computational 2D materials database (C2DB). 2D Mater 8(4):044002. https://doi.org/10.1088/2053-1583/ac1059
30. Haastrup S, Strange M, Pandey M, Deilmann T, Schmidt PS, Hinsche NF, Gjerding MN, Torelli D, Larsen PM, Riis-Jensen AC, Gath J, Jacobsen KW, Mortensen JJ, Olsen T, Thygesen KS (2018) The computational 2D materials database: high-throughput modeling and discovery of atomically thin crystals. 2D Mater 5(4):042002. https://doi.org/10.1088/2053-1583/aacfc1
31. Schütt KT, Gastegger M, Tkatchenko A, Müller K-R, Maurer RJ (2019) Unifying machine learning and quantum chemistry with a deep neural network for molecular wavefunctions. Nat Commun 10(1):5024. https://doi.org/10.1038/s41467-019-12875-2
32. Brockherde F, Vogt L, Li L, Tuckerman ME, Burke K, Müller K-R (2017) Bypassing the kohn-sham equations with machine learning. Nat Commun 8(1):872. https://doi.org/10.1038/s41467-017-00839-3
33. Rasmussen A, Deilmann T, Thygesen KS (2021) Towards fully automatized GW band structure calculations: what we can learn from 60.000 self-energy evaluations. NPJ Comput Mater 7(1):22. https://doi.org/10.1038/s41524-020-00480-7
34. Riemelmoser S, Verdi C, Kaltak M, Kresse G (2023) Machine learning density functionals from the random-phase approximation. J Chem Theory Comput 19(20):7287. https://doi.org/10.1021/acs.jctc.3c00848
35. Wilkins DM, Grisafi A, Yang Y, Lao KU, DiStasio RA, Ceriotti M (2019) Accurate molecular polarizabilities with coupled cluster theory and machine learning. Proc Natl Acad Sci 116(9):3401. https://doi.org/10.1073/pnas.1816132116

36. Hansen K, Biegler F, Ramakrishnan R, Pronobis W, von Lilienfeld OA, Müller K-R, Tkatchenko A (2015) Machine learning predictions of molecular properties: accurate many-body potentials and nonlocality in chemical space. J Phys Chem Lett 6(12):2326. https://doi.org/10.1021/acs.jpclett.5b00831
37. Unke OT, Chmiela S, Sauceda HE, Gastegger M, Poltavsky I, Schütt KT, Tkatchenko A, Müller K-R (2021) Machine learning force fields. Chem Rev 121(16):10142. https://doi.org/10.1021/acs.chemrev.0c01111
38. Dick S, Fernandez-Serra M (2021) Highly accurate and constrained density functional obtained with differentiable programming. Phys Rev B 104(16):161109. https://doi.org/10.1103/PhysRevB.104.L161109
39. Li L, Snyder JC, Pelaschier IM, Huang J, Niranjan U-N, Duncan P, Rupp M, Müller K-R, Burke K (2015) Understanding machine-learned density functionals. Int J Quantum Chem 116(11):819. https://doi.org/10.1002/qua.25040
40. Kirkpatrick J, McMorrow B, Turban DHP, Gaunt AL, Spencer JS, Matthews AGDG, Obika A, Thiry L, Fortunato M, Pfau D, Castellanos LR, Petersen S, Nelson AWR, Kohli P, Mori-Sánchez P, Hassabis D, Cohen AJ (2021) Pushing the frontiers of density functionals by solving the fractional electron problem. Science 374(6573):1385. https://doi.org/10.1126/science.abj6511
41. Perdew JP, Parr RG, Levy M, Balduz JL (1982) Density-functional theory for fractional particle number: derivative discontinuities of the energy. Phys Rev Lett 49(23):1691–1694. https://doi.org/10.1103/PhysRevLett.49.1691
42. Kuisma M, Ojanen J, Enkovaara J, Rantala TT (2010) Kohn-sham potential with discontinuity for band gap materials. Phys Rev B 82(11):115106. https://doi.org/10.1103/PhysRevB.82.115106

Generative Deep Learning for the Inverse Design of Materials

Yixuan Zhang, Teng Long, and Hongbin Zhang

Abstract Beyond the forward inference of materials properties using machine learning, generative deep learning techniques applied on materials science allow the inverse design of materials, i.e., assessing the composition—processing—(micro-)structure—property relationships in a reversed way. In this review, we focus on the (micro-)structure—property mapping, i.e., crystal structure—intrinsic property and microstructure—extrinsic property, and summarize comprehensively how generative deep learning can be performed. Three key elements, i.e., the construction of latent spaces for both the crystal structures and microstructures, generative learning approaches, and property constraints, are discussed in detail. A perspective is given outlining the challenges of the existing methods in terms of computational resource consumption, data compatibility, and yield of generation.

1 Introduction

Developing and employing advanced structural, functional, and quantum materials is of significant importance helping us tackle the challenges like energy shortage and information explosion [1]. Conventional materials science research relies mostly on the trial-and-error experiments and individual domain expertise, leading to resource- and time-costly materials development and exploitation. Recently, the data-driven approaches based on machine learning have emerged unlocking the fourth paradigm of materials science, in addition to the approaches based on empirical experimentation, phenomenological theory, numerical simulations using quantum mechanics [2]. For instance, focusing on the core problem of materials science, *i.e.*, to map out the composition—processing—(micro-)structure—property (CPSP) relationships, machine learning can be applied to quantify each link, *e.g.*, identifying synthesis

Y. Zhang · T. Long · H. Zhang (✉)
Institute of Materials Science, Technische Universität Darmstadt, Darmstadt, Germany
e-mail: hongbin.zhang@tu-darmstadt.de

T. Long
School of Materials Science and Engineering, Shandong Universiy, Jinan, China

© The Author(s) 2026
M. te Vrugt (ed.), *Artificial Intelligence and Intelligent Matter*, Machine Intelligence for Materials Science, https://doi.org/10.1007/978-3-032-04129-6_8

recipe [3], engineering microstructure in additive manufacturing [4], and statistical modelling of various physical properties [5]. However, despite many progresses enabling the acceleration of materials design [2, 6–8], the CPSP relationships covering the whole compositional and functional space are still far from quantitatively revealed [9, 10], which can be attributed to the high dimensionality of the underlying design space [11]. Therefore, even with the help of machine learning, the current materials design practices are mostly following the many-to-one pattern along the CPSP chain, i.e., predict and synthesize unreported compounds, perform the measurements, and select the one with optimal properties [2, 12, 13].

To further accelerate materials design, it is essential to access the CPSP chain in a reversed way, *e.g.*, predicting the compounds with target properties together with the processing conditions for the desired microstructure. This leads to the concept of inverse design, which can be carried out based on high-throughput (HTP) combinatorial screening, global optimization, and generative models [14]. Taking the crystal structure – intrinsic property (as given by the crystal structures) as an example, the HTP computational workflows usually consist of three steps: (1) generating new crystal structures by substituting possible atoms of typical prototypes [10]; (2) conducting density functional theory (DFT) computation on the hypothetical structures [15]; (3) screening the compositions with desired properties. In this regard, the HTP method follows again the massive generation and screening scheme, which is still resource-demanding, though it can reduce the cost of materials design by providing guidance to experimental investigations. For solid state materials, HTP screening is performed mostly for the mapping of crystal structure – intrinsic property link based on DFT calculations [16–18], which rely on the implementation of automatized workflows based on platforms such as atomic simulation environment [19], atomate [20], AiiDa [21], etc. Such workflows can nowadays be constructed for many classes of advanced materials such as magnetic materials [22]. It is interesting to get the other simulation tools such as molecular dynamics, phase fields, and finite element (FE) modelling integrated into operative HTP workflows, so that the other links in the CPSP chain can be explored, as well as experimental workflows. Furthermore, the global optimization method relies on extracting effective descriptors that affect material properties based on existing databases, and constructing reliable surrogate models in order to automatically adjusts these descriptors to achieve better properties for the newly discovered materials [23–31]. Such methods include but are not limited to Bayesian optimization and genetic algorithms, which avoid extensive calculations and thus reduce the cost of the design process [32]. An obvious advantage of the global optimization approach formulated based on Bayesian optimization is to implement and execute the so-called closed-loop adaptive design strategy, which can be directly integrated with experimental investigations and allows an iterative optimization based on active learning [33]. Such a closed-loop adaptive design strategy comprises database curation, surrogate modelling, balanced exploration (searching new regions of the design space) and exploitation (refining known optimal regions), and guided experimental synthesis/characterization, and has been successfully applied to optimize the compositions of NiTi-based shape memory alloys [34], $BaTiO_3$-based piezoelectric ceramics [35], ferroelectric perovskite with higher Curie temperature [36], high entropy alloys with enhanced hardness [37] and as Invar alloys with reduced thermal expansion [38].

In machine learning, the concept of vector space is fundamental and crucial. A vector space is a collection of vectors that can be linearly combined through addition and scalar multiplication. For regression problems in materials science, each vector in the vector space typically represents a sample with multiple features, such as the chemical composition and physical properties of a compound. In this space, machine learning algorithms aim at finding a function (i.e., a distribution in terms of statistics) to fit the relationship between sample data and target outcomes [39–41]. However, the actual chemical or physical parameter space is often high-dimensional and complex. Directly learning and optimizing in a high-dimensional space can encounter efficiency and accuracy issues. This challenge can be mitigated by introducing the latent space, which is ideally a continuous low-dimensional representation used to describe data in the original high-dimensional vector space [39, 40] and is valuable to better assess the distribution of physical properties. Through an appropriate encoder, every original descriptor vector can be mapped to a unique vector in the latent space; Whereas through a decoder, vectors in the latent space can be decoded back into descriptor vectors with explicit physical meanings. Such a latent space greatly facilitates the applications of Bayesian optimization for materials design, where sampling can be efficiently performed with acquisition functions balancing between exploration and exploitation.

Importantly, with the help of the properly constructed latent spaces, generative deep learning can be valuable to go beyond the known design space to further promote inverse design. On the one hand, there is one unique vector in the latent space corresponding to the original descriptor vector as inputs and hence one specific compound, and all the vectors in the latent space can be decoded into descriptor vectors with explicit physical meaning. On the other hand, such a latent space should be constructed in such a way that the joint distribution $P(x, y)$ of target physical properties y and descriptors x should be faithfully reproduced, which distinguishes it from deterministic models focusing on the conditional distribution $P(y|x)$ [11]. Given such a challenge to construct a robust latent space, inverse design based on generative deep learning models has not been extensively done. Furthermore, there are three major machine learning approaches which can be used for generative models, i.e., generative adversarial nets (GAN), variational autoencoder (VAE), and diffusion-based generative model (DGM), including various types such as denoising diffusion probabilistic models (DDPM), denoising score matching (DSM) and discrete denoising diffusion probabilistic model (D3PM) [42–45]. The implementation and application of such models are a bit more technical involved, in comparison to those frequently used forward inference models, such as random forest [46], support vector machine [47], neural networks [48], and transformer [49]. Additionally, another challenging task is how to perform the optimization of physical properties, which is best done in a multi-objective manner and hence entails exploring the usually large chemical/parameter/latent space efficiently to recommend proper candidates fulfilling the requirements [50].

In this review, we focus on the (micro-)structure – property relationships, i.e., crystal structure—intrinsic property and microstructure – extrinsic property (as defined by microstructure), and summarize the current approaches realizing the inverse design of materials based on the generative deep learning. For each section (*i.e.*, crystal structure and microstructure), we elaborate on three critical aspects, namely,

(1) how to achieve efficient communication between machines and humans, i.e., how to generate machine readable descriptors informed with domain expertise,
(2) how to improve the performance of machine learning models,
(3) how to apply constraints so that materials with desired properties will be predicted[11, 14, 51–56].

Specifically, advances in inverse design are discussed in terms of descriptors or representations for materials, deep learning algorithms, and the integration of property constraints, with related studies summarized in Tables 1 (for crystal structures) and Table 2 (for microstructure), respectively. Finally, the pending challenges will be outlined, together with a bird-view outlook.

2 Section I: Inverse Design of Crystal Structures

The generative deep learning-based inverse design method actively optimizes the non-linear relationship between material structures and properties without external intervention, thus attracting intensive attention [51, 76, 77]. In general, such a method extracts knowledge from the existing structure–property datasets and applies the learned knowledge in designing new materials. One of the main challenges of applying the inverse design method for crystal structures is the descriptors of crystalline materials. In Table 1, we summarize the state-of-the-art crystal structure generative models based on the representations of crystal structures, the model architectures, whether there are constraints, the reversibility of the used representations, the application scope of the models, and their generation efficiency.

Representations, also known as descriptors or features, are a set of parameters that represent the structural features of materials. In this section, they refer to the parameters corresponding to the crystal structures that are used as input for the machine learning models. It is noted that crystalline parameters as specified in standard .cif files are often not sufficient for machine learning because the three-dimensional (3D) periodicity is not automatically encoded, and also that the crystal structures in real space are bidirectionally mapped to the descriptors in the latent space.

Models are deep learning generative models used to predict new crystal structures. These models are a type of machine learning model that learns to generate new data similar to the data used for training. From the mathematical perspective, these models attempt to mimic the distribution of the training data through various approaches and metrics.

Table 1 List of generative deep learning-based approaches for the inverse design of crystal structures in terms of representation, machine learning model, constraint, reversibility, scope, and efficiency (cf. the main text for details)

Work	Representations	Models	Constraints	Reversibility	Scope	Efficiency
CrystalGAN [57]	Vector	GAN	None	Low	Binary and ternary	Low
IMatGen [58]	Voxels	VAE	Stable or not	High	Binary	High
CGCNN [59]	Graphs	None	None	None	Multicomponent	High
iCGCNN [60]	Graphs	None	None	None	Multicomponent	High
CCDCGAN [61, 62]	Voxels	GAN	Formation energy	High	Multicomponent	High
SmVAE [63]	Graphs	VAE	Four textural properties, three properties related to natural gas separation and three properties related to flue gas separation	Middle	Multicomponent	High
CDVAE [64, 65]	Vectors	VAE	Stability	High	Multicomponent	High
Cond-DFC-VAE [66]	Voxels	VAE	Formation energy, bandgap, bulk/shear modulus, etc	Middle	Multicomponent	High
FTCP [67, 68]	Vectors	VAE	Formation energy, bandgap, Thermoelectric power factor	High	Multicomponent	Low
ZeoGAN [69]	Voxels	GAN	Heat absorption	Middle	Ternary	High
Composition Conditioned Crystal GAN [70]	Vectors	GAN	Pourbaix stability and the band gaps	High	Ternary	Low
PGCGM [71, 72]	Vectors	GAN	Formation energy	High	Multicomponent	High
DP-CDVAE [73]	Vectors	DDPM & VAE	None	High	Multicomponent	High

(continued)

Table 1 (continued)

Work	Representations	Models	Constraints	Reversibility	Scope	Efficiency
Scalable Diffusion using UniMat [74]	Vectors	DDPM	Composition	High	Multicomponent	High
MatterGen [75]	Vectors	DDPM & DSM & D3PM	Composition, symmetry, magnetic, electronic, mechanical properties and supply-chain-risk	High	Multicomponent	High

Table 2 List of deep learning-based microstructure generation methods, their features, advantages and disadvantages

Work	Models	Constraints(properties)	Database	Dimension	Comments
Yang et al. [113]	GP-Hedge + GAN	Style loss Optical property	Synthetic microstructure images by GRF (5000)	2D	GP-Hedge as constrain for effective property searching
Iyer et al. [114]	ACWGAN-GP	different cooling methods	UHCSDB (172)	2D	Investigated the processing-structure linkage
Lambard et al. [115]	StyleGAN2 with ADA	None	Ferrite-martensite SEM (3000)	2D	Low data requirement
Chun et al. [116]	GAN	None	HMX SEM image (1)	2D	Seamlessly generation across dimensions
Ma et al. [117]	pg-GAN Pix2Pix GAN	Different processing methods	U-10Mo SEM-BSE (272)	2D	Spatial exploration by GAN needs careful evaluation
Fokina et al. [118]	StyleGAN	None	SEM of Alporas aluminum foam (1); Micro-CT of Berea sandstone (1); Micro-CT of Ketton limestone (1)	2D	multiple image resolutions
Squires et al. [119]	DCGAN	None	FIB-SEM of a three-phase solid oxide fuel cell anode (1); hypoeutectoid steel image from DoITPoMS (1)	2D	Using microstructural inpainting to recover the defects and unwanted artefacts

(continued)

Table 2 (continued)

Work	Models	Constraints(properties)	Database	Dimension	Comments
Lee et al. [120]	DCGAN Cycle GAN Pix2Pix	None	OM (9216) and SEM (10,000) images of steel surface; OM (3045) and SEM (3186) images with × 2000 magnification of steel surface; SEM of Li-battery cathode/anode with spherical (3045) and wiry (3186) morphologies; Hand-drawn sketches (1200) and martensitic (440) micrographs	2D	Using large amount of dataset Image style changing
Kench et al. [121]	SliceGAN	None	Micrographs meet special criteria from DoITPoMS (87)	2D3D	Construct 3D microstructure from 2D images
Hsu et al. [122]	Wasserstein-GAN	None	SOFC anode microstructure PFIB-SEM Synthetic 3D (1)	3D	Generated structures outperform Dream.3D results in electrochemical simulations
Henkes et al. [123]	Wasserstein-GAN	None	Synthetic spherical inclusions microstructures; Micro-CT scanned wood-plastic composite microstructure (1)	3D	Investigate the influence of network topology, filter number, and geometrical and physical inductive biases
Gayon-Lombardo et al. [124]	DC-GAN	None	Li-ion battery cathode XCT (1); SOFC anode XCT (1)	3D	Generate arbitrarily large synthetic microstructural volumes and the periodic boundaries
Cang et al. [125]	Constrained-VAE	Style loss	Sandstone microstructures (200)	2D	Incorporate a style loss as constrain into the model training

(continued)

Table 2 (continued)

Work	Models	Constraints(properties)	Database	Dimension	Comments
Kim et al. [126]	GPR-VAE	Ultimate tensile strength; uniform elongation toughness	Synthetic dual-phase steels (4000)	2D	Investigate the structure–property linkages in continues space
Düreth et al. [127]	Diffusion model	Microstructure class	NFFA-Europe, UHCSDB (13,000); Fiber Composite (36)	2D	High model stability
Azqadan et al. [128]	Diffusion model	Processing parameters	AZ80 magnesium alloy components with different casting-forgings (27)	2D	Investigate the process-structure linkage
Lyu et al. [129]	Diffusion model	Permeability	A 2D database comprising various types of structural features (8000) and 3D porous material database (3000)	2D & 3D	Achieve 3D microstructure generation with property control
Lee et al. [130]	Diffusion model	Volume fraction Effective elastic modulus Light intensity change ratio	Micro-CT images of SAOED composites (300) and stress fields calculated by FEM (1920)	2D	Employ multiple target material properties into conditioned DGM

The term *constraint* comes from nonlinear programming problems in operations research, referring to additional conditions that must be satisfied while achieving the main objective. In inverse design, the main objective is the generation of new crystal structures, whereas desired intrinsic properties can be considered as constraints when generating crystal structures. A constraint that can be integrated into the objective function or as a tolerance for screening after the generation process.

Reversibility refers to the ability to achieve bidirectional mapping between crystal structures in real space and descriptors in latent space. Specifically, it includes two aspects (1) whether the model can correctly recognize the known crystal structures, i.e., uniquely encoding and decoding the existing crystal structures, and (2) whether the newly designed crystal structure can be correctly recognized, i.e., sampling the latent space for reasonable descriptors which can be decoded to crystal structures in the real space. Therefore, this indicator is generally characterized by the reproducibility rate of the lattice parameters and atomic positions in the crystal structures during the transformation between real and latent spaces.

The scope indicates the compositional space covered by the inverse design model, e.g., binary/ternary means that the model can only be used in a specific binary/ternary system (sometimes transferrable for other binary/ternary compositions), whereas a multicomponent model can generate new crystal structures with more than three elements.

Efficiency refers to the ability to generate new materials effectively, i.e., the ratio of the number of crystal structures fulfilling the constraints with respect to the total number of generated structures via decoding.

3   Crystal Structure Descriptors

In order to enable the inverse design of crystalline materials, it is necessary to create descriptors mapping the crystal structures for the deep learning models [78, 79]. Descriptors of crystalline materials can be divided into three categories: (1) elemental descriptors that describe the chemical information of materials constituents [80–82], (2) structural descriptors that describe the geometric information of crystal structures [83, 84], and (3) combined descriptors describing both the chemical and structural information [59, 85]. Either the elemental descriptors or the structural descriptors highlight only partial information of crystalline materials, which are not sufficient to specify the crystalline materials. Even though it is possible to combine them, the heterogeneity and poor connectivity make the description of crystalline materials inefficient. In comparison, the combined descriptors integrate the essential information of crystalline material as a whole and therefore have the potential to reconstruct the crystalline materials. However, not all combined descriptors can be used in the deep learning methods, as inverse design methods also require that identifiable crystalline materials should be extracted via decoding from the corresponding descriptor [86]. Since all crystalline materials are a periodic arrangement of atoms in the 3D space, the unit cells are good starting points for generating descriptors [87]. In general, a unit cell is composed of lattice parameters, atomic coordinates, and atomic species. The lattice parameters describe the size and shape of the unit cell, the atomic coordinates indicate the atomic position in the unit cell, and the atomic species mark the elements at the corresponding positions. Thus, the descriptors of the crystalline material used in generative models should allow the extraction of all such information to reconstruct the unit cell. There are deep learning models focusing on the chemical information [88–90] and its extracting [91, 92], the structural information is indispensable. In this regard, generative deep learning to predict crystal structures remains a significant challenge [61].

Nevertheless, there are a few existing solutions which have been applied successfully for generative deep learning, including the voxel method [58], the vector method [64, 65, 73–75], and the graph theory method [63], as demonstrated in Fig. 1. The voxel method uses 3D grids to voxelize the crystal structures, i.e., creating 3D voxels to record the atomic species at the corresponding atomic positions [86], and either the voxels themselves or the vectors obtained by encoding the voxels are used to get the latent space [93–95]. Due to usually required dense grids for voxelization and comparatively limited number of atoms in unit cells, the voxels are significantly sparse (if only the atomic positions are labeled). Thus, severe information loss will occur during the reconstruction process, i.e., the positions of the atoms cannot be reproduced accurately. In order to avoid sparsity, atomic positions can be transformed into atomic densities by introducing a Gaussian smearing with element types being scale parameters. Such a representation scheme can achieve a reconstruction ratio of about 70%, making it an acceptable solution for generative deep learning of crystal structures.

Subsequently, Noh et al. proposed an improvement to this method [58], with three essential modifications: (1) the lattice parameters are also treated using voxels, i.e., one extra voxelization explicitly created to store the shape and size of the unit cell, dubbed as the cell voxel hereafter, (2) The scale parameters for Gaussian smearning are considered as hyper-parameters, instead of using the element types, and 3) individual voxels are created for different elements (called the basis voxels) and stored in different channels. The improved voxel method was successfully applied for the binary V–O system, with a reconstruction ratio of more than 90%. The voxel approach was extended to multicomponent systems by Court et al. [66] by abandoning using multiple channels to store elemental information and resolving back to describing different elemental species by scale parameters, while retaining the lattice parameter voxelization. Such a scheme enables the design of multicomponent crystal structures [66] with the cost of a reduced reconstruction ratio. Long et al. further improved the voxel approach by decomposing different elements into different channels, fixing the scale parameters, and introducing empty channels to maintain the consistency of the voxel data, resulting in an overall reconstruction ratio above 85% [62]. It is noted that Kim et al. successfully used the voxel method to describe porous materials, demonstrating the potential of the voxel approach for microstructure representation [69].

The voxel approach has become one of the most successful solutions in crystal structure representation, with a high reconstruction ratio and the potential to extend to microstructure. However, it exhibits a few shortcomings: (1) it does not reflect the

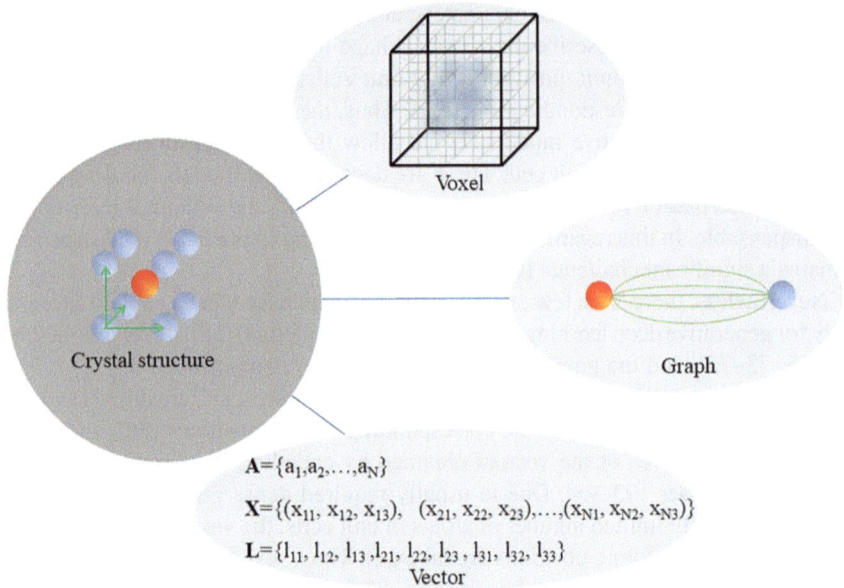

Fig. 1 Common descriptors of crystal structures, with the crystal structure on the left and the voxel descriptors, graph descriptors and vector descriptors from top to bottom on the right

symmetry information, which is vital in crystallography, (2) it cannot identify the similarity of the structures after rotation, i.e., it cannot ensure the rotational symmetry of the descriptors, and (3) it requires a significant amount of computational resources for data processing.

The graph theory approach uses nodes and edges to describe atoms and interatomic connectivity, as well as channels to reflect the elemental information weighted based on the interconnections between the individual atoms [60, 96, 97]. In this regard, it accounts for the translational and rotational symmetry properly. For example, the properties of the corresponding atoms and bonds are added by the channels in crystal graphs, giving rise to the necessary description of the crystal structures. The graph-based descriptors have become one of the most attractive methods [59]. Xie et al. used the one-hot features to store information for both vertices and edges and found that such features suffered from poor reversibility to reconstruct the crystal structures. This dramatically limits the application of the graph theory approach in generative deep learning. Even though the method has been further developed by Park et al., there is still no report about reconstructing crystal structures from graphs [60, 96, 97]. Yao et al. have solved the problem from a different perspective, i.e., using nodes to represent a group of atoms in the graph and edges to represent the interactions between groups [63]. This approach enables the description of massive systems with the cost of relying on other computational methods to achieve reconstruction. The method has shown good performance in the design of materials such as MOF. Therefore, we believe the graph theory approach can accurately capture the symmetry and periodicity of crystalline materials, but its poor reversibility limits its applications in generative models.

The vector method uses vectors to directly describe the atomic species and positions and the lattice parameters. Kim et al. proposed a method to incorporate symmetry and robustness through data augmentation, such as adding perturbations and rotations to the original data during the training process, and successfully applied it for the Mg-Mn–O system [70]. However, this did not solve the problem of the low reconstruction ratio of the vector method. Given that the atomic position information is complicated to reconstruct accurately, Ren et al. enhanced the description of atomic positions by describing the structures of crystalline materials using both Cartesian and Fourier transformed coordinates [67]. After encoding and decoding, such two types of information can be cross-validated, thus significantly improving the reconstruction ratio of the crystal structures. However, the generation process has extremely low efficiency due to the redundant representation. In contrast, Xie et al. proposed a solution by using elemental species, atomic coordinates, and lattice constants as descriptors. Assuming that the lattice parameters can be accurately reconstructed, the scale of the noise of elemental species and atomic coordinates can be measured by scoring neural networks with data argumentation [64]. As a result, a more accurate reconstruction of elemental species and atomic coordinates can be achieved. At the same time, they proposed a pre-optimization scheme to determine the corresponding atomic coordinates and elemental species during the generation process, which guaranteed the stability of the generated structures. Furthermore, Zhao et al. improved the validity of the generation by introducing the element properties, atomic pairwise distance constraints and structural symmetry into the generation model, resulting in a higher success rate and more symmetrized structures in generation [71, 72]. Recently, Yang et al. have defined a four-dimensional scalable vectorized material representation UniMat based on the periodic table of elements utilizing the a priori knowledge of the periodic table [74], with also good performance for both reconstruction and generation. In short, the vector method turns out to be a good solution for describing crystalline materials and becomes extremely promising after the solution proposed by Xie et al. In particular, the rise of DGMs based on Markov diffusion processes can improve the reconstruction rate significantly, which guarantee robust forward and backward mappings and hence reliable reconstruction and generation of crystal structures.

4 Crystal Structure Generative Models

Generative models are the key to inverse design, in order to learn existing crystal structures and design new ones. Three typical generative deep learning models, namely, variational autoencoder (VAE), generative adversarial network (GAN), and diffusion model, are illustrated in Fig. 2.

The VAE model consists of two parts, i.e., the encoder and the decoder. The former takes descriptors as discussed above as inputs and maps them into vectors in the latent

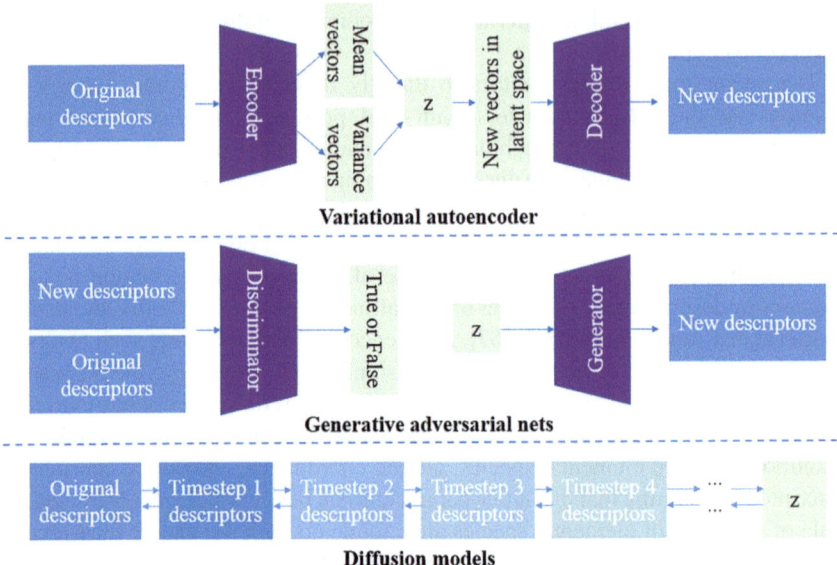

Fig. 2 The architectures of commonly used generative deep learning models, *e.g.*, VAE, GAN, and diffusion model, where z denotes the latent space

space. The vectors in the latent space should satisfy a specific distribution, and they contain all necessary information to reconstruct the inputs. The decoder uses the latent space vectors as input to perform the reconstruction. After training the VAE model, new vectors are randomly sampled in the latent space according to the learned distribution, hence generating new crystal structures [42]. Decoders and encoders are generally neural network models, specifically common types are convolutional neural networks (CNNs) [98] and graphical neural networks (GNNs) [99].

Gomez-Bombarelli et al. applied the VAE model to design molecules [100]. The first successful application of the VAE model to crystal structure design was conducted by Noh et al. [58], where a 2D CNN-based VAE model was used to design new V–O materials with high efficiency. It was demonstrated that the VAE model is more efficient in generating new V–O materials than the global optimization model. The VAE model was subsequently used by Court et al. to generate Heusler alloys, chalcogenides, and binary alloys with various compositions [66]. They adopted the idea of using voxel descriptors, but they constructed a 3D CNN model directly on the voxels without using crystal images [77]. However, this study also pointed out the shortcomings of the VAE model, i.e., the generated structures were mostly not stable and needed to be re-optimized by further DFT calculations.

Since descriptors based on the vector method can be applied to better specify the crystal structures, Ren et al. applied the VAE model with vector descriptors using CNN constructs[68]. Due to a large amount of noise in the vector descriptors generated with the VAE method, only a few portion of the generated descriptors can be transformed into reasonable crystal structures [101]. The efficiency of the VAE model in generating new materials is therefore low, with a success rate of less than 1% [102]. Xie et al. proposed the crystal diffusion VAE (CDVAE) model to address this problem[64]. They assumed that there was a deviation of the structures generated by the VAE model from their stable state, so the feasibility of finding the target materials could be increased by analyzing the noise distribution [103–105]. Such analysis relies on the noise conditional score network (NCSN) model. Besides, CDVAE also takes the stability of the materials into account by exhausting the possibilities of the generated materials, so only optimal structures can be selected [103]. The CDVAE model also demonstrates the possibility of using GNN constructions in the VAE model.

In general, the VAE model can generate complex images and control the hidden space distribution. It is believed that the choice of distribution functions integrates the domain knowledge on the crystalline materials, which is the key to accelerate model training [52]. However, the choice of the distribution functions in VAE significantly affects the model performance, which may lead to a significant drop in the performance of such models.

Conceptually, the GAN model consists of a generator and a discriminator where the generator aims to generate new descriptors from a set of random arrays, while the discriminator will determine whether the distribution of the generated descriptors is statistically consistent with the descriptors of the known crystal material [43, 106]. Thus, the generator tries to ensure that the generated descriptors cannot be identified by the discriminator as generated descriptors by improving the similarity to the existing structures, while the discriminator will be trained to find the difference between the generated and the actual descriptors. The generator and the discriminator are trained to compete against each other simultaneously, and eventually, the generator will be capable of generating descriptors of crystalline materials that are sufficiently good.

The GAN model was first applied to generate crystal structures in the CrystalGAN model by Nouira et al. [57]. They used vectors to describe the structures of binary materials, i.e., A-H and B-H, and used the GAN model to generate ternary structures, A-B-H. Although the generation ratio demonstrated in this work was not high, CrystalGAN successfully generated Ni-Pd-H crystal structures using Ni–H and Pd-H crystal structures as inputs, enabling the generation of heterogeneous crystal structures. Subsequently, Long et al. used 2-dimensional crystal images based on voxelization as inputs and outputs of a GAN model [61]. This method was successful when acting on a specific binary system, Bi-Se, and was able to find new Bi-Se structures that did not exist in the training set, with a decent generation ratio. Kim et al. successfully generated new structures of Mg-Mn–O using the GAN method with higher generation rates compared to the VAE models. This demonstrates the advantages of the GAN method in exploration [70]. Long et al. then embarked on an attempt to use the GAN model for multicomponent systems, successfully generating crystalline materials with various compositions [62]. Nevertheless, the generation ratio dropped in its application of multicomponent systems, mainly due to the limited number of crystal structures in the training dataset for a vast space of compositions.

The GAN model has demonstrated advantages in generating new structures using image-based descriptors, and also has the potential to be combined with vector-based descriptors (no research has been reported). It is worth noting that GAN models learn the distribution of the latent space by themselves implicitly and therefore are not immune to mode collapse. This leads to a decrease in the structural generation ratio of such models.

Diffusion models, which are inspired by non-equilibrium thermodynamics, use denoising to generate data. The process of learning by denoising involves two Markov Chain processes: the forward process and the reverse process. During the forward process, random noise is gradually added to the data at a series of time steps from t_1 to t_n, with samples at the current time step drawn from a Gaussian distribution conditioned on the samples at the previous time step, and the variance of the distribution following a predetermined schedule. After long enough forward time steps, the samples become standard Gaussian distributions. In the reverse process, starting from a standard Gaussian distribution, the noise is reduced at each time step in the backward direction from t_n to t_1 [44, 107]. After training, we can use the DGM to generate data by simply passing randomly sampled noise through the learned denoising process. In fact, the Markov chain process in the algorithm shifts the goal from matching distributions in the data space and the potential space to finding a strategy that describes the direction in which the noisy samples at the current time step are most likely to change towards a steady state within every time step, *i.e.*, what the DGM actually does can be seen as finding a hypothetical potential energy surface which, by inputting a noisy sample and the corresponding timestep, will output a more stabilized state at the next timestep.

Xie et al. employed the score-based network NCSN as a decoder to denoise the perturbed structures to the stable crystal structures [64, 65]. Pakornchote et al. added an extra DDPM model on the top of CDVAE, where CDVAE is used to predict the lattice parameters and the number of atoms in the unit cell, and the DDPM model helps to denoise the fractional coordinates and predict the atomic coordinates [73]. Yang et al. directly applied the DDPM on their UniMat representation for both unconditional and conditional crystal structure generation [74]. Based on the CDVAE descriptors, Xie et al. applied the D3PM, DSM and DDPM with limit distribution for atom type denoising, fractional coordinate denoising and lattice distribution denoising, respectively, where the matching of these three denoising models is by score evaluation using an SE(3)-equivariant GNN named GemNet-dT [75].

The current trends show that crystal structure generation is a complex problem and requires multiple models acting together. In addition, the consideration of whether a crystalline material can be synthesized is an interesting direction, as it has been mostly discussed for molecules [108, 109]. Currently, crystal structure generation models are mostly CNN-/GNN-based VAE and GAN models, but diffusion models are also starting to enter the mainstream. Nevertheless, if transformers can be applied to construct generative deep learning models, their capability will possibly be further enhanced [110, 111].

5   Crystal Structure Constraints

While current generative deep learning models have been mostly applied on generating new crystal structures, integrating the physical property prediction in such models has started to attract more and more attention in recent years. In this regard, the property constraints can be incorporated either as a screening approach or as an optimizing approach, as shown in Fig. 3. The former involves two steps: (1) generation of the new structures and (2) screening structures with desired properties. Noh et al. applied this approach by firstly generating many V_xO_y structures using the VAE model and then selecting those with good thermodynamic stability with formation energies below -0.5 eV/atom [58]. Such an approach allows for exploring a larger chemical space, but it cannot be applied to optimize the target properties during the design process, which dramatically slows down the design of functional crystalline materials. This problem was ameliorated in the work by Court et al. [66], where they generated a large number of crystal structures using the VAE model and used a forward prediction model, i.e., the CGCNN model developed by Xie et al. to predict the properties (including bandgap, mechanical properties, and formation energy) of the generated structures. Such a scheme improved the efficiency of screening functional materials by selecting only crystal structures exhibiting target properties in the generation process [59], without considering the material properties during the training process though.

To address such an issue, Long et al. added material properties in the loss function of the GAN model as a back propagator to achieve joint optimization of material properties and crystal structures [61]. It is noted that modifying the loss function will affect the discriminator's evaluation of the generated structures. For instance, if the formation energies of the generated structures are too high, the value of the loss function will increase and therefore prevents the generation of such structures. Detailed analysis reveals that this approach improves the efficiency of generating the structures with target properties by 100 times, which can significantly improve

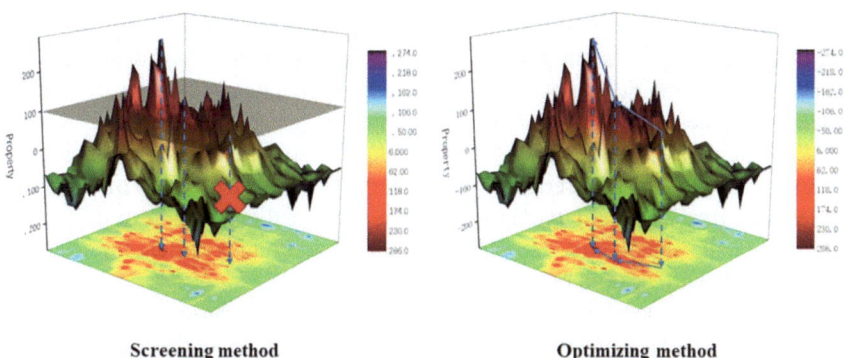

Screening method   **Optimizing method**

Fig. 3 Two methods of adding constraints to the generative models: the screening method (left panel) and the optimizing method (right panel)

the generation ratio of functional crystalline materials [62]. Kim et al. added heat absorption as a constraint in their GAN model to design porous materials, enabling the prediction of porous crystalline materials with heat absorption in a specific range [69]. Yao et al. also added other functional properties to the loss function in the VAE model, leading to the direct design of functional MOF materials [63]. Taking this a step further, Xie et al. introduced the process of structure optimization directly into the generative model. By generating lattice parameters, atomic positions, and element species individually and exhausting the possibilities based on the above constraints, they could predict materials with desired properties [64]. Unlike the cases of GAN and VAE, in diffusion models, the cross-attention layers enable the incorporation of constraints into the joint distribution, thus controlling the direction of structure generation towards distributions with target properties. The corresponding control nets can be added during the initial model training process as done in UniMat [74], or later to fine-tune the pre-trained model as implemented in Mattergen [75].

In short, the essence of property constraints is to balance exploration and exploitation. Unconstrained models can be considered as explorative models, focusing on the design of new crystal structures in the latent space that are distinct from the existing ones [112]. In contrast, constrained models explore the latent space to identify crystal structures with desired properties, emphasizing the design of novel structures based on the available information. According to our research on adding constraints, the balance of exploration and exploitation will significantly affect the performance of generative models, which could be a future research direction for generative deep learning models.

6   Section II: Inverse Design of Microstructure

Microstructure is a general term that describes the geometrical features and topological arrangements at a particular length-scale. e.g., nanoscale or macroscale. That is, it refers to the arrangement of a material's constituent components, including atoms, molecules, grains, and phases, at a microscopic level, which can be formulated as the size, shape, orientation, and distribution of microstructural units including grains, phase boundaries, defects, inclusions, and precipitates. Microstructure plays a crucial role in determining the extrinsic properties of materials, such as their strength, ductility, toughness, hardness, conductivity, magnetization, and corrosion resistance. For example, a material's strength depends on the size and distribution of the grains, while its thermal conductivity depends on the shape and orientation of the grains., Therefore, understanding the morphology of microstructure is critical for designing new materials with desired properties.

Microstructure morphology can be analyzed using both experimental and computational methods. These methods help to identify key features of the microstructure, which can be used to investigate material properties and design new materials with desirable properties. Experimental techniques used for microstructure characterization include optical microscopy, scanning electron microscopy (SEM), transmission electron microscopy (TEM), X-ray diffraction (XRD), electron backscatter diffraction (EBSD), atomic force microscopy (AFM), X-ray computer tomography (XCT), micro computed tomography (micro-CT) scans, and Four-dimensional scanning transmission electron microscopy (4D-STEM), etc. These techniques produce 2D or 3D microstructure images with varying levels of resolution and depth, as summarized in Table 3 (see [131, 132] for more details). Computational methods can be used to further analyze microstructure morphology, including quantification and reconstruction of statistically equivalent microstructures. The features considered in such theoretical methods can be classified into several categories, including statistical functions (e.g., two-point correlation function, linear-path function, and two-point cluster correlation function), deterministic physical descriptors (e.g., volume fraction, total surface area, and number of clusters), statistical physical descriptors (e.g., cluster's nearest center distance and orientation angle of a cluster's principle axis), spectral density function, texture synthesis and multiple-point statistics [133].

Once the explicit or implicit features of microstructure have been characterized, it is possible to reconstruct the microstructure using different methods. For example, if statistical descriptors are used for the characterization, reconstruction can be achieved through optimization using methods such as simulated annealing (SA), genetic algorithms, or other gradient-based algorithms. Alternatively, reconstruction of texture synthesis or multiple-point statistics can be done through Gaussian pyramids [133]. There are also established software tools, such as Dream3D[134] and OptiMic [135], which incorporate these statistical and computational algorithms for simulating microstructure. These tools are designed to facilitate the reconstruction process and provide a platform for analyzing and designing materials with specific microstructures and properties.

While statistical and physical features can be useful for characterizing microstructure morphology, they are limited to specific microstructure systems or morphologies. Additionally, they may overlook important features or relationships within the microstructure, such as small-scale variations or complex interdependencies. Machine learning generative models offer a more comprehensive and flexible approach to microstructure representation and design. These models can learn from large datasets of microstructures and generate new ones that conform to specific constraints or objectives. Moreover, they can uncover complex relationships between the microstructure features and material properties, enabling the design of microstructures with optimized and novel properties. In recent years, generative deep learning models have shown a great promise in advancing microstructural materials science, and are becoming increasingly important for designing new microstructures with tailored properties. Therefore, the development and application of generative deep learning models for microstructure design are a critical area of research that will have significant impact on a wide range of technological fields. In Table 2, we summarize the state-of-the-art microstructure generative models in terms of the representations of microstructure (mainly the data types and feature scales used by the model), the model architectures, the constraints applied during the generation, and the accuracy and interpretability of the models.

Table 3 The experimental measuring approaches of microstructures

Experimental approach	Resolution (typical)	Data Type	Merits	Drawbacks	Comments
Optical microscopy	Micrometers	2D	Simple, widely used	Low resolution	Useful for observing overall morphology and microstructure
Scanning electron microscopy (SEM)	Nanometers	2D	High resolution, surface imaging, compositional analysis	Requires vacuum environment, sample preparation	Widely used in materials science
Transmission electron microscopy (TEM)	Nanometers	2D	High resolution, internal structure and composition, crystal defects	Requires thin samples, complex sample preparation	Useful for studying microstructure and composition at the nanoscale
X-ray diffraction (XRD)	Angstroms	2D	Determines crystal structure and composition, non-destructive	limited to crystalline materials	Useful for identifying and characterizing different materials and their properties
Atomic Force microscopy (AFM)	Nanometers	2D/3D	High resolution, surface topography and roughness, mechanical properties, non-destructive, operates in air and liquids	Slow scanning	Useful for studying surface properties and topography at the nanoscale
Electron backscatter diffraction (EBSD)	Nanometers	2D	Determines crystallographic orientation, non-destructive	Surface analysis only	Useful for studying crystallographic orientation and deformation history
X-ray computer tomography (XCT)	Micrometers	3D	High resolution 3D imaging, internal structure, non-destructive	Requires high dose of X-ray radiation	Useful for imaging biological and porous materials

(continued)

Table 3 (continued)

Experimental approach	Resolution (typical)	Data Type	Merits	Drawbacks	Comments
Micro computed tomography (micro-CT) scan	Micrometers	3D	High resolution 3D imaging, internal structure	Requires X-ray radiation, limited to relatively small samples	Useful for imaging biological and porous materials
Four-dimensional scanning transmission electron microscopy (4D-STEM)	Nanometers	2D	High resolution imaging, electron diffraction	Requires complex data analysis, limited temporal resolution	Useful for studying dynamics at the nanoscale

7 Microstructure Representations

In contrast to the representation of crystal structures, which only requires a quantitative description of atomic positions, atomic species, and crystal lattices, the construction of microstructure descriptors is impeded by the numerous metrics that entail intricate intercorrelations with the desired properties and the vast visual phase space of microstructures. As microstructural morphology becomes more complex, traditional microstructure representations such as volume fractions, Minkowski tensors and spatial correlations [133, 136] become less effective. Contemporary microstructure generation tends to train a machine learning model directly using real or synthetic microstructure images. However, the preparation of microstructure samples is complex and expensive, the existing experimental images are insufficient to comprehensively cover the vast visual phase space. Consequently, a complete construction of microstructure features as microstructure descriptors is currently not feasible [137]. At present, the assessment of microstructural features is largely based on human intuition and varies depending on the specific system being analyzed and the method of measurement employed. However, with the advent of computer vision (CV) technology, it has become possible to quantitatively detect and analyze abstract information within images [138, 139], CV algorithms can be employed to digitize experimental microstructure images and then processed to achieve various research goals. Consequently, utilizing experimental or synthetic digitized microstructure images directly to describe the microstructures is a more feasible and dependable approach compared to constructing statistical microstructure descriptors. This approach addresses the first issue mentioned earlier in this paragraph. In order to gain a better understanding of the visual phase space of microstructures, CV algorithms can be combined with unsupervised generative machine learning models, as discussed above for crystal structure generative deep learning models. The resulting algorithms can extract abstract features from digital crystal images, comprehending the distribution of microstructures, and such generative deep learning models with the learned distributions can be applied to generate new microstructure images in the reversed way with desired physical properties optimized.

To build a high-quality uniform generative model from scratch, a large and diverse dataset with over 50,000 samples is essential. Although established microstructure image databases, such as the NFFA-EUROPE SEM Dataset (~21,169), ASM Micrograph Database (~4100), UHCSDB [123] (~961) and DoITPoMS(~818) provide tens of thousands of high-quality microstructure images and are wildly used for benchmarking new analysis techniques or training machine learning models, the current scale of existing micrograph datasets falls short of the threshold required for a uniform generative model. Therefore, proper pre-processing and data enrichment are vital as they help machine learning models capture and understand the underlying features in the images. Due to the limitations of experimental approaches, there is a deficiency of intermediate data that covers the vast visual phase space. As a result, the current generative model tends to focus on local features by segmenting a large image into small ones or on a small region of the visual phase space with a specialized dataset. To choose between a local generative model trained on specific experimental micrographs or a uniform generative model trained on simulated synthetic augmented micrographs, one must weigh the benefits and trade-offs. The localized generative model provides a precise description of the local data distribution, leading to an accurate understanding of the microstructure-property relationships in the specific system. This approach can help to uncover system-specific mechanisms and theories. In contrast, the uniform generative model provides more opportunities to explore unexplored regions in the visual phase space, increasing the probability of finding potential new materials with desired properties. The following paragraphs will describe specific examples or applications of each approach, combining the summarized information in Table 2 to illustrate their benefits and trade-offs further.

Firstly, for the generative models that focus on local features, the key characteristic is that only a few microstructures (less than 10) exist in the database. The main idea behind local feature generation models is to use convolutional layers to capture the underlying statistical features of the images and then reconstruct or generate the microstructures for specific use. It is worth mentioning that the reconstructed microstructures may differ from the original structures, but the morphological statistical distribution remains the same. Therefore, these models are mainly used for constructing larger microstructures for simulation purposes or recovering defects or blurred regions caused by experiments.

Ref. [116, 118, 119] constructed generation models by randomly sampling regions in a single image, focusing solely on the local features of a specific material compound. These models are primarily aimed at microstructure modeling for simulation calculations or image inpainting to recover blurred regions. Particularly, Chun et al. used a size 3000 × 3000 pixels SEM image of a class V cyclotetramethylenetetranitramine pressed energetic material, and generated a training dataset containing 12,500 images with a size of 161 × 161 by cropping the original image in random position. Fokina et al. tested two systems during their investigation, the SEM image of Alporas aluminum foam and the digital rock by micro-CT. For Alporas aluminum foam, 16,000 size 128×128 images were random cut from the original 751×751 SEM images; Whereas for the digital rock, 10,000 400×400 2D slices of Berea sandstone and 10,240 256×256 2D slices of Ketton limestone were randomly cut from the original micro-CT images separately. Squires et al. used FIB-SEM of a three-phase solid oxide fuel cell anode and a hypoeutectoid steel image from DoIT-PoMS as their benchmark tests for image inpainting, an occluded region was first set and in each training iteration a batch of training images were sampled from the unoccluded region.

Ref. [121–124] also focused on capturing local features but in 3D reconstruction. The same strategy of conducting random segmentations on microstructure was employed to build the training database. The segmented 2D or 3D samples were used to train the generative models, which were then utilized for 3D structure reconstruction for simulation purposes. Specifically, Kench et al. [121] selected 87 out of 818 2D micrographs in the DoITPoMS dataset based on a set of exclusion criteria, and the inpainting technic was applied to remove the scale bars in the images, to maintain the data features to the great extent under the extremely data-scarce situation. Hsu et al. [122] trained their 3D Wasserstein-GAN (WGAN) model using the 3D data of the solid oxide fuel cell (SOFC) anode containing yttria-stabilized zirconia, nickel, and pore phases measured by Xe plasma focused ion beam combined with SEM (Xe PHIB-SEM), with the volume of $110 \times 124 \times 8 \ \mu m^3$, during the training a subvolume of $65 \times 65 \times 65 \ nm^3$ were randomly sampled. Henkes et al. [123] used two datasets to train and test their model performance, one was the dataset of synthetic spherical inclusions microstructure with 32^3 voxels, another was a micro-CT scanned wood-plastic composite microstructure of $950 \times 240 \times 850$ voxels, and then 1000 samples of $64 \times 64 \times 64$ voxels were picked out via Latin-Hypercube sampling for model training. Gayon-Lombardo et al. [124] also used two datasets for model benchmarking, the first was a Li-ion battery cathode measured by XCT with a size of $100.7 \times 100.3 \times 100.3 \ \mu m^3$, the second one is the same SOFC anode dataset as in Ref.[122]. During the training of each dataset, more than 10,000 subvolumes were extracted using an 8-voxel-stride sampling function. It is noteworthy that all current generative models for 3D microstructures focus on local features. This could be attributed to the challenges posed by the scarcity of 3D data and the significant information demand required for the reconstruction process during the model's training.

Secondly, for generative models that focus on small regions of visual phase space, the dataset size lies between hundreds to thousands across several morphologies of a specific compound system. These models can well describe the local morphological distribution of the system and thus often have good interpretive ability in describing the process-structure–property linkage.

Among the papers listed in Table 2, Ref. [115, 117, 120] investigated the performance of generative models in describing the local morphological distribution of microstructures. Lambard et al. [115] used a datasets of 3000 SEM images of 30 ferrite-martensite dual-phase steels with different martensite fractions to train a StyleGAN2 model. Ma et al. [117] used 272 SEM-BSE images of depleted U-10Mo alloy from 10 classes, each class denoted a different processing history. The original images were cropped into 10,080 512 × 512 pixels images with the augmentation including horizontal shift, rotations and horizontal/vertical flipping. The classes information was used for training a random forest (RF) classification model. Lee et al. [120] collected four distinct datasets to evaluate the performance of different GAN models in microstructure generation, including one large dataset containing 19,216 256 × 256 pixels steel micrographs with different magnifications (9216 from OM and 10,000 from SEM), one dataset containing 6231 256 × 256 pixels steel micrographs with × 2000 magnifications (3186 from OM and 3045 from SEM), one dataset of 971 spherical and 1130 wiry SEM micrographs of size 128 × 128 pixels, and one dataset containing 1200 hand-drawn sketches of steel SEM images and another 440 martensite steel SEM images.

Ref. [113, 114, 125, 126] took a step further, they used constrained model to investigate these local regions where more promising properties are expected. Yang et al. [113] used a 5000 synthetic micrograph database of size 128 × 128 pixels generated by GRF model, where the parameters in GRF model were carefully controlled to guarantee the dispersity of the data. Iyer et al. [114] cropped the original 172 steel SEM images in UHCSDB into 7000 size 128 × 128 pixels micrographs. Cang et al. [125] used a small dataset of 200 sandstone micrographs and considered the properties including Young's modulus, diffusivity and fluid permeability to test the performance of their morphology constrained VAE model. Kim et al. [122] used 4000 statistically synthetic dual-phase steel micrographs for their GPR-VAE model performance benchmark. Azqadan et al. [128] utilized 27 AZ80 magnesium alloy components with different casting-forgings to train a DDPM conditioned on different processing parameters. Lee et al. [130] used 300 SAOED composites Micro-CT images to train a unconditioned DGM and generated 5760 random samples. The target properties of 1920 out of 5760 samples were calculated using FE modeling and were used to train a U-Net surrogate model. The target properties of the rest samples were predicted by this U-net model, where the conditioned DGM model was trained based on such a dataset with 5760 samples.

Finally, for uniform generative models, besides the requirement of large datasets, it is also essential to have a stable model to ensure the reliable understanding of global data distribution. In Düreth et al.'s [127] work, they trained their state-of-the-art DDPM on a large dataset consisting of more than 13,000 raw data in 13 different classes, sourced from NFFA-Europe and UHCSDB. Lyu et al. [129] used a 2D database comprising various types of structural features consists of 8000 micrographs and 3D porous material database consisting 3000 microstructures to train their class-conditioned DDPM. These large and diverse datasets on the one hand ensured good coverage of the visual phase space and on the other hand met the large data requirements due to the huge scale and complex diffusion process of the DDPM model.

8 Microstructure Generative Models

According to the analysis above about using VAE and GAN models for the inverse design of crystalline materials, it is clearly that researchers have roughly the same preference for VAE or GAN models. However, the situation changes dramatically for microstructure generative deep learning models. As shown in Table 2, there are only 2 out of 15 investigated papers using VAE models. This is partially because the generated images from VAE are always blurred than those obtained using GAN. Such an issue can be straightforwardly solved in crystalline material inverse design via an extra procedure, for example, DFT or molecular dynamic structural relaxations which can be done in a HTP way. Similar extra treatments are challenging for microstructure, because there are no reliable physical simulations available which can reproduce experimentally available microstructures. Nevertheless, to understand the reason for the blurry generations of VAE, we need to start with the essential difference between the VAE and GAN models in terms of how they describe their data density function.

VAE is designed to learn an explicit density function in the latent space, which serves as an effective feature representation to approximate the true distribution in addition to enabling the generation of new data. That is, VAE also expects to model the data distribution explicitly and obtains the well-defined latent representation which can be used for inference. This is achieved by optimizing the loss function of VAE, which involves maximizing the evidence lower bound of likelihood. The optimization process ensures that the encoded posterior distribution of latent vectors, conditioned on the current samples, is as close as possible to the normal distribution with the help of reparameterization (calculated by the Kullback–Leibler (KL) divergence), and that the decoded results from these vectors with added noises are as close as possible to the original data. However, VAE assumes that all the data obeys a multivariate Gaussian distribution with independent components, which may not hold for real arbitrarily complex distributions. Consequently, this assumption can lead to suboptimal sample quality in the latent space and correspondingly blurred reconstructed results in the physical space.

In Table 2, investigations that utilized VAE were primarily interested in its capacity to construct effective latent representations. Cang et al. [125] introduced a morphology constraint during the training of the VAE model by implementing an additional morphology style loss penalty generated by a pre-trained VGG net. Despite the VGG model being trained on image training sets rather than VAE latent representations, the structural-morphology distribution learned by VGG prevented the mismatching of neighboring latent samples to significantly deviated images. Besides, to prevent the model from generating a cluster distribution, a model collapse loss was also incorporated. Furthermore, the added losses resulted in a reduction in the weight of the KL divergence loss in the total loss, indicating a decrease in the uncertainty considered during training. These two factors combined to guarantee a higher efficiency of training and higher clarity of the micrograph generation of this constrained VAE than normal VAE. Kim et al. [122] employed Gaussian process regression (GPR) to depict the associations between the latent space vectors and mechanical properties. Additionally, they utilized the trained GPR to identify the highest uncertainty points, which served two purposes: firstly to enhance the diversity of the dataset, and secondly to swiftly identify the microstructures with desired target properties.

In contrast to VAE, GANs do not operate using an explicit distribution function. Instead, they learned to generate samples from the training distribution through a zero-sum game between the generator and discriminator. This gives rise to two fundamental differences between GAN and VAE. Firstly, there is no uncertainty in the mapping from the latent space to the samples, and secondly, the inference using latent space representation is problematic. Furthermore, optimizing the zero-sum game using Jensen-Shannon (JS) divergence can lead to instability during the optimization process and often ends up in saddle points. Various attempts have been proposed to address this instability by modifying the model structure, latent space and loss function. This trend can also be seen implicitly in the development of microstructural GAN models.

In the case of 2D microstructural GAN models, Ref. [113, 114, 116, 119] modified the loss function and latent space of the GAN model to tackle their specific problems. Yang et al. [113] incorporated the same style loss and model collapse loss used in Cang et al. [125] into the GAN model, and they also introduced a GP-Hedge Bayesian optimization (BO) to control the sampling in latent space in order to generate microstructures with more promising optical properties. Iyer et al. [114] proposed an auxiliary classifier Wasserstein GAN with gradient penalty (ACWGAN-GP) model for synthesizing steel microstructures under specific processing conditions. To achieve this, they trained an additional classifier alongside the GAN model to ensure the model could differentiate between different processing conditions. As a result, the classifier loss and a gradient penalty loss (to prevent model from collapse during training) were incorporated into the loss function, and an extra processing condition vector was introduced into the latent space. Chun et al. [116] employed the modified latent space and loss function to a simple GAN model, they achieved the morphology generating control of HMX microstructures by extending the loss of GAN into patch-based loss, and applying a combined latent vector (the global morphology parameter sampled from an uniform distribution and the local stochasticity parameter sampled from an uniform random distribution) of each grid. These modifications resulted in a scalable mapping of each grid to some specific region within the model perception in the micrograph, consequently enabled tractable control of the microstructure morphology generation. Squires et al. [119] proposed a deep convolutional GAN (DCGAN) model for the inpainting tasks of recovering defects and unwanted artifacts in micrographs via generator optimization approach or seed optimization approach. To address the requirement of matching boundaries during inpainting, for the generator optimization approach, a content loss function for annulus region was included in the loss function; And for the seed optimization approach, an extra seed optimization was carried out separately when generating new micrographs.

The remaining 2D microstructural generation models primarily addressed their target problems through modifications to the model structure. For example, Fokina et al. [118] utilized a StyleGAN architecture in combination with image quilting technique for their microstructure reconstruction, which resulted in the generation of high-resolution and high-quality micrographs. In the StyleGAN model, a multi-layer perceptron (MLP) was used to learn an affine transform to project the latent vectors into an intermediate latent space that is disentangled from the data distribution. This transform was then applied to each convolutional block through a normalization algorithm called adaptive instance normalization to achieve better control over the generation process. Similarly, Lambard et al. [115] employed an updated StyleGAN2 architecture on a small dataset consists of 3000 dual-phase steel micrographs. An adaptive discriminator augmentation (ADA) was also implemented to stabilize the training of GAN model within limited data regime. This allowed for the generation of microstructures with good quality and good interpolations between microstructures. Ma et al. [117] tested the performance and interpretability of two different GAN models: the progressively growing GAN (pg-GAN) and the pix2pix GAN. The pg-GAN model trains progressively from low to high resolution data layer by layer, using reliable weights obtained from previous layer as a weighted residual during the training of the next layer. The pix2pix GAN uses a labeled image as a constraint to a U-net type generator to achieve point-to-point mapping. Finally, Lee et al. [120] employed DCGAN for generating new virtual micrographs, resulting in highly realistic and visually appealing microstructures. They also utilized Cycle GAN and Pix2Pix GAN to conduct style transform between OM and SEM or sketches and SEM images.

In 3D microstructure generation, due to the further reduction of data amount compared to that of 2D microstructure, the aims were mainly focused on reconstructing larger 3D microstructure from the data of a single microstructure, where only the local features were required. Therefore, the original GAN architecture was sufficient for handling such tasks. For example, Gayon-Lombardo et al. [124] applied a DCGAN in their study, and Hsu et al. [122] and Henkes et al. [123] used the WGAN with the Wasserstein distance as the loss function to guarantee the smooth gradient during the model training. They also explored the performance of an S4 equivariant network that considers rotational equivariance. While the S4 CNN showed better quality in generating microstructure, it required a significant amount of computational time and memory.

As we can see from previous part, generative models have advanced considerably in the last few years, with efforts focused either on improving the performance of VAE within reasonable computational limits via better variational posteriors, or on improving the stability of GAN through better loss functions and discriminators. However, a fundamental question arises: Can we reap the benefits of both VAE and GAN, rather than being forced to choose between Scylla and Charybdis? Specifically, is it possible to create a generative deep learning model that trains a simple objective function and is compatible with highly expressive neural networks? The DDPM model proposed by Ref. [44, 107] modified the goal of generator from 'mapping standard Gaussian distributions to data distributions' to 'fitting the inverse process of a defined Gaussian Markov Chain which maps standard Gaussian distributions to data distributions'. In this way, the generator only needs to match each small step of the inverse process corresponding to the forward Markov Chain, rather than optimizing the generator and the variational posterior/discriminator simultaneously. The work in Ref. [127–130] has demonstrated the great potential of DDPM on capturing complex microstructural morphologies and controllable new micrograph generations.

9 Microstructure Constraints

Microstructure plays a critical role in determining the mechanical, thermal, electrical, and magnetic properties of materials. The morphology properties of microstructures, such as grain size, shape, orientation, distribution, and phase composition, as well as porosity, connectivity, and tortuosity of the pore network, which have significant influence on material properties such as strength, ductility, electrical and magnetic behavior, can be characterized using a variety of experimental techniques. For instance, the mechanical properties of microstructures, such as strength, ductility, toughness, and fatigue resistance, can be characterized using techniques like nanoindentation, tensile testing, and fatigue testing. The thermal properties of microstructures, such as thermal conductivity, specific heat capacity, and thermal expansion coefficient, can be characterized using techniques such as laser flash analysis, differential scanning calorimetry (DSC), and thermomechanical analysis (TMA). The electrical properties of microstructures, such as electrical conductivity, resistivity, and dielectric constant, can be characterized using techniques such as impedance spectroscopy and dielectric spectroscopy. The magnetic properties of microstructures, such as magnetization, coercivity, and remanence, can be characterized using magnetic measurements techniques, such as vibrating sample magnetometry (VSM) and magnetic force microscopy (MFM). Understanding the relationship between microstructure and material properties is crucial for optimizing the performance of materials in various applications, including energy storage, catalysis, and biomaterials.

In addition to experimental methods, simulation methods can also be used to evaluate extrinsic properties driven by microstructure, usually multiscale simulations in order to capture the underlying mechanisms and to bridge to device performance. For instance, FE analysis is a widely used simulation method for evaluating mechanical properties of microstructures, such as stress distribution and deformation behavior. Molecular dynamics (MD) simulations can be used to study thermal and mechanical properties of microstructures at the atomistic level, providing insights into properties such as thermal conductivity and strength. Phase-field simulations can be applied to model the evolution of microstructure during various processing routes, such as solidification or annealing, and predict the resulting microstructure properties. Monte Carlo simulations can be used to simulate the behavior of a system with randomly distributed variables, such as the distribution of pores in a microstructure, and predict the resulting properties. These simulation methods are increasingly being used in conjunction with experimental techniques to gain a more comprehensive understanding of extrinsic properties and their relationship to microstructure.

Despite the plethora of experimental and simulation methods available for evaluating such microstructure-derived extrinsic properties, there is still a lack of a comprehensive process-microstructure-property database comparable to those available for crystalline materials, such as the Materials Project or the Inorganic Crystal Structure Database (ICSD). The diversity and complexity of microstructures, as well as the vast number of morphologies that can influence their properties, make it challenging to develop a complete and accurate database. As a consequence, apart from the screening method or optimizing method like crystal constrains, microstructure generative models also developed active learning type constrains on the top of these two methods to generate new microstructures with desired properties. The development of such microstructure constraint generative models represents a promising avenue for advancing the field of microstructure design and understanding the relationships between microstructure and extrinsic properties.

For microstructure-property constraints, Cang et al. [125] added a style penalty evaluated by a pre-trained VGG net and Gram matrices to the VAE. It is showed that adding a meaningful additional loss can greatly improve the performance of the VAE, providing theoretical support for the application of physical constraints in generative models. Yang et al. [113] transplanted the same type loss function to a GAN model and added GP-Hedge evaluation to actively generate new structures with desired optical properties. Kim et al. [126] used the same GPR active learning strategy on a VAE model to establish the relationship between microstructure and mechanical properties. Lyu et al. [129] used a class-conditioned DDPM to generate porous 3D microstructures with different permeabilities. Lee et al. [130] embedded volume fraction, effective elastic modulus and light intensity change ratio as three controlling properties into conditional DGM for the inverse design of mechanoluminescence particle composites.

In addition to the microstructure-property linkage, the processing-microstructure mapping can also be used as a constraint to the microstructure generative model. This relationship describes how the microstructure is influenced by the processing history, such as the thermal treatment conditions. By incorporating such processing information into generative models, a more accurate prediction of the resulting microstructure and its properties can be achieved. This can be done by treating the processing conditions as different classes to the microstructures and introducing a classifier into the generative model. The relationship between the processing conditions and resulting microstructures can be learned from a large dataset collected through either experimental or multi-physics simulation approaches, such as FE analysis or MD. The resulting generative model can then be used to predict new microstructures under different processing conditions and to optimize processing conditions for optimal microstructures.

There have been various attempts to incorporating processing-structure constraints into generative models. For screening-type constraints, Ma et al. [117] used a random forest classifier to predict processing conditions and evaluate microstructure representations. In contrast, Iyer et al. [114] incorporated processing constraints in an optimizing way, predicting different cooling methods using an additional classifier built into the discriminator and introducing corresponding scoring loss and latent vectors in the GAN architecture. The work of Düreth et al. [114] found that DDPM could generate different classes of microstructures by comparing descriptors such as spatial three-point correlations or Gram matrices, without the need for an explicit classifier in the DDPM model. Azqadan [128] applied the same class-conditioned DDPM structure to AZ80 magnesium alloy to achieve the processing parameter control without any explicit classifier. This is because the MC process used in DDPM naturally ensures a healthy and traceable distribution mapping, which provides good classification ability. These methods enable microstructure generation under different processing conditions and complete major parts in the closed-loop of CPSP inverse design, providing a more comprehensive understanding of the mechanisms behind microstructure formation and its impact on material properties.

10 Section III: Challenges and Outlook

Given the strong interest of the materials science and relevant communities and significant progresses on data-driven acceleration of materials design, the inverse design strategy offers an opportunity to tackle the CPSP relationships in an innovative way, boosted by the application of generative deep learning and Bayesian optimization. However, there are still many challenges to be properly handled.

For the crystal structure inverse design, the low success rate in designing crystalline materials can be attributed to several factors. Firstly, the exploration space of materials that humans can investigate is limited. Traditional experimental methods for discovering new materials are time-consuming and resource-costly, thereby restricting the number of materials that can be explored. Additionally, the complex nature of crystal structures makes it challenging to synthesize new materials even with predictions made available based on automated inverse design methods. Secondly, an essential challenge in crystal structure inverse design is the development of effective descriptors. The existing descriptors may not capture all the critical information needed, e.g., how to properly describe and impose the crystallographic symmetries including the point group and the translational symmetry (i.e., 3D periodicity) during the crystal structure generation. Thirdly, the efficiency of generative models is a critical factor in crystal structure inverse design. Generative models must be able to sample a vast space of possible crystal structures, identify promising candidates with desired properties, and fine tune themselves with the examined new structures efficiently. Therefore, the updating of generative models that can better describe the latent-real space distribution is of vital importance.

Similar challenges hold for the microstructure inverse design such as limited exploration space and high demand for efficient models. Although the use of synthetic micrographs can help to expand the database and overcome the shortage of data, the subtle differences between the real and synthetic microstructures and the complex parameter settings during the machine learning simulations can lead to biased distributions. From the physical point of view, the formation and evolution of various microstructures is driven by the thermodynamic and diffusive processes depending on the initial compositions and heat treatments. That is, it is critical to evaluate the composition – processing – (micro-)structure links in the CPSP chain by either systematic experimental investigations or quantitative simulations of such processes, e.g., via phase field modelling [140]. This entails the establishment of an inverse design paradigm for the composition – processing – (micro-)structure links. Such a paradigm is valuable so that extra physical constraints can be applied in the generative models, giving rise to better efficiency when dealing with deficient datasets. On the other hand, it makes a lot of sense to curate digitized microstructure database following the FAIR principles [141], so that generative models can be developed to minimize the differences between the real and synthetic microstructures.

Thus, to address the challenges in the development of generative deep learning models for both crystal structures and microstructures, it is crucial to improve the quality and diversity of the available data. While experimental data collection remains an essential approach, the high resource- and time–cost involved in syntheses and measurements make it impractical to rely solely on the experiments to increase the amount of data. As an alternative, one possible solution is to establish a synthetic-real data linkage using style-transfer or multi-fidelity ML models. This approach allows for the modelling of more real structures using synthetic ones, or the simulation of more properties using real structures. Additionally, data mining of previous results using generative pre-trained models, such as GPT4 [142], has proven to be a useful tool in collecting and analyzing data and images from papers and technical reports, leading to the development of a homogeneous and extensive database. Ultimately, improving the quality and diversity of data will enhance the performance and accuracy of generative models, allowing for more efficient and effective applications in crystal and microstructure inverse design. But in contrast to fields like images or natural language, where vast amounts of data are generated daily on the internet, the natural sciences often struggle with limited data availability, even with the aid of the previously mentioned data augmentation methods. Therefore, in the short term, relying solely on enhancing the dataset to improve the expressive power of structural generation models is not a practical solution. Instead, we may need to explore better self-supervised learning methods to overcome this challenge.

In terms of descriptors and representations, crystal structures present challenges due to their large number of atoms and lattice sizes, which exceed the limits for voxel-based descriptors. Including more materials without reducing reconstruction accuracy can lead to memory issues. Vector-based descriptors also face challenges due to their inherent heterogeneity, requiring significant zero padding. Additionally, the meaningful atomic positions due to the periodicity in crystal structures make vector-based descriptors difficult to use. The graph theory methods have not fully addressed the challenge of reconstructing crystal structures, but optimization during reconstruction may be a possible solution. For both types of generative models, physics-informed descriptors that map between the structures and properties are crucial for success. These descriptors should be explainable and understandable, providing insights into the underlying physical mechanisms governing processing-structure–property relationships.

From the modelling perspective, the learning speed of generative models, being VAE or GAN, is not yet fast enough. In order to improve the efficiency of generative model structures, it is necessary to incorporate more physics. Quantum machine learning shows promising acceleration effects for unsupervised learning model training, which may help to solve the problem of slow training [143]. The diffusion models simplify the training objective function, make the generative models much more stable and compatible with highly expressive neural networks. In addition to the advancements in model structures, current transformer architectures have demonstrated promising results in terms of enhancing data comprehension. For instance, the Graphormer [144] applies the transformer architecture to the GNN to tackle over-smoothing problems, leading to a significant improvement in the GNN's expressive ability. Therefore, there is still a considerable scope for further improvements through the combination and reconfiguration of model architectures. Furthermore, multiscale models that can connect crystal structures and microstructures can provide a more comprehensive understanding of the processing-structure–property relationships. One possible approach to achieving this is to use the weight from the crystal generative model as additional information in the low-depth layer of the microstructure generative model.

All of the aforementioned approaches provide an opportunity to develop a large, general model similar to GPT and generalize all tasks within this model with the assistance of domain knowledge models. Furthermore, with a quantitative mapping of the CPSP relationships, generative deep learning models can make efficient and reliable predictions, resulting in an effective surrogate model which can be combined with the experimental investigations to further improve the closed-loop adaptive design approach. In particular, the resulting inverse design strategy is promising for the realization of future autonomous experimentation for inorganic solid state materials [145], which will significantly accelerate the materials discoveries and utilizations. Such a promising paradigm involves leveraging a generic generative model to predict possible new materials with desired properties and the processing conditions to synthesize and optimize the microstructure, as well as using robotic systems to conduct high-throughput experimentation, enabling the rapid exploration of a vast search space of synthesis conditions and material compositions. By analyzing the experimental data, the generative model can be trained and improved, forming a closed loop that guides subsequent rounds of autonomous experimentation. In this regard, high-throughput DFT can be further accelerated using ML to acquire sufficient training data (see the chapter by Thorsten Deilmann in this volume). Moreover, with the emergent quantum machine learning (see chapter by Ivana Nikoloska), it is hoped that the generative deep learning approaches can be further accelerated to explore a more comprehensive latent space. It is also noted that the generic inverse problem can all be tackled using such approaches, including the classical modelling such as classical DFT theory (see chapter by Alessandro Simon/Martin Oettel). Therefore, we believe the potential for the generative deep learning approaches to revolutionize the field of materials discovery and engineering is enormous and disruptive, as they explore the latent space beyond the known design space, hence facilitating the development and employment of novel materials with unprecedented properties for various applications.

Acknowledgements This work is funded by the Deutsche Forschungsgemeinschaft (DFG, German Research Foundation)—Project-ID 405553726—TRR 270 and Project-ID 443703006—CRC 1487 Iron, upgraded. The authors gratefully acknowledge the computing time provided to them on the high-performance computer Lichtenberg at the NHR Centers NHR4CES at TU Darmstadt.

References

1. Chen A, Zhang X, Zhou Z (2020) Machine learning: Accelerating materials development for energy storage and conversion. InfoMat 2:553–576
2. Agrawal A and Choudhary A 2016 Perspective: Materials informatics and big data: Realization of the "fourth paradigm" of science in materials science *APL Materials* **4** 053208
3. Huo H, Bartel CJ, He T, Trewartha A, Dunn A, Ouyang B, Jain A, Ceder G (2022) Machine-Learning Rationalization and Prediction of Solid-State Synthesis Conditions *Chem.* Mater 34:7323–7336
4. Jannesari Ladani L (2021) Applications of artificial intelligence and machine learning in metal additive manufacturing. J. Phys. Mater. 4:042009
5. Schmidt J, Marques M R G, Botti S and Marques M A L 2019 Recent advances and applications of machine learning in solid-state materials science *npj Computational Materials* **5** 83
6. Matouš K, Geers MGD, Kouznetsova VG, Gillman A (2017) A review of predictive nonlinear theories for multiscale modeling of heterogeneous materials. J Comput Phys 330:192–220
7. Liu Z, Wang Y, Cai L, Cheng Q, Zhang H (2016) Design and manufacturing model of customized hydrostatic bearing system based on cloud and big data technology. Int J Adv Manuf Technol 84:261–273
8. Butler KT, Davies DW, Cartwright H, Isayev O, Walsh A (2018) Machine learning for molecular and materials science. Nature 559:547–555
9. de Pablo J J, Jones B, Kovacs C L, Ozolins V and Ramirez A P 2014 The Materials Genome Initiative, the interplay of experiment, theory and computation *Current Opinion in Solid State and Materials Science* **18** 99–117
10. de Pablo JJ, Jackson NE, Webb MA, Chen L-Q, Moore JE, Morgan D, Jacobs R, Pollock T, Schlom DG, Toberer ES, Analytis J, Dabo I, DeLongchamp DM, Fiete GA, Grason GM, Hautier G, Mo Y, Rajan K, Reed EJ, Rodriguez E, Stevanovic V, Suntivich J, Thornton K, Zhao J-C (2019) New frontiers for the materials genome initiative npj Comput Mater 5:1–23
11. Sanchez-Lengeling B, Aspuru-Guzik A (2018) Inverse molecular design using machine learning: Generative models for matter engineering. Science 361:360–365
12. Kaufmann K, Zhu C, Rosengarten AS, Maryanovsky D, Harrington TJ, Marin E, Vecchio KS (2020) Paradigm shift in electron-based crystallography via machine learning. Science 367:564–568
13. Tolle K M, Tansley D S W and Hey A J G 2011 The Fourth Paradigm: Data-Intensive Scientific Discovery [Point of View] *Proceedings of the IEEE* **99** 1334–7
14. Noh J, Gu G H, Kim S and Jung Y 2020 Machine-enabled inverse design of inorganic solid materials: promises and challenges *Chem. Sci.* **11** 4871–81
15. Kirklin S, Saal J E, Meredig B, Thompson A, Doak J W, Aykol M, Rühl S and Wolverton C 2015 The Open Quantum Materials Database (OQMD): assessing the accuracy of DFT formation energies *npj Computational Materials* **1** 1–15
16. Marzari N (2021) Electronic-structure methods for materials design Nature Materials 20:14

17. Gjerding M N, Taghizadeh A, Rasmussen A, Ali S, Bertoldo F, Deilmann T, Knøsgaard N R, Kruse M, Larsen A H, Manti S, Pedersen T G, Petralanda U, Skovhus T, Svendsen M K, Mortensen J J, Olsen T and Thygesen K S 2021 Recent progress of the Computational 2D Materials Database (C2DB) *2D Mater.* **8** 044002
18. Haastrup S, Strange M, Pandey M, Deilmann T, Schmidt P S, Hinsche N F, Gjerding M N, Torelli D, Larsen P M, Riis-Jensen A C, Gath J, Jacobsen K W, Mortensen J J, Olsen T and Thygesen K S 2018 The Computational 2D Materials Database: high-throughput modeling and discovery of atomically thin crystals *2D Mater.* **5** 042002
19. Hjorth Larsen A, Jørgen Mortensen J, Blomqvist J, Castelli I E, Christensen R, Dułak M, Friis J, Groves M N, Hammer B, Hargus C, Hermes E D, Jennings P C, Bjerre Jensen P, Kermode J, Kitchin J R, Leonhard Kolsbjerg E, Kubal J, Kaasbjerg K, Lysgaard S, Bergmann Maronsson J, Maxson T, Olsen T, Pastewka L, Peterson A, Rostgaard C, Schiøtz J, Schütt O, Strange M, Thygesen K S, Vegge T, Vilhelmsen L, Walter M, Zeng Z and Jacobsen K W 2017 The atomic simulation environment—a Python library for working with atoms *Journal of Physics: Condensed Matter* **29** 273002
20. Mathew K, Montoya J H, Faghaninia A, Dwarakanath S, Aykol M, Tang H, Chu I, Smidt T, Bocklund B, Horton M, Dagdelen J, Wood B, Liu Z-K, Neaton J, Ong S P, Persson K and Jain A 2017 Atomate: A high-level interface to generate, execute, and analyze computational materials science workflows *Computational Materials Science* **139** 140–52
21. Pizzi G, Cepellotti A, Sabatini R, Marzari N and Kozinsky B 2016 AiiDA: automated interactive infrastructure and database for computational science *Computational Materials Science* **111** 218–30
22. Zhang H (2021) High-throughput design of magnetic materials *Electron.* Struct 3:033001
23. Mockus J 2012 *Bayesian Approach to Global Optimization: Theory and Applications* (Springer Science & Business Media)
24. Li Q, Zhou D, Zheng W, Ma Y, Chen C (2013) Global Structural Optimization of Tungsten Borides *Phys.* Rev Lett 110:136403
25. Oganov A R, Ma Y, Lyakhov A O, Valle M and Gatti C 2010 Evolutionary Crystal Structure Prediction and Novel High-Pressure Phases *High-Pressure Crystallography* NATO Science for Peace and Security Series B: Physics and Biophysics ed E Boldyreva and P Dera (Dordrecht: Springer Netherlands) pp 293–323
26. Oganov AR, Lyakhov AO, Valle M (2011) How Evolutionary Crystal Structure Prediction Works—and Why *Acc.* Chem Res 44:227–237
27. Oganov AR, Pickard CJ, Zhu Q, Needs RJ (2019) Structure prediction drives materials discovery. Nat Rev Mater 4:331–348
28. Allahyari Z and Oganov A R 2020 Coevolutionary search for optimal materials in the space of all possible compounds *npj Computational Materials* **6** 1–10
29. Lyakhov AO, Oganov AR (2011) Evolutionary search for superhard materials: Methodology and applications to forms of carbon and TiO$\{\}\_\{2\}$ *Phys.* Rev B 84:092103
30. Simon A and Oettel M 2024 Machine Learning approaches to classical density functional theory
31. Simon A, Weimar J, Martius G, Oettel M (2024) Machine Learning of a Density Functional for Anisotropic Patchy Particles. J Chem Theory Comput 20:1062–1077
32. Nikoloska I and Simeone O 2022 Bayesian Active Meta-Learning for Black-Box Optimization *2022 IEEE 23rd International Workshop on Signal Processing Advances in Wireless Communication (SPAWC)* 2022 IEEE 23rd International Workshop on Signal Processing Advances in Wireless Communication (SPAWC) pp 1–5
33. Lookman T, Balachandran P V, Xue D and Yuan R 2019 Active learning in materials science with emphasis on adaptive sampling using uncertainties for targeted design *npj Computational Materials* **5**
34. Xue D, Balachandran PV, Hogden J, Theiler J, Xue D, Lookman T (2016) Accelerated search for materials with targeted properties by adaptive design. Nat Commun 7:11241
35. Yuan R, Liu Z, Balachandran P V, Xue D, Zhou Y, Ding X, Sun J, Xue D and Lookman T 2018 Accelerated Discovery of Large Electrostrains in BaTiO₃ -Based Piezoelectrics Using Active Learning *Advanced Materials* **30** 1702884

36. Balachandran P V, Kowalski B, Sehirlioglu A and Lookman T 2018 Experimental search for high-temperature ferroelectric perovskites guided by two-step machine learning *Nature Communications* **9**

37. Wen C, Zhang Y, Wang C, Xue D, Bai Y, Antonov S, Dai L, Lookman T, Su Y (2019) Machine learning assisted design of high entropy alloys with desired property. Acta Mater 170:109–117

38. Rao Z, Tung P-Y, Xie R, Wei Y, Zhang H, Ferrari A, Klaver T P C, Körmann F, Sukumar P T, Kwiatkowski da Silva A, Chen Y, Li Z, Ponge D, Neugebauer J, Gutfleisch O, Bauer S and Raabe D 2022 Machine learning–enabled high-entropy alloy discovery *Science* **378** 78–85

39. Zhou Z-H 2021 *Machine Learning* (Springer Nature)

40. El Naqa I and Murphy M J 2015 What Is Machine Learning? *Machine Learning in Radiation Oncology: Theory and Applications* ed I El Naqa, R Li and M J Murphy (Cham: Springer International Publishing) pp 3–11

41. Carleo G, Cirac I, Cranmer K, Daudet L, Schuld M, Tishby N, Vogt-Maranto L, Zdeborová L (2019) Machine learning and the physical sciences *Rev.* Mod Phys 91:045002

42. Kingma D P and Welling M 2014 Auto-Encoding Variational Bayes arXiv:1312.6114 *[cs, stat]*

43. Goodfellow I, Pouget-Abadie J, Mirza M, Xu B, Warde-Farley D, Ozair S, Courville A and Bengio Y 2014 Generative Adversarial Nets *Advances in Neural Information Processing Systems* vol 27 (Curran Associates, Inc.)

44. Ho J, Jain A and Abbeel P 2020 Denoising Diffusion Probabilistic Models

45. Austin J, Johnson D D, Ho J, Tarlow D and Berg R van den 2023 Structured Denoising Diffusion Models in Discrete State-Spaces

46. Ho TK (1998) The random subspace method for constructing decision forests. IEEE Trans Pattern Anal Mach Intell 20:832–844

47. Bhavsar H and Panchal M H 2012 A review on support vector machine for data classification *Int. J. Adv. Res. Comput. Eng. Technol* 185–9

48. Ding S, Li H, Su C, Yu J, Jin F (2013) Evolutionary artificial neural networks: a review. Artif Intell Rev 39:251–260

49. Vaswani A, Shazeer N, Parmar N, Uszkoreit J, Jones L, Gomez A N, Kaiser Ł and Polosukhin I 2017 Attention is All you Need *Advances in Neural Information Processing Systems* vol 30 (Curran Associates, Inc.)

50. Deb K and Deb K 2014 Multi-objective Optimization *Search Methodologies: Introductory Tutorials in Optimization and Decision Support Techniques* ed E K Burke and G Kendall (Boston, MA: Springer US) pp 403–49

51. Choudhary K, DeCost B, Chen C, Jain A, Tavazza F, Cohn R, WooPark C, Choudhary A, Agrawal A, Billinge S J L, Holm E, Ong S P and Wolverton C 2021 Recent Advances and Applications of Deep Learning Methods in Materials Science arXiv:2110.14820 *[cond-mat, physics:physics]*

52. Lopez-Alvis J, Laloy E, Nguyen F, Hermans T (2021) Deep generative models in inversion: a review and development of a new approach based on a variational autoencoder. Comput Geosci 152:104762

53. Zunger A (2018) Inverse design in search of materials with target functionalities Nat Rev Chem 2:1–16

54. Lu Z (2021) Computational discovery of energy materials in the era of big data and machine learning: A critical review. Materials Reports: Energy 1:100047

55. Lee S, Byun H, Cheon M, Kim J and Lee J H 2021 Machine learning-based discovery of molecules, crystals, and composites: A perspective review *Machine learning* **38** 12

56. Alberi K, Nardelli MB, Zakutayev A, Mitas L, Curtarolo S, Jain A, Fornari M, Marzari N, Takeuchi I, Green ML, Kanatzidis M, Toney MF, Butenko S, Meredig B, Lany S, Kattner U, Davydov A, Toberer ES, Stevanovic V, Walsh A, Park N-G, Aspuru-Guzik A, Tabor DP, Nelson J, Murphy J, Setlur A, Gregoire J, Li H, Xiao R, Ludwig A, Martin LW, Rappe AM, Wei S-H, Perkins J (2018) The 2019 materials by design roadmap. J Phys D: Appl Phys 52:013001

57. Nouira A, Sokolovska N and Crivello J-C 2019 CrystalGAN: Learning to Discover Crystallographic Structures with Generative Adversarial Networks arXiv:1810.11203 *[cs, stat]*
58. Noh J, Kim J, Stein HS, Sanchez-Lengeling B, Gregoire JM, Aspuru-Guzik A, Jung Y (2019) Inverse Design of Solid-State Materials via a Continuous Representation Matter 1:1370–1384
59. Xie T, Grossman JC (2018) Crystal Graph Convolutional Neural Networks for an Accurate and Interpretable Prediction of Material Properties *Phys.* Rev Lett 120:145301
60. Park CW, Wolverton C (2020) Developing an improved crystal graph convolutional neural network framework for accelerated materials discovery *Phys.* Rev. Materials 4:063801
61. Long T, Fortunato NM, Opahle I, Zhang Y, Samathrakis I, Shen C, Gutfleisch O, Zhang H (2021) Constrained crystals deep convolutional generative adversarial network for the inverse design of crystal structures npj Comput Mater 7:1–7
62. Long T, Zhang Y, Fortunato N M, Shen C, Dai M and Zhang H 2021 Inverse design of crystal structures for multicomponent systems arXiv:2104.08040
63. Yao Z, Sánchez-Lengeling B, Bobbitt NS, Bucior BJ, Kumar SGH, Collins SP, Burns T, Woo TK, Farha OK, Snurr RQ, Aspuru-Guzik A (2021) Inverse design of nanoporous crystalline reticular materials with deep generative models Nat Mach Intell 3:76–86
64. Xie T, Fu X, Ganea O-E, Barzilay R and Jaakkola T 2021 Crystal Diffusion Variational Autoencoder for Periodic Material Generation arXiv:2110.06197 *[cond-mat, physics:physics]*
65. Wines D, Xie T and Choudhary K 2023 Inverse design of next-generation superconductors using data-driven deep generative models
66. Court CJ, Yildirim B, Jain A, Cole JM (2020) 3-D Inorganic Crystal Structure Generation and Property Prediction via Representation Learning. J Chem Inf Model 60:4518–4535
67. Ren Z, Tian SIP, Noh J, Oviedo F, Xing G, Li J, Liang Q, Zhu R, Aberle AG, Sun S, Wang X, Liu Y, Li Q, Jayavelu S, Hippalgaonkar K, Jung Y, Buonassisi T (2022) An invertible crystallographic representation for general inverse design of inorganic crystals with targeted properties Matter 5:314–335
68. Ren Z, Noh J, Tian S, Oviedo F, Xing G, Liang Q, Aberle A, Liu Y, Li Q, Jayavelu S, Hippalgaonkar K, Jung Y and Buonassisi T 2020 Inverse design of crystals using generalized invertible crystallographic representation arXiv:2005.07609 *[cond-mat, physics:physics]*
69. Kim B, Lee S and Kim J 2020 Inverse design of porous materials using artificial neural networks *Science Advances*
70. Kim S, Noh J, Gu G H, Aspuru-Guzik A and Jung Y 2020 Generative Adversarial Networks for Crystal Structure Prediction *ACS Cent. Sci.* **6** 1412–20
71. Zhao Y, Al-Fahdi M, Hu M, Siriwardane EMD, Song Y, Nasiri A, Hu J (2021) High-Throughput Discovery of Novel Cubic Crystal Materials Using Deep Generative Neural Networks. Advanced Science 8:2100566
72. Zhao Y, Siriwardane EMD, Wu Z, Fu N, Al-Fahdi M, Hu M, Hu J (2023) Physics guided deep learning for generative design of crystal materials with symmetry constraints npj Comput Mater 9:1–12
73. Pakornchote T, Choomphon-anomakhun N, Arrerut S, Atthapak C, Khamkaeo S, Chotibut T, Bovornratanaraks T (2024) Diffusion probabilistic models enhance variational autoencoder for crystal structure generative modeling Sci Rep 14:1275
74. Yang M, Cho K, Merchant A, Abbeel P, Schuurmans D, Mordatch I and Cubuk E D 2023 Scalable Diffusion for Materials Generation
75. Zeni C, Pinsler R, Zügner D, Fowler A, Horton M, Fu X, Shysheya S, Crabbé J, Sun L, Smith J, Nguyen B, Schulz H, Lewis S, Huang C-W, Lu Z, Zhou Y, Yang H, Hao H, Li J, Tomioka R and Xie T 2024 MatterGen: a generative model for inorganic materials design
76. Debnath A, Krajewski A M, Sun H, Lin S, Ahn M, Li W, Priya S, Singh J, Shang S, Beese A M, Liu Z-K and Reinhart W F 2021 Generative deep learning as a tool for inverse design of high-entropy refractory alloys arXiv:2108.12019
77. Sohn K, Lee H and Yan X 2015 Learning Structured Output Representation using Deep Conditional Generative Models *Advances in Neural Information Processing Systems* vol 28 (Curran Associates, Inc.)

78. Ghiringhelli LM, Vybiral J, Levchenko SV, Draxl C, Scheffler M (2015) Big Data of Materials Science: Critical Role of the Descriptor *Phys.* Rev Lett 114:105503
79. Seko A, Togo A and Tanaka I 2017 Descriptors for Machine Learning of Materials Data arXiv:1709.01666
80. Ward L, Agrawal A, Choudhary A and Wolverton C 2016 A general-purpose machine learning framework for predicting properties of inorganic materials *npj Computational Materials* 2 1–7
81. Stanev V, Oses C, Kusne A G, Rodriguez E, Paglione J, Curtarolo S and Takeuchi I 2018 Machine learning modeling of superconducting critical temperature *npj Computational Materials* 4 1–14
82. Nelson J, Sanvito S (2019) Predicting the Curie temperature of ferromagnets using machine learning *Phys.* Rev. Materials 3:104405
83. Behler J (2011) Atom-centered symmetry functions for constructing high-dimensional neural network potentials. J Chem Phys 134:074106
84. Bartók AP, Kondor R, Csányi G (2013) On representing chemical environments *Phys.* Rev B 87:184115
85. Himanen L, Jäger M O J, Morooka E V, Federici Canova F, Ranawat Y S, Gao D Z, Rinke P and Foster A S 2020 DScribe: Library of descriptors for machine learning in materials science *Computer Physics Communications* **247** 106949
86. Hoffmann J, Maestrati L, Sawada Y, Tang J, Sellier J M and Bengio Y 2019 Data-Driven Approach to Encoding and Decoding 3-D Crystal Structures arXiv:1909.00949 *[cond-mat, physics:physics, stat]*
87. Monaco H L, Artioli G, Viterbo D, Ferraris G, Gilli G, Zanotti G and Catti M 2002 *Fundamentals of Crystallography* (Oxford University Press)
88. Dan Y, Zhao Y, Li X, Li S, Hu M and Hu J 2020 Generative adversarial networks (GAN) based efficient sampling of chemical space for inverse design of inorganic materials *npj Comput Mater* **6** 84
89. Sawada Y, Morikawa K and Fujii M 2019 Study of Deep Generative Models for Inorganic Chemical Compositions arXiv:1910.11499 *[cond-mat, physics:physics]*
90. Anon Deep learning enabled inorganic material generator - Physical Chemistry Chemical Physics (RSC Publishing)
91. Yang W, Dilanga Siriwardane EM, Hu J (2022) Crystal Structure Prediction Using an Age-Fitness Multiobjective Genetic Algorithm and Coordination Number Constraints. J Phys Chem A 126:640–647
92. Hu J, Yang W, Dong R, Li Y, Li X, Li S, Siriwardane EMD (2021) Contact map based crystal structure prediction using global optimization. CrystEngComm 23:1765–1776
93. Wu J, Zhang C, Xue T, Freeman B and Tenenbaum J 2016 Learning a Probabilistic Latent Space of Object Shapes via 3D Generative-Adversarial Modeling *Advances in Neural Information Processing Systems* vol 29 (Curran Associates, Inc.)
94. Zhu J-Y, Zhang Z, Zhang C, Wu J, Torralba A, Tenenbaum J B and Freeman W T 2018 Visual Object Networks: Image Generation with Disentangled 3D Representation arXiv:1812.02725 *[cs, stat]*
95. Valsesia D, Fracastoro G and Magli E 2018 Learning Localized Generative Models for 3D Point Clouds via Graph Convolution
96. O'Keeffe M, Yaghi OM (2012) Deconstructing the crystal structures of metal-organic frameworks and related materials into their underlying Nets Chem. Rev 112:675–702
97. Delgado-Friedrichs O, Hyde ST, O'Keeffe M, Yaghi OM (2017) Crystal structures as periodic graphs: the topological genome and graph databases. Struct Chem 28:39–44
98. Albawi S, Mohammed T A and Al-Zawi S 2017 Understanding of a convolutional neural network *2017 International Conference on Engineering and Technology (ICET)* pp 1–6
99. Scarselli F, Gori M, Tsoi AC, Hagenbuchner M, Monfardini G (2009) The Graph Neural Network Model. IEEE Trans Neural Networks 20:61–80
100. Gómez-Bombarelli R, Wei J N, Duvenaud D, Hernández-Lobato J M, Sánchez-Lengeling B, Sheberla D, Aguilera-Iparraguirre J, Hirzel T D, Adams R P and Aspuru-Guzik A 2018 Automatic Chemical Design Using a Data-Driven Continuous Representation of Molecules *ACS Cent. Sci.* **4** 268–76

101. Gu J, Wang Z, Kuen J, Ma L, Shahroudy A, Shuai B, Liu T, Wang X, Wang G, Cai J, Chen T (2018) Recent advances in convolutional neural networks. Pattern Recogn 77:354–377
102. Liu R, Lehman J, Molino P, Such F P, Frank E, Sergeev A and Yosinski J 2018 An Intriguing Failing of Convolutional Neural Networks and the CoordConv Solution arXiv:1807.03247 *[cs, stat]*
103. Song Y and Ermon S 2020 Generative Modeling by Estimating Gradients of the Data Distribution arXiv:1907.05600 *[cs, stat]*
104. Dhariwal P and Nichol A 2021 Diffusion Models Beat GANs on Image Synthesis arXiv:2105.05233 *[cs, stat]*
105. Cai R, Yang G, Averbuch-Elor H, Hao Z, Belongie S, Snavely N and Hariharan B 2020 Learning Gradient Fields for Shape Generation *Computer Vision – ECCV 2020* Lecture Notes in Computer Science vol 12348, ed A Vedaldi, H Bischof, T Brox and J-M Frahm (Cham: Springer International Publishing) pp 364–81
106. Gui J, Sun Z, Wen Y, Tao D and Ye J (2020) A Review on Generative Adversarial Networks: Algorithms, Theory, and Applications arXiv:2001.06937 *[cs, stat]*
107. Sohl-Dickstein J, Weiss E, Maheswaranathan N, Ganguli S (2015) Deep Unsupervised Learning using Nonequilibrium Thermodynamics In: Proceedings of the 32nd international conference on machine learning international conference on machine learning (PMLR) pp 2256–65
108. Szczypiński FT, Bennett S, Jelfs KE (2021) Can we predict materials that can be synthesised? Chem Sci 12:830–840
109. Xia Y, Zabaras N (2021) Bayesian multiscale deep generative model for the solution of high-dimensional inverse problems arXiv:2102.03169 [cs, stat]
110. Fuchs FB, Wagstaff E, Dauparas J and Posner I (2021) Iterative SE(3)-Transformers arXiv:2102.13419 [cs, stat]
111. Fuchs FB, Worrall DE, Fischer V and Welling M (2020) SE(3)-Transformers: 3D Roto-translation equivariant attention networks arXiv:2006.10503 [cs, stat]
112. Jung D, Choi Y (2021) Systematic review of machine learning applications in mining: exploration. Exploit, Reclam Miner 11:148
113. Yang Z, Li X, Catherine Brinson L, Choudhary AN, Chen W, Agrawal A (2018) Microstructural materials design via deep adversarial learning methodology. J Mech Des 140
114. Iyer A, Dey B, Dasgupta A, Chen W, Chakraborty A (2019) A conditional generative model for predicting material microstructures from processing methods
115. Lambard G, Yamazaki K, Demura M (2023) Generation of highly realistic microstructural images of alloys from limited data with a style-based generative adversarial network Sci Rep 13:566
116. Chun S, Roy S, Nguyen YT, Choi JB, Udaykumar HS, Baek SS (2020) Deep learning for synthetic microstructure generation in a materials-by-design framework for heterogeneous energetic materials Sci Rep 10:13307
117. Ma W, Kautz E, Baskaran A, Chowdhury A, Joshi V, Yener B, Lewis D (2020) Image-driven discriminative and generative machine learning algorithms for establishing microstructure-processing relationships. J Appl Phys 128:134901
118. Fokina D, Muravleva E, Ovchinnikov G, Oseledets I (2020) Microstructure synthesis using style-based generative adversarial network Phys. Rev E 101:043308
119. Squires I, Cooper SJ, Dahari A, Kench S (2022) Two approaches to inpainting microstructure with deep convolutional generative adversarial networks
120. Lee J-W, Goo NH, Park WB, Pyo M, Sohn K-S (2021) Virtual microstructure design for steels using generative adversarial networks. Eng Rep 3:e12274
121. Kench S, Squires I, Dahari A and Cooper SJ (2022) MicroLib: A library of 3D microstructures generated from 2D micrographs using SliceGAN
122. Hsu T, Epting WK, Kim H, Abernathy HW, Hackett GA, Rollett AD, Salvador PA, Holm EA (2021) Microstructure generation via generative adversarial network for heterogeneous, topologically complex 3D materials JOM 73 90–102

123. Henkes A, Wessels H (2022) Three-dimensional microstructure generation using generative adversarial neural networks in the context of continuum micromechanics. Comput Methods Appl Mech Eng 400:115497
124. Gayon-Lombardo A, Mosser L, Brandon NP, Cooper SJ (2020) Pores for thought: generative adversarial networks for stochastic reconstruction of 3D multi-phase electrode microstructures with periodic boundaries. npj Comput Mater 6 1–11
125. Cang R, Li H, Yao H, Jiao Y, Ren Y (2018) Improving direct physical properties prediction of heterogeneous materials from imaging data via convolutional neural network and a morphology-aware generative model Comp Mater Sci 150:212–221
126. Kim Y, Park HK, Jung J, Asghari-Rad P, Lee S, Kim JY, Jung HG, Kim HS (2021) Exploration of optimal microstructure and mechanical properties in continuous microstructure space using a variational autoencoder. Mater Des 202:109544
127. Düreth C, Seibert P, Rücker D, Handford S, Kästner M, Gude M (2023) Conditional diffusion-based microstructure reconstruction Mater Today Commun 35:105608
128. Azqadan E, Jahed H, Arami A (2023) Predictive microstructure image generation using denoising diffusion probabilistic models. Acta Mater 261:119406
129. Lyu X, Ren X (2024) Microstructure reconstruction of 2D/3D random materials via diffusion-based deep generative models. Sci Rep 14 5041
130. Lee K-H, Lim HJ, Yun GJ (2024) A data-driven framework for designing microstructure of multifunctional composites with deep-learned diffusion-based generative models. Eng Appl Artif Intell 129:107590
131. Hawkes PW, Spence JCH (2007) Science of microscopy (New York, NY: Springer)
132. Goodhew PJ, Humphreys J, Humphreys J (2014) Electron Microscopy and Analysis. CRC Press, London
133. Bostanabad R, Zhang Y, Li X, Kearney T, Brinson LC, Apley DW, Liu WK, Chen W (2018) Computational microstructure characterization and reconstruction: Review of the state-of-the-art techniques. Prog Mater Sci 95:1–41
134. Groeber M A and Jackson M A 2014 DREAM.3D: a digital representation environment for the analysis of microstructure in 3D. Integr Mater Manuf Innov 3 56–72
135. Serrao PH, Sandfeld S, Prakash A (2021) OptiMic: A tool to generate optimized polycrystalline microstructures for materials simulations. SoftwareX 15 100708
136. Schröder-Turk GE, Mickel W, Kapfer SC, Schaller FM, Breidenbach B, Hug D, Mecke K (2013) Minkowski tensors of anisotropic spatial structure New. J Phys 15:083028
137. Holm EA, Cohn R, Gao N, Kitahara AR, Matson TP, Lei B, Yarasi SR (2020) Overview: computer vision and machine learning for microstructural characterization and analysis Metall Mater Trans A 51 5985–99
138. Szeliski R (2022) Computer vision: algorithms and applications. Springer International Publishing, Cham
139. Chai J, Zeng H, Li A, Ngai EWT (2021) Deep learning in computer vision: A critical review of emerging techniques and application scenarios. Machine Learning with Applications 6:100134
140. Steinbach I (2013) Phase-Field model for microstructure evolution at the mesoscopic scale. Annu Rev Mater Res 43:89–107
141. Wilkinson MD, Dumontier M, Aalbersberg Ij J, Appleton G, Axton M, Baak A, Blomberg N, Boiten J-W, da Silva Santos L B, Bourne P E, Bouwman J, Brookes AJ, Clark T, Crosas M, Dillo I, Dumon O, Edmunds S, Evelo CT, Finkers R, Gonzalez-Beltran A, Gray AJG, Groth P, Goble C, Grethe JS, Heringa J, 't Hoen PAC, Hooft R, Kuhn T, Kok R, Kok J, Lusher SJ, Martone ME, Mons A, Packer AL, Persson B, Rocca-Serra P, Roos M, van Schaik R, Sansone S-A, Schultes E, Sengstag T, Slater T, Strawn G, Swertz MA, Thompson M, van der Lei J, van Mulligen E, Velterop J, Waagmeester A, Wittenburg P, Wolstencroft K, Zhao J, Mons B (2016) The FAIR Guiding Principles for scientific data management and stewardship. Sci Data 3 160018
142. OpenAI 2023 GPT-4 Technical Report

143. Lloyd S, Mohseni M, Rebentrost P (2013) Quantum algorithms for supervised and unsupervised machine learning arXiv:1307.0411
144. Ying C, Cai T, Luo S, Zheng S, Ke G, He D, Shen Y, Liu T-Y (2022) Do transformers really perform badly for graph representation? advances in neural information processing systems
145. Stach E, DeCost B, Kusne AG, Hattrick-Simpers J, Brown KA, Reyes KG, Schrier J, Billinge S, Buonassisi T, Foster I, Gomes CP, Gregoire JM, Mehta A, Montoya J, Olivetti E, Park C, Rotenberg E, Saikin SK, Smullin S, Stanev V, Maruyama B (2021) Autonomous experimentation systems for materials development: A community perspective Matter 4:2702–2726
146. Thakre S, Karan V, Kanjarla AK (2023) Quantification of similarity and physical awareness of microstructures generated via generative models. Comput Mater Sci 221:112074
147. Attari V, Khatamsaz D, Allaire D, Arroyave R (2023) Towards inverse microstructure-centered materials design using generative phase-field modeling and deep variational autoencoders. Acta Mater 259:119204
148. Sase K, Shibuta Y (2023) Prediction of microstructure evolution at the atomic scale by deep generative model in combination with recurrent neural networks. Acta Mater 259:119295

Machine Learning for Identifying Dynamical Phases in Topological Lasers

Stephan Wong, Doris E. Reiter, and Sang Soon Oh

Abstract Identifying phases and analyzing the stability of dynamic states are common and important problems that appear in a variety of physical systems. However, drawing a phase diagram in high-dimensional and large parameter spaces has proven to be challenging. In this chapter, we will look at a data-driven method to obtain the phase diagram of lasing modes in photonic topological insulator lasers. The classification is based on the temporal behaviour of the topological modes obtained via numerical integration of the rate equation. An unsupervised learning method is used and an adaptive library is constructed in order to distinguish the different topological modes present in the generated parameter space. We start by introducing photonic topological lasers and Su-Schrieffer-Heeger lattices with saturable gain. Then, we look at different dynamic mode decomposition methods for a parameter space defined as the gain and loss parameters. Finally, we classify the topological phases of the topological lasing modes using the library automatically determined by top-down and bottom-up classification approaches.

1 Introduction

A laser is a light emitting device composed of an optical resonator, called optical cavity, and gain medium which provides optical gain. Nowadays lasers are widely used in our life as well as in various fields of scientific research. The optical cavities of those lasers are made up of one optical resonator or an array of coupled optical

S. Wong (✉)
Center for Integrated Nanotechnologies, Sandia National Laboratories, Albuquerque, NM, USA
e-mail: stephan_wong@hotmail.fr

D. E. Reiter
Condensed Matter Theory, TU Dortmund University, Dortmund, Germany
e-mail: doris.reiter@tu-dortmund.de

S. S. Oh
School of Physics and Astronomy, Cardiff University, Cardiff, UK
e-mail: ohs2@cardiff.ac.uk

© The Author(s) 2026
M. te Vrugt (ed.), *Artificial Intelligence and Intelligent Matter*, Machine Intelligence
for Materials Science, https://doi.org/10.1007/978-3-032-04129-6_9

167

resonators. The laser is referred to as a photonic topological insulator (PTI) laser when the optical resonators are coupled to each other to form a topological insulator and lase with topological modes. They are also called topological lasers. These PTI lasers are particularly interesting because they can lase with topological edge modes [1] or corner modes [2] which are robust to structural defects and show phase-locking between the optical resonators. Therefore, significant research efforts have been made on PTI lasers especially focusing on spatial stability, i.e., the robustness of lasing modes to geometrical defects. However, lasers are open systems and thus non-Hermitian systems in which energy is exchanged with the surrounding external systems making the lasing modes dynamic with time-varying optical fields and carrier populations. Moreover, they are nonlinear due to the optical gain that depends on the mode intensities and complex light-matter interactions in the gain medium. Consequently, without taking into account their dynamic nature, a description of lasing is incomplete.

To explore the dynamics and temporal stability of lasing modes, one can use the linear stability analysis to solutions of coupled rate equations [3]. However, it becomes challenging when we apply it to more complex systems such as lasing modes in coupled lasers. This is because the analytical approach easily fails due to the lack of analytical solutions for high-dimensional parameter space and the emergence of complex and diverse dynamic phases [4]. Yet, numerical methods can offer solutions for the coupled rate equations enabling us to classify the lasing modes based on simulation results. For a high-dimensional and large parameter space, however, the simulation can be very costly and drawing the phase diagram is not a straightforward task. It is even more challenging when we do not know all the possible phases present in the laser system. Unfortunately, this is true for most of dynamic systems in nature.

Machine learning (ML) can be advantageous in tackling such problems that require repeated numerical simulations, since it can significantly reduce the number of simulations. Depending on the type of dataset, i.e., whether the data are labelled or not, different strategies of ML can be used: a supervised learning and an unsupervised learning. On one hand, the supervised learning strategy relies on labelled data, a dataset of input-output pairs. This strategy has been utilized in topological photonics [5] to draw topological phase diagrams [6], calculate topological invariants [7], or explore topological band structures [8]. On the other hand, in the unsupervised learning strategy, we extract information from the dataset which does not have labels. This strategy is useful for dimensional reductions which keep only the main features of the high-dimensional structure of the dataset, or for clustering problems where the data is classified into different types [9]. For instance, this has been successful in obtaining the phase transition in the Ising model [10], and clustering Hamiltonians that belong to the same symmetry classes [11].

In this chapter, we introduce a representation classification method to study the spatio-temporal dynamics of nonlinear topological systems. The results will be analyzed based on the phase diagram of the Su-Schrieffer-Heeger (SSH) lattice [12] with a domain wall and with saturable gain [4, 13]. Without involving any detailed knowledge on the complex system, the algorithm constructs an appropriate library of the different phases automatically. To build the library, we will employ two approaches:

a top-down approach in which the library has numerous phases that are merged into the equivalent phases, and a bottom-up approach in which the library is constructed on the fly to obtain the most accurate classification.

2 Physics of Photonic Topological Insulator Lasers

The field of PTI lasers has sparked interest since the first topological lasers were demonstrated [1, 14], resulting in a great success story [15–17]. While we refer to recent overview articles on the latest developments on PTI lasers [18, 19], we here briefly highlight the main concepts of a PTI laser.

2.1 A Brief Introduction to Photonic Topological Insulator Lasers

PTI lasers combine the advantages of topological photonics with laser physics by an optical cavity designed with a topological array. A topological array or a topological insulator is a special photonic crystal structure that features unique edge modes that are topologically protected. Such edge modes form when combining two materials that differ in their properties regarding their wave function, mostly quantified by a winding number. The topological array can be one-dimensional (1D) or two-dimensional (2D) and the arrangement can be of different types depending on how different coupling coefficients are arranged.

The 2D photonic topological structures are interesting because it is possible to create waveguides which guide the light without losses in arbitrary paths, in particular around edges. In contrast, a perfect bending of light is not possible in conventional waveguides, where bending always leads to scattering losses. Designing such topological waveguides is achieved by topological edge modes. Examples of 2D photonic topological structures are the perturbed honeycomb lattice [21] and the perturbed kagome lattice [16]. The honeycomb lattice features a photonic Dirac cone, but when slightly distorted, i.e., when the hexagons are stretched or compressed, bands with distinct topologies appear [21]. Combining a stretched with a compressed honeycomb lattice results in the occurrence of topological modes at the interface. The topological modes are topologically protected and as such also robust against defects and disorders.

Examples of 2D photonic structures are shown in Fig. 1. In Fig. 1a, a combination of a compressed Kagome lattice (upper part in red) and a stretched Kagome lattice (lower part, blue) are shown. Then, a topological edge mode is formed at the interface between the two parts which due to construction has sharply curved corners. In the lower panel the light field amplitude of the edge mode travelling along the interface is displayed. Due to its spatial symmetry, bending the light is possible without scat-

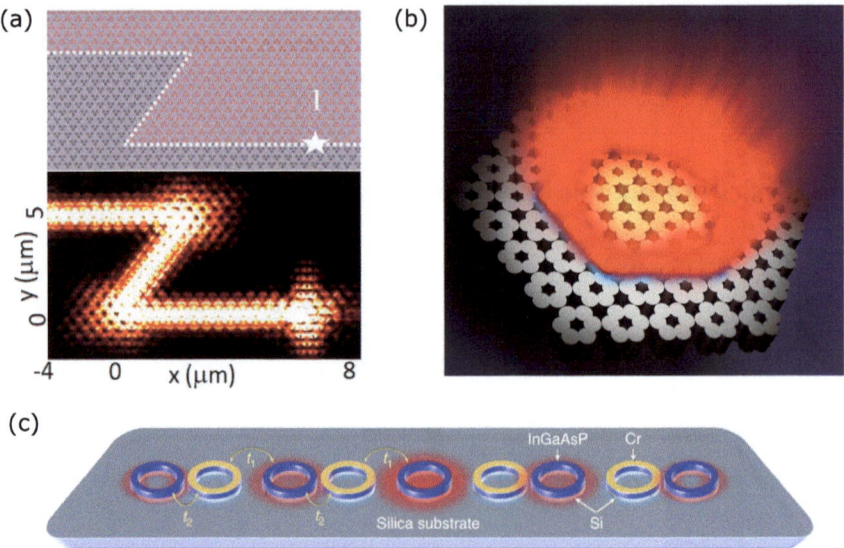

Fig. 1 **Structures for PTI lasers**: **a** A combination of two Kagome lattices supporting a topo-logical edge mode between them. Light can travel along the edge mode with minimal losses, even though there exist strong bendings as shown in the simulated electric field. Kagome lattices can be used for PTI lasers. Reprinted (adapted) with permission from [16]. Copyright 2020 American Chemical Society. All Rights Reserved. **b** Artistic view of two honeycomb lattices featuring an edge mode between them. Each site is a VCSEL made from semiconductor nanostructures. Due to their arrangements, the VCSELs lase into the topological mode, forming a PTI laser. Figure provided by Sebastian Klembt, a similar figure can be found in Ref. [17]. **c** Schematic of a one-dimensional topological array on a hybrid silicon platform. Each ring is a nanostructure of InGaAsP quantum well layers on top of silicon. Every other ring has an additional layer of Cr. This structure is an implementation of the 1D-SSH model. Figure reproduced from [20], published under a CC BY 4.0 license

tering and reflection losses, which in a regular waveguide would be associated with significant losses [16].

In a PTI laser, topological edge modes or topological cavities are combined with lasing materials to make use of the great advantage of photonic topological modes. Figure 1b shows a PTI laser based on the deformed honeycomb lattice. In the outer part the hexagons are compressed, in the inner part the hexagons are stretched. Each site is a vertical microcavity featuring a gain medium between distributed Bragg mirrors, thus, forming a vertical-cavity surface-emitting laser (VCSEL). When pumped, in total 30 of the VCSEL structures lase coherently into the topological mode [17].

PTI lasers also exists in 1D, as shown in Fig. 1c. The lattice is now formed by the combination of two nano-rings, each composed of a layered semiconductor structure. Every other ring has an additional Cr layer. On the right and left side the arrangement changes, such that in the center a topological mode is formed, which can feature lasing. This 1D structure is interesting, because it is a realization of the SSH model [20], which is a well-established theoretical model in the condensed matter physics.

2.2 Su-Schrieffer-Heeger Model for PTI Lasing

The SSH model describes a one-dimensional chain lattice with topological features [12]. Originally it was derived to describe electrical conductivity and soliton physics, while nowadays it is a standard model in topological photonics. A sketch of the SSH lattice is shown in Fig. 2a. Two lattice sites A and B form the unit cell. Within that unit cell the sites are coupled via the parameter t_1, while the coupling between unit cells is given by the parameter t_2. In the center, the lattice changes and accordingly a topological mode is formed.

To include lasing in the SSH model, at each site n a linear gain g_n and a linear loss γ_n is introduced. This leads to a dynamics at each lattice site, which can be described by solving the equations of motion

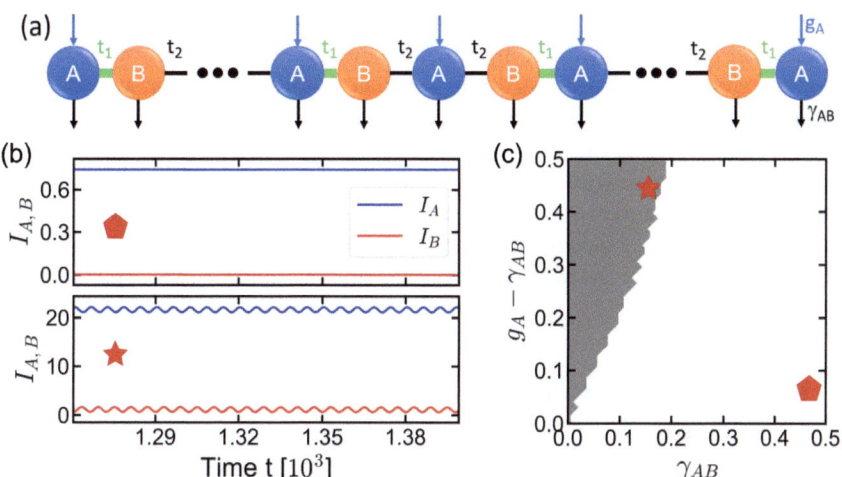

Fig. 2 **Phase diagram of the domain-wall-type Su-Schrieffer-Heeger (SSH) lattice with saturable gain. a** Sketch of the SSH lattice feature unit cells composed of site A and site B. The coupling within the unit cell is of strength t_1 and between cells t_2. In the center the lattice changes leading to a topological mode. To introduce lasing, gain of the strength g_A acts on each site A and losses γ_{AB} act on all lattice sites. **b** Representative time-evolution of the total intensity I_A (and I_B) of the A (and B) sublattice for the non-oscillating (top) and oscillating (bottom) topological lasing mode. **c** Phase diagram of the SSH lattice shown in panel (a). The red pentagon and star mark the parameter space position of the time-evolutions plotted in panel (b). The oscillating (red star) and non-oscillating (red pentagon) topological modes displayed are chosen at $(\gamma_{AB}, g_A - \gamma_{AB}) = (0.16, 0.44)$ and $(0.48, 0.06)$, respectively

$$i\frac{da_p}{dt} = i\left(\frac{g_A}{1+|a_p|^2} - \gamma_A\right)a_p + t_{1,2}b_p + t_{2,1}b_{p-1}, \tag{1}$$

$$i\frac{db_p}{dt} = i\left(\frac{g_B}{1+|b_p|^2} - \gamma_B\right)b_p + t_{1,2}a_p + t_{2,1}a_{p+1}, \tag{2}$$

where a_p and b_p are respectively the amplitudes of the A and B sites on the p-th unit cell, g_σ and γ_σ are the linear gain and linear loss at the site $\sigma = A, B$, and the couplings $t_{1,2}$ (and $t_{2,1}$) are either t_1 or t_2 (and t_2 or t_1) depending on the lattice sites [see Fig. 2a]. If the dynamics is coherent, lasing occurs. We quantify this by considering the dynamics of the amplitude I_A of the sum of all lattice sites A (and I_B respectively).

In our calculations, we set the gain just to occur at the A sites with g_A, while the gain at all B sites is zero $g_B = 0$. The losses are assumed to be the same for all sites with $\gamma_A = \gamma_B = \gamma_{AB}$, as in Refs. [4, 13]. We take a lattice consisting of $N_s = 21$ sites. Depending on the parameters, we can find two different behaviours/modes as shown in Fig. 2b: In the non-oscillating mode (top) the amplitudes $I_{A,B}$ stay constant for all times and no dynamics occurs. In the oscillating mode (bottom), the amplitudes $I_{A,B}$ oscillate in a steady state.

The parameters to distinguish between the modes are given in the phase diagram shown in Fig. 2c with the examples in Fig. 2b marked by red symbols. For one-dimensional cases like the SSH model with gain/loss, the equations of motion can be solved and the phase diagram can be calculated exactly. However, as soon as we consider higher dimensions obtaining the phase diagram is not as simple and new methods have to be sought.

In the following, we will thus explore whether ML learning techniques are helpful to draw the phase diagram of the PTI laser. For this, we need to classify the dynamical behaviour of the lasing modes. The solutions presented in Fig. 2c, from Ref. [4], will act as a benchmark for the ML results, by showing the shaded areas in our results.

3 Dynamical Behavior Identified by Representation Classification

Given a set of samples with unknown dynamical behaviors, we aim to identify their dynamical behaviors using representation classification. The fundamental idea of the representation classification method consists of identifying the samples based on a constructed library $\mathcal{L}$. The manual construction of the library constitutes the supervised learning part of the representation classification method where the dynamical behaviors of interest are utilized to construct the library for subsequent identification of their dynamics. Figure 3 shows the outcomes of this classification method. While it fails to correctly reproduce the phase diagram in the literature [4], leveraging the supervised learning part in an adaptive construction of the library gives more reasonable results [22], as we will see later in this chapter.

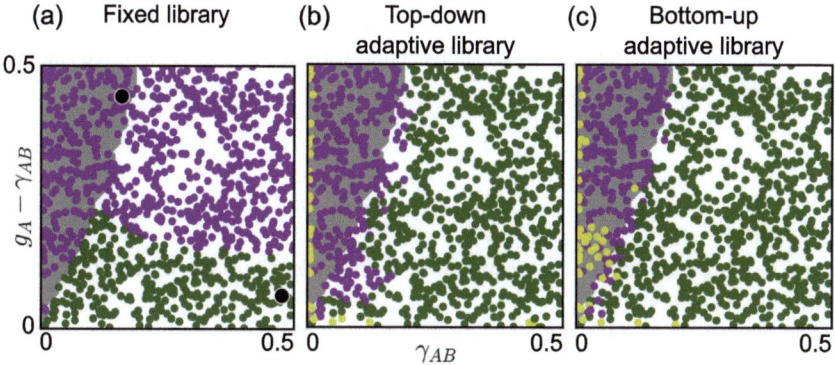

Fig. 3 Outcome of the representation classification. Phase diagram obtained with representation classification (**a**) from a fixed library composed two regimes (one oscillating and one non-oscillating), **b** from a top-down adaptive library, **c** from a bottom-up adaptive library. The purple, green and yellow dots correspond respectively to the identified oscillating, non-oscillating and transient regimes. The black dots represent the regimes used for the construction of the library. The fixed library is composed of the black dots located at $(\gamma_{AB}, g_A - \gamma_{AB}) = (0.48, 0.06)$ and $(0.16, 0.44)$, γ_{AB} and g_A are, respectively, the linear loss and gain on the A sites. The white and grey areas are overlays of the referenced phase diagram obtained in Fig. 2

The process of constructing the library $\mathcal{L}$ in the representation classification method relies on the dynamical behaviors we want to identify. If we assume that the dynamical behavior of the topological laser evolves on a low D-dimensional attractor, meaning that the system will behave the same over time once the transient is passed, then the underlying behaviors of the system can be accurately approximated by a reduced-order model. In particular, this means that the dynamics of the system can be described by a low D-dimensional vector space, where the basis vectors $\Phi = \{\phi_i\}_{i=1,\dots,D}$ are used to approximate the system's spatio-temporal dynamics $x(t)$ (or the measured state at time t) close to the attractor as:

$$x(t) \approx x_D(t) := \sum_{i=1}^{D} \phi_i \beta_i(t) = \Phi \beta(t) \tag{3}$$

where β_i are the weighted coefficients in the above linear combination of basis states ϕ_i. The library will therefore be composed of the bases representing each nonlinear regime of interest in order to identify the dynamical regimes of each samples, namely to classify the samples into different phases. Specifically, the library $\mathcal{L}$ is a set of bases each of them spanning the appropriate desired dynamical behaviors:

$$\mathcal{L} = \{\Phi_1, \dots, \Phi_J\} = \{\phi_{j,i}\}_{j=1,\dots,J\,,\,i=1,\dots,D}, \tag{4}$$

where J is the number of regimes, Φ_j are the bases of each of the dynamical regime j, and $\phi_{j,i}$ are the corresponding basis states. Therefore, the spatio-temporal dynamics $x(t)$ can now be approximated as:

$$x(t) \approx x_{J_D}(t) := \sum_{j=1}^{J} \sum_{i=1}^{D} \phi_{j,i} \beta_{j,i}(t) = \sum_{j=1}^{J} \Phi_j \beta_j(t) \tag{5}$$

where $\beta_{j,i}$ are coefficients corresponding to the contributions of the dynamical regimes in the library.

The dynamical regime of each sample is identified by finding the basis with the highest contribution to the approximated reconstruction dynamics [Eq. (5)]. More precisely, the classification strategy consist of finding the vector space within each dynamics in the library that has the highest projection measurement at a particular time t_i (or time window $[t_i : t_{i+N_w}]$):

$$j^* = \arg \max_{j=1,\dots,J} \|P_j x(t_i : t_{i+N_w})\|_2, \tag{6}$$

where $x(t_i : t_{i+N_w})$ is the notation for the state measured within the time window $[t_i, t_{i+N_w}]$ with N_w the time step window size, namely the vector $[x(t_i), \dots, x(t_{i+N_w})]$, and P_j is the projection operator onto the bases of the regime j in the library $\mathcal{L}$:

$$P_j = \Phi_j \Phi_j^{+} \tag{7}$$

with Φ_j^{+} being the pseudo-inverse of Φ_j. In Eq. (6), $\|\cdot\|_2$ is the ℓ^2-norm of a vector defined as $\|v\|_2 := \sqrt{\sum_i |v_i|^2}$, and arg max is the function that returns the index corresponding to the maximum value.

3.1 Decomposition Method

The basis vectors Φ_j approximating the dynamical regime j can be found via different decomposition methods. Depending on the decomposition methods, the classification will yield different results as it focuses on different features (see Fig. 4). In this section, three methods will be covered with an highlight on the features translated into the bases: The proper orthogonal decomposition (POD), the dynamical mode decomposition (DMD), and the augmented dynamical mode decomposition (aDMD).

Common to all the decomposition methods is the use of the data matrix X built from the dynamical data at hand. The data matrix is a $(N_s \times N_t)$-matrix that gathers the N_t data snapshots $x(t_i)$ into columns:

$$X = \left[(t_1), \dots, x(t_{N_t}) \right]. \tag{8}$$

Here, N_s is the number of sites in the system, namely $N_s = 21$. In the remaining of the chapter, the vectors $x(t_i)$ are chosen to be the complex-valued amplitudes of the modes at the A and B sites. Exploring alternative "observables", such as the absolute values or the total intensities per sublattice, is reserved for a future investigation and is beyond the scope of this chapter.

3.1.1 Proper Orthogonal Decomposition

The proper orthogonal decomposition (POD) [23] method is a common decomposition method based on the singular value decomposition (SVD) of the data matrix.

The data matrix is first decomposed via SVD:

$$X = U \Sigma V^\dagger \tag{9}$$

where U and $V^\dagger$ are $(N_s \times N)$ and $(N \times N_t)$ unitary matrices, respectively, and Σ is a diagonal $(N \times N)$-matrix $\mathrm{diag}(\sigma_1, \ldots, \sigma_N)$, with $N = \min(N_s, N_t)$. The diagonal entries of Σ are the so-called singular values, and are ordered in ascending order $\sigma_1 > \sigma_2 > \ldots > \sigma_N \geq 0$. Because the singular values can be obtained from the eigenvalue of $X X^T$, they can be interpreted as the variance of the data matrix. The columns of U, called the singular vectors, are thus ordered according to the variance σ_i they capture in the data matrix.

Intuitively, the POD method can be seen as performing a space-time separation of the data matrix as the SVD of X can explicitly be written as:

$$X_{im} = \sum_{n=1}^{N} U_{in} \sigma_n V_{nm}^\dagger, \tag{10}$$

where the columns of U contain the spatial information, while the rows of $V^\dagger$ have the temporal information at each spatial grid point.

The POD method consists of choosing the singular vectors, also called the POD modes, as being the basis Φ_j used for approximating the spatio-temporal dynamics. Yet, given the data matrix is typically large, the resulting size of the POD basis is correspondingly large as well, making the computation of the projector in the classification [Eq. (7)] impractical. To reduce the number of basis vectors used to approximate the dynamics, the POD basis U is truncated according to a cut-off value r while retaining the main information of the data matrix. More precisely, this is realized by keeping only the r highest terms in the decomposition [Eq. (10)], namely by keeping the POD modes with the r highest variance they capture in the data matrix X:

$$X_{im} \simeq \sum_{n=1}^{r} U_{in} \sigma_n V_{nm}^\dagger. \tag{11}$$

In a matrix form, the truncation reads:

$$X \simeq U_r \Sigma_r V_r^\dagger, \tag{12}$$

where U_r, Σ_r and $V_r^\dagger$ are the truncated matrix of U, Σ and $V^\dagger$, respectively. Although the cut-off value r can be chosen based on different criteria [24], r implicitly depends on the dimension of the vector space needed to approximate the attractor and is typically chosen so that the POD modes retain a certain amount of the variance σ_X in the data, namely:

$$\sum_{n=1}^{r} \sigma_i > \sigma_X. \tag{13}$$

The vectors used to build the basis Φ_j are therefore the reduced POD basis, namely $\Phi_j^{(\text{POD})} = U_r$.

However, the POD modes represent a static picture of dynamics and do not explicitly model the temporal dynamics of the time series. This method will therefore most likely fail to identify the correct dynamical regime in the classification step [see discussion in Sect. 3.2]. Nonetheless, for didactic purposes, it is important to introduce this concept.

3.1.2 Dynamical Mode Decomposition

The dynamical mode decomposition (DMD) [23, 25, 26] method is an alternative to the POD method for learning the dynamics of nonlinear systems.

Indeed, the DMD method can extract the spatio-temporal patterns of the data matrix by considering the linear mapping, namely the matrix A, between the dynamics starting at time t_1 and at time t_2. Explicitly, the linear mapping A is defined by

$$X_2 = A X_1, \tag{14}$$

where X_1 is the data matrix starting at some time steps t_1, and X_2 is the data matrix starting at the next time step t_2, namely:

$$X_1 = \left[x(t_1), x(t_2), \ldots, x(t_{N_t - 1}) \right] \tag{15}$$

and

$$X_2 = \left[x(t_2), x(t_3), \ldots, x(t_{N_t}) \right]. \tag{16}$$

The definition of the matrix A [Eq. (14)] is similar to the equation for a linear stability analysis of discrete maps. Solving the eigenvalue problem for the matrix A will therefore give us information about both the spatial profiles and the temporal evolutions with its eigenvectors and eigenvalues, respectively. In particular, if the eigenvalue problem is written as:

$$A\Psi = \Psi\Lambda \qquad (17)$$

where the columns of Ψ are called the DMD modes ψ_i and the corresponding DMD eigenvalues λ_i are the diagonal entry of Λ, then the DMD modes ψ_i give us the spatial profile of the eigenmodes while their corresponding eigenvalues λ_i have their temporal information. Because the DMD eigenvalues λ_i are eigenvalues of an evolution operator, they are of the form $\lambda_i = e^{\Omega_i \Delta t}$, with $\Delta t = t_2 - t_1$. The temporal evolution quantities of interest are thus given by Ω_i:

$$\Omega_i = \frac{\ln(\lambda_i)}{\Delta t} = i\omega_i + \mu_i \qquad (18)$$

where ω_i is the oscillation frequency of the DMD modes ψ_i, and μ_i is its growth (or decay) rate if $\mu_i > 0$ (or $\mu_i < 0$).

At its core, the DMD method decomposes the data into a set of coupled spatio-temporal modes. Conceptually, the DMD combines the POD method in the spatial domain with the discrete Fourier transform in the time domain for the oscillating behavior. Notably, DMD goes even beyond these comparisons by additionally providing an estimation of the growth (or decay) rate in time via $\mu_i > 0$ (or $\mu_i < 0$).

In the DMD method, the basis vectors used for the construction of the library $\mathcal{L}$ are the DMD modes Ψ. However, the eigendecomposition is not feasible by direct computation as the size of the matrix A is typically large, and all the DMD modes are not needed for similar reasons to the POD method. Instead, the DMD modes and eigenvalues can be obtained via the truncated data matrices of X_1 and X_2. The algorithm starts with the truncated SVD of $X_1 \approx U_r \Sigma_r V_r^\dagger$ in which Eq. (14) becomes:

$$X_2 \approx A U_r \Sigma_r V_r^\dagger. \qquad (19)$$

The matrix A is then projected onto the truncated POD subspace:

$$A_r := U_r^\dagger A U_r \approx U_r^\dagger X_2 V_r \Sigma_r^{-1}. \qquad (20)$$

The eigenvalue problem for A_r is solved with:

$$A_r W = W\Lambda, \qquad (21)$$

from which the truncated DMD modes can now be written as [27]:

$$\Psi_r = X_2 V_r \Sigma_r^{-1} W, \qquad (22)$$

and still approximately satisfy the relation in Eq. (17). The basis vectors used in the library construction are therefore $\Phi_j^{(\text{DMD})} = \Psi_r$.

Nevertheless, it is the DMD eigenvalues λ_i that carry the temporal information. Akin to the POD method, the DMD basis captures only the spatial information, leading outcomes comparable to those of the POD method [see discussion in Sect. 3.2].

3.1.3   Time-Augmented Dynamical Mode Decomposition

The time-augmented DMD (aDMD) consists in incorporating the temporal information into the DMD basis [28].

Indeed, although the DMD captures the temporal evolution of the dynamics, the temporal information is not directly incorporated in the DMD modes used for constructing the basis of the attractor's vector space. Instead, the set of vectors $\{(\psi_{r,i}), (\lambda_i \psi_{r,i}), (\lambda_i^2 \psi_{r,i}), \ldots, (\lambda_i^{N_w} \psi_{r,i})\}$ gives an idea of the evolution of the DMD mode $\psi_{r,i}$, as λ_i is similar to a time-evolution operator, i.e., multiplying by λ_i is equivalent to shifting by one time step. The temporal information is therefore included in the DMD by constructing a time-augmented basis [28]:

$$\psi_{r,i}^{(N_w)} = \begin{bmatrix} \psi_{r,i} \\ \lambda_i \psi_{r,i} \\ \vdots \\ \lambda_i^{N_w} \psi_{r,i} \end{bmatrix}. \tag{23}$$

Because the time-augmented vectors Eq. (23) exhibit the time-evolution of the DMD vectors, these vectors constitute the basis used for constructing the library $\mathcal{L}$, namely $\Phi_j^{(\text{aDMD})} = \Psi_{r,i}^{(N_w)}$, with $\Psi_{r,i}^{(N_w)} = [\psi_{r,1}^{(N_w)}, \psi_{r,2}^{(N_w)}, \ldots]$.

The identification is now slightly different than for the POD and DMD, as the projection measurement is realized over a time window $[t_i, t_i + N_w]$, as shown in Eq. (6). In consequence, the outcome of the aDMD method gives now better results, even though this is far from being accurate as we will see in the next section.

3.2   Classification

After acquiring the basis composing the library $\mathcal{L}$, we can now proceed with classifying the samples using the basis derived from the representative dynamics. The representative dynamics are here randomly chosen within each known dynamical phases, namely the oscillating and non-oscillating modes [see Fig. 2], and they are depicted by the black dots in Fig. 4. In order to avoid the transient regime and to get closer to the attractor's dynamics, the bases are constructed from the time series after starting at the 1800-th time step.

The phase diagram results from the identification of the sample with respect to the representative dynamics as shown in Eq. (6). In the phase diagrams, we color-coded the different identified regimes, where the purple (and pink) dots always mark the oscillating regime, and the green dots the non-oscillating regime. As we will see later, the yellow dots correspond to the transient regime and the orange dots to the transition regime. The white and grey areas are overlays of the referenced phase diagram obtained in Fig. 2.

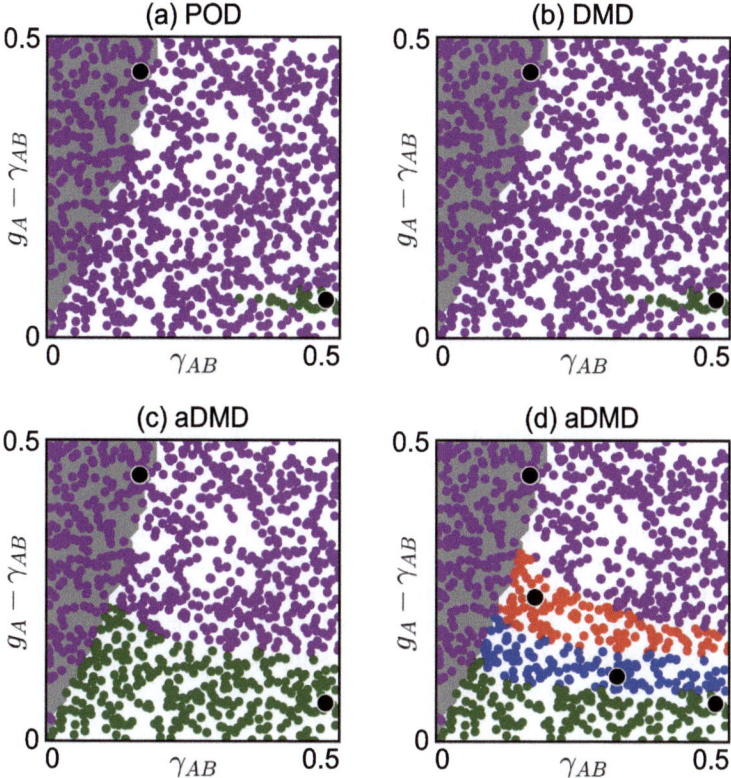

Fig. 4 Representation classification using different decomposition methods. Phase diagrams derived using a fixed library composed of two regimes (one in the oscillating and one in the non-oscillating phase), and with the bases in the library generated using the **a** POD, **b** DMD, **c** aDMD with $N_w = 25$. **d** Same as in panel (c) but with four regimes in the fixed library (one in the oscillating and three in the non-oscillating phase) The purple and green dots correspond respectively to the identified oscillating and non-oscillating regimes. The black dots represent the regimes used for the construction of the library. These black dots are located at $(\gamma_{AB}, g_A - \gamma_{AB}) = (0.16, 0.44)$ and $(0.48, 0.06)$ for panel (a), (b) and (c); and additionally at $(0.31, 0.11)$ and $(0.17, 0.24)$ for panel (d). The white and grey areas are overlays of the referenced phase diagram obtained in Fig. 2

Figure 4 shows the phase diagrams obtained from the different decomposition methods described in the previous section. In Fig. 4a, b and c, the purple (green) dots are the identified regime j^* [see Eq. (6)] from the oscillating (non-oscillating) regime. On one hand, Fig. 4a, b display the phase diagrams obtained from the POD and DMD method, respectively. As anticipated, the POD method fails to accurately reproduce the reference phase diagram [see Fig. 2] as the bases do not contain any temporal information: We observe that many time series are not correctly identified [Fig. 4a]. Similarly, the DMD method does not correctly identify the dynamics [Fig. 4b] as the bases from the DMD are similar to the POD bases. On the other hand, Fig. 4c illustrates that the aDMD (with $N_w = 25$) has better classification results: Less samples are being misidentified as the aDMD basis contains the temporal information of the dynamics.

While one might consider utilizing a different set of representative dynamics in the parameter space to construct the library $\mathcal{L}$ for potential improved results, such an approach would likely results in only marginal improvement of the classification. Instead, Fig. 4d demonstrates that redundantly increasing the number of representative dynamics in the library $\mathcal{L}$–in this case, four in total–yields much better results in correctly identifying the oscillating to the non-oscillating regime. In this context, the blue and red dots are the identified samples from the two new added representative dynamics: The identified oscillating and non-oscillating regimes now have a better fitting with the referenced phase diagram [Fig. 2], despite belonging to distinct regimes, namely having different index j^*. Therefore being able to identify the three colored for the oscillating regimes (green, blue, red) as corresponding to a single oscillating regime would give a more accurate diagram. This is the rationale of the methods, termed top-down and bottom-up adaptive classification, developed in the next sections: Adding redundant representative dynamics in the library $\mathcal{L}$ and being able to merge them corresponding to the oscillating or non-oscillating regime could potentially bring us closer to the desired phase diagram. The implications of these adaptive classification methods are two-fold: Firstly, prior knowledge of the system would not be needed, only measurement of its dynamics; secondly, the phase diagram would be constructed automatically from the samples.

4 Top-Down Adaptive Representation Classification

The top-down adaptive representation classification consist of adaptively refining the library initially composed of a redundant set of representative dynamics. This eliminates the manual process of selecting representative dynamics known from prior knowledge of the complex system considered. In practice, this top-down approach starts with many samples for the construction of the library, and then reduce the library size by merging some of them. From the previous section, this would mean that we merge the three phases in the non-oscillating region in Fig. 4d, and consider them as a single regime. Based on some measures in the decision process, this automated construction of the library thus removes the manual construction of the regimes.

The regimes are merged when considered to be equivalent, which is defined based on the dissimilarity between the subspace of different regimes. More precisely, the equivalence relation relied on the subspace alignment γ_{ij} that measures the dissimilarity between two subspaces i and j by projecting one subspace onto the other:

$$\gamma_{ij} := \frac{\|P_i P_j\|_F^2}{\|P_i\|_F \|P_j\|_F}, \tag{24}$$

with $\|\cdot\|_F$ the Frobenius norm of a matrix, $\|M\|_F := \sqrt{\sum_{ij} |M_{ij}|^2}$, and the projection operator defined as $P_j = \Phi_j \Phi_j^+$ [Eq. (7)]. Therefore, the regimes i and j are said to be equivalent, denoted by $i \sim j$, if the fraction of information retained after the projection onto each other, $\gamma_{ij} \in [0, 1]$, is high enough:

$$\gamma_{ij} > \gamma_{th}, \tag{25}$$

where $\gamma_{th} \in [0, 1]$ is a hyper-parameter deciding the threshold value for merging different regimes. Essentially, the relation Eq. (25) is numerically computed with a crucial emphasis on preserving the transitivity property of the equivalence relation, meaning that if $i \sim j$ and $j \sim k$ then $i \sim k$. The relation Eq. (25) is then indeed an equivalence relation because the reflexive ($i \sim i$) and symmetric ($i \sim j \Rightarrow j \sim i$) property of the relation is automatically satisfied from the definition of γ_{ij} [Eq. (24)]. Figure 5a illustrates the algorithm of the top-down approach. With an initial library $\mathcal{L}$ composed of a large number of representative regimes, the subspace alignment between all the bases is calculated, which can be viewed as being the (i, j) entries of a symmetric matrix γ [left of panel of Fig. 5a]. Then, the bases having a subspace alignment higher than γ_{th} are grouped together, while ensuring the transitivity relation is satisfied. Visually, this grouping procedure resemble a block diagonalization procedure of the subspace alignment matrix [middle panel of Fig. 5a]. Finally, for each of the blocks, one basis is selected and considered as the representative of the dynamical regime, which are plotted Fig. 5c.

In its core, the top-down representation classification entails in identifying the time series based on a comprehensive library of bases. It is only after this classification that equivalent identified regimes are merge using alignment subspace γ_{ij} and the equivalence relation specified in Eq. (25). Figure 5 shows the phase diagram resulting from the top-down algorithm with an initial library composed of $J = 60$ randomly chosen regimes. With this top-down approach, the resulted phase diagram is closed to the reference one [see Fig. 2] as we can see that the oscillating (purple dots) and non-oscillating (green dots) regimes can be distinguished. We can also see that a third regime is being distinguished from the oscillating and non-oscillating dynamics present in the literature. This is the transient regime (yellow dots) situated close to the $\gamma_{AB} = 0$ or $g_A - \gamma_{AB} = 0$ axis. Because we start looking at the dynamics after a time t_0, the transient regimes is an indication that a longer starting time and therefore longer simulation might be needed to consider them being in either one of the oscillating or non-oscillating regimes. In general, varying the parameter values also changes the relaxation times of the system, and the transient regime may then be the dominant regime. If that is the case, this would mean that the simulation has not run long enough to observe the main dynamical regimes.

The threshold parameter γ_{th} controlling the merging process is a critical quantity in the algorithm, as a low threshold γ_{th} will easily merge regimes while a high γ_{th} will still yield a relatively large size of the library as demonstrated in Fig. 5d. As a consequence, the two extreme cases will either consider all the dynamical regimes as being equivalent and identify all of them as being the same, or will not be able

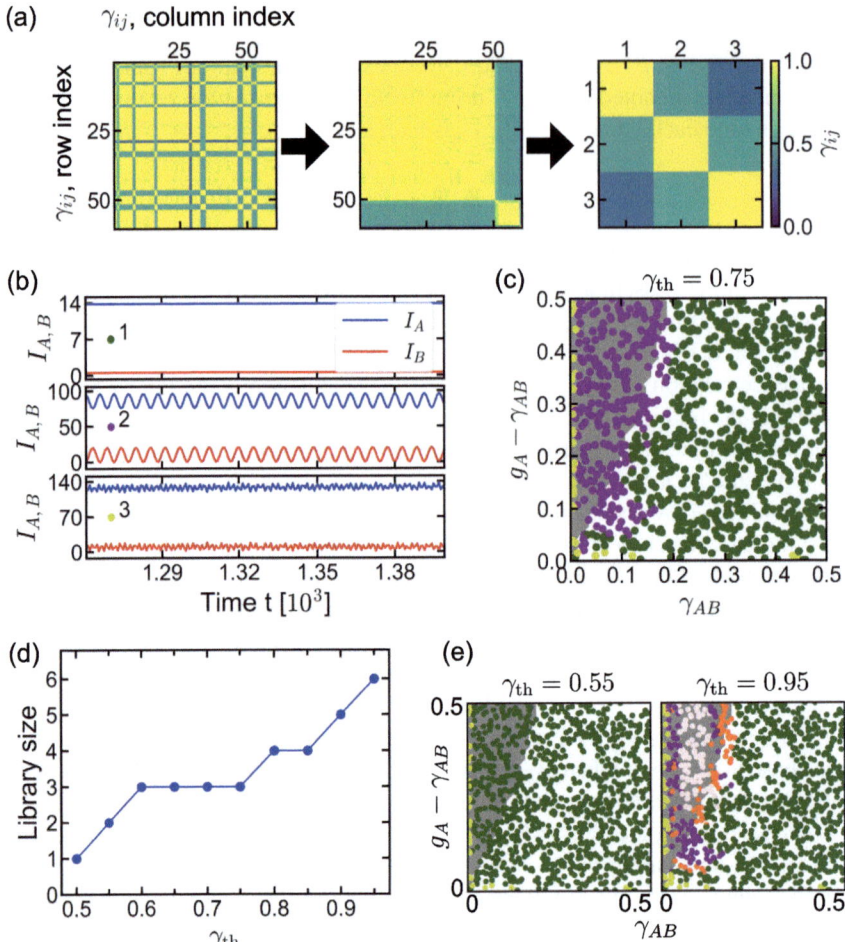

Fig. 5 Principle of the top-down adaptive representation classification. a Schematic of the top-down library generation: Starting with an initial library composed of $J = 60$ regimes, the subspace alignment matrix γ_{ij} is plotted (left panel), grouped into equivalent regimes with $\gamma_{th} = 0.75$ (middle panel), then the subspace alignment matrix from the reduced library is plotted after the equivalent regimes are merged (right panel). **b** Representative time-evolution of the total intensity I_A (and I_B) of the A (and B) sublattice for the non-oscillating (top), oscillating (middle) topological lasing mode, and for the transient mode (bottom). **c** Phase diagram obtained using the top-down representation classification approach with an initial library composed of randomly chosen $J = 60$ regimes, with the hyper-parameter threshold $\gamma_{th} = 0.75$. **d** Library size against the hyper-parameter threshold γ_{th}, with an initial library composed of randomly chosen $J = 60$ regimes. **e** Phase diagrams obtained using the top-down classification approach with $\gamma_{th} = 0.55$ (left) and $\gamma_{th} = 0.95$ (right). The purple, green, yellow and orange dots correspond respectively to the identified oscillating, non-oscillating, transient and transition regimes. The white and grey areas are overlays of the referenced phase diagram obtained in Fig. 2.

to see similarities between the time series, respectively. Therefore, the threshold is arbitrarily chosen based on the refinement of the desired library [see Fig. 5d, e]. Figure 5e shows examples of the phase diagram obtained with two different γ_{th} values. With $\gamma_{th} = 0.55$, the threshold value is too low as all the oscillating and non-oscillating regimes are being considered as equivalent, despite being able to distinguish the transient regimes with the others. However, using the same initial library but with a finer threshold value $\gamma_{th} = 0.95$, the top-down classification method is able to resolve six regimes grouped into four main regimes: The oscillating and non-oscillating known from the reference phase diagram [see Fig. 2], along with the regimes corresponding to the transient regime and to the transition between the two topological phases. Out of the six regimes, two are grouped into the oscillating phases, and two are grouped into the transient phases. The transition regime is different than the transient regimes, as the dynamics has already been reached the stationary regimes, and corresponds to the parameter values along the boundary between the two topological phases [see grey and white shaded areas in Fig. 5e]: therefore the method cannot correctly classify the regimes in this region as either oscillatory or non-oscillatory. Notably, this finer threshold provides distinct sets of modes in the oscillating parameter space region [see the purple and pink dots in Fig. 5e]. Those distinct oscillating modes were not present in the reference diagram [Fig. 2], and may have been overlooked as only the dynamics of the total intensity were considered in the literature [4].

Despite the success of the classification using the top-down approach, we can still see misclassification of the samples in the low γ_{AB} and low $g_A - \gamma_{AB}$ region (bottom-left region of the present phase diagram [Fig. 5b]), where some time series are interpreted as oscillating instead of non-oscillating regime. This is demonstrative of the limitation of this top-down automatic method where the initially constructed library may not encompass certain paths connecting similar bases. For example, the regimes i and k may not be similar enough to be considered as equivalent with respect to γ_{th}, but an "intermediate" regime j missing in the initial library and equivalent to both regimes i and k, i.e., $i \sim j$ and $j \sim k$, might make the regimes i and k equivalent thanks to the transitivity relation. This issue is dubbed here as the library having missing path, and the natural strategy to overcome this problem is to increase the initial library size to fill all the missing paths, as we will see in the next section.

5 Bottom-Up Adaptive Representation Classification

As an alternative approach to the top-down method, we propose the bottom-up classification which adaptively adds samples to the library on the fly. In practice, this bottom-up approach starts with few samples in the library and add bases to the library whenever the library is consider to be not good enough. The library is then reduced using the top-down approach to merge equivalent regimes. The key advantage of this bottom-up approach is the automatic construction of a library based on its quality. Therefore, the method does not rely on a more or less good choice of samples composing the library and has the potential to overcome the missing issue present in the top-down classification.

A sample is added to the library whenever the library cannot retrieve the dynamics of the given sample using its bases. This construction procedure is assessed via the maximal projection of the measurement onto the regimes' subspace ϵ:

$$\epsilon := \max_{j=1,\ldots,J} \frac{\|P_j y(t) - y(t)\|_2}{\|y(t)\|_2}, \tag{26}$$

with $\|\cdot\|_2$ the L_2-norm of a vector, and P_j is the projector defined in Eq. (7). In other words, the library is said to be good enough if the worst relative reconstruction error, ϵ, is low enough:

$$\epsilon < \epsilon_{\text{th}}, \tag{27}$$

where ϵ_{th} is the hyper-parameter deciding the threshold quality of the library. Figure 6a depicts the bottom-up approach algorithm. The library begins with a single sample, and is adaptively constructed based the relative reconstruction error [see left of panel Fig. 6a]. Once the library is established [see middle of panel Fig. 6a], the top-down approach is employed in order to merge equivalent regimes [see right of panel Fig. 6a]. Figure 6c shows the representative of the dynamical regimes after the top-down method is employed.

Fundamentally, the bottom-up representation classification scheme consists in initially classifying time series based on a given library or incorporating the sample into the library if the library is not good enough. Subsequently, the different phases are merged into equivalent regimes using the top-down method. Starting with a library composed of a single randomly chosen sample, the phase diagram obtained from the bottom-up classification is depicted in Fig. 6c. Comparable to the top-down approach shown in Fig. 5c, three distinct regimes are identified, corresponding to the oscillating (purple dots), non-oscillating (green dots) and transient (yellow dots) regimes. Notably, the phase diagram obtained from the bottom-up classification demonstrates an improved accuracy in identifying the regimes. The misclassifications previously observed of the oscillating and non-oscillating regimes, which were attributed to missing paths in the library, are now considerably reduced. Only few samples near the topological transition boundary are not correctly identified. Similarly to the top-down approach, the transient regimes indicates the need of a longer simulation time to correctly classify them.

Together with the hyper-parameter γ_{th}, the threshold hyper-parameter ϵ_{th} plays a crucial role in the bottom-up approach. Indeed, the decision of whether to add a sample to the library is a critical step in the algorithm. Setting a low threshold parameter will add too many samples, while a high ϵ_{th} will not add samples to the library at all, as shown in Fig. 6d. Therefore, the threshold value ϵ_{th} is, again, arbitrarily chosen based on the desired quality of the library [Fig. 6d, e]. Figure 6e illustrates two examples with different threshold parameter value, given $\gamma_{\text{th}} = 0.95$. Using $\epsilon_{\text{th}} = 0.05$, namely with a library that gives less than 5% of the reconstruction error of the measurement, the resulting phase diagram has four main regimes corresponding to the oscillating,

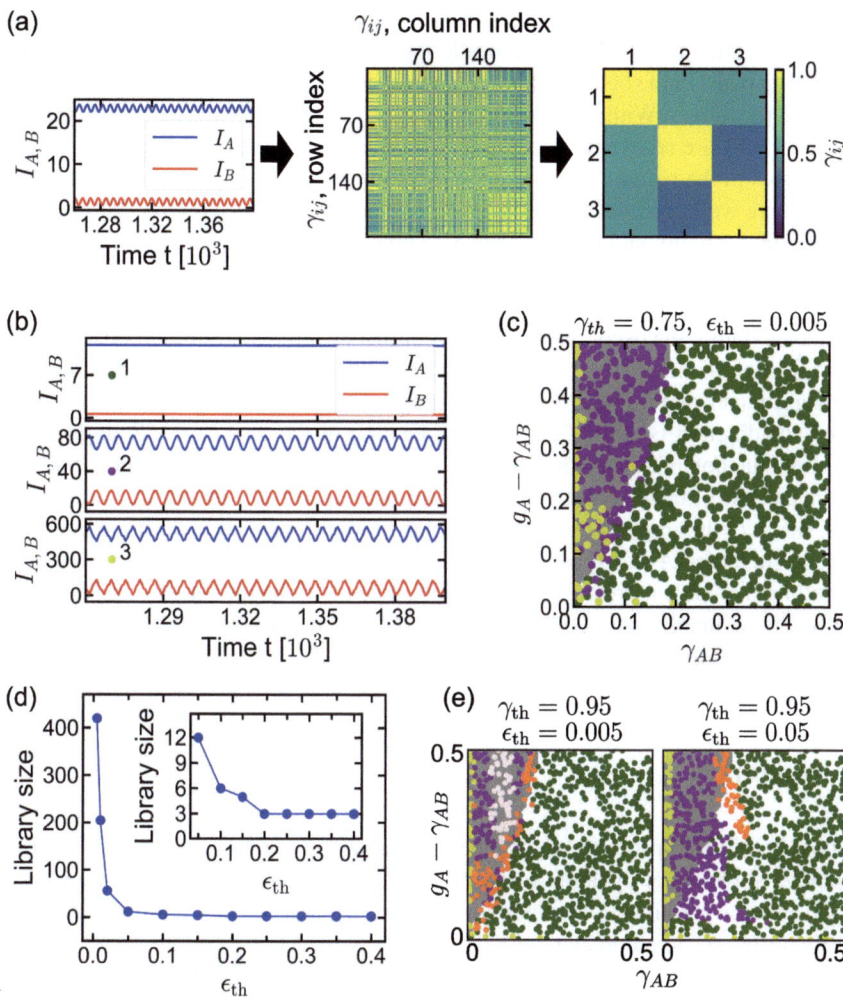

Fig. 6 Principle of the bottom-up adaptive representation classification. a Schematic of the bottom-up library generation: Starting with an initial library composed of a single randomly chosen regime (left panel), the library is increased according to the library quality where the subspace alignment matrix γ_{ij} of the final library is plotted (middle panel), then the subspace alignment matrix from the reduced library via the top-down approach is plotted (right panel). **b** Representative time-evolution of the total intensity I_A (and I_B) of the A (and B) sublattice for the non-oscillating (top), oscillating (middle) topological lasing mode, and for the transient mode (bottom). **c** Phase diagram obtained using the bottom-up representation classification approach with an initial library composed of single randomly chosen regime, with the hyper-parameters threshold $\epsilon_{th} = 0.005$, $\gamma_{th} = 0.75$. **d** Library size against the hyper-parameter threshold ϵ_{th}, with $\gamma_{th} = 0.95$ and with an initial library composed of a single randomly chosen regime. **e** Phase diagrams obtained using the bottom-up classification approach with $\gamma_{th} = 0.95$, and with $\epsilon_{th} = 0.005$ (left) and $\epsilon_{th} = 0.05$ (right). The purple, green, yellow and orange dots correspond respectively to the identified oscillating, non-oscillating, transient and transition regimes. The white and grey areas are overlays of the referenced phase diagram obtained in Fig. 2

non-oscillating, transition and transient regimes. However, there are instances of misidentification between the two oscillating and non-oscillating phases, likely due to missing paths in the obtained library, and suggesting that some samples need to be added in the library. With a better library quality, specifically setting $\epsilon_{th} = 0.005$, the missing paths are recovered, leading to a more accurate prediction of the topological phases in the resulting phase diagram: Both the oscillating and non-oscillating regimes are well positioned in their respective parameter space region. Moreover, a better library quality yields a more detailed phase diagram with the transition points not being able to be considered as oscillatory or non-oscillatory as this regime correspond to the transition boundary between the two topological phases, although some points are within the oscillatory regime. Significantly, the bottom-up representation classification distinguishes two oscillating modes [see the purple and pink dots in the right panel of Fig. 6e], instead of a single oscillating regime as shown in the reference diagram [see Fig. 2]. The identification of the distinct oscillating modes highlight the potential novel insights given by the data-driven classification method. In the context of this chapter, the classification method takes into account the complex values of the amplitudes of $x(t)$, rather than solely focusing on the total intensity of each sublattice A and B as done in Ref. [4, 13]. The use of complex valued here therefore enables a more detailed description of the dynamic pattern based on the entire lattice, incorporating information such as the relative phase difference of the sites or the absolute value of amplitudes.

6   Summary

In this chapter, we introduced an unsupervised machine-learning approach to identify topological phases of dynamic systems. In principle, this identification process is a representation classification, which finds a library that associates a finite number of labels for phases to parameters of a dynamic system. As an example, we explained how to draw the phase diagrams of the domain-wall-type SSH lattice with saturable gain.

As a mean to characterize the dynamic property, we introduced different decomposition methods including the POD, the DMD and the aDMD. By comparing the representation classification results using a fixed library based on the different methods, we observed that the aDMD works best for the domain-wall-type SSH lattice which has spatial and temporal variations in its field amplitudes.

In constructing the library, one can employ two schemes: top-down and bottom-up approaches that merge similar phases in a library or adaptively construct a library according to its quality, respectively. It is noteworthy that the libraries are constructed automatically and it does not require any detailed knowledge on the system. The libraries are optimized to represent all possible phases in a given parameter space by considering different measures defined on the libraries' bases. In the SSH laser example, we found the bottom-up adaptive scheme was the best approach to tackle the problem of drawing the phase diagram.

The bottom-up adaptive scheme can avoid pitfalls that can be encountered when using reverse engineering, such as missing paths in the equivalence relations, and can be used as a strategy to extend the method to more complex systems. While maybe not all phases might be identified on the first try, it is capable of clustering similar behaviour and gives a first classification of the different modes in the system. It should be complemented by a thorough analysis of these modes. Nonetheless, because of its different approach to drawing the phase diagram and its capability of clustering similar behaviour, reverse engineering holds the potential to find novel topological lasing modes, which could have been overlooked in other approaches.

Additionally, the machine-learning approach can be applied to many other dynamic systems composed of nonlinear resonators, for example, an array of quantum emitters and coupled mechanical resonators.

References

1. Bandres MA, Wittek S, Harari G, Parto M, Ren J, Segev M, Christodoulides DN, Khajavikhan M (2018) Topological insulator laser: experiments. Science 359(6381):eaar4005
2. Wu J, Ghosh S, Gan Y, Shi Y, Mandal S, Sun H, Zhang B, Liew TC, Su R, Xiong Q (2023) Higher-order topological polariton corner state lasing. Sci Adv 9(21):eadg4322
3. Longhi S, Kominis Y, Kovanis V (2018) Presence of temporal dynamical instabilities in topological insulator lasers. EPL 122(1):14004
4. Malzard S, Schomerus H (2018) Nonlinear mode competition and symmetry-protected power oscillations in topological lasers. New J Phys 20(6):063044
5. Yun J, Kim S, So S, Kim M, Rho J (2022) Deep learning for topological photonics. Adv Phys X 7(1):2046156 Mar
6. Araki H, Mizoguchi T, Hatsugai Y (2019) Phase diagram of a disordered higher-order topological insulator: a machine learning study. Phys Rev B 99(8):085406
7. Zhang P, Shen H, Zhai H (2018) Machine learning topological invariants with neural networks. Phys Rev Lett 120(6):066401
8. Peano V, Sapper F, Marquardt F (2021) Rapid exploration of topological band structures using deep learning. Phys Rev X 11(2):021052
9. Pedregosa F, Weiss R, Brucher M, Varoquaux G, Gramfort A, Michel V, Thirion B, Grisel O, Blondel M, Prettenhofer P, Weiss R, Dubourg V, Vanderplas J, Passos A, Cournapeau D, Brucher M, Perrot M, Duchesnay É (2011) Scikit-learn: machine learning in Python. J Mach Learn Res 12:2825–2830
10. Wang L (2016) Discovering phase transitions with unsupervised learning. Phys Rev B 94(19):195105
11. Scheurer MS, Slager R-J (2020) Unsupervised machine learning and band topology. Phys Rev Lett 124(22):226401
12. Su W-P, Schrieffer JR, Heeger AJ (1979) Solitons in polyacetylene. Phys Rev Lett 42(25):1698–1701
13. Malzard S, Cancellieri E, Schomerus H (2018) Topological dynamics and excitations in lasers and condensates with saturable gain or loss. Opt Express 26(17):22506
14. Bahari B, Ndao A, Vallini F, El Amili A, Fainman Y, Kanté B (2017) Nonreciprocal lasing in topological cavities of arbitrary geometries. Science 358(6363):636–640
15. Harari G, Bandres MA, Lumer Y, Rechtsman MC, Chong YD, Khajavikhan M, Christodoulides DN, Segev M (2018) Topological insulator laser: theory. Science 359(6381):eaar4003
16. Gong Y, Wong S, Bennett AJ, Huffaker DL, Oh SS (2020) Topological insulator laser using valley-Hall photonic crystals. ACS Photonics 7(8):2089–2097

17. Dikopoltsev A, Harder TH, Lustig E, Egorov OA, Beierlein J, Wolf A, Lumer Y, Emmerling M, Schneider C, Höfling S, Segev M, Klembt S (2021) Topological insulator vertical-cavity laser array. Science 373(6562):1514–1517
18. Yan Q, Xiaoyong H, Yulan F, Cuicui L, Fan C, Liu Q, Feng X, Sun Q, Gong Q (2021) Quantum topological photonics. Adv Opt Mater 9(15):2001739
19. Li Z, Luo X-W, Gu Q (2023) Topological on-chip lasers. APL Photonics 8(7):070901
20. Zhao H, Miao P, Teimourpour MH, Malzard S, El-Ganainy R, Schomerus H, Feng L (2018) Topological hybrid silicon microlasers. Nat Commun 9(1):981
21. Long-Hua W, Xiao H (2015) Scheme for achieving a topological photonic crystal by using dielectric material. Phys Rev Lett 114(22):223901 June
22. Wong S, Olthaus J, Bracht TK, Reiter DE, Oh SS (2023) A machine learning approach to drawing phase diagrams of topological lasing modes. Commun Phys 6(1):1–7
23. Proctor JL, Brunton SL, Brunton BW, Kutz JN (2014) Exploiting sparsity and equation-free architectures in complex systems. Eur Phys J Spec Top 223(13):2665–2684
24. Gavish M, Donoho DL (2014) The optimal hard threshold for singular values is $4/\sqrt{3}$. IEEE Trans Inf Theory 60(8):5040–5053
25. Schmid PJ (2010) Dynamic mode decomposition of numerical and experimental data. J Fluid Mech 656:5–28
26. Tu JH, Rowley CW, Luchtenburg DM, Brunton SL, Kutz JN (2014) On dynamic mode decomposition: theory and applications. J Comput Dyn 1(2):391–421
27. Kutz JN, Fu X, Brunton SL (2016) Multiresolution dynamic mode decomposition. SIAM J Appl Dyn Syst 15(2):713–735
28. Kramer B, Grover P, Boufounos P, Nabi S, Benosman M (2017) Sparse sensing and DMD-based identification of flow regimes and bifurcations in complex flows. SIAM J Appl Dyn Syst 16(2):1164–1196

Artificial Intelligence Reshaping the Semiconductor Metrology

Md. Ashiqur Rahman Laskar, Srijan Chakrabarti, and Umberto Celano

Abstract As technology advances, the challenges in characterization of nanoelectronics devices and materials also increase. To overcome the challenges, relying on only cutting-edge hardware doesn't help the metrologist in this new generation. Computational power of machine learning (ML), hence artificial intelligence (AI) has come forward as a strong support to modern metrology tools and data analysis. Here we review how AI is reshaping the metrology power for the nanoelectronics manufacturing community, especially in failure analysis and materials characterization. In this chapter, diverse contributions of AI in major metrology tools such as experiment automation, sample identification, data denoising, fast and robust data analysis etc. are discussed with their impact on the characterization and advancement of nanoelectronics.

1 Introduction

At the time of writing, the semiconductor industry is undergoing a period of massive changes. The ubiquity of electronics in our lives is more profound than ever, and it's accelerating. Wearables, edge computing, smart lenses, and novel computing paradigms are just the tip of the iceberg. Driving this revolution, as always, are advancements in complementary metal-oxide semiconductor (CMOS) process development for logic and memory devices. But unlike the past 50 years, we are witnessing a simultaneous eruption of innovative chip architectures. Fin-based transistors are yielding to nanosheet technologies, memories are stacking skyward, and advanced packaging techniques are enabling heterogeneous integration of hyper-complex systems on chip (SoC). This confluence creates a monumental shift in chip manufacturing, brimming with exciting opportunities and daunting challenges.

One foundational pillar for advancing chip manufacturing has traditionally been metrology. Metrology, a niche field with deep roots in material science and solid-state physics, has always been a cornerstone of progress in chipmaking. Its focus: to

Md. A. R. Laskar · S. Chakrabarti · U. Celano (✉)
Computer and Energy Engineering, Arizona State University, Tempe, USA
e-mail: Umberto.Celano@asu.edu

© The Author(s) 2026
M. te Vrugt (ed.), *Artificial Intelligence and Intelligent Matter*, Machine Intelligence for Materials Science, https://doi.org/10.1007/978-3-032-04129-6_10

characterize, at scale, the intricate dance of manufacturing processes, materials, and their integration. But in this era of unprecedented changes, metrology must move just as fast. Chip requirements are continuously evolving, demanding ever-smaller dimensions, higher accuracy, and unmatched sensitivity. Metrology can only fulfill its vital role in the silicon transformation by keeping pace with the rapid innovations driving the industry.

Historically an essential but understated component of the semiconductor industry, metrology, has now gained significant recognition for its critical role. The recent white paper "Metrology Gaps in the Semiconductors Ecosystem" from the National Institute of Standards and Technology (NIST), issued under the CHIPS & Science Act, brings this critical field into sharp focus [1]. For students, scholars, and anyone interested in the core processes of chip manufacturing, this document is an invaluable resource. However, its most compelling insight is the transformative potential of AI to reshape semiconductor metrology on several fronts. The white paper paints a clear picture: AI isn't just playing in the sandbox; it's poised to transform the very landscape of how we measure and understand chip behavior. From automating complex characterization processes to unlocking new frontiers in nanomaterial analysis, AI promises to propel metrology to unprecedented heights.

While an extensive treatment of the measurement science and methods is beyond the scope of this chapter. Here, we focus on reviewing the state-of-the-art application of AI for materials characterization and semiconductor metrology. The reader will find an overview of the field, with references to methods and ongoing research for material quality assessment, in-line metrology for high-throughput manufacturing, and novel metrology concepts. The chapter is structured with seven main sections. The introduction offers an overview of the field with details of the main concepts of neural networks (NNs) and deep learning (DL) applied to metrology offering a detailed overview of some of the most recent work in the semiconductor metrology community. Later, we introduce some of the advancements in the use of AI for scanning probe microscopies, a class of techniques in which where a nanosized high-aspect ratio probe is used to sense the surface properties, including physical dimensions with Angstrom precision and various electronic and magnetic properties. The third section discusses the role of virtual metrology in which the knowledge of experimental data and simulated processes are combined to predict the properties of a metrological measurement for an unknown structure. Fourth, we introduce the major signs of progress obtained for electron beam imaging, including scanning electron microscopy (SEM) and transmission electron microscopy (TEM). Later, we introduce the role of AI-assisted data analysis for X-ray-based and optical methods, including X-ray diffraction, ellipsometry, interferometry, and surface profilometry. These represent some of the most commonly available, fully automated, and high-throughput measurements in chip manufacturing. Finally, we provide a general overview of the possible usages of AI for data analysis Raman spectroscopy and nuclear Magnetic Resonance spectroscopy (NMR).

2 Machine Learning for Metrology in Nanoelectronics

The revolution of nanoelectronics is advancing very fast and transforming our day-to-day lives. Following Moore's law, [2] the device density has increased as the scaling down continues. Here metrology is very crucial to achieve the ultimate device performance as it ensures quality and critical parameters during the fabrication process. However, the nanoelectronics' evaluation and headway are pushing existing metrology techniques' limitations [3, 4]. In addition, the requirements of huge data collection, faster data processing, and complex analysis have made the situation more challenging for characterization purposes. Therefore, researchers adopted ML algorithms widely with various metrology tools and techniques in the last two decades [5–7]. ML is a combination of computer science and artificial intelligence that learns from the training data set and can perform specific tasks with unseen data sets, just like imitating the way of a human. Broadly speaking, ML enables the researchers to acquire the metrology data autonomously/effectively, classify them, and then interpret and even predict the correlative consequences. An appropriate ML model and proper training data sets can maximize the accuracy of the expected results. Investigation reveals that researchers are commonly using different ML architectures in scanning probe microscopy (SPM), [8, 9] electron microscopy (EM), [10, 11] different spectroscopies [12–14] and x-ray diffraction (XRD) [15, 16] for pattern and structure recognition, data classification, materials identification, artifacts removal, phase validation, defects investigation, inverse problem, roughness calculation, fringe analysis, auto-alignment, carrier concentration, image segmentation, crystal characteristics, peak position and spectra analysis etc. that significantly impact the overall capability of nanoelectronics characterization.

Researchers have successfully developed neural networks (NNs) and deep learning (DL) models that can accomplish many tasks as an alternative to human operators. For instance, trained NNs have been reported to quantify lattice strain and electric field-driven sample navigation in 4D scanning transmission electron microscopy (STEM) [17–19]. Classification of carbon nanomaterials by VGG-16 network, [20] characterization of carbon nanotubes by multi-layer perception model, [21] inspection of wafer defect by unsupervised NNs [22], etc. have also been possible so far. Tremendous advancement has been observed in SPM metrology capabilities, such as the elimination of image distortions, [23] automatic probe conditioning, [24] enhanced signal-to-noise ratio (SNR), [25] surface reconstruction [26], etc. In nanoelectronics fabrication, thin film quality, thickness measurement, and materials property characterization are critical. Interestingly, ML has shown outstanding performance in these aspects. For example, complex optical modeling was replaced by ML algorithms to calculate film thickness by ellipsometry, [27, 28]. Fringe analysis algorithms analyzed interferometric measurements for the surface topography reconstruction, [29] convolutional neural network (CNN) was introduced to identify and quantify the chemical elements[30], etc. Thus, the contributions of machine learning to nanoelectronics metrology are innumerable.

3 ML Assisted Scanning Probe Microscopy for Nanoelectronics

As the shrinkage of electronic devices continues with time, the importance of precise and localized study of the material's properties becomes dominant. SPM techniques have become very popular in this aspect [31–35]. Despite the groundbreaking capabilities of Scanning Probe Microscopy (SPM) at the nanoscale, traditional SPM techniques are constrained by several factors, including data acquisition speed, noise levels, dynamic processing, and data analysis efficiency [36–40]. Given the scenario, the researchers recently introduced ML in SPM technology to overcome the conventional limitations and improve the overall performance [9, 41–44]. In addition, measurement automation, tip conditioning, 3D imaging, sample identification and classification. Figure 1a shows different areas of application where advancements are possible with the help of ML [45]. It does not only contribute to reduction in experimental time but also enhance the data accuracy by avoiding uncertainty. So far various ML models have shown promising performance to help SPM characterize nanoelectronics devices and materials [8].

For example, ML frameworks have been utilized to study ferroelectric materials in terms of switching characteristics, domain walls, dynamic growth, polarization, etc. [46–51]. Huang et al. reported a support vector machine approach, in short SVM, to identify ferroelectric domains [52]. In this approach, domain walls are subjected to locate hysteresis behavior in piezo response force microscopy (PFM). It helps avoid confusion while identifying ferroelectric regions through switching spectroscopy mode. In a recent work, Liu et. al introduced a hypothesis learning model to investigate the switching behavior in ferroelectric thin films [53]. Here they emphasized on

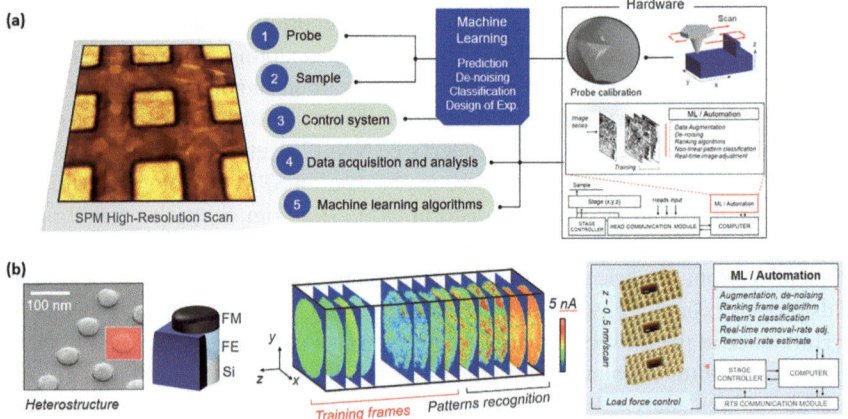

Fig. 1 **a** Diverse applications of ML in scanning probe microscopy for nanoelectronics characterization. **b** Autonomous 3D conductive AFM on metal–oxide–semiconductor sample powered by features recognition and material scalpel control [8]. Laskar et al. APL Machine Learning 1.4 (2023). Licensed under the terms of Creative Commons CC BY

threshold based processing to find the domain changes. At first, domains are written by pulse voltage on $BaTiO_3$ (BTO) sample and then corresponding PFM image data are used to predict the domain switching behavior by applying iterative cycle. On the other hand, researchers used deep neural network to minimize the noise and hyperspectral data in PFM of band excitation [25]. Thus, it allows to characterize a material (usually which provides weak signal response) more robustly. The DNN is quite fast and can also be applied to KPFM and MFM.

With the aim of monitoring tip-induced leakage and defect identification in thin oxide films, Liu et al. proposed a workflow to study the conductive spots autonomously using C-AFM [54]. Based on their controlled trajectory model, the ML framework can explore specific features under a wide field of view and then can carry out localized measurements. For instance, it could identify conductive spots on $Hf_{0.54}Zr_{0.48}O_2$ film and analyze the structure and characteristics of those hot spots that are directly related to the quality, and reliability of the dielectric layer. Object segmentation, background thresholding and field programmable gate array (FPGA) powered control were the fundamental keys of this method. Prior that, ANN model was designed for Scanning Tunneling Microscopy (STM) technique to investigate complex electronic states of quantum matter [55]. Hence it opened a new door for liquid crystal and fractionalized electronic materials characterization. The supervised ANN could automatically recognize modulation characteristics from the STM images throgh d-symmetry. In that study, $Bi_2Sr_2CaCu_2O_8$ and CuO_2 were used as reference samples. Besides, ML has contributed to analysis of complex heterostructure interfaces. For example, Strelcov et al. studied the $BiFeO_3$-$CoFe_2O_4$ heterostructure samples by using the first-order reversal curve technique in I-V [56]. In the I-V mapping, they employed data mining analysis and various equations of electronic transport. That in turn provided significant information on interface oxygen vacancies, Fowler Nordheim tunneling current, Poole Frenkel transport etc.

Measurement assistance and automation are other fields of study that can be widely improved through ML applications in SPM [57–59]. Consequently, enhanced data acquisition, artifacts removal, high accuracy data interpretation etc. have been possible for nanoelectronics samples. For example, Krull et al. introduced the DeepSPM model that can recognize not only good or bad topography images but also navigate the probe to a good area within the sample [60]. Thus it makes the system as self-dependent to record some acceptable AFM images with more than 80% accuracy. Other groups of researchers addressed the issues associated with tip contamination and tip geometric deviation, however, at the same time, they came up with the solution to halt these limitations with the help of different ML models [24, 60]. Some of them utilized the CNN model to identify the double tip effects that commonly appear in the used probes when the tip has multiple vertex points due to degradation. On the other hand, CNN was also used for pulse-based tip conditioning when tip contamination became dominant. Surprisingly, assessment of probe state and corresponding tip condition have achieved more than 90% accuracy through CNN and reverse image checking models. In addition, the findings on image distortion elimination [23], noise reduction [61], and artifact (Z drift) compensation [62] are praiseworthy to advance the nanoelectronics technology one step forward in metrology.

Investigation of grain boundaries (GBs) in photovoltaic materials is very crucial. Researchers commonly rely on conductive AFM and KPFM to investigate the GBs effectively [63–67]. Nonetheless, the process becomes slow because of identifying the grain boundaries manually and executing manual spectroscopy. Liu et. al, for the first time, proposed an ML framework that can identify the GBs and assist in I-V spectroscopy [68]. In this work, they chose hybrid perovskite solar cells as a reference.

4 ML Supported Virtual Metrology for Semiconductor Manufacturing

In the last few years, virtual metrology (VM) has become very popular in the semiconductor industry to achieve highly efficient process control and monitoring [69–71]. VM refers to a way of predicting the properties of something without actually measuring it. VM can effectively estimate the wafer or product quality during the production process without going through each wafer individually. Clearly, VM involves heavy use of machine learning algorithms and previous in-line metrology data, which can provide shortened root cause analysis and enhanced yield [72, 73]. Several ML algorithms are commonly used in different steps of VM process, such as principal component analysis (PCA), partial least squares (PLS), CNN etc. for feature extraction;[74] multiple linear regression (MLR), multiple layer perceptron (MLP), recursive neural network (RNN), gaussian process regression (GPR), Bayesian neural network (BNN) etc. for quality estimation [75].

Greeneltch et. al recently proposed a design oriented VM technique to predict the oxide deposition thickness [76]. It provides metrology augmentation with in-fab data and supports the processing steps. However, it requires a full wafer history to predict the process thickness with high accuracy. In this research, gradient-boosted tree architecture (GBT) is built for chemical vapor deposition (CVD), where a high aspect ratio oxide deposition process was targeted. The GBT data driven approach not only computes design features but also recommends appropriate process recipes. In another work, Choi et al. proposed a neural network (NN) algorithm to predict amorphous carbon film thickness [77]. Initially, in-situ plasma chemistry was augmented for the plasma-enhanced chemical vapor deposition (PECVD) and then applied to the modeling of the plasma process. Their predictive model includes Box-Behnken statistical approach for the experiment design at three levels considering four factors. Among their three predictive models, the model trained with optical emission spectroscopy sensor data and equipment status could show 99.5% accuracy to estimate the carbon layer thickness. Later, researchers introduced advanced VM known as feature based virtual metrology (FVM) to overcome the limitations (non-synchronized trajectory, non-equal step duration etc.) of frequently used VM techniques [78]. Here it doesn't require any data pre-processing and emphasizes on batch features to under-stand batch process dynamic behavior in semiconductor industry. Their FVM model is based on the correlative information of metrology experiments and process features whereas statistics pattern analysis (SPA) [79] plays the most important role. Besides, the authors claim that MLR and Principal component regression (PCR) approaches are not suitable when too many variables (independent) and multilinearity exist.

To achieve the precise nanostructure goals in nanoelectronics fabrication, auto-mated critical Dimension SEM (CD-SEM) plays a significant role in robust and accurate VM [81, 82]. CD-SEM is a system that assists in the precision dimension measurement of nanostructures or patterns on a wafer. Ding et al. demonstrated that their deep learning (DL) architecture is the potential for post-etch metrology compen-sating the background noise, edge low contrast, etc., in CD-SEM images [83]. The proposed architecture includes double U-Net segmentation and binarization which resolves the challenges associated with unbalanced dataset. Figure 2b shows the VM schematic for post etching CD-SEM image. Conversely, chemical mechanical polishing (CMP) has also been studied under virtual metrology. Although CNN, RNN etc. are suitable for CMP-VM, researchers commonly have used BNNs in this case [74]. Moreover, DL models, for example, least squares generative adver-sarial network (LS-GAN), were introduced in CMP for unrealized target outputs and capturing input distribution [84]. Eventually, process drift control was improved, and run-to-run process variation was minimized.

5 ML Aided Electron Microscopy for Nanoelectronics

Machine learning has added an exceptional ability to electron microscopy that directly or indirectly boosts the characterization power of nanoelectronics. Auto-mated experiments, crystal phase or interface identification, image denoising, atomic structure validation, defects classification, composition analysis etc. are the major contributions of ML for scanning electron microscopy (SEM), transmission electron microscopy TEM, scanning TEM (STEM) [85–91]. Figure 3 illustrates few examples of these contributions.

Detection and classification of defects are very important for any type of nano-electronics device, particularly as defects can tune, modify, and compromise the reli-ability and functionality of the device. Therefore, the ability to detect, measure, and tune the materials defectivity is key in the CMOS manufacturing process. Several ML algorithms have been developed and tested to support the electron microscopy for this purpose [11, 92]. For example, a supervised learning based neural network model was proposed for defects identification in $Mo_{1-x}W_xSe_2$ and monolayer graphene samples. Initially they trained the model with exclusive simulated images and then applied to the experimental images to classify a variety of defects [93]. Another group of researchers constructed a deep learning architecture that could recognize not only vacancy defects but also Si impurity in graphene samples [94]. Besides, in doped

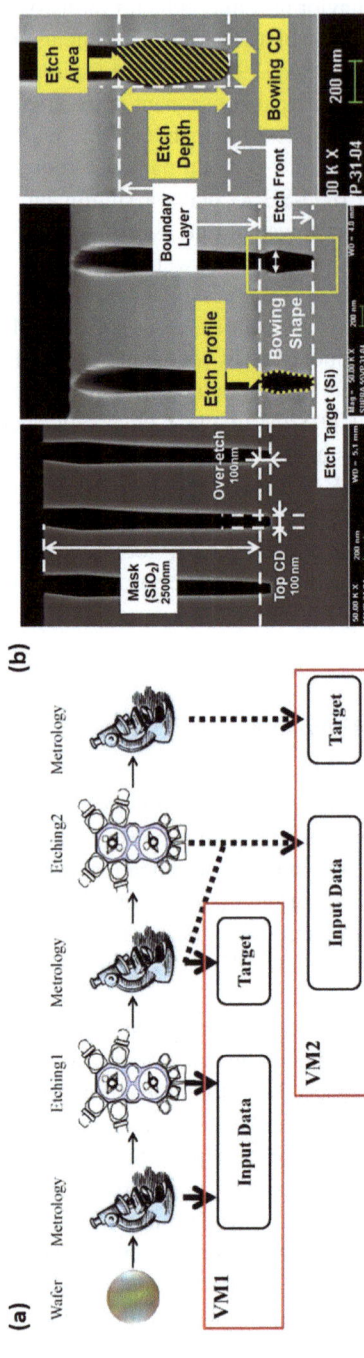

Fig. 2 **a** A generalized schematic of machine learning based VM framework for wafer etching [70]. Reproduced with permission from Kang et al. Expert Systems with Applications 36.10 (2009). Copyright 2009, Elsevier, All Rights Reserved. **b** Quantification of Si etching profile from SEM image through VM process [80]. Kwon et al. Materials 14.11 (2021). Licensed under the terms of Creative Commons CC BY

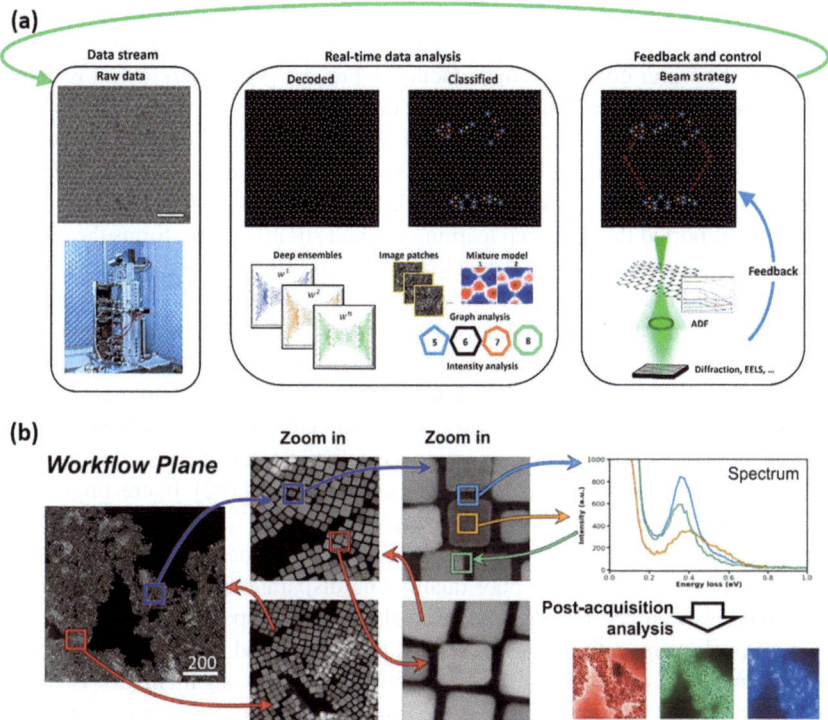

Fig. 3 a A workflow based on ensemble learning and iterative training (ELIT) model for defects evaluation and atomic manipulation [105]. Reproduced with permission from Roccapriore et al. ACS nano. 16,10 (2022). Copyright 2022, American Chemical Society, All Rights Reserved. **b** Example of a ML reinforced workflow for image acquisition, spectra capture and data analysis [106]. Kalinin et al. npj Computational Materials 9.1 (2023). Licensed under the terms of Creative Commons CC BY

WS_2 sample, phase evolution was successfully studied by classifying defect structure and applying density function theory (DFT) [95]. Lee et. al introduced Residual U-Net classifier to identify isolated defects from the annular dark field (ADF) image of $WSe_{2-2x}Te_{2x}$ monolayer sample [96]. It also facilitated the measurement of atomic spacing and strain-related oscillations in proximity to atomic vacancies. On the other hand, Dennler et. al proposed a VGG16 convolution neural network architecture for automated analysis of crystal defects in group III-V semiconductor materials deposited on top of Si substrate [97]. In this work, STEM images were segmented at multiple steps, and various regions were identified based on defectivity, non-defectivity, symmetric fault, etc. Thus, VGG16 CNN was able to outperform the classical Bragg filter method. Earlier, Kirschner et al. developed a neural network model for local composition analysis and thickness measurement in compound semiconductor samples such as AlGaAs [98]. They utilized high resolution transmission electron microscopy (HREM) images of the sample and quantitative image processing using the neural network. Cheon et. al reported a CNN-based defect classification system that can be particularly efficient in classifying various surface defects on the wafer during the manufacturing process [99]. Therefore, it is helpful for process engineers to decide whether the defective wafer can be repaired or not to continue the fabrication process. More than two thousand SEM images were used to train the CNN algorithm and later tested it with 30 images having unknown defects. Eventually, the algorithm model could achieve higher than 90% accuracy. Other researchers reported that the deep learning method slightly outperforms the conventional ML algorithm for wafer defect classification [100]. With a large data set of 12 k augmented images, the test accuracy results show that CNN (95.3%) provides one percent higher accuracy in defect detection and classification than the random forest method (94.2%). In addition, the K-means algorithm is also used to quantify and classify defects from the SEM images [101]. In addition, Jared et. al investigated Inception-V1 algorithm-based CNN architecture for defect classification in semiconductors. Although it outperformed the automatic defect classification (ADC) model, however, the resulting accuracy was just over 85% [102]. Interestingly, some other investigators observed low test accuracy of 60–70% when they used the same datasets, respectively, in ResNet-50 and Inception-V3 models for a fair comparison [103]. It was assumed that the poor quality and disparateness of the initial dataset caused the low accuracy. Conversely, Su et. al reported 3 separate CNN architectures based on back propagation, linear vector quantization and radial basis algorithms for semiconductor wafer post-sawing inspection [104]. Among them, first two could reach almost 100% accuracy whereas the last one ended its performance at 90% of test accuracy.

The necessity of denoising the TEM or STEM data is inevitable as only clean data can provide reliable, meaningful, and reproducible data analysis [107, 108]. The researcher's preferred algorithms are fully convolutional Neural Networks (FCNN), generative adversarial Networks (GAN), sparse coding, ANN, etc., for this purpose. For instance, Lee et. al mentioned FCMM based model to improve the signal to noise ratio (SNR) of ADF-STEM images [96]. Surprisingly, the model could help to determine the strain field around a single atom defect after reducing the noise level effectively. Wang et al. proposed a DL model called Noise2Atom to minimize the noise in HADF-STEM and consecutive structural similarity (CSS) architecture to assess the image quality after the denoising process [109]. A deep neural network consisting of a trained atrous convolutional encoder-decoder was also demonstrated by another group of researchers to decrease signal attenuation [110]. For low and high electron doses, it demonstrated superior performance while denoising the electron micrograph. Based on the CycleGAN approach, Quan et al. showed how an asymmetric model could enhance the electron microscopy images and outperform the traditional Noise2Noise model [111]. It was a semi-supervised learning model without paired training data, which could effectively remove the artifacts. Prior to that, Buban et al. investigated auto-encoder (AE) architecture for noise reduction. With the help of the AE, they could successfully apply on $SrTiO_3$ sample image, indeed, lattice defects were visible from the noise-free reconstructed images [112]. It is also noteworthy to mention the works of Vincent [113] and Mohan [114] for their attempts at deep denoising of TEM images. While one of them developed a supervised deep CNN, another proposed simulation based denoising architecture to study the atomic-level defects. A slightly different approach was observed when convolutional dual-decoder autoencoder (CDDAE) and exit-wave reconstruction were proposed to denoise the atomic resolution TEM images of CVD-grown graphene monolayers [115]. Suveer et al. presented a CNN technique named TSRNet (TEM Super-Resolution Net) to convert low-resolution noisy data into four times higher-resolution data [116]. According to the presented methodology, it consists of three modules such as feature encoder, residual mapping, and upscale processor. Here, the residual mapping enhances the reconstruction of edges. In addition, trained sparse coding has also been studied with the aim of image denoising and achieving high SNR [117, 118]. The result could demonstrate improved phase imaging, hence the precision electron holography.

ML assisted automation can save the electron microscopy operational time and make things much easier for the operator. A couple of investigations on instrument control, alignment, workflows, data acquisition etc. have come to light despite some limitations [10, 106, 119, 120]. Moreno et al. presented a framework using Python machine learning scripts that automates the image capture and reconstruction [121]. Other work shows that sparse data analytics based a closed loop model can trigger the autonomous image acquisition in JEOL STEM. In fact, it demonstrates that data can be collected from a large area of interest on the sample and classify different features [122]. ML-based software package was reported by Schorb et al. for automated tasks of feature detection and image acquisition [123]. Previously, Ghosh et al. explored an iterative deep-learning model for reproducible feature extraction and recognition [124]. Graphene and NiO-lanthanum strontium manganite samples' STEM images were utilized to test the model in this work. The author claimed that U-Net DL can be incorporated to this to improve the uncertainty quantification and robustness of the prediction. Followed by this work, researchers proposed ensemble learning and iterative training (ELIT) CNN to investigate the effects of electron beam irradiation on atomic structure and bonding during live data stream [105]. Here they identified single vacancy lines defect on MoS_2 and topological defect in graphene sample using the ELIT. On the other hand, automated microscopic alignment, aberration correction and converge angle maximization have been also considered through the power machine learning [125, 126]. Recently, Roccapriore et. al established a deep kernel learning (DKL) algorithm that can help in automated navigation and discovering the plasmonic functionality on 2D $MnPS_3$ [127]. Rotunno et al. successfully demonstrated that their designed CNN could determine the misalignment of the orbital angular momentum (OAM) sorter, and it can be potentially used for spherical aberration correction [128, 129]. To push the performance of direct electron detectors (DED) for a nanoscale semiconductor chips, Correa et al. introduced ML, which can provide sub-pixel super-resolution by counting the electrons in terms of beam energy and depth of the diode [130].

It is evident that ML is also a benefaction to nanoelectronics metrology if we talk about the chemical composition evaluation techniques such as energy dispersive X-ray (EDX) and electron energy loss spectroscopy (EELS) embedded in electron microscopy. For example, Chatzidakis et al. demonstrated a deep convolutional neural network to resolve the issue of zero loss peak (ZLP) drift of the EELS spectrometer [131]. It is very important to compensate for the drift. Otherwise, absolute peak position becomes complicated in the spectra. In a different work, ML approach was deployed for the low-loss energy region of the spectra by applying the strategy of ZLP subtraction, thus allowing the opportunity to study local electronic property of the WS_2 nanoflowers [132]. In addition to the contribution of ML algorithms in 3D tomography reconstruction for quantum dots, [133] phase information retrieval by SegNet network, [134] and segmenting, and characterization of carbon nanotubes (CNTs) from the SEM or TEM images [21] are also praiseworthy.

6 Machine Learning in X-Ray Diffraction

XRD allows researchers to analyze the structural and compositional features of crystalline materials and other microstructural attributes. The method was developed when Max Von Laue discovered that crystals diffracted X-rays and experience constructive and destructive interference due to the periodicity in atomic arrangement in the lattice structure. The diffraction pattern is discernible and drawn out from the very interference phenomenon. The diffraction pattern is then captured by a point detector [135], a commonly used detector type to lay out the inferred structural information. This method is, therefore, a fundamental aspect of crystallographic research across the fields of physics, chemistry, and materials science. Analysis of XRD data is crucial in the determination of crystallographic structures, with some of the distinctive features being peak morphology, amplitude, and spatial placement. Well established methodologies such as Sonneveld and Visser technique for preprocessing of data by determining background noise levels followed by refinement of peak positions and intensities. Noise is isolated from the peak contributions by evaluating the noise component's standard deviation and mean value. In addition to the previous, Rietveld [136] and Pawley [137] refinement have traditionally been applied for phase identification and crystallographic analysis. Nevertheless, when we are dealing with novel compounds in thin-film form, traditional methods like Rietveld refinement often encounter constraints owing to the absence of reference patterns in databases and the probability of encountering unknown film textures. For instance, the direct-space method, statistical methodologies, and cultivation of single crystals have been utilized to extract information related to the symmetry of novel materials [138]. Though these methods are effective for the said purposes, they entail significant iteration time, feature engineering, and rely on human intervention, which results in inefficiency, particularly in the context of high throughput experiments. The intricacies of these methods make them less viable for fast acquisition of data, which finally leads to the idea of implementing machine learning and deep neural network models to negate these hinderances. In the following subsection we will introduce Deep Neural Network and Machine Learning concepts briefly and finally explore through some case studies of XRD data analysis derived from recent years. In this section, we will cover a few applications of Machine Learning algorithms in XRD data analysis. These studies are drawn from various literature published in reputed journals. The techniques will give the reader an overview of advancements made in XRD data analysis through ML.

In a study on XRD, Wang et al. used CNN [139] to interpret limited XRD pattern data. To augment the data for training, noise was extracted from experimental spectra to get combined with the highlight peaks of theoretical spectra that were established. The CNN model showed a very high identification rate of 96.7% [140]. The researchers collected theoretical data and experimental data of MOFs. They collected noise from the experimental data, randomized and merged it with mean peaks of the theoretical data to form a new augmented dataset for model training. A Convolutional Neural Network of one input layer, four convolution layers, three fully connected layers, and one output layer was built [141]. Leveraging the CNN models, the main highlight of their work was the one-by-one identification of XRD patterns, which was previously attainable through traditional machine learning algorithms. Their noise-based data augmentation technique expanded the dataset, increasing the robustness of the model against experimental variations like peak shifts, intensity scaling, and FWHM broadening. To further elucidate the decision-making process, CAM [142] or Class Activation maps of convolution layers were used, which in addition also highlighted the inner workings of the model, preventing the inner model from being coined as a black-box [140].

Szymanski et al. demonstrated an autonomous and adaptive XRD [143]. The autonomous workflow of adaptive XRD is represented in Fig. 4. The researchers called the measurement process adaptive characterization. ML enables changes to measurement parameters such as re-scanning parts of XRD diffractograms with higher resolution in a closed loop manner without human intervention, increasing precision and negating the iterative approach. While many recent adaptive techniques have been researched and developed, most of them have been applied to samples which showcased unchanging properties. According to the researcher, the highlight of this method was its implementation on transitional samples. It observed intermediate phases of materials such as Lithium Lanthanum Zirconium Oxide(LLZO), a promising solid electrolyte for lithium batteries [144, 145]. The method effectively reduced scan times from the conventional method's 10 min to less than 4- 6 min while increasing precision in peak detections. Scan times are automatically adjusted to compensate for reduced intensity due to the absence of a synchrotron light source. The phase identification was done using the XRD-Autoanalyzer algorithm [146], a previously published work from the author. XRD-Auto analyzer used CNN along with dynamic time-warping technique to fit spectrum into 2Θ peak position typically in the range of $10°$ to $80°$.

A microstructural characterization study by Boulle et al. demonstrated the use of CNN to invert the XRD dataset to retrieve depth-resolved strain profiles in disordered materials. The CNN model exhibits robust performance in predicting simple

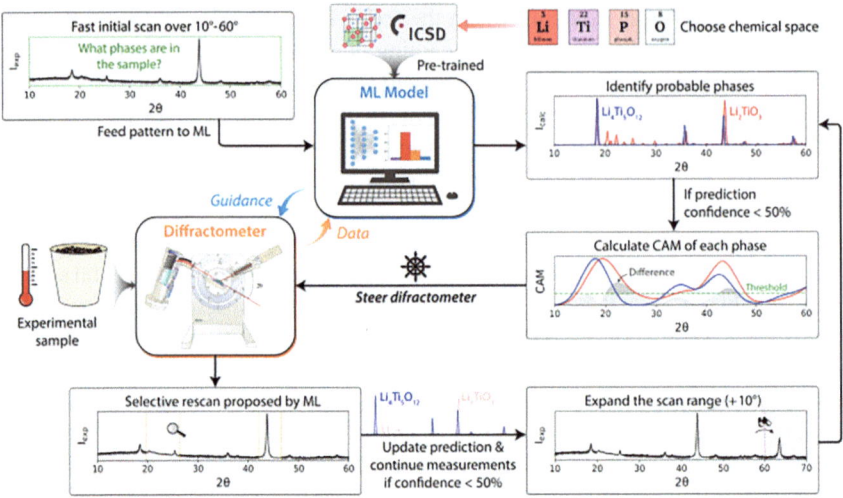

Fig. 4 Shows the workflow of autonomous and Adaptive XRD. Based on retrieved confidence metrics instructions are carried to perform rescans for phase identification and scan range is accordingly adjusted. Szymanski et al. *Npj Comput Mater* **9**, 31 (2023). Licensed under the terms of Creative Commons CC BY

descriptors of strain distribution with high accuracy while also predicting complex strain shapes when insufficient information is present in the XRD dataset [147].

The maximum accuracy of the CNN model was high for predicting strain with an error range superseding that of traditional human-driven analysis. This result again highlights the key advantage of deploying machine learning in XRD data analysis, which pertains to a highly efficient and less iterative approach compared to the human-based method.

Machine learning methods such as support vector machines(SVM) [148], Random Forest (RF), and Gradient Boosting were used by Yanxon et al. to identify artifacts in powder diffraction data. An exemplary powder dataset should have a diffraction pattern characterized by concentric Debye–Scherrer rings, each exhibiting homogenous distribution. The research group referred here investigated methods where machine learning is implemented to identify and separate diffraction spots in XRD images with more than 8 million pixels [149]. Prior to training the machine learning model feature engineering where all the features were normalized to a standard scale. Their finding suggested that gradient boosting is more efficient than gradient boosting for K-Nearest Neighbors (KNN), RF, etc. The yielding TP rates of gradient boosting are over 90%. Though this method has certain drawbacks where it failed to identify spots with strong pixel intensity [149]. While another study from Oviedo et al. [138] showed how augmentation of scarce XRD data can help predict crystallographic dimensions and space groups. They implemented a one-dimensional CNN, which could classify dimensionality with 93% accuracy while having 89% accuracy for space group classification. Negating misclassification of data allows

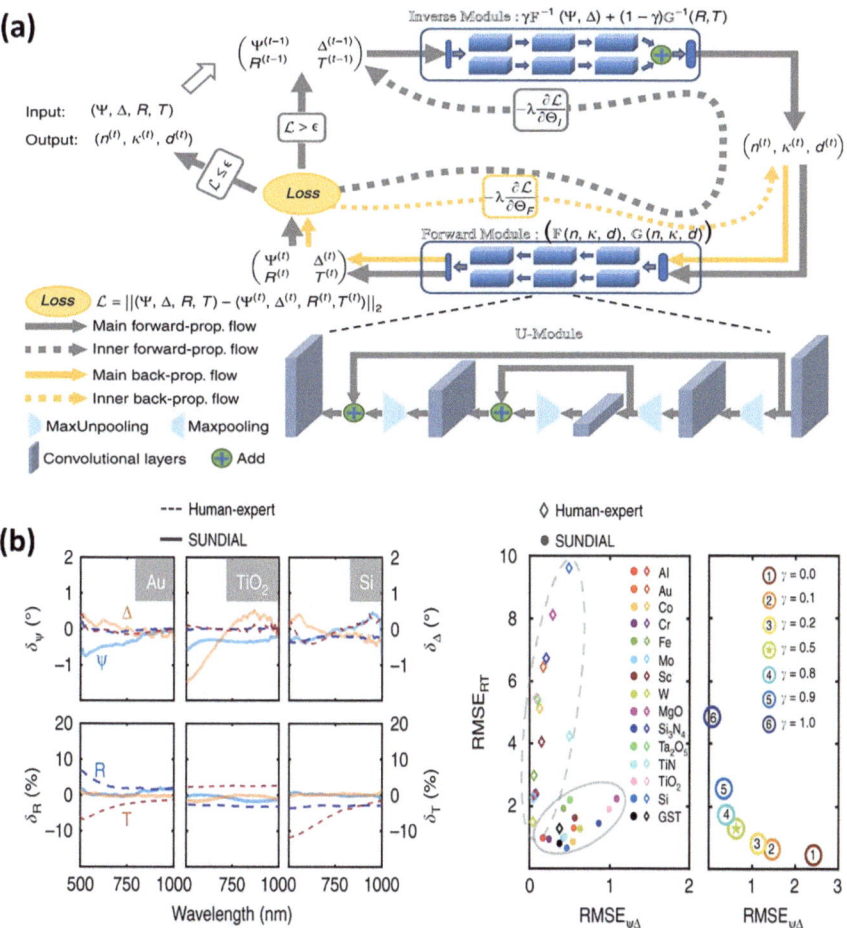

Fig. 5 a Schematic view of SUNDIAL algorithm process flow. Forward and inverse modules of the deep neural network are represented. **b** Comparison of SUNDIAL approach vs traditional fitting method shows accuracy of SUNDIAL. Solid dots represent SUNDIAL RMSE values. Liu et al. *Light Sci Appl* **10**, 55 (2021). Licensed under the terms of Creative Commons CC BY

for high interpretability, for this they used a weak regularizer by using average class activation maps from a global pooling layer. In addition their predictive technique allowed a XRD pattern to be classified in less than 6 min by completing evaluating maximum pattern step size before loss in accuracy takes place[138]. To minimize loss, the Adam optimizer algorithm was implemented; their model had a smaller number of convolution layers compared to previously implemented models. This approach allowed their CNN model to be faster and less susceptible to overfitting.

7   Machine Learning in Optical Metrology

Ellipsometry is a century-old technique to determine changes in polarization states of light when reflected at an acute or obtuse angle from the sample surface. This technique is non-destructive in nature, has a sensitivity of a few nanometers, and can be used to measure thickness at varying environmental conditions such as elevated temperature and pressure, thereby making it suitable as a robust measurement technique in semiconductor process manufacturing. To date, this method and all its variations play a crucial role in ensuring the quality and performance of modern semiconductor devices by measuring the uniformity, roughness, and quality of thin films and patterned structures.

The interpretation of data in this technique is based on inference of motive from the results, which leads it to be categorized as an inverse technique [150]. Generally, the use of various fitting techniques is used to set initial parameters based on the material properties for multiple films or substrates which requires input from an expert. This process is an open loop process based on a trial-and-error method, which makes it inefficient [28]. To negate the involvement of human–machine learning algorithms, such ANN was deployed as a data preprocessor to guess the initial values of (n,k,d) later used for the fitting process by the group at Urban et al. [151]. This method was reported to be fast, but it produced wrong initial parametric inputs, resulting in a bad fit in the data. The group Liu et al. deployed a deep neural network-based model in ellipsometry to solve the previously discussed issues. The SUNDIAL approach is an optimizing framework with two neural modules, the inverse and forward[28], which leads it to be a closed-loop feedback method.

Surface roughness is traditionally measured using stylus profilometry or contact profilometry. In this method a stylus is traced along the surface to form a profile measuring the roughness metrics of the sample. This technique is limited by the area of measurement, one of the potential alternatives is photographic profilometry. Implementation of machine learning techniques and image processing allows this method to thrive and allows for real-time acquisition [152, 153]. CNN is implemented to find surface roughness from raw images with 9% error[154]. Wang et al. reported the use of heterogenous CNN which lowered the error to 4% [155]. In the study from Rifai et al., the group proposed the use of deep learning models for vision-based surface roughness measurement. Their approach did not require feature extraction as they used CNN to automate the feature extraction process. This decreased the prediction time and complexity while making it easier to be implemented in the production workflow. MAPE (Mean Absolute Percentage error) loss function reportedly achieved the accuracy required.

Cooper et al. discussed some of the drawbacks of using CNN to measure surface roughness, one highlighted as CNN not suitable enough for profiling in the perpendicular direction. They addressed this issue by presenting a study on texture-aware surface roughness prediction based on ridgelet transform [156]. They used RF for predicting surface roughness, this approach negated possibilities of instability while computing data of large scale as data normalization was not required while also asserting robustness and accuracy from varying data. The key highlight of using ridgelets was their rotation, translation, and scaling features, which were well suited for surface roughness evaluation measurement technique. They reported an error of 0.5% for mean roughness prediction, which considerably wins over the use of Convolutional Neural Networks discussed earlier [156].

Deep Learning algorithms have also been utilized in interferometry, another optical metrology technique. In this technique, multiple waves are superimposed to form a unique output wave [157]. An optical instrument known as an interferometer is used to measure changes in the wavefront precisely [158]. While looking at various stages throughout the workflow of interferometry, the use of deep learning in data preprocessing, analysis, and post-processing is highlighted by most research groups. Preprocessing data has several benefits. Some of the crucial advantages are denoising, negating geometrical distortions by correction of sub-pixel through classical mathematical models, enhancement of fringe pattern, and separation of color channel for faster acquisition speed.

8 Machine Learning in Spectroscopy

Spectrometry is a general term encompassing techniques that measure the interactions between any kind of radiation and matter. Probing how matter interacts with the radiation differently, absorbing, reflecting, or scattering it in specific ways. It can be thought of as a form of interrogation of the sample where we send a type of radiation to a specimen and listen to its "response" to learn about its properties. Machine Learning and Deep Learning algorithms have also found their usage in spectroscopy. Its application ranges from data optimization, enhancement, automation, and quantification to highlight a few. In this section, we will draw examples from the literature on the implementation of ML in a few spectroscopy techniques like Raman Spectroscopy, XPS (X-ray photoelectron Spectroscopy), NMR (Nuclear Magnetic Resonance spectroscopy), and EELS.

Raman Spectroscopy is a non-destructive, robust, sensitive and versatile analytical method to obtain chemical inference of materials [159]. The principle of Raman spectroscopy lies in the interaction of matter with electromagnetic radiation. This interaction is conducted through absorption, scattering, and transmission. The Raman effect occurs when electromagnetic radiation or light interacts with electrons in the chemical bond of a material, which causes the excitation of molecules and results in a frequency shift of the interacted light. In addition to that, the Raman effect is also observed in inelastic scattering in excited target material due to the inherent lattice vibration of the material. As the lattice of vibration of all materials can be quantified, the presence or traces of such in a sample can be easily identified [160]. The method was used to identify strains in two-dimensional chalcogenides by Li et al.; they measured local strain on molecular beam epitaxially grown thin film chalcogenides [161]. Raman was particularly useful as other methods, such as XRD, have rather large spot sizes to accommodate device-level study. With the current increase of manufacturing the complex and often processing time of spectral data from Raman spectroscopy can be a hindrance. Therefore, the application of machine learning algorithms makes the workflow more efficient as evident from its application in other metrology areas. Qi et al. published a review of recent studies on machine-assisted Raman spectroscopy, discussing progress made with the implementation of established ML methods such as KNN, RF, SVM, and neural network-based algorithms like CNN, RNN, GAN, and ANN [162]. The efficiency of ML implementation was highlighted while analyzing twist angle and vibrational metrics of twisted bilayer graphene(tBLG) by Sheremetyeva et al. They implemented machine learning regressor models for correlating Raman spectral data and the twist angle of tBLG [163].

In a study published by Zhang et al. on high-speed identification of carbon nanotubes using Raman Spectroscopy with implementation of CNN [164]. When performing Raman scattering, signals can get masked by noise, which hinders high-speed imaging by inducing noise. The group achieved high speed Raman imaging capability with minimal signal to noise ratio of 0.9. Their reported accuracy was more than 90% for labelling spectral. Mao et al. published an analysis using machine learning assisted Raman of MoS_2, a two-dimensional material with potential implementation in device fabrication [165]. They implemented an RF algorithm for extracting metrics from the Raman spectra for the classification of MoS_2 film. The use of RF solved the inefficiency of sample identification using a single variable and categorized and predicted defects that would otherwise be ignored. To further add to the research of implementing ML in Raman, Lu et al. correlated Raman maps with photoluminescence features from monolayer MoS_2 using DenseNet, gradient-boosted trees, and Shapley additive explanation (SHAP) models [166]. They also deployed SVM and Gaussian Mixture Model (GMM) to relate features across Raman and Photoluminescent attributes. Figure 6a shows their results along with an illustration of the DenseNet model. Their study paved a new characterization methodology for two-dimensional materials [166]. Another area where the integration of ML is used significantly is data enhancement. The spectral data from Raman spectroscopy is often insufficient to implement deep learning techniques. Kim et al. discussed data enhancement using GAN. The group used data augmentation in spectrum data using federated learning, which consists of models such as GAN. This method was reportedly more effective than existing methods [167].

XPS is another robust analytical technique for compositional and structural knowledge. The probing depth of XPS is a few nanometers. XPS data is hindered by quantification error due to the presence of noise and change in the resolution of energy spectra. The consideration of fitting the model depends on these factors, making the process tedious for inexperienced users. Automation of XPS analysis is thus required, and in recent years, many research groups have implemented machine learning and deep learning to automate the process[168, 169]. In a recent study by Pielsticker et al., supervised machine learning framework-based CNN was implemented to automate transitional metal XPS data quantification. The training dataset was composed using artificial spectral data from XPS [170].

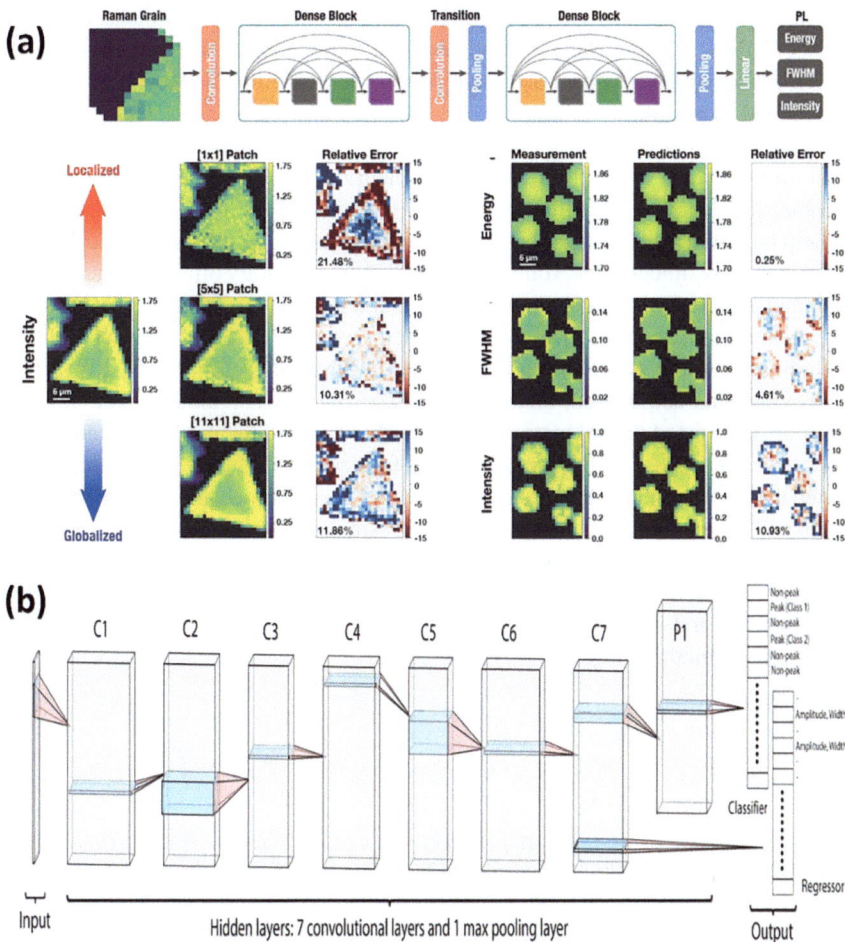

Fig. 6 a Shows the prediction results from DenseNet. The schematic of the model is represented along with measured data set, predicted results and relative error and input dataset in clockwise order. Lu et al., Adv. *Mater.* 2022, 34, 2,202,911. Licensed under the terms of Creative Commons CC BY. **b** Figurative schematic of DEEPPicker used in 2d NMR spectra. Seven CNN layers are represented along with max pooling layer. Output schematic shows data falling under classifier and regressor. Li et al. *Nat Commun* **12**, 5229 (2021). Licensed under the terms of Creative Commons CC BY

Nuclear Magnetic Response has often been a go to technique for researchers to study structural and chemical identities of materials. The basic principle behind this technique is the use of magnetic field to transfer energy from a lower level to higher energy level in a nuclei which inherently has multiple spins and is charged electrically[171]. Implementing machine learning algorithms in NMR has a lot of potential, and recent studies have shown the same. Li et al. proposed a deep neural network for NMR spectral peak picking coined as DEEP picker, which essentially allows automation in analyzing 2D NMR spectra [172], Fig. 6b shows the schematic view of DEEP Picker. Their method resolved the challenge of overlapping peaks which earlier involved a trial-and-error process to determine whether a point could be labelled as a peak.

In another work NMR shift prediction was performed where data points were less than optimal. Rull et al. proposed a graph neural network based message passing Geometrical learning algorithm where edges were utilized for passing information to help build data locally at nodes while utilizing edge aggregation function [173].

In summary, integrating AI into semiconductor metrology offers enhanced precision and efficiency in the characterization and analysis of today's nanoelectronics. ML techniques have been successfully applied across various metrological methods, including scanning probe microscopy, electron microscopy, and X-ray diffraction, improving data acquisition, analysis, and experimental automation. The development of AI-assisted virtual metrology has further enabled the prediction and control of critical parameters during semiconductor fabrication, addressing challenges associated with traditional metrology approaches. As device dimensions continue to scale down and fabrication processes become increasingly complex, the role of AI in optimizing metrology flow will become even more critical. This chapter thus explored the transformative role of AI in semiconductor metrology, highlighting its current contributions and potential for future advancements and positioning the role of AI in metrology methodologies.

References

1. Commerce, U.D.o., *Metrology Gaps in the Semiconductor Ecosystem*, C.R.a.D. Office, Editor. 2023, National Institute of Standards and Technology (NIST)
2. CIRCUITS, I., *CRAMMING MORE*. Understanding Moore's Law: Four Decades of Innovation, 2006: p. 55.
3. Lundstrom M et al (2007) Nanoelectronics: metrology and computation. In: AIP conference proceedings. 2007. American Institute of Physics
4. Ma Z, Seiler DG (2017) Metrology and diagnostic techniques for nanoelectronics. CRC Press
5. Brown KA et al (2019) Machine learning in nanoscience: big data at small scales. Nano Lett 20(1):2–10
6. Bishop CM, Nasrabadi NM (2006) Pattern recognition and machine learning. 4. Springer
7. Extance A (2018) How AI technology can tame the scientific literature. Nature 561(7722):273–275
8. Rahman Laskar MA, Celano U (2023) Scanning probe microscopy in the age of machine learning. APL Mach Learn 1(4)
9. Gordon OM, Moriarty PJ (2020) Machine learning at the (sub) atomic scale: next generation scanning probe microscopy. Mach Learn: Sci Technol 1(2):023001
10. Botifoll, M., I. Pinto-Huguet, and J. Arbiol, *Machine learning in electron microscopy for advanced nanocharacterization: current developments, available tools and future outlook*. Nanoscale Horizons, 2022.
11. Treder, K.P., et al., *Applications of deep learning in electron microscopy*. Microscopy, 2022. **71**(Supplement_1): p. i100-i115.
12. Zuo, C., et al., *Deep learning in optical metrology: a review*. Light: Science & Applications, 2022. **11**(1): p. 39.

13. Kwak H, Kim J (2023) Semiconductor Multilayer Nanometrology with Machine Learning. Nanomanufacturing and Metrology 6(1):15
14. Lansford JL, Vlachos DG (2020) Infrared spectroscopy data-and physics-driven machine learning for characterizing surface microstructure of complex materials. Nat Commun 11(1):1513
15. Surdu V-A, György R (2023) X-ray Diffraction Data Analysis by Machine Learning Methods—A Review. Appl Sci 13(17):9992
16. Zhao X et al (2023) Machine Learning Automated Analysis of Enormous Synchrotron X-ray Diffraction Datasets. The Journal of Physical Chemistry C 127(30):14830–14838
17. Bruefach A, Ophus C, Scott MC (2022) Analysis of interpretable data representations for 4D-STEM using unsupervised learning. Microsc Microanal 28(6):1998–2008
18. Munshi, J., et al., *Disentangling multiple scattering with deep learning: application to strain mapping from electron diffraction patterns.* npj Computational Materials, 2022. **8**(1): p. 254.
19. Roccapriore KM et al (2022) Automated experiment in 4D-STEM: exploring emergent physics and structural behaviors. ACS Nano 16(5):7605–7614
20. Luo Q, Holm EA, Wang C (2021) A transfer learning approach for improved classification of carbon nanomaterials from TEM images. Nanoscale Advances 3(1):206–213
21. Al-Khedher M et al (2007) Quality classification via Raman identification and SEM analysis of carbon nanotube bundles using artificial neural networks. Nanotechnology 18(35):355703
22. Chang C-Y et al (2009) An unsupervised neural network approach for automatic semiconductor wafer defect inspection. Expert Syst Appl 36(1):950–958
23. Kocur V et al (2023) Correction of AFM data artifacts using a convolutional neural network trained with synthetically generated data. Ultramicroscopy 246:113666
24. Rashidi M, Wolkow RA (2018) Autonomous scanning probe microscopy in situ tip conditioning through machine learning. ACS Nano 12(6):5185–5189
25. Borodinov, N., et al., *Deep neural networks for understanding noisy data applied to physical property extraction in scanning probe microscopy.* npj Computational Materials, 2019. **5**(1): p. 25.
26. Villarrubia JS (1997) Algorithms for scanned probe microscope image simulation, surface reconstruction, and tip estimation. J Res Nat Inst Stand Technol 102(4):425
27. Kwak, H., et al., *Non-destructive thickness characterisation of 3D multilayer semiconductor devices using optical spectral measurements and machine learning.* Light: Advanced Manufacturing, 2021. **2**(1): p. 9–19.
28. Liu, J., et al., *Machine learning powered ellipsometry.* Light: Science & Applications, 2021. **10**(1): p. 55.
29. Wu D, Fang F (2021) Development of surface reconstruction algorithms for optical interferometric measurement. Front Mech Eng 16:1–31
30. Drera G, Kropf CM, Sangaletti L (2020) Deep neural network for x-ray photoelectron spectroscopy data analysis. Machine Learning: Science and Technology 1(1):015008
31. Hui F, Lanza M (2019) Scanning probe microscopy for advanced nanoelectronics. Nature electronics 2(6):221–229
32. Celano, U., *The Atomic Force Microscopy for Nanoelectronics.* Electrical Atomic Force Microscopy for Nanoelectronics, 2019: p. 1–28.
33. Celano U et al (2014) Three-dimensional observation of the conductive filament in nanoscaled resistive memory devices. Nano Lett 14(5):2401–2406
34. Laskar MAR et al (2024) Adaptive Scalpel Scanning Probe Microscopy for Enhanced Volumetric Sensing in Tomographic Analysis. Adv Mater Interfaces 11(21):2400187
35. Chakrabarti, S., *Study of Dielectric Degradation Using Self-Sensing and Optical Conductive Probes.* 2024, Arizona State University.
36. Gołek F et al (2014) AFM image artifacts. Appl Surf Sci 304:11–19
37. Hla S-W et al (2004) Single-atom extraction by scanning tunneling microscope tip crash and nanoscale surface engineering. Nano Lett 4(10):1997–2001
38. Straton JC et al (2015) Removal of multiple-tip artifacts from scanning tunneling microscope images by crystallographic averaging. Advanced Structural and Chemical Imaging 1(1):1–12

39. Tewari S et al (2017) Robust procedure for creating and characterizing the atomic structure of scanning tunneling microscope tips. Beilstein J Nanotechnol 8(1):2389–2395

40. Wagner R et al (2011) Uncertainty quantification in nanomechanical measurements using the atomic force microscope. Nanotechnology 22(45):455703

41. Kalinin, S.V., et al., *Big, deep, and smart data in scanning probe microscopy.* 2016, ACS Publications.

42. Nguyen LTP, Liu BH (2022) Emerging machine learning strategies for diminishing measurement uncertainty in SPM nanometrology. Surf Topogr Metrol Prop 10(3):033002

43. Vasudevan RK et al (2023) A processing and analytics system for microscopy data workflows: the Pycroscopy ecosystem of packages. Advanced Theory and Simulations 6(11):2300247

44. Liu, Y., et al., *Harnessing Automation and Machine Learning in Scanning Probe Microscopy to Accelerate Physics Discovery.* Bulletin of the American Physical Society, 2024.

45. Gongora, A.E., et al., *A Bayesian experimental autonomous researcher for mechanical design.* Science advances, 2020. **6**(15): p. eaaz1708.

46. Agar JC et al (2019) Revealing ferroelectric switching character using deep recurrent neural networks. Nat Commun 10(1):4809

47. Holstad, T.S., et al., *Application of a long short-term memory for deconvoluting conductance contributions at charged ferroelectric domain walls.* npj Computational Materials, 2020. **6**(1): p. 163.

48. Li, L., et al., *Machine learning–enabled identification of material phase transitions based on experimental data: Exploring collective dynamics in ferroelectric relaxors.* Science advances, 2018. **4**(3): p. eaap8672.

49. Liu Y et al (2020) High-speed piezoresponse force microscopy and machine learning approaches for dynamic domain growth in ferroelectric materials. ACS Appl Mater Interfaces 12(8):9944–9952

50. Zhang F et al (2021) Maximizing Information: a machine learning approach for analysis of complex nanoscale electromechanical behavior in defect-rich PZT films. Small Methods 5(12):2100552

51. Gobeljic D et al (2016) Nanoscale mapping of heterogeneity of the polarization reversal in lead-free relaxor–ferroelectric ceramic composites. Nanoscale 8(4):2168–2176

52. Huang B, Li Z, Li J (2018) An artificial intelligence atomic force microscope enabled by machine learning. Nanoscale 10(45):21320–21326

53. Liu, Y., et al., *Autonomous scanning probe microscopy with hypothesis learning: Exploring the physics of domain switching in ferroelectric materials.* Patterns, 2023. **4**(3).

54. Liu, Y., et al., *Exploring leakage in dielectric films via automated experiments in scanning probe microscopy.* Applied Physics Letters, 2022. **120**(18).

55. Zhang Y et al (2019) Machine learning in electronic-quantum-matter imaging experiments. Nature 570(7762):484–490

56. Strelcov E et al (2014) Deep data analysis of conductive phenomena on complex oxide interfaces: physics from data mining. ACS Nano 8(6):6449–6457

57. Kelley KP et al (2020) Fast scanning probe microscopy via machine learning: non-rectangular scans with compressed sensing and gaussian process optimization. Small 16(37):2002878

58. Liu, Y.-L., et al. *An on-line variable speed scanning method with machine learning based feedforward control for atomic force microscopy.* in *2019 12th Asian Control Conference (ASCC).* 2019. IEEE.

59. Liu, Y., et al., *Explainability and human intervention in autonomous scanning probe microscopy.* Patterns, 2023. **4**(11).

60. Krull A et al (2020) Artificial-intelligence-driven scanning probe microscopy. Communications Physics 3(1):54

61. Roy, P., et al. *Adaptive thresholding: A comparative study.* in *2014 International conference on control, Instrumentation, communication and Computational Technologies (ICCICCT).* 2014. IEEE.

62. Cerbu, D., et al. *Deep learning-enabled vertical drift artefact correction for AFM images.* in *Metrology, Inspection, and Process Control XXXVI.* 2022. SPIE.

63. Bahrami B et al (2020) Nanoscale spatial mapping of charge carrier dynamics in perovskite solar cells. Nano Today 33:100874

64. Laskar MAR et al (2020) Phenylhydrazinium iodide for surface passivation and defects suppression in perovskite solar cells. Adv Func Mater 30(22):2000778

65. Reza KM et al (2021) Grain Boundary Defect Passivation in Quadruple Cation Wide-Bandgap Perovskite Solar Cells. Solar RRL 5(4):2000740

66. Laskar, M.A.R. and Q. Qiao. *Sodium Hydroxide Assisted Work Function Tuning of Hole Transport Material for Efficient Wide-Bandgap Perovskite Solar Cells.* in *Electrochemical Society Meeting Abstracts prime2020.* 2020. The Electrochemical Society, Inc.

67. Laskar, M.A.R. and Q. Qiao. *Chemically aromatic novel additive material for short circuit current density improvement in organic-inorganic hybrid perovskite solar cells.* in *Organic, Hybrid, and Perovskite Photovoltaics XXI.* 2020. SPIE.

68. Liu, Y., et al., *Disentangling electronic transport and hysteresis at individual grain boundaries in hybrid perovskites via automated scanning probe microscopy.* arXiv preprint arXiv:2210. 14138, 2022.

69. Gu, J.M. and S.J. Hong. *Virtual metrology for tsv etch depth measurement using optical emission spectroscopy.* in *2015 IEEE Electrical Design of Advanced Packaging and Systems Symposium (EDAPS).* 2015. IEEE.

70. Kang P et al (2009) A virtual metrology system for semiconductor manufacturing. Expert Syst Appl 36(10):12554–12561

71. Kim, T.-G., et al. *In-line metrology for atomic resolution local height variation.* in *2017 28th Annual SEMI Advanced Semiconductor Manufacturing Conference (ASMC).* 2017. IEEE.

72. Chen, P., et al. *Virtual metrology: A solution for wafer to wafer advanced process control.* in *ISSM 2005, IEEE International Symposium on Semiconductor Manufacturing, 2005.* 2005. IEEE.

73. Lee KB, Kim CO (2020) Recurrent feature-incorporated convolutional neural network for virtual metrology of the chemical mechanical planarization process. J Intell Manuf 31(1):73–86

74. Dreyfus, P.-A., et al., *Virtual metrology as an approach for product quality estimation in Industry 4.0: a systematic review and integrative conceptual framework.* International Journal of Production Research, 2022. **60**(2): p. 742–765.

75. Vallejo M et al (2019) Soft metrology based on machine learning: a review. Meas Sci Technol 31(3):032001

76. Greeneltch, N.G., et al. *Design-aware virtual metrology and process recipe recommendation.* in *DTCO and Computational Patterning II.* 2023. SPIE.

77. Choi, J.E. and S.J. Hong, *Machine learning-based virtual metrology on film thickness in amorphous carbon layer deposition process.* Measurement: Sensors, 2021. **16**: p. 100046.

78. Suthar K et al (2019) Next-generation virtual metrology for semiconductor manufacturing: A feature-based framework. Comput Chem Eng 127:140–149

79. Wang J, He QP (2010) Multivariate statistical process monitoring based on statistics pattern analysis. Ind Eng Chem Res 49(17):7858–7869

80. Kwon J-W et al (2021) Development of virtual metrology using plasma information variables to predict Si etch profile processed by SF6/O2/Ar capacitively coupled plasma. Materials 14(11):3005

81. Orji NG et al (2018) Metrology for the next generation of semiconductor devices. Nature electronics 1(10):532–547

82. Chien, K.-C., C.-H. Chang, and D. Djurdjanovic, *Virtual metrology modeling of reactive ion etching based on statistics-based and dynamics-inspired spectral features.* Journal of Vacuum Science & Technology B, 2021. **39**(6).

83. Ding, S., et al. *Double U-Net based Virtual Metrology on Plasma-Etch CD-SEM Images: AM: Advanced Metrology.* in *2023 34th Annual SEMI Advanced Semiconductor Manufacturing Conference (ASMC).* 2023. IEEE.

84. Kim S, Jang J, Kim CO (2021) A run-to-run controller for a chemical mechanical planarization process using least squares generative adversarial networks. J Intell Manuf 32:2267–2280

85. Dey, B., *Deep Learning based SEM image Denoising Approaches for Improving Metrology and Inspection of Thin Resists.* 2023.
86. Bonnet N (1998) Multivariate statistical methods for the analysis of microscope image series: applications in materials science. J Microsc 190(1–2):2–18
87. Bosman M et al (2006) Mapping chemical and bonding information using multivariate analysis of electron energy-loss spectrum images. Ultramicroscopy 106(11–12):1024–1032
88. Jesse S et al (2016) Big data analytics for scanning transmission electron microscopy ptychography. Sci Rep 6(1):26348
89. Kalinin SV, Sumpter BG, Archibald RK (2015) Big–deep–smart data in imaging for guiding materials design. Nat Mater 14(10):973–980
90. Torruella P et al (2018) Clustering analysis strategies for electron energy loss spectroscopy (EELS). Ultramicroscopy 185:42–48
91. Kim, M., et al. *Frequency-informed deep-learning denoising method supporting sub-nm metrology for high NA EUV lithography.* in *DTCO and Computational Patterning II.* 2023. SPIE.
92. López de la Rosa, F., et al., *A review on machine and deep learning for semiconductor defect classification in scanning electron microscope images.* Applied Sciences, 2021. **11**(20): p. 9508.
93. Ziatdinov M et al (2017) Deep learning of atomically resolved scanning transmission electron microscopy images: chemical identification and tracking local transformations. ACS Nano 11(12):12742–12752
94. Ziatdinov, M., et al., *Building and exploring libraries of atomic defects in graphene: Scanning transmission electron and scanning tunneling microscopy study.* Science advances, 2019. **5**(9): p. eaaw8989.
95. Maksov, A., et al., *Deep learning analysis of defect and phase evolution during electron beam-induced transformations in WS2.* npj Computational Materials, 2019. **5**(1): p. 12.
96. Lee C-H et al (2020) Deep learning enabled strain mapping of single-atom defects in two-dimensional transition metal dichalcogenides with sub-picometer precision. Nano Lett 20(5):3369–3377
97. Azimi SM et al (2018) Advanced steel microstructural classification by deep learning methods. Sci Rep 8(1):2128
98. Kirschner H, Hillebrand R (2000) Neural networks for HREM image analysis. Inf Sci 129(1–4):31–44
99. Cheon S et al (2019) Convolutional neural network for wafer surface defect classification and the detection of unknown defect class. IEEE Trans Semicond Manuf 32(2):163–170
100. Lei, H., et al. *Automated wafer defect classification using a convolutional neural network augmented with distributed computing.* in *2020 31st Annual SEMI Advanced Semiconductor Manufacturing Conference (ASMC).* 2020. IEEE.
101. Halder, S., et al. *SEM image analysis with K-means algorithm: AM: Advanced metrology/ DI: Defect inspection.* in *2018 29th Annual SEMI Advanced Semiconductor Manufacturing Conference (ASMC).* 2018. IEEE.
102. O'Leary J, Sawlani K, Mesbah A (2020) Deep learning for classification of the chemical composition of particle defects on semiconductor wafers. IEEE Trans Semicond Manuf 33(1):72–85
103. Imoto K et al (2019) A CNN-based transfer learning method for defect classification in semiconductor manufacturing. IEEE Trans Semicond Manuf 32(4):455–459
104. Su C-T, Yang T, Ke C-M (2002) A neural-network approach for semiconductor wafer post-sawing inspection. IEEE Trans Semicond Manuf 15(2):260–266
105. Roccapriore KM et al (2022) Probing electron beam induced transformations on a single-defect level via automated scanning transmission electron microscopy. ACS Nano 16(10):17116–17127
106. Kalinin, S.V., et al., *Machine learning for automated experimentation in scanning transmission electron microscopy.* npj Computational Materials, 2023. **9**(1): p. 227.

107. De Backer A et al (2013) Atom counting in HAADF STEM using a statistical model-based approach: Methodology, possibilities, and inherent limitations. Ultramicroscopy 134:23–33

108. De Backer A et al (2016) StatSTEM: An efficient approach for accurate and precise model-based quantification of atomic resolution electron microscopy images. Ultramicroscopy 171:104–116

109. Wang F et al (2020) Noise2Atom: unsupervised denoising for scanning transmission electron microscopy images. Applied Microscopy 50(1):1–9

110. Ede JM, Beanland R (2019) Improving electron micrograph signal-to-noise with an atrous convolutional encoder-decoder. Ultramicroscopy 202:18–25

111. Quan, T.M., et al. *Removing imaging artifacts in electron microscopy using an asymmetrically cyclic adversarial network without paired training data.* in *2019 IEEE/CVF International Conference on Computer Vision Workshop (ICCVW).* 2019. IEEE.

112. Buban JP, Choi S-Y (2017) Auto-encoders for noise reduction in scanning transmission electron microscopy. Microsc Microanal 23(S1):130–131

113. Vincent JL et al (2021) Developing and evaluating deep neural network-based denoising for nanoparticle TEM images with ultra-low signal-to-noise. Microsc Microanal 27(6):1431–1447

114. Mohan S et al (2022) Deep denoising for scientific discovery: A case study in electron microscopy. IEEE Transactions on Computational Imaging 8:585–597

115. Lee J et al (2020) Contrast transfer function-based exit-wave reconstruction and denoising of atomic-resolution transmission electron microscopy images of graphene and cu single atom substitutions by deep learning framework. Nanomaterials 10(10):1977

116. Suveer, A., et al. *Super-resolution reconstruction of transmission electron microscopy images using deep learning.* in *2019 IEEE 16th International Symposium on Biomedical Imaging (ISBI 2019).* 2019. IEEE.

117. Anada S et al (2019) Sparse coding and dictionary learning for electron hologram denoising. Ultramicroscopy 206:112818

118. Anada S et al (2020) Simulation-trained sparse coding for high-precision phase imaging in low-dose electron holography. Microsc Microanal 26(3):429–438

119. Al-Najjar, A., et al. *Enabling autonomous electron microscopy for networked computation and steering.* in *2022 IEEE 18th International Conference on e-Science (e-Science).* 2022. IEEE.

120. Kalinin SV et al (2021) Automated and autonomous experiments in electron and scanning probe microscopy. ACS Nano 15(8):12604–12627

121. Moreno XC et al (2023) An open-source microscopy framework for simultaneous control of image acquisition, reconstruction, and analysis. HardwareX 13:e00400

122. Olszta M et al (2022) An automated scanning transmission electron microscope guided by sparse data analytics. Microsc Microanal 28(5):1611–1621

123. Schorb M et al (2019) Software tools for automated transmission electron microscopy. Nat Methods 16(6):471–477

124. Ghosh, A., et al., *Ensemble learning-iterative training machine learning for uncertainty quantification and automated experiment in atom-resolved microscopy.* npj Computational Materials, 2021. **7**(1): p. 100.

125. Zhang C et al (2021) Aberration corrector tuning with machine-learning-based emittance measurements and Bayesian optimization. Microsc Microanal 27(S1):810–812

126. Sagawa R et al (2021) Aberration measurement and correction in scanning transmission electron microscopy using machine learning. Microsc Microanal 27(S1):814–816

127. Roccapriore KM, Kalinin SV, Ziatdinov M (2022) Physics discovery in nanoplasmonic systems via autonomous experiments in scanning transmission electron microscopy. Advanced Science 9(36):2203422

128. Rotunno E et al (2021) Convolutional neural network as a tool for automatic alignment of electron optical beam shaping devices. Microsc Microanal 27(S1):822–824

129. Rotunno E et al (2021) Alignment of electron optical beam shaping elements using a convolutional neural network. Ultramicroscopy 228:113338

130. Correa G, Muller D (2020) Machine Learning for Sub-pixel Super-resolution in Direct Electron Detectors. Microsc Microanal 26(S2):1932–1934
131. Chatzidakis M, Botton G (2019) Towards calibration-invariant spectroscopy using deep learning. Sci Rep 9(1):2126
132. Roest LI et al (2021) Charting the low-loss region in electron energy loss spectroscopy with machine learning. Ultramicroscopy 222:113202
133. Han Y et al (2021) Deep learning STEM-EDX tomography of nanocrystals. Nature Machine Intelligence 3(3):267–274
134. Cherukara MJ, Nashed YS, Harder RJ (2018) Real-time coherent diffraction inversion using deep generative networks. Sci Rep 8(1):16520
135. Brügemann L, Gerndt EKE (2004) Detectors for X-ray diffraction and scattering: a user's overview. Nucl Instrum Methods Phys Res, Sect A 531(1):292–301
136. Rietveld HM (1969) A profile refinement method for nuclear and magnetic structures. J Appl Crystallogr 2(2):65–71
137. Pawley GS (1981) Unit-cell refinement from powder diffraction scans. J Appl Crystallogr 14(6):357–361
138. Oviedo, F., et al., *Fast and interpretable classification of small X-ray diffraction datasets using data augmentation and deep neural networks*. npj Computational Materials, 2019. **5**(1): p. 1–9.
139. LeCun Y, Bengio Y, Hinton G (2015) Deep learning. Nature 521(7553):436–444
140. Wang, H., et al., *Rapid Identification of X-ray Diffraction Patterns Based on Very Limited Data by Interpretable Convolutional Neural Networks*. Journal of Chemical Information and Modeling, 2020. **60**(4).
141. Srivastava, N., et al., *Dropout: A Simple Way to Prevent Neural Networks from Overfitting*.
142. Zhou, B., et al. *Learning Deep Features for Discriminative Localization*. in *2016 IEEE Conference on Computer Vision and Pattern Recognition (CVPR)*. 2016. IEEE.
143. Szymanski, N.J., et al., *Adaptively driven X-ray diffraction guided by machine learning for autonomous phase identification*. npj Computational Materials, 2023. **9**(1): p. 1–8.
144. Hu Y et al (2022) Probing the Phase Transition during the Formation of Lithium Lanthanum Zirconium Oxide Solid Electrolyte. ACS Appl Mater Interfaces 14(37):41978–41987
145. Manthiram A, Yu X, Wang S (2017) Lithium battery chemistries enabled by solid-state electrolytes. Nat Rev Mater 2(4):1–16
146. Szymanski NJ et al (2021) Probabilistic Deep Learning Approach to Automate the Interpretation of Multi-phase Diffraction Spectra. Chem Mater 33(11):4204–4215
147. Boulle A, Debelle A (2023) Convolutional neural network analysis of x-ray diffraction data: strain profile retrieval in ion beam modified materials. Machine Learning: Science and Technology 4(1):015002
148. Cortes C, Vapnik V (1995) Support-vector networks. Mach Learn 20(3):273–297
149. Yanxon H et al (2023) Artifact identification in X-ray diffraction data using machine learning methods. J Synchrotron Radiat 30(Pt 1):137–146
150. Rosa R (1988) The inverse problem of ellipsometry: a bootstrap approach. Inverse Prob 4(3):887
151. Urban FK, Tabet MF (1994) Real time, in-situ ellipsometry solutions using artificial neural network pre-processing. Thin Solid Films 245(1):167–173
152. Everton SK et al (2016) Review of in-situ process monitoring and in-situ metrology for metal additive manufacturing. Mater Des 95:431–445
153. Liu Y et al (2022) Machine vision based condition monitoring and fault diagnosis of machine tools using information from machined surface texture: A review. Mech Syst Signal Process 164:108068
154. Rifai AP et al (2020) Evaluation of turned and milled surfaces roughness using convolutional neural network. Measurement 161:107860
155. Wang P et al (2019) Heterogeneous data-driven hybrid machine learning for tool condition prognosis. CIRP Ann 68(1):455–458
156. Cooper C et al (2022) Texture-Aware Ridgelet Transform and Machine Learning for Surface Roughness Prediction. IEEE Trans Instrum Meas 71:1–10

157. *Principles of Optics - Google Books.*
158. Abdelsalam, D.G., et al., *Interferometry and its Applications in Surface Metrology*, in *Optical Interferometry*. 2017, IntechOpen.
159. *Pelletier: Analytical applications of Raman spectroscopy - Google Scholar.*
160. Das RS, Agrawal Y (2011) Raman spectroscopy: Recent advancements, techniques and applications. Vib Spectrosc 57(2):163–176
161. Li Q et al (2020) Localized strain measurement in molecular beam epitaxially grown chalcogenide thin films by micro-Raman spectroscopy. ACS Omega 5(14):8090–8096
162. Qi Y et al (2023) Recent Progresses in Machine Learning Assisted Raman Spectroscopy. Advanced Optical Materials 11(14):2203104
163. Sheremetyeva N et al (2020) Machine-learning models for Raman spectra analysis of twisted bilayer graphene. Carbon 169:455–464
164. Zhang J et al (2022) High-speed identification of suspended carbon nanotubes using Raman spectroscopy and deep learning. Microsyst Nanoeng 8(1):1–9
165. Mao Y et al (2020) Machine Learning Analysis of Raman Spectra of MoS2. Nanomaterials 10(11):2223
166. Lu A-Y et al (2022) Unraveling the Correlation between Raman and Photoluminescence in Monolayer MoS2 through Machine-Learning Models. Adv Mater 34(34):2202911
167. Kim Y, Lee W (2022) Distributed Raman Spectrum Data Augmentation System Using Federated Learning with Deep Generative Models. Sensors 22(24):9900
168. Ghosh K et al (2019) Deep Learning Spectroscopy: Neural Networks for Molecular Excitation Spectra. Advanced Science 6(9):1801367
169. Rankine CD, Madkhali MMM, Penfold TJ (2020) A Deep Neural Network for the Rapid Prediction of X-ray Absorption Spectra. J Phys Chem A 124(21):4263–4270
170. Pielsticker L et al (2023) Convolutional neural network framework for the automated analysis of transition metal X-ray photoelectron spectra. Anal Chim Acta 1271:341433
171. Singh, M.K. and A. Singh, *Chapter 14 - Nuclear magnetic resonance spectroscopy*, in *Characterization of Polymers and Fibres*, M.K. Singh and A. Singh, Editors. 2022, Woodhead Publishing. p. 321–339.
172. Li D-W et al (2021) DEEP picker is a deep neural network for accurate deconvolution of complex two-dimensional NMR spectra. Nat Commun 12(1):5229
173. Rull H, Fischer M, Kuhn S (2023) NMR shift prediction from small data quantities. Journal of Cheminformatics 15(1):114

Machine Learning for Active Matter

Giovanni Volpe

Abstract The integration of machine learning, and particularly deep learning, with active matter research provides new possibilities for research and innovation. Active matter systems, which encompass a broad range of natural and artificial entities that consume energy to perform mechanical work, present unique challenges due to their intrinsic out-of-equilibrium dynamics. Recent advancements in deep learning, offer unprecedented opportunities to analyze, model, and understand these complex systems. By addressing both the opportunities and challenges, including the need for physics-informed models and the reality gap between simulations and real-world applications, this chapter highlights the mutual benefits of combining machine learning with active matter research.

1 Introduction

Active matter refers to a class of materials composed of individual units that consume energy to generate movement and exert forces on their surroundings [1]. Unlike passive materials, which rely on external forces and energy inputs, active matter systems are characterized by their intrinsic ability to self-organize and exhibit collective dynamics. This is observed in both natural systems, such as flocks of birds, bacterial colonies, and cellular tissues, and in artificial systems, including synthetic self-propelled particles and robotic swarms—some of these systems are shown in Fig. 1.

The study of active matter provides insights into fundamental processes of life and paves the way for innovative applications in technology and medicine. By understanding how active matter systems operate, researchers can develop new materials with adaptive properties, design efficient robotic systems, and create novel methods for drug delivery and tissue engineering.

G. Volpe (✉)
Department of Physics, University of Gothenburg, Gothenburg, Sweden
e-mail: giovanni.volpe@physics.gu.se

© The Author(s) 2026   217
M. te Vrugt (ed.), *Artificial Intelligence and Intelligent Matter*, Machine Intelligence
for Materials Science, https://doi.org/10.1007/978-3-032-04129-6_11

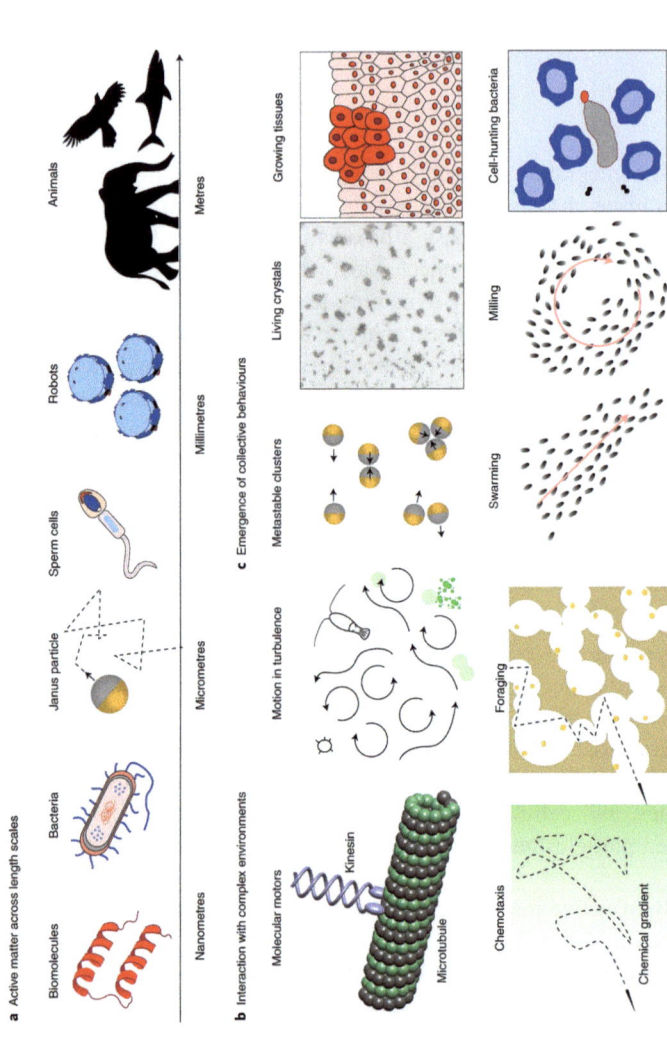

Fig. 1 Active-matter systems and phenomena. **a.** Examples of active particles range in size from micrometres to metres (for example, biomolecular motors, motile bacteria, sperm cells, artificial microscopic particles, fish, birds, mammals and robots). **b.** Active particles react to environmental signals and optimize their behaviour to attain certain goals—for example, biomolecular motors move along microtubules, microorganisms swim in turbulent flows, motile cells respond to chemotactic gradients, and animals look for food (foraging). **c.** Interactions between active particles may lead to complex collective behaviours, such as the growth of metastable clusters of particles (living crystals), and to the emergence of collective dynamics such as swarming and milling. Reproduced from Ref. [2]

The rise of machine learning in scientific research marks a transformative period in the way complex systems are studied and understood. Machine learning, a subset of artificial intelligence (AI), began gaining prominence in the mid-20th century with the development of algorithms capable of learning from data. The advent of more powerful computers and the accumulation of large datasets in the 1990s and 2000s further propelled its growth. Key milestones, such as the creation of the backpropagation algorithm for training neural networks and the introduction of support vector machines, laid the groundwork for modern machine learning techniques. In recent years, breakthroughs in deep learning have enabled unprecedented capabilities in pattern recognition and prediction, making machine learning an indispensable tool in scientific research [3].

For active matter, machine learning provides powerful methods to analyze the complex, often non-linear behaviors exhibited by these systems [2]. By using vast amounts of experimental and simulation data, machine learning algorithms can uncover hidden patterns, predict future states, and optimize the design and control of active matter systems. This integration provides tools to enhance our understanding of natural phenomena and drive innovation in creating synthetic active materials with tailored properties, ultimately bridging the gap between theory and practical applications.

2 Recent Advancements in Deep Learning

The evolution of AI has seen remarkable advancements [4], particularly with the advent of deep learning, which has revolutionized the study and manipulation of complex systems, including active matter [2]. The journey began in the mid-20th century with the development of simple neural networks, but it was not until the 1980s and 1990s, with the introduction of backpropagation and more powerful computers, that significant progress was made. An overview of the main currently available machine learning tools is shown in Fig. 2.

In the early 2000s, the resurgence of interest in neural networks, coupled with the availability of large datasets and enhanced computational power, led to breakthroughs in deep learning [3]. In particular, the emergence of Convolutional Neural Networks (CNNs) marked a crucial moment to seed the ongoing deep learning revolution. CNNs, inspired by the visual cortex of the brain, revolutionized computer vision by efficiently processing image data through layered structures that capture spatial hierarchies. For active matter research, CNNs have become essential in analyzing visual data from experiments, enabling the identification and classification of patterns in the organization of active matter systems.

Recurrent Neural Networks (RNNs) and their advanced variants, Gated Recurrent Unit (GRU) and Long Short-Term Memory (LSTM) networks, emerged as crucial tools to handle sequential data. Their ability to capture temporal dependencies made them useful in studying the dynamic behaviors of active matter over time. These net-

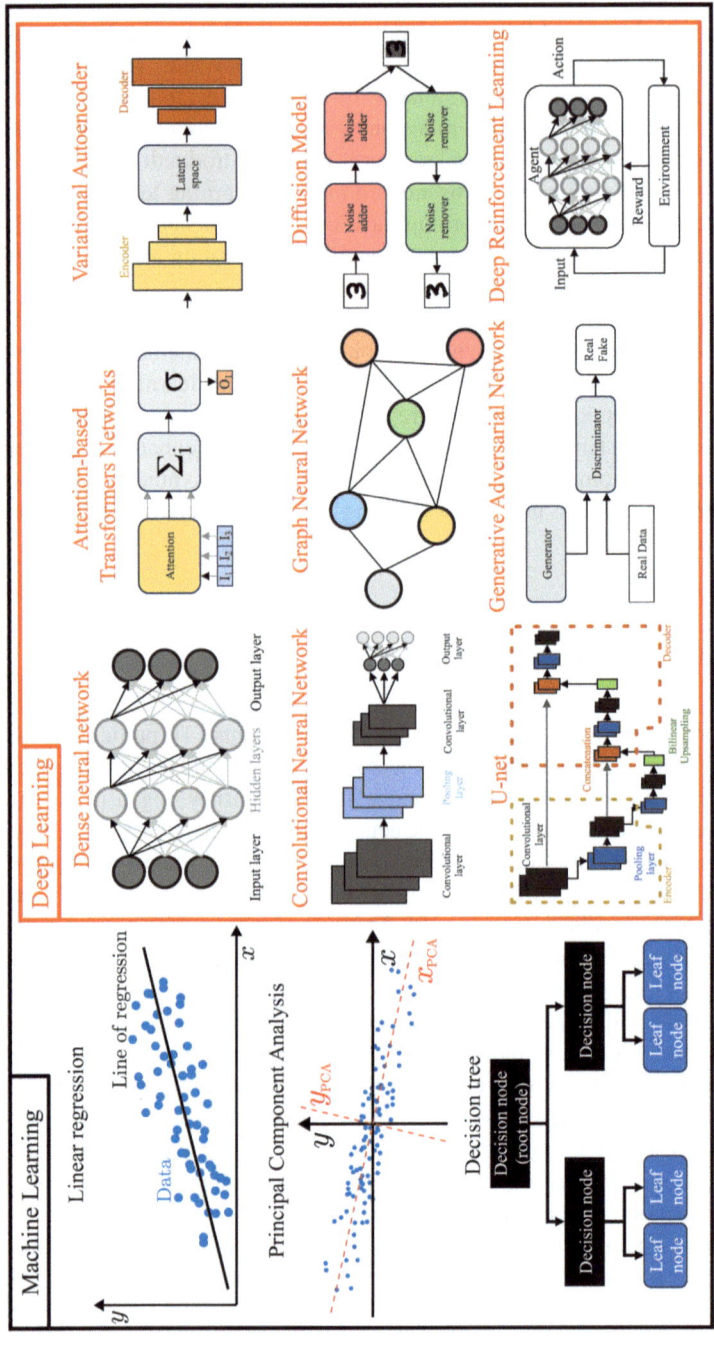

Fig. 2 Machine learning and deep learning. Deep learning (orange rectangle) is a subset of machine learning (black rectangle). Typical machine learning approaches include linear regression, principal component analysis, and decision trees. Typical deep learning approaches include Dense Neural Networks (DNN), Convolutional Neural Networks (CNN), U-Nets, Attention-Based Transformer Networks, Graph Neural Networks (GNNs), Generative Adversarial Networks (GANs), Variational Autoencoders, Diffusion Models, and Deep Reinforcement Learning. Reproduced from Ref. [5]

works facilitate the modeling of temporal evolution in systems like bacterial colonies or synthetic self-propelled particles, offering insights into their collective dynamics.

The introduction of the Attention-Based Transformer Networks in 2017 by Vaswani et al. [6] represented a major leap forward in handling data sequences. Transformers, with their self-attention mechanism, have enabled better handling of contextual information, which is crucial for understanding the complex interactions within active matter systems. These models have been adapted to predict and analyze the behavior of active particles by capturing dependencies across different scales and timeframes.

Generative Adversarial Networks (GANs), introduced by Ian Goodfellow in 2014 [7], brought a new dimension to deep learning by enabling the generation of realistic data, arguably starting the generative AI revolution. In the context of active matter, GANs can be used to simulate realistic scenarios of particle interactions and behaviors, for example providing valuable synthetic data for training and testing other AI models. This innovation has profound implications for studying rare or difficult-to-observe phenomena in active matter systems.

Deep Reinforcement Learning is another significant development. It combines deep learning with reinforcement learning principles. Pioneered by researchers at DeepMind, this approach has enabled the development of agents that can learn to control and optimize active matter systems through interactions with their environments. Applications include the autonomous control of robotic swarms and the optimization of synthetic active materials for desired behaviors.

Most recently, Diffusion Models have emerged as a powerful generative AI modeling framework. These models, which learn to generate data by iteratively refining random noise and represent the cutting edge of generative modeling. For active matter, they can be employed to create high-quality simulations of active matter dynamics, probably eventually displacing GANs.

3   Machine Learning Applications in Active Matter

The integration of machine learning and AI into the study of active matter presents unprecedented and unexpected opportunities [2]. Active matter often exhibits emergent phenomena that are challenging to analyze and predict using traditional methods. AI offers powerful tools to enhance experimental techniques, develop data-driven models, and optimize control strategies. For example, using AI, researchers can improve the resolution and accuracy of video microscopy, automate particle tracking, and gain deeper insights into the collective dynamics of active matter. Furthermore, AI-driven approaches enable the prediction of complex patterns, optimization of navigation and search strategies, and accurate control of synthetic and biological systems. Here, we will review some of the fields where AI has had a transformative role within active matter research.

3.1 Data Acquisition and Analysis

The integration of machine learning into data acquisition and analysis techniques has significantly advanced the study of active matter. One of the primary areas of impact is in video microscopy and particle tracking [8, 9], where traditional methods often struggle with the sheer volume and complexity of data generated by active matter systems.

Machine learning algorithms have introduced new ways to enhance the quality and efficiency of video microscopy. Techniques such as deep-learning-based image enhancement can significantly improve the resolution and clarity of microscopic images, allowing researchers to observe finer details. These improvements are particularly useful to study dynamic processes in biological systems, such as cell motility and bacterial colony formation, where high-resolution imaging is crucial.

In particle tracking, machine learning has enabled more accurate and automated tracking of individual particles. Traditional tracking methods often require extensive manual intervention and can be prone to errors, especially in densely packed or highly dynamic environments. Machine learning algorithms, particularly those based on CNNs, can automatically identify and track particles with high precision, even in challenging conditions [10–12]. This automation increases the accuracy of tracking, while also speeding up the analysis process, allowing researchers to handle larger datasets and uncover more detailed insights into active particle behaviors.

Several studies have demonstrated the effectiveness of machine learning in improving experimental data analysis in active matter research. For instance, deep learning models have been employed to analyze the motion patterns of synthetic self-propelled particles, revealing complex behaviors that were previously difficult to quantify [10]. By training on extensive datasets, these models can predict particle trajectories and identify subtle patterns in their movement, providing a deeper understanding of the underlying driving mechanisms.

In biological systems, machine learning has been used to analyze time-lapse microscopy data of cell cultures, enabling the identification of distinct phases of cell behavior and the prediction of future states [13]. This capability is particularly valuable in biomedical research, where understanding cell dynamics can inform the development of new treatments and therapies.

Another notable application is in the study of collective behaviors, such as swarming and flocking. Machine learning algorithms have been utilized to analyze video data of these phenomena, identifying individual agents' roles and interactions within the group [14, 15]. This analysis has provided new insights into how collective behaviors emerge and evolve, offering potential strategies to control and optimize such behaviors in synthetic systems.

3.2 Pattern Recognition

The classification of patterns in active matter systems often relies on subjective methods like visual inspection, which can introduce bias and inconsistencies. This "black box problem" has existed long before the advent of AI. In this case, AI, and particularly machine learning algorithms, can provide a more objective and data-driven approach to pattern recognition. These algorithms can be trained on large datasets to identify patterns based on statistical evidence, ensuring a more reliable and reproducible classification process.

One significant advantage of AI is its ability to handle the complexity of 3D pattern recognition, a task that is challenging using traditional methods. In 3D, spatial relationships and depth information add difficulty. 3D CNNs have been developed to process volumetric data effectively. These models extract essential features for accurate pattern recognition, making them invaluable in various applications.

For instance, in medical imaging, AI has been employed to identify tumors in MRI scans with high accuracy, surpassing human performance in some cases. Autonomous driving systems use AI to recognize and classify objects in 3D space, ensuring safe navigation through complex environments. In robotics, AI enables the precise manipulation and navigation of objects by understanding their 3D structure.

3.3 Data-Driven Models

The application of deep learning to infer models from time-series data can revolutionize the study of active matter. Deep learning models, particularly RNNs, GRUs and LSTMs, can be trained on time-series data collected from experiments or simulations to capture temporal dependencies and dynamic patterns, making them ideal to study the evolution of active matter systems over time. By learning from the temporal sequences of particle positions, velocities, and interactions, these models can infer the underlying rules governing the system's dynamics. This enables the identification of key parameters and interactions that drive the observed behaviors, providing a deeper understanding of the mechanisms at play.

Moreover, deep learning models can predict future states of the system based on current and past data. This predictive capability is particularly valuable for exploring how active matter systems evolve, allowing researchers to anticipate emergent phenomena.

Several case studies highlight the power of deep learning in predicting complex patterns and behaviors in active matter systems. For example, in the study of bacterial colonies, deep learning models have been used to analyze time-lapse microscopy data, predicting the formation of spatial patterns such as biofilms and swarms [13]. These predictions help researchers understand how bacterial communities organize and respond to environmental changes, informing strategies for controlling harmful biofilms or promoting beneficial microbial activities.

In synthetic active matter, deep learning has been applied to predict the collective dynamics of self-propelled particles. By training models on experimental data, researchers have been able to forecast the emergence of collective behaviors such as clustering, phase separation, and vortex formation [16, 17]. By enhancing our theoretical understanding of the phenomena, these predictions can guide the design of synthetic systems with desired properties, such as optimized swarm robotics or targeted drug delivery mechanisms.

Another notable application is in the field of active nematics, where deep learning models have been used to predict the formation and evolution of topological defects [18, 19]. These defects, which play a crucial role in the behavior of liquid crystals and other active materials, can be challenging to study due to their complex and dynamic nature. Deep learning models trained on time-series data have successfully predicted the locations and trajectories of these defects, providing new insights into their role in active matter systems.

3.4 Navigation and Search Strategies

The optimization of navigation and search strategies in fluctuating environments is a critical aspect of active matter research. Reinforcement learning, a type of machine learning where agents learn to make decisions by receiving rewards or penalties, has proven to be an effective tool to enhance these strategies [20, 21]. For a comprehensive exploration of AI-based navigation and communication problems of active particles, refer to the chapter by Hartmut Löwen and Benno Liebchen.

Reinforcement learning algorithms are particularly suited for dynamic and unpredictable environments, where traditional navigation strategies may fail. In active matter systems, agents such as self-propelled particles or robotic swarms must navigate through environments that can change rapidly and unpredictably. Reinforcement learning algorithms enable these agents to learn optimal navigation strategies through trial and error, continuously improving their performance based on feedback from their interactions with the environment.

In practice, reinforcement learning involves training an agent using a reward-based system where successful actions are reinforced, and unsuccessful ones are discouraged. This approach allows the agent to develop an understanding of the environment and adapt its navigation strategy accordingly. For instance, in fluctuating environments, a reinforcement-learning-trained agent can learn to identify and exploit transient pathways, avoid obstacles, and optimize its route to reach a target efficiently.

Reinforcement learning has already demonstrated significant success in various applications involving biological systems and robotics. One notable example is the use of reinforcement learning in optimizing the navigation of robotic swarms [22, 23]. Researchers have developed reinforcement learning algorithms that enable robots to coordinate their movements, avoid collisions, and achieve collective goals such as search-and-rescue missions or environmental monitoring. These algorithms have shown remarkable adaptability, allowing robots to perform effectively even in complex and changing environments.

In biological systems, reinforcement learning has been applied to study and influence animal behavior. For instance, researchers have used reinforcement learning to train artificial agents that interact with live animals, guiding their movements and studying their responses to different stimuli [24]. This approach has provided valuable insights into the navigation strategies of animals and has potential applications in conservation efforts, where guiding animals away from hazardous areas could be crucial.

Another exciting application is in the field of synthetic biology, where reinforcement learning is used to optimize the behavior of bio-hybrid systems. These systems, which combine biological components with reinforcement learning algorithms that learn to control biological processes with precision. For example, reinforcement learning has been employed to optimize the motility of bio-hybrid micro-robots, enhancing their ability to navigate through complex biological environments for tasks such as targeted drug delivery [25].

3.5   Collective Dynamics

The study of collective dynamics in active matter systems encompasses a wide range of phenomena, including swarming, flocking, and other forms of coordinated behavior [26, 27]. These behaviors arise from local interactions among individual agents, leading to complex and often unpredictable global patterns. Multi-agent systems and deep reinforcement learning provide useful tools to understand and manipulate these collective behaviors, offering insights that can be applied in both natural and synthetic systems. For a detailed discussion on how systems of simple motile agents can be used to emulate intelligent systems through emergent behavior, see the chapter by Julian Jeggle and Raphael Wittkowski.

Multi-agent systems simulate the interactions of numerous autonomous agents, each following simple rules, to study the emergence of collective behaviors [28]. These systems are particularly effective in modeling swarming and flocking behaviors observed in nature, such as bird flocks, fish schools, and insect swarms. By adjusting the rules governing individual agent behavior, researchers can explore how different interaction parameters influence the overall dynamics of the system.

For example, in swarming systems, agents might follow rules that dictate alignment with neighbors, attraction to nearby agents, and repulsion from those that are too close. These simple rules can lead to complex behaviors, such as the formation of stable swarms, dynamic flocks, or even chaotic movement patterns. Multi-agent simulations help identify the key factors that drive these behaviors and provide a platform for testing hypotheses about natural systems.

In synthetic systems, multi-agent models can be used to design and control robotic swarms. These models guide the development of algorithms that enable robots to coordinate their actions, adapt to changing environments, and achieve collective objectives. Applications range from environmental monitoring and disaster response to precision agriculture and autonomous exploration.

Deep reinforcement learning further enhances the study of collective dynamics by enabling agents to learn optimal behaviors through interactions with their environment. Unlike traditional multi-agent systems that rely on predefined rules, deep reinforcement learning agents use neural networks to develop strategies based on rewards and penalties, allowing for more adaptive and robust behaviors.

In simulated scenarios, deep reinforcement learning has been used to model and understand a variety of collective behaviors. For deep reinforcement learning to simulate the flocking behavior of birds, discovering new insights into how local interactions can lead to global coordination [15, 29]. These simulations reveal the underlying principles of flock formation and maintenance, providing valuable information that can be applied to both biological studies and the design of robotic systems.

Real-world applications of deep reinforcement learning in collective dynamics include the development of robotic swarms capable of complex, coordinated tasks [22]. For example, deep reinforcement learning algorithms have been used to train swarms of drones to navigate through cluttered environments, avoid obstacles, and work together to achieve common goals [30]. These capabilities are crucial for applications such as search-and-rescue operations, where coordination and adaptability are essential.

3.6   Control and Manipulation of Active Matter

AI-driven control systems and machine learning techniques offer unprecedented precision and flexibility in managing the behavior of active matter systems, enabling innovative solutions across various fields. The development of AI-driven control systems has the potential to revolutionize the way we can manipulate active matter, particularly synthetic active particles and bio-hybrid systems.

AI-driven control systems utilize advanced algorithms to manipulate synthetic active particles. For example, reinforcement learning can be employed to develop control policies that adjust the propulsion mechanisms of synthetic particles. This precise control is crucial for applications such as micro-robotics, where active particles need to perform tasks in confined or dynamic spaces.

In addition, computer vision and deep learning techniques can be used to monitor and adjust the behavior of these particles in real-time. By processing visual data from microscopes or cameras, AI systems can dynamically steer particles towards desired trajectories, ensuring accurate performance even in the presence of environmental fluctuations.

Bio-hybrid systems, which integrate biological components with synthetic elements, can also benefit significantly from AI-driven control. Machine learning algorithms can be used to manipulate the motility and interactions of biological entities, such as engineered cells or microorganisms. For instance, AI can control the direction and speed of bacteria equipped with magnetic or light-responsive elements, enabling targeted movement and task execution.

One of the most promising applications of AI in active matter is targeted drug delivery within the context of precision medicine [31]. Machine learning models can optimize the navigation of drug-carrying active particles to specific locations within the body. By learning from biological data and patient-specific conditions, AI systems can control external fields (such as magnetic or acoustic fields) to steer these particles accurately. This process might involve real-time tracking of the particles' positions (possibly at an ensemble level) and adjusting the fields accordingly to guide the particles to efficiently reach and release their therapeutic payloads at targeted sites, minimizing side effects and improving treatment efficacy.

Active matter systems equipped with AI capabilities are also being developed for environmental sensing [32]. For instance, swarms of active particles or drones can be deployed to monitor environmental conditions, such as pollution levels or hazardous substances. Machine learning algorithms optimize their search patterns and data collection strategies, ensuring comprehensive and efficient environmental monitoring. For an in-depth overview of these methods and their applications, see the chapter by Oliver Kamps, Tim W. Kroll, and Oliver Mai.

Beyond these examples, AI-driven optimization of active matter behaviors has potential in numerous other fields, including materials science, where active matter can be used to create smart materials that respond to environmental stimuli, and bioengineering, where bio-hybrid systems can be designed to perform complex biological functions.

3.7 Integration with Experimental Platforms

Finally, the integration of AI with experimental platforms can significantly enhance the capabilities of researchers in the field of active matter. By combining machine learning techniques with experimental setups, scientists can achieve real-time data analysis, feedback, and control, leading to more precise and insightful experiments. Hybrid approaches that merge AI with traditional methods can also be highly effective in advancing the study of active matter, leading also to AI-enabled scientific discovery.

In real-time experimental setups, AI can analyze the collected data, offering immediate feedback and adjustments, similar to what is being done in the field of optical manipulation [5]. For instance, in video microscopy of active particles, AI algorithms can track particle movements, detect anomalies, and adjust imaging parameters on the fly to enhance data quality. This real-time capability allows for more accurate monitoring of active matter behaviors and enables researchers to make informed decisions during experiments.

AI-driven control systems can also adapt experimental conditions based on real-time data analysis. For example, in experiments involving synthetic active particles, machine learning algorithms can adjust the environmental conditions, such as temperature or chemical gradients, to observe how these changes affect particle behavior. This adaptive control can help optimize experimental conditions and potentially discover new phases and behaviors.

In bio-hybrid systems, researchers have integrated AI with traditional biological experiments to control and study the behavior of living organisms. For instance, by combining AI-driven imaging with traditional cell culture techniques, scientists have been able to monitor and manipulate the behavior of cell colonies in real-time. This integration can lead to breakthroughs in understanding cellular dynamics and optimizing conditions for tissue engineering.

In studies of synthetic active particles, hybrid approaches have been used to optimize navigation and control. Researchers have combined machine learning algorithms with traditional experimental setups to guide the movement of particles in complex environments [33]. By training AI models on experimental data, they have developed strategies that allow particles to navigate through obstacles and reach target destinations efficiently.

Hybrid approaches have also been employed in environmental monitoring applications. For example, researchers have used AI to enhance the capabilities of traditional sensor networks deployed in aquatic environments. Machine learning algorithms analyze data from sensors in real-time, identifying patterns and providing insights into water quality and pollutant distribution. This integration has improved the accuracy and responsiveness of environmental monitoring systems.

4   Opportunities and Challenges

The intersection of machine learning and active matter research provides a landscape rich with opportunities and challenges. As AI continues to evolve, its integration with active matter systems promises to revolutionize data acquisition, analysis, and manipulation, enabling unprecedented advancements in both fundamental research and practical applications. However, this convergence also brings forth a set of complex challenges that must be addressed to fully realize the potential of these technologies.

This section explores the opportunities that AI-driven approaches offer in enhancing experimental techniques, understanding biological and synthetic systems, and developing innovative applications. It also discusses the significant challenges that researchers face, including the interpretability of machine learning models, the integration of physical principles, and the bridging of the gap between simulations and real-world scenarios.

4.1 Opportunities

The integration of machine learning with active matter research opens up a vast array of opportunities for scientific advancement and technological innovation.

Improvement of data acquisition and analysis. The integration of AI into data acquisition and analysis has already greatly enhanced the capabilities of active matter research. Advanced machine learning algorithms already enable real-time data analysis and feedback control, transforming traditional experimental approaches. These enhanced methods allow for the continuous monitoring and adjustment of experimental conditions, ensuring optimal data quality and experimental accuracy. AI-driven systems can process vast amounts of data almost instantaneously, identifying patterns and anomalies that might be missed by manual analysis. This real-time feedback loop improves the precision of experiments while also accelerating the discovery process. Furthermore, the development of adaptive experimental setups that incorporate AI allows for dynamic adjustments based on ongoing data analysis. These setups can optimize experimental parameters on-the-fly, leading to more efficient and insightful investigations, also within the framework of AI-enabled scientific discovery.

Swarm robotics. The principles and techniques developed in swarm robotics offer valuable inspiration for the study and control of active matter systems. Genetic algorithms, behavior trees, and GANs from the field of swarm robotics can be effectively applied to active matter to optimize and control its behavior. Genetic algorithms mimic natural selection processes to evolve optimal solutions for complex problems, making them ideal for tuning the parameters and interactions within active matter systems to achieve desired collective behaviors. Behavior trees, widely used in robotics for decision-making and task execution, can be adapted to manage the complex dynamics of active matter, ensuring coherent and efficient responses to environmental changes. GANs, which generate new data based on training examples, can simulate realistic scenarios and explore the potential behaviors of active matter under various conditions.

Furthermore, building on the principles of swarm intelligence [26, 27], where simple rules and local interactions lead to complex global behaviors, can significantly enhance the design and control of active matter systems. By mimicking the decentralized and adaptive strategies observed in natural swarms, researchers can develop robust and scalable active matter systems capable of performing complex tasks. These systems can be used in applications ranging from environmental sensing to medical interventions, where coordinated and adaptive behaviors are crucial.

Systematization of active matter. Machine learning offers powerful tools for the systematic classification and understanding of diverse active matter systems. By applying advanced algorithms to the vast and complex data generated by these systems, researchers can uncover underlying patterns and categorize behaviors with unprecedented precision. This classification is crucial for making sense of the diverse manifestations of active matter, from bacterial colonies and cell motility to synthetic self-propelled particles and robotic swarms. Machine learning enables the identification of key properties and interactions that define different types of active matter, facilitating a deeper understanding of their dynamics and mechanisms.

Furthermore, the creation of comprehensive databases and taxonomies for active matter behaviors and properties can be greatly enhanced by machine learning techniques. These databases can serve as valuable resources for researchers, providing structured and accessible information on the various behaviors observed in active matter systems. By organizing data into detailed taxonomies, researchers can systematically compare different systems, identify commonalities and differences, and develop generalized theories and models.

Biological insights. Reinforcement learning and other AI techniques are revolutionizing the study of sensory inputs and interaction strategies in biological systems, providing profound insights into the complex behaviors and processes of living organisms. Reinforcement learning, which involves training agents to make decisions based on rewards and penalties, is particularly effective in decoding how biological systems, such as animals or cells, process sensory information and interact with their environment. By simulating these processes and optimizing behavioral strategies, researchers can unravel the sophisticated mechanisms underlying navigation, foraging, communication, and other activities. For example, understanding how cancer cells interact with their microenvironment can lead to the development of more effective treatments. Similarly, insights into animal behavior can inform conservation strategies and improve animal welfare.

In addition to reinforcement learning, other AI analysis methods are being employed to gain a deeper understanding of biological processes. Machine learning algorithms can analyze large-scale biological data, identifying patterns and correlations that might be invisible to human observers. This capability is crucial to study complex systems like neural networks in the brain, gene regulatory networks, and cellular signaling pathways, which provides a bridge from active matter to variosu omics fields and tools, such as those developed to study brain connectivity [34]. AI can also integrate diverse types of biological data, from genetic sequences to imaging data, providing a holistic view of biological phenomena.

Evolutionary perspectives. The exploration of evolutionary perspectives in active matter systems is greatly enhanced by AI-driven models and multi-objective optimization techniques. These advanced methods allow researchers to simulate and analyze the evolutionary processes that shape the behavior and morphology of active matter, providing insights into how these systems adapt and evolve over time.

Multi-objective optimization, inspired by natural selection, is used to explore the vast landscape of possible configurations and behaviors in active matter systems. By optimizing multiple objectives simultaneously, such as efficiency, robustness, and adaptability, researchers can identify the trade-offs and synergies that drive evolutionary success. This approach is particularly useful for understanding how complex behaviors and structures emerge in active matter, from the cooperative dynamics of microbial colonies to the movement patterns of synthetic autonomous agents.

Moreover, studying the evolutionary dynamics of active matter can inform the design of bio-inspired materials and systems. By understanding the adaptive strategies that have evolved in nature, engineers can create synthetic systems that mimic these strategies, leading to more resilient and efficient designs. This intersection of evolutionary biology and materials science holds great promise for the development of advanced technologies, from adaptive materials that change properties in response to environmental stimuli to self-healing materials that repair themselves over time.

Embodied intelligence. The development of smart, autonomous active materials with integrated computational capabilities is at the frontier of active matter research. These materials, imbued with embodied intelligence, should be able to sense, process, and respond to environmental stimuli autonomously, opening up a plethora of innovative applications across various fields.

The integration of computational capabilities into active materials will allow them to perform complex tasks without external control. These smart materials can process sensory information from their surroundings, make decisions based on that information, and adapt their behavior in real time. For example, materials embedded with microprocessors and sensors can monitor environmental conditions such as temperature, pressure, and chemical composition, and autonomously adjust their properties or actions accordingly. This level of autonomy is crucial for applications in remote or hazardous environments where human intervention is limited or impossible.

The design of materials that can autonomously respond to environmental stimuli involves a combination of material science, robotics, and AI. These materials will likely be composed of responsive polymers, bio-hybrid structures, or nanomaterials that change their physical properties in reaction to external triggers. AI algorithms play a critical role in managing these responses, ensuring that the material's actions are appropriate and efficient. For instance, AI can enable a material to optimize its configuration for load-bearing tasks in real-time or to self-heal when damaged by triggering specific chemical reactions.

The implications of embodied intelligence in active materials are vast. In biomedical engineering, smart materials can lead to the development of advanced prosthetics that adapt to the user's movements and environmental conditions, improving functionality and comfort. In environmental monitoring, autonomous materials can continuously assess and respond to pollution levels, providing real-time data and mitigation strategies. In robotics, these materials can enhance the capabilities of soft robots, enabling them to navigate complex terrains and perform delicate tasks with greater precision.

Technological applications. The integration of AI and active matter opens up a wide range of technological innovations, particularly in the fields of medicine, environmental monitoring, and materials science. These applications exploit the advanced capabilities of AI-driven systems to optimize performance, enhance functionality, and provide real-time adaptive responses.

One of the most promising applications of AI in active matter is in the development of targeted drug delivery systems. These systems will utilize optimized navigation and release mechanisms to deliver therapeutic agents directly to specific sites within the body. By integrating machine learning algorithms (possibly using some properties of the particles as sensors for the local environment), these systems will learn to navigate complex biological environments, avoiding obstacles and homing in on target cells or tissues with high precision. This targeted approach will improve the efficacy of treatments and minimize side effects, as the drug is concentrated at the site of need rather than being distributed throughout the body. AI can also optimize the timing and dosage of drug release, ensuring maximum therapeutic effect in the context of precision medicine.

AI-enhanced active matter systems can also revolutionize environmental monitoring and sensing. These technologies can provide real-time adaptive responses to changing environmental conditions, making them highly effective for applications such as pollution detection, climate monitoring, and ecosystem management. For instance, swarms of autonomous sensors equipped with AI algorithms can dynamically adjust their deployment strategies based on real-time data, ensuring comprehensive coverage and accurate measurement of environmental parameters. These systems can detect and respond to pollutants or other hazards more quickly and effectively than traditional methods, providing valuable data for environmental protection and management.

The development of smart materials and responsive systems is another area where AI and active matter can converge to deliver groundbreaking innovations. Smart materials can change their properties in response to external stimuli such as temperature, light, or chemical signals. By integrating AI, these materials can be designed to respond more intelligently and autonomously. For example, AI-driven smart materials can self-heal when damaged, adjust their stiffness or flexibility based on load conditions, or change their thermal conductivity in response to temperature variations. These responsive capabilities make them ideal for a wide range of applications, including adaptive building materials, wearable technology, and advanced manufacturing processes.

Interdisciplinary research. The fusion of AI with active matter research is inherently interdisciplinary, drawing from fields such as physics, biology, engineering, and computer science. This confluence of disciplines fosters synergies that drive innovative solutions and new discoveries, as each field brings its unique perspectives, methodologies, and expertise to the table. For instance, physicists and biologists working together can apply AI to unravel the complex interactions within biological systems, such as cellular processes or animal behaviors. Engineers can then use these insights to create bio-inspired materials or devices that mimic these natural systems. AI researchers can enhance these efforts by providing advanced algorithms for data analysis and predictive modeling, ensuring that the solutions are both innovative and grounded in solid scientific understanding.

4.2 Challenges

Despite the promising opportunities described in the previous section, the integration of machine learning and active matter research is fraught with significant challenges.

Black-box models. One of the significant challenges in integrating machine learning with active matter research is the interpretability of black-box models (and in fact one of the key challenges for the whole field of AI). While these models, particularly deep learning algorithms, can achieve high levels of accuracy and predictive power, their decision-making processes often remain opaque. This lack of transparency can hinder trust and acceptance, particularly in fields where understanding the underlying mechanisms is crucial. Addressing this challenge involves developing methods to explain and interpret the decisions made by AI models. Techniques such as explainable AI (XAI) aim to make the inner workings of machine learning models more transparent, providing insights into how they process data and arrive at their conclusions. By elucidating these processes, researchers can ensure that the models' predictions are reliable and based on understandable principles. Enhancing the interpretability of AI models will build trust and facilitate the validation and refinement of these models, ultimately leading to more robust and scientifically sound applications in active matter research.

Generalization and extrapolation. Ensuring that machine learning models can generalize and perform well outside their training data ranges is another critical challenge. Models often excel within the scope of the data they were trained on but struggle when confronted with new, unseen conditions. This limitation can undermine their usefulness in real-world applications, where variability and unpredictability are the norm. To address this, researchers should enhance the robustness and adaptability of AI models. Techniques such as transfer learning, domain adaptation, and robust training methods are being employed to improve models' ability to extrapolate and maintain performance across diverse and novel scenarios. By ensuring that AI models can generalize effectively, researchers can develop more reliable and versatile tools for predicting and controlling active matter systems, ultimately leading to broader and more impactful applications.

Physics-informed models. Incorporating physical principles and conservation laws into machine learning models is essential to enhance their accuracy and reliability. Traditional data-driven models, while powerful, can sometimes produce results that violate known physical laws or lack interpretability. By integrating fundamental physics into these models, researchers can ensure that the predictions are physically meaningful. This approach leads to the development of hybrid models that combine the strengths of data-driven techniques with the rigor of physics-based methods. This synergy will enhance the robustness and credibility of the models, making them more applicable to real-world scenarios. Moreover, such hybrid models will facilitate a deeper understanding of active matter systems by providing insights that are grounded in established physical principles, ultimately advancing both theoretical and applied research in the field.

Reality gap. Bridging the gap between simulation and real-world performance, often referred to as the "reality gap", is a significant challenge in the application of AI models and algorithms. While simulations provide a controlled environment for developing and testing AI-driven models, the complexities and unpredictabilities of real-world conditions can lead to discrepancies in performance. Ensuring that AI models translate effectively from virtual environments to practical applications requires rigorous validation and testing in diverse, real-world scenarios. Researchers are developing techniques such as domain randomization and transfer learning to enhance the robustness of models, allowing them to adapt to variations that occur outside of the simulated settings. Additionally, incorporating real-world data into the training process can help bridge this gap, making the models more resilient and accurate when deployed.

Data quality and quantity. Ensuring high-quality, representative datasets for training AI models is a foundational requirement because the performance and reliability of machine learning models heavily depend on the data they are trained on. Therefore, high-quality datasets that accurately reflect the diversity and complexity of active matter systems are crucial for developing robust and generalizable models. Addressing the limitations of small or noisy datasets is particularly important in this field, as insufficient or poor-quality data can lead to overfitting, bias, and inaccurate predictions. Researchers are employing various strategies to overcome these challenges, such as data augmentation, synthetic data generation, and advanced noise reduction techniques. Additionally, the collection of larger and more diverse datasets through collaborative efforts and improved experimental techniques is essential.

Computational complexity. Managing the high computational demands of advanced machine learning algorithms is another significant challenge. Addressing this challenge involves developing more efficient algorithms that can deliver high performance without excessive computational costs. Techniques such as model compression, pruning, and quantization can reduce the size and complexity of models, making them more suitable for real-time applications. Additionally, high-performance computing resources, such as GPUs and cloud computing, can help manage the intensive computational requirements.

Ethical and societal implications. Ethical concerns arise particularly in the context of biological research, where AI-driven experiments can impact living organisms. It is essential to establish clear ethical guidelines and regulatory frameworks to govern the use of AI in these settings, ensuring that research practices respect the intrinsic value of life and avoid unintended harm.

One significant ethical consideration is the potential for AI to exacerbate existing biases or introduce new ones, particularly when models are trained on unrepresentative datasets. This can lead to skewed results and potentially harmful consequences in applications such as healthcare or environmental monitoring. Ensuring transparency and accountability in AI model development and deployment is vital to mitigate these risks.

Furthermore, the societal impact of AI-driven technologies in active matter must be carefully considered. These technologies hold the promise of significant benefits, such as improved medical treatments and enhanced environmental protection. However, they also pose risks, including job displacement due to automation and the potential for misuse in surveillance or control mechanisms. Engaging with a broad range of stakeholders, including ethicists, policymakers, and the public, is essential to navigate these complex issues and ensure that the development and deployment of AI technologies are aligned with societal values and goals.

5   Conclusions

The integration of machine learning and active matter research has created a symbiotic relationship that drives advancements in both fields. Machine learning provides powerful tools for analyzing and predicting the complex behaviors of active matter systems, enhancing our understanding and control of these dynamic phenomena. In return, active matter research offers rich and diverse datasets that challenge and refine machine learning algorithms, pushing the boundaries of what these technologies can achieve. This interplay accelerates scientific discovery and technological innovation, as improvements in machine learning techniques directly contribute to more sophisticated and effective studies of active matter, and vice versa. The mutual benefits of this integration highlight the transformative potential of combining AI and active matter research, leading to breakthroughs that would be unattainable within the confines of a single discipline.

The future of integrating AI and active matter research is ripe with potential breakthroughs and applications. Emerging technologies such as quantum computing and neuromorphic engineering promise to further enhance the capabilities of machine learning models, enabling them to tackle even more complex and large-scale active matter systems. Interdisciplinary collaborations will continue to be a driving force, bringing together expertise from physics, biology, engineering, and computer science to address the multifaceted challenges in this field.

Promising areas for future research include the development of more advanced and interpretable AI models that can provide deeper insights into the mechanisms of active matter; the exploration of bio-inspired materials and systems that mimic the adaptive and autonomous behaviors observed in nature; and the application of AI-driven active matter technologies in medicine, environmental monitoring, and robotics to create innovative solutions to real-world problems.

References

1. Bechinger C, Di Leonardo R, Löwen H, Reichhardt C, Volpe G, Volpe G (2016) Rev Mod Phys 88(4):045006
2. Cichos F, Gustavsson K, Mehlig B, Volpe G (2020) Nat Mach Intell 2(2):94
3. LeCun Y, Bengio Y, Hinton G (2015) Nature 521(7553):436
4. Midtvedt B, Pineda J, Moberg HK, Bachimanchi H, Pereira JB, Manzo C, Volpe G (2025) Deep Learning Crash Course. No Starch Press, San Francisco, CA
5. Ciarlo A, Ciriza DB, Selin M, Maragò OM, Sasso A, Pesce G, Volpe G, Goksör M (2024) Nanophotonics (0)
6. Vaswani A, Shazeer N, Parmar N, Uszkoreit J, Jones L, Gomez AN, Kaiser Ł, Polosukhin I (2017) Advances in neural information processing systems 30
7. Goodfellow I, Pouget-Abadie J, Mirza M, Xu B, Warde-Farley D, Ozair S, Courville A, Bengio Y (2024) Advances in neural information processing systems 27
8. Rivenson Y, Göröcs Z, Günaydin H, Zhang Y, Wang H, Ozcan A (2017) Optica 4(11):1437
9. Midtvedt B, Helgadottir S, Argun A, Pineda J, Midtvedt D, Volpe G (2021) Appl Phys Rev 8(1)
10. Newby JM, Schaefer AM, Lee PT, Forest MG, Lai SK (2018) Proc Natl Acad Sci 115(36):9026
11. Hannel MD, Abdulali A, O'Brien M, Grier DG (2018) Opt Express 26(12):15221
12. Helgadottir S, Argun A, Volpe G (2019) Optica 6(4):506
13. Jeckel H, Jelli E, Hartmann R, Singh PK, Mok R, Totz JF, Vidakovic L, Eckhardt B, Dunkel J, Drescher K (2019) Proc Natl Acad Sci 116(5):1489
14. Gazzola M, Tchieu AA, Alexeev D, De Brauer A, Koumoutsakos P (2016) J Fluid Mech 789:726
15. Verma S, Novati G, Koumoutsakos P (2018) Proc Natl Acad Sci 115(23):5849
16. McDermott D, Reichhardt C, Reichhardt C (2020) Phys Rev E 101(4):042101
17. Zvyagintseva D, Sigurdsson H, Kozin VK, Iorsh I, Shelykh IA, Ulyantsev V, Kyriienko O (2022) Commun Phys 5(1):8
18. Doostmohammadi A, Ignés-Mullol J, Yeomans JM, Sagués F (2018) Nat Commun 9(1):3246
19. Zhou Z, Joshi C, Liu R, Norton MM, Lemma L, Dogic Z, Hagan MF, Fraden S, Hong P (2021) Soft Matter 17(3):738
20. Colabrese S, Gustavsson K, Celani A, Biferale L (2017) Phys Rev Lett 118(15):158004
21. Bachimanchi H, Pinder MI, Robert C, De Wit P, Havenhand J, Kinnby A, Midtvedt D, Selander E, Volpe G (2024) Limnol Oceanogr Lett
22. Hüttenrauch M, Šošić A, Neumann G (2019) J Mach Learn Res 20(54):1
23. Blais MA, Akhloufi MA (2023) Cognitive Robotics
24. Bierbach D, Lukas J, Bergmann A, Elsner K, Höhne L, Weber C, Weimar N, Arias-Rodriguez L, Mönck HJ, Nguyen H et al (2018) Front Robot AI 5:3
25. Dong H, Lin J, Tao Y, Jia Y, Sun L, Li WJ, Sun H (2024) Lab Chip 24(5):1419
26. Vicsek T, Zafeiris A (2012) Phys Rep 517(3–4):71
27. Argun A, Callegari A, Volpe G (2021) Simulation of Complex Systems. IOP Publishing Ltd, Bristol, UK
28. Schweitzer F (2007) Brownian agents and active particles: Collective dynamics in the natural and social sciences. Springer Verlag, Heidelberg, Germany
29. P. Sunehag, G. Lever, S. Liu, J. Merel, N. Heess, J.Z. Leibo, E. Hughes, T. Eccles, T. Graepel, in *Artificial life conference proceedings* (MIT Press One Rogers Street, Cambridge, MA 02142-1209, USA journals-info ..., 2019), pp. 103–110
30. Hodge VJ, Hawkins R, Alexander R (2021) Neural Comput Appl 33(6):2015
31. Johnson KB, Wei WQ, Weeraratne D, Frisse ME, Misulis K, Rhee K, Zhao J, Snowdon JL (2021) Clin Transl Sci 14(1):86
32. Ye Z, Yang J, Zhong N, Tu X, Jia J, Wang J (2020) Sci Total Environ 699:134279
33. M. Praeger, Y. Xie, J.A. Grant-Jacob, R.W. Eason, B. Mills, Machine Learning: Science and Technology **2** (2021)
34. Mijalkov M, Kakaei E, Pereira JB, Westman E, Volpe G, Initiative ADN (2017) PLoS ONE 12(8):e0178798

Implementations of Artificial Intelligence in Nanosystems

A Primer on Neuromorphic Hardware: Underlying Principles and Approaches—From Sensors to Processors

Akhilesh Jaiswal, Md. Abdullah-Al Kaiser, and Maryam Parsa

Abstract Brain-like energy efficiency and brain-like cognitive ability has intrigued researchers to mimic the fundamental computing principles of the brain on hardware computing platforms. Such neuromorphic systems present a paradigm shift in hardware computing technology and leverage both CMOS and beyond-CMOS emerging technologies to create large-scale neuromorphic systems. This chapter presents spike-based computations and spike-based communications as key fundamental principles underlying neuromorphic systems and discusses several approaches spanning sensors and processors that implements these fundamental principles into scalable, energy efficient and intelligent neuromorphic hardware solutions.

1 Introduction

The brain is the most remarkable intelligent system known to mankind. While digital computers spanning simple hand calculators to supercomputers have massive computing abilities, the unprecedented cognitive ability of the brain is by far largely unmet by the state-of-the-art silicon computing systems. Further, despite the unmatched cognitive abilities of the brain, the brain consumes orders of magnitude less power compared to existing computing solutions [1]. Thus, it comes as no surprise that the brain has been a constant source of inspiration for modern computing systems since its early inception [2]. The emerging field of *neuromorphic computing* is at the forefront of translating brain-inspiration into a novel computing paradigm. While the scope of neuromorphic computing is expanding day by day due to immense

A. Jaiswal (✉) · Md. A.-A. Kaiser
University of Wisconsin-Madison, Madison, WI, USA
e-mail: akhilesh.jaiswal@wisc.edu

Md. A.-A. Kaiser
e-mail: mkaiser8@wisc.edu

M. Parsa
George Mason University, Fairfax, VA, USA
e-mail: mparsa@gmu.edu

© The Author(s) 2026   241
M. te Vrugt (ed.), *Artificial Intelligence and Intelligent Matter*, Machine Intelligence
for Materials Science, https://doi.org/10.1007/978-3-032-04129-6_12

research interest across the community, on a very broad level, neuromorphic computing encompasses both hardware and software innovations inspired closely by the brain. The goal of hardware aspects of neuromorphic computing is to create solid-state systems that mimic the fundamental computing principles of the brain. The goal of software aspects of neuromorphic computing is to enable learning algorithms inspired by the learning ability of the brain. This chapter will focus on the hardware aspect of neuromorphic computing. Neuromorphic hardware can either be sensors or processors. Both the neuromorphic sensors and processors are based on two fundamental key attributes—*Spike Computation* and *Spike Communication*.

1.1 Spike Computation Using Simplified Neuronal Models

Neurons are the basic computing units in the brain. While several neuronal models are of interest to the neuromorphic community, a well-known simplified model of key interest to the hardware community is the leaky-integrate fire (LIF) model of the neuron, see Fig. 1. A typical *spiking neural network* (SNN) is composed of pre(input) and post(output) neurons interconnected through synaptic weights. The input preneuronal spikes are modulated by the weights W_i, and summed up as depicted in the figure. The summed up output signal alters the membrane potential (V_{mem}) of the postneuron in a leaky-integrate fashion i.e. the membrane potential increases in response to input spikes modulated by the weights, and leaks otherwise. The postneuron generates a spike if the membrane potential crosses a certain neuron threshold. The membrane potential is then reset, and the postneuron is prevented from spiking for a certain period of time known as the *refractory period*. The generated spikes then form the input or preneuron spike for the next layer of the spiking neural network. Thus, an SNN is composed of layers of neurons (LIF or other neuronal models) connected through synaptic weights. For each layer, input spike is multiplied by associated weights and then summed together. The summation signal based on the specific neuron model is then used to generate the output spikes for the specific layer of the SNN.

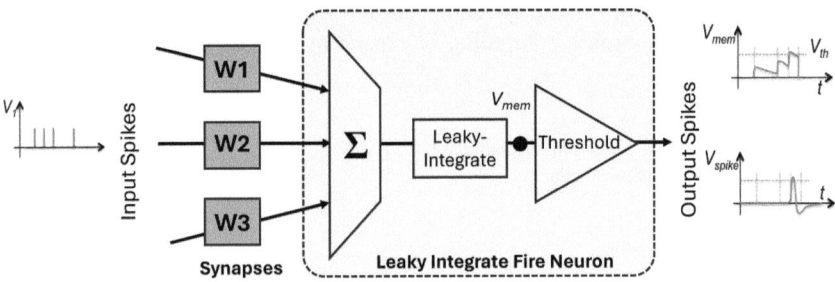

Fig. 1 Spike processing through a simplified model consisting of synapses and Leaky-Integrate-Fire Neurons. Inspired from Ref. [3]

This multiply and accumulate operation (MAC operation) followed by thresholding is a key computing aspect of SNNs. Note, since inputs to each layer of SNNs are spikes—which in a digital circuit implementation is typically represented as 0 or 1—multiply operations in an SNN is essentially multiplication by 1, thereby largely simplifying the hardware cost to implement the MAC operation. Thus, as we will see later, typical neuromorphic processors implement these simplified MAC operations in a distributed parallel manner followed by LIF neuronal thresholding at the output.

1.2 Spike Communication Using Address Event Representation

In a typical digital computing system information is transferred from one computing block to another in form of series of digital bits called *words* (for example, 32-bit or 64-bit words). In a neuromorphic system, information is transferred as addresses of the neurons that wish to exchange spikes (or events). The address-event-representation (AER) protocol facilitates communication of sparse, asynchronous, and random events or spikes resembling biological synaptic transmission, primarily employed in neuromorphic vision chips [4, 5] and neural processors [6]. Typically, the transmitter block encodes and sends the event using row and column addresses in two-dimensional neuromorphic systems [7–9]. In contrast, the receiver decodes and recreates the spike based on the received addresses [10, 11]. The transmitter and receiver communicate through a common data bus using a hand-shaking protocol. A simplified block diagram of the AER protocol is illustrated in Fig. 2.

In a two-dimensional system, the neurons are arranged in rows and columns. When a neuron generates an event spike, it initiates a row request (RR) signal and awaits the receiver's row acknowledge (RA) signal. An asynchronous row arbitration system ensures that only one row is acknowledged depending on the received buffer state (ready to latch new data). Upon receiving the row acknowledgment (RA) signal, the sending neuron activates its column request (CR) signal and awaits a reply on

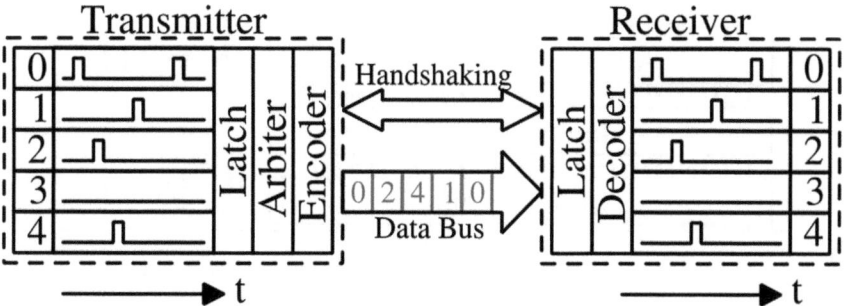

Fig. 2 Simplified AER protocol diagram

the column acknowledge (CA) line. Another column arbitration system guarantees the selection of one column at a time, and the column acknowledge signal (CA) is sent. Once both row and column acknowledgment signals are received, the activated neuron resets both request signals. The peripheral address encoder converts the spike's row and column addresses (x, y) based on RA and CA, transmitting them over the data bus. In a two-dimensional system, the neurons share the RR and RA signals along rows and CR and CA signals along columns. The transmitter can send addresses for activated neurons individually or in burst-mode [9]. In burst mode, the row arbiter selects the entire row, and column lines are latched in the periphery for concurrent reading, exhibiting high throughput. The arbitration system ensures that a new row will not be read until all the event spikes of the acknowledged row are serviced. An AER transmitter architecture for a two-dimensional neuromorphic system is shown in Fig. 3. On the receiver side, the addresses from the data bus are latched, decoded, and sent to the recipient neurons.

AER is a bio-inspired communication protocol widely used in neuromorphic systems that leverages sparse spike-based communications for energy efficiency.

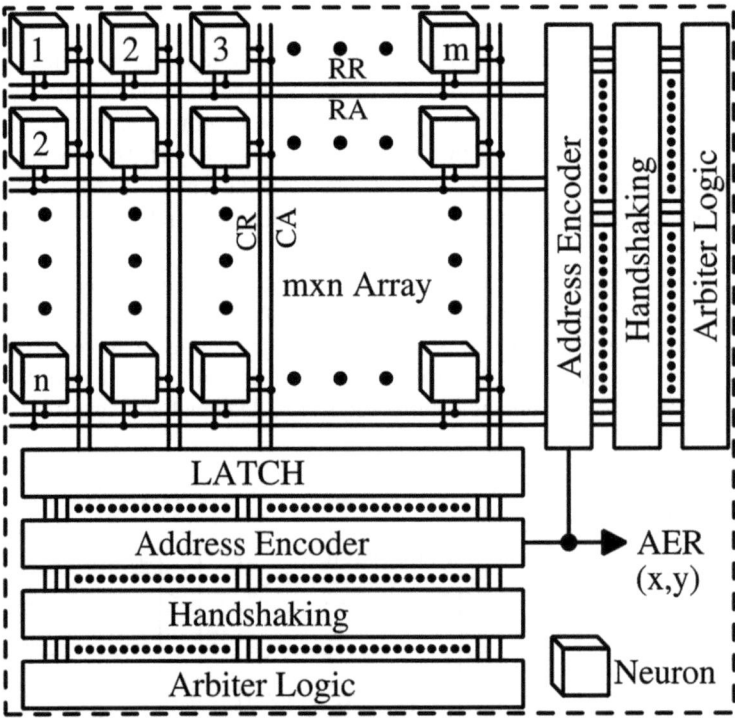

Fig. 3 AER transmitter architecture for a two-dimensional neuromorphic system

2 Neuromorphic Sensors

Rapid and robust response to environmental stimuli is a hallmark of animal intelligence and is crucial for survival, for example, escaping a predator or attacking a prey, in the ever changing surroundings. Sensory modalities like vision, olfactory, touch etc. play a critical role in driving these stereotypical animal responses to stimuli. Sensory systems, as such, have long been looked upon as a source of inspiration to process external stimuli and extract useful information [12–14]. In fact, the field of neuromorphic hardware has its inception in attempts to mimic the animal vision system in pioneering works from Mead and Mowald [15]. Recently, several neuromorphic works on various sensory modalities can be found in the literature including vision [16, 17], olfactory [18–20] and touch [21–23]. Among these, vision systems have been studied the most and applied to several interesting applications spanning computer vision [24] to robotic control [25, 26], etc. We will focus on vision based neuromorphic sensors as an example neuromorphic sensing system elaborating the overall approach, underlying circuits, and future works in the field.

Today's camera systems, which work as engineered eyes for machines, work very differently than biological eyes. A typical camera is based on CMOS image sensors [27], which consists of millions of photodiodes converting light into electric charges in a synchronous fashion, essentially capturing the information from external world in a frame-by-frame format. These frame-based cameras are currently the workhorse for almost all artificial vision-based systems. However, irrespective of the significance of information present in a visual scene, these cameras capture all the pixels in every frame thereby generating enormous volumes of data. Such data deluge fueled by frame-based cameras lead to slow frame speeds, high energy-consumption and latency in processing millions of pixels per frame. Neuromorphic cameras also called event-based Dynamic Vision Sensors (DVS), on the other hand, are retina-inspired (retinomorphic) systems that mimic the change detection attribute of biological retina. Pixels in a DVS camera monitor temporal changes in light intensity and generate ON/OFF signal, called spikes (or events) when they encounter a change in light intensity greater than a pre-defined threshold. This leads to asynchronous, sparse spike generation in response to light intensity changes as opposed to traditional cameras that generate continuous stream of frame-based data for all pixels in the focal plane array.

The two key aspects that drive the design of DVS sensors are—(1) spike-generation in response to temporal intensity-change and (2) spike-based AER communication. To see how these aspects are inspired by the biological eye, consider the fact that the retina is arranged in a layered fashion consisting of many different kind of cells. Bipolar cells in the retina are known to generate electrical signals in response to changes in light intensity [28]. There are ON and OFF bipolar cells in the retina. The ON (OFF) bipolar cells generate signals when the light intensity incident on a particular cell increases (decreases). The signals generated from the retinal cells are then processed through various layers of the retina and are eventually sent to the brain through optical nerve as spikes or sharp pulses of voltage. The behavior of bipo-

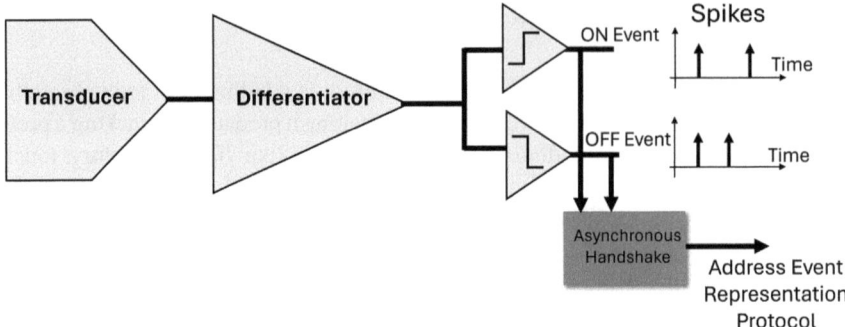

Fig. 4 A representative figure showing key circuit blocks of a neuromorphic dynamic vision sensor

lar cells is encapsulated in VLSI circuits that function as ON/OFF spike-generators in response to temporal intensity-change, and the communication of retinal outputs to the brain through spikes is encapsulated using the AER representation discussed earlier. Together the spike-generation and AER representation leads to highly sparse data generated from DVS cameras. Spike are generated only for those pixels that experience change in light intensity, as opposed to frame-based cameras that generate data for all pixels irrespective of changes in light intensity. Further, the data generated from DVS cameras is asynchronous i.e. a DVS pixel generates an ON/OFF spike as and when the pixel experiences a change in light intensity, as opposed to synchronous operation of frame-based camera, where pixel response is synchronized with respect to a global clock signal.

A circuit diagram showing a typical DVS pixel is depicted in Fig. 4. It consists of a photodiode (transducer) usually coupled with a logarithmic circuit. The logarithmic circuit helps to increase the dynamic range of the pixel. The logarithmic pixel output is fed into an analog differentiator circuit. The differentiator circuit detects temporal changes in the pixel output. If the output of the differentiator circuit crosses a pre-defined threshold a spike is generated using thresholding or comparator circuit. There are typically two thresholding circuits, one which generates a spike for positive changes in light intensity (ON spike) and other that generates spikes for negative changes in light intensity (OFF spikes). DVS cameras were first proposed in [16]. Subsequently, other camera systems inspired by the DVS cameras have been proposed, a notable camera system among them is the DAVIS camera. DAVIS pixels generate conventional frame-based data as well as DVS data, thus combining the best of both the traditional and neuromorphic cameras [29]. Recently, several works employing DVS cameras with neural networks have been used for a wide range of applications. These include hand-gesture recognition, etc. In most of these works, an event-based DVS camera is used alongside a conventional frame-based camera for decision making.

2.1 Recent Advances and Future Works in Neuromorphic Sensing

The current landscape in neuromorphic vision sensors is dominated by CMOS based circuits implementing DVS or DAVIS pixels. Recently, several interesting works have been proposed that either use emerging materials for DVS pixels or embed more complex computations into neuromorphic cameras. Spike generation in response to light intensity change is one aspect of retinal computation. Over the past decade, retinal neuroscience has identified several more complex retinal computations performed by various layers of the retinal cells beyond the simple light-change detection aspect, a feature mostly related to the functioning of bipolar cells. Different cells in various layers of the retina work in unison to generate highly specific feature-spikes that encode motion, shape and other aspects of objects present in the visual scene. Some examples of these feature-spikes are Object Motion Sensitivity (OMS) and Looming Detection (LD) functions. OMS is the ability of the retina to identify moving objects in moving background, while LD is the capability of the retina to identify approaching threats. Integrated Retinal Functionality in Image Sensors (IRIS) [30] is a recent proposal to leverage 3D integration of chips to mimic various computations of retinal layers for generating feature-spikes that represent OMS and LD functionality of retinal circuits. These spikes differ from traditional DVS spikes in that they carry highly specific motion-based information as opposed to light change intensity detection. 2D materials have also been used for mimicking aspects of retinal computation in [31, 32]. These works signify that emerging materials, 3D integration and newer discoveries in retinal neuroscience are underexplored areas and can significantly improve the adoption as well as capabilities of neuromorphic sensors in the near future.

Neuromorphic cameras represent a paradigm shift in imaging technology, offering unique advantages due to their high temporal resolution, very high dynamic range, low power consumption, and high pixel bandwidth (on the order of kHz) [33]. These cameras facilitate continuous, real-time processing in scenarios where traditional frame-based cameras fall short, enabling advancements in motion segmentation, recognition, neuromorphic control, space imaging, and onboard robotics. Primarily utilized in robotics and computer vision, the applications of event-based cameras range from low-level vision tasks such as feature detection and tracking to high-level applications including 3D reconstruction, pose estimation, real-time gaze estimation [34] and Simultaneous Localization and Mapping (SLAM) [33]. Specifically, they have found wide use in real-time monitoring systems, autonomous driving, and video surveillance [35], hypersonic research applications [36], real-time autofocusing for microscopic imaging [37], and space situational awareness and imaging [38].

3   Neuromorphic Processors

3.1   *Basic Principles for Neuromorphic Processors*

An abstract view of a conventional central processing unit (CPU) is shown in Fig. 5. It has two distinct sub-systems, a processing unit, and a memory unit. The processing and the memory unit are connected through a bus that transfers data back and forth from the memory unit to the processing unit. A typical operation of a CPU can be understood as follows—based on input commands (instructions) data is loaded (*load operation*) into the processing unit from the memory unit, the processing unit then applies basic arithmetic functions to the data and the result is subsequently stored (*store operation*) back into the memory unit. Such load-store based operations form the fundamental primitives for computing in a digital system. These systems characterized by two segregated units, the processing and the memory unit belong to a class of computer architecture called the von-Neumann architecture, named after the famed mathematician John von-Neumann who played a pioneering role in design of modern computers. A critical drawback of the von-Neumann architecture is the existence of segregated processing and memory units which leads to energy and latency expensive frequent transfer of data from the processing unit to the memory unit and vice-versa. Commonly known as the 'von-Neumann' or the 'memory' bottleneck, the segregated system design, limits the energy and throughput efficiency of current computing systems.

Neuromorphic computing systems, on the other hand, are based on principles of computing employed by the biological brain. Three key principles stand out in contrast to the von-Neumann architecture. First, processing and storage in the brain is accomplished by complex networks of neurons and synapses distributed throughout the brain in a closely intertwined fashion. In other words, unlike today's computing systems, the brain does not consist of physically segregated processing and memory units. The highly distributed and collocated processing (neurons) and memory (synapses) in the brain inevitably implies no (or minimal) bottleneck associated with transfer of data between the processing and the memory units. Thus, co-located and tightly connected processing and memory units are a key characteristic of neuromorphic hardware. As we will see later, processing-in-memory using emerging memory devices takes the concept of bringing processing and memory closer to each other to another level, wherein memory sub-systems are designed such that they can store data as well as perform massively parallel computing simultaneously. Second, spike-based communication is a common characteristic across neuromorphic systems and has direct inspiration from the biological brain. Neurons and networks of neurons in the brain communicate with each other through short pulses of voltages called spikes, similarly, blocks of neurons with associated synapses inside a neuromorphic hardware communicate with each other using spikes and AER. Thus, instead of transferring floating point data between different blocks, neuromorphic systems transfer addresses using AER, that are cheap to transfer, leading to significant saving in data transfer cost. Third, similar to the brain, neuromorphic hardware exhibits a high-level

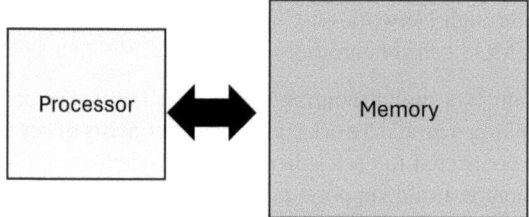

Von-Neumann Machines

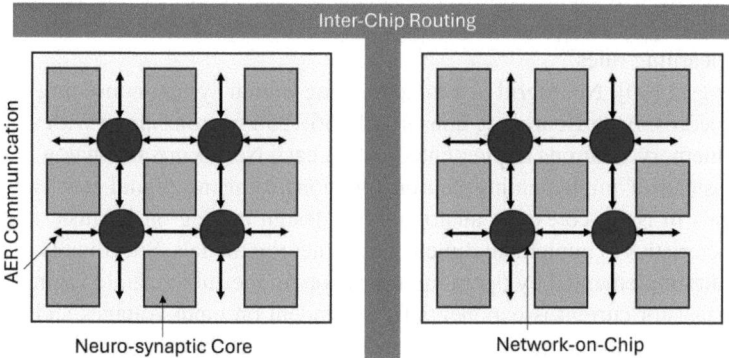

Neuromorphic Processor

Fig. 5 (Top) A von-Neumann computing machines with segregated processing and computing. (Bottom) A neuromorphic processor with tightly-intertwined computing and processing. Inspired from Ref. [1]

of sparsity i.e. only a small fraction of total neurons on a hardware generate a spike at any given point of time. Thus, the neuromorphic system can be designed to be spike-driven or event-driven, wherein a particular block is activated only when a spike is received for processing. This in turn leads to significant energy savings. In summary, co-located memory and processing, spike-based communication and event-driven sparsity are some of the key aspects of neuromorphic systems found across various hardware platforms. A high-level block diagram of a typical neuromorphic system is shown in Fig. 5. At a broad level neuromorphic systems can be divided into those based on CMOS technology and those that utilize emerging memory technologies.

3.2 CMOS Based Neuromorphic Computing Hardware

Both analog and digital CMOS hardware have been used to construct neuromorphic systems. In general, neuromorphic chips contain 'blocks' or 'cores' where each block or core consists of neurons with local memory. The cores in turn interact with each other using AER through Network-on-Chip (NoC). Below is a brief description of two well-known CMOS based neuromorphic systems.

1. Loihi [39]: Loihi is a digital neuromorphic chip fabricated on Intel 14nm for accelerating spiking neural networks in silicon. It consists of several (128) neuro-synaptic cores connected to each other through asynchronous NoC implementing a mesh structure. Loihi supports mapping of various spiking neural network topologies including those that require dense, all-to-all connectivity as well as those that need only sparse connectivity between neurons. Learning in a neuromorphic system usually consists of updating synaptic memory. Loihi includes a learning engine allowing changes to synaptic state variable to implement different local learning rules.

2. Neurogrid [40]: Neurogrid is a mixed-analog-digital system consisting of many Neurocores. Each Neurocore consists of 256×256 neurons along with some on-chip memory. Neurons implemented within each Neurocore are analog in nature i.e. instead of implementing neuron functionality using digital Boolean gates, neurons in Neurocores use analog circuit design relying on intrinsic transistor characteristics to implement dimensionless neuron models. Specifically, the neurons are implemented by operating transistors in the subthreshold regime where the transistor current is exponentially dependent on input voltages and helps in implementing the characteristic neuron functions. 16 Neurocores chips are connected using a tree structure to form a Neurogrid system.

Other notable CMOS neuromorphic systems are Spinnaker [41] and BrainScale-2 [42].

3.3 Emerging Memory Based Neuromorphic Computing Hardware

The sustained dimensional scaling of transistors has driven continued improvements in the performance of digital computing systems including on-chip memory storage in accordance with Moore's Law [43]. However, with the recent slowdown of Moore's Law, alternate devices are being explored to improve the storage density and power efficiency of existing memory sub-systems. Non-volatile memory (NVM) devices have emerged as a promising class of memory devices in recent times [44]. In general, these non-volatile devices consist of two-terminal devices. The resistance of the device can be modulated between high and low resistance levels using either external current or voltage stimulus. Interestingly, many NVMs allow programming the resistance level not just between the high (max) and low (min) levels but also at intermediate values between the high and low level. This behavior in turn allows NVM to store more than one bit of data in a single device. For example, an NVM that can be programmed to 16 different levels stores 4-bit data. This multi-bit storage coupled with the small formfactor of NVM devices leads to very high-density mem-

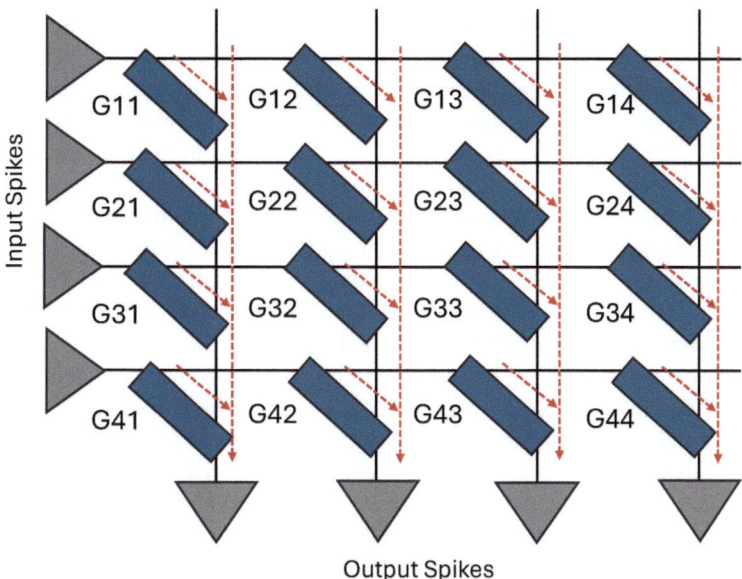

Fig. 6 An NVM based crossbar array capable of massively parallel MAC operations using analog computing

ory storage. Some of the most widely explored NVM devices are Resistive-RAM (RRAM), Phase Change Memory (PCM), Magnetic-RAM (MRAM), etc.

In general, the switching of RRAM device is due to growth of conductive filament between the two electrodes of the RRAM device [45]. Low resistance results when a filament is formed between the top and bottom electrode by application of appropriate voltage. In absence of such a conductive filament the RRAM device exhibits high resistance. By controlling the growth of the conductive filament, RRAM devices can exhibit multiple resistance levels and thus can store multiple bits. Based on the composition of the conductive filament RRAM devices can be classified in two broad categories—conductive bridge RAM (CBRAM) based on metal ion filaments and OxRAM based on oxygen vacancy filaments.

Phase Change Memory (PCM) [46] works by altering the material property of a small active volume formed of phase change material embedded within two electrodes. Data is stored in a PCM device by changing the active region from crystalline to amorphous state and vice-versa. When the device is in crystalline (amorphous) state, it exhibits low (high) resistance. The PCM device can be switched from crystalline to amorphous state and vice-versa by application of appropriate current pulses. The transition from amorphous to crystalline phase and vice-versa is driven by intricate control of temperature of the active region. Current pulses applied to the PCM device generate heat allowing an electrical means to program the PCM device. Like RRAMs, PCM devices also exhibit multi-bit storage by appropriately controlling the applied current pulses.

MRAMs [47] belong to the family of spintronic devices. A typical MRAM device is composed of a Magnetic Tunnel Junction (MTJ). MTJ consists of two nano-magnets separated by a non-magnetic oxide space. When the spin direction of two nano-magnets point in same direction (parallel state), the MTJ exhibits low resistance state, while the nano-magnets pointing in opposite directions (anti-parallel state) leads to high resistance state. Typically, an MTJ has only two stable states—low and high resistance. Unlike RRAMs and PCMs MRAMs do not exhibit multi-bit storage. Note, other spintronic devices like domain-wall magnets [48] can exhibit multi-bit storage, however, the resistance range is usually much smaller compared to RRAMs and PCMs. MRAMs also have higher endurance compared to RRAMs and PCM and is an important differentiator for MRAM technology.

NVMs like RRAM, PCM, MRAM can be arranged in a crossbar fashion enabling massively parallel MAC operations. Referring to Fig. 6 input spikes are applied on the horizontal wires and the resultant analog MAC operation is obtained on the vertical wires. An NVM is stationed between each horizontal and vertical wire. Current flowing through each NVM performs a multiplication operation between its conductance value and applied input voltage (input spike) in accordance with Ohm's Law. The multiplicative current flowing through all the NVM devices in the same column add up based on Kirchoff's current law, resulting in addition operation. Thus, the output current on each column represents analog MAC operation. A neuron circuit, for example, an LIF neuron, can be placed on the output of each column thereby generating an output spike based on the analog MAC current. Note, MAC operation (computing) is performed in a massively parallel manner inside the memory devices, hence such schemes are called processing-in-memory [49] schemes and are currently an area of intense research investigation. Finally, processing-in-memory operations can also be performed using CMOS memories [50].

3.4 Recent Advances and Future Works in Neuromorphic Computing Using NVMs

Recent works in neuromorphic computing using NVMs can be categorized into three distinct efforts [51]. On the device side, high precision NVM devices that can store several different levels of resistive values are being actively explored. On the circuits and system level larger arrays and holistic systems are being build using NVM based neuromorphic platforms. Finally, algorithm-hardware co-design, wherein hardware non-ideal behavior can be mitigated through algorithmic innovations are taking center stage in processing-in-memory based neuromorphic systems.

4 Summary

This chapter presented an overview of neuromorphic hardware from sensors to processors, identifying basic underlying principles and specific approaches that have been used by the research community in realizing those principles. The chapter discusses spikes based computations and spike based communications as key distinctive features of neuromorphic systems.

References

1. Roy K, Jaiswal A, Panda P (2019) Towards spike-based machine intelligence with neuromorphic computing. Nature 575(7784):607–617
2. Von Neumann J, Kurzweil R (2012) The computer and the brain. Yale university Press
3. Jaiswal A, Roy S, Srinivasan G, Roy K (2017) Proposal for a leaky-integrate-fire spiking neuron based on magnetoelectric switching of ferromagnets. IEEE Trans Electron Devices 64(4):1818–1824
4. Lichtsteiner P, Posch C, Delbruck T (2006) A 128 x 128 120db 30mw asynchronous vision sensor that responds to relative intensity change. In: 2006 IEEE international solid-state circuits conference-digest of technical papers, pp 2060–2069, IEEE
5. Son B, Suh Y, Kim S, Jung H, Kim J-S, Shin C, Park K, Lee K, Park J, Woo J, et al. (2017) 4.1 a 640× 480 dynamic vision sensor with a $9\mu m$ pixel and 300meps address-event representation. In: 2017 IEEE International Solid-State Circuits Conference (ISSCC), pp 66–67, IEEE
6. Akopyan F, Sawada J, Cassidy A, Alvarez-Icaza R, Arthur J, Merolla P, Imam N, Nakamura Y, Datta P, Nam G-J et al (2015) Truenorth: Design and tool flow of a 65 mw 1 million neuron programmable neurosynaptic chip. IEEE Trans Comput Aided Des Integr Circuits Syst 34(10):1537–1557
7. Lazzaro J, Wawrzynek J, Mahowald M, Sivilotti M, Gillespie D (1992) Silicon auditory processors as computer peripherals. Advances in Neural Information Processing Systems, vol 5
8. Boahen KA (2000) Point-to-point connectivity between neuromorphic chips using address events. IEEE Trans Circuits Syst II Analog Digital Signal Proc 47(5):416–434
9. Boahen KA (2004) A burst-mode word-serial address-event link-i: transmitter design. IEEE Trans Circuits Syst I Regul Pap 51(7):1269–1280
10. Boahen KA (2004) A burst-mode word-serial address-event link-ii: receiver design. IEEE Trans Circuits Syst I Regul Pap 51(7):1281–1291
11. Lin J, Boahen K (2009) A delay-insensitive address-event link. In: 2009 15th IEEE symposium on asynchronous circuits and systems, pp 55–62, IEEE
12. Mead CA (1987) Neural hardware for vision. Eng Sci 50(5):2–7
13. Mead C (1990) Neuromorphic electronic systems. Proc IEEE 78(10):1629–1636
14. Koickal TJ, Hamilton A, Tan SL, Covington JA, Gardner JW, Pearce TC (2007) Analog vlsi circuit implementation of an adaptive neuromorphic olfaction chip. IEEE Trans Circuits Syst I Regul Pap 54(1):60
15. Mead CA, Mahowald MA (1988) A silicon model of early visual processing. Neural Netw 1(1):91–97
16. Leñero-Bardallo JA, Serrano-Gotarredona T, Linares-Barranco B (2011) A 3.6us latency asynchronous frame-free event-driven dynamic-vision-sensor. IEEE J Solid-State Circuits 46(6):1443–1455
17. Delbruck T et al (2008) Frame-free dynamic digital vision. In: Proceedings of international symposium on secure-life electronics. Advanced electronics for quality life and society, vol 1, pp 21–26, Citeseer

18. Hu X, Khanzada S, Klütsch D, Calegari F, Amin H (2022) Implementation of biohybrid olfactory bulb on a high-density cmos-chip to reveal large-scale spatiotemporal circuit information. Biosens Bioelectron 198:113834

19. Datta-Chaudhuri T, Araneda RC, Abshire P, Smela E (2016) Olfaction on a chip. Sens Actuators, B Chem 235:74–78

20. Koickal TJ, Hamilton A, Pearce TC, Tan S-L, Covington JA, Gardner JW (2006) Analog vlsi design of an adaptive neuromorphic chip for olfactory systems. In: 2006 IEEE international symposium on circuits and systems, pp 4–pp, IEEE

21. Abou Khalil A, Valle M, Chible H, Bartolozzi C (2018) Cmos event-driven tactile sensor circuit. Integration 63:315–322

22. Rahiminejad E, Parvizi-Fard A, Iskarous MM, Thakor NV, Amiri M (2021) A biomimetic circuit for electronic skin with application in hand prosthesis. IEEE Trans Neural Syst Rehabil Eng 29:2333–2344

23. Liu F, Deswal S, Christou A, Sandamirskaya Y, Kaboli M, Dahiya R (2022) Neuro-inspired electronic skin for robots. Sci Robot 7(67):eabl7344

24. E. Piątkowska, A. N. Belbachir, S. Schraml, and M. Gelautz, "Spatiotemporal multiple persons tracking using dynamic vision sensor," in *2012 IEEE Computer Society Conference on Computer Vision and Pattern Recognition Workshops*, pp. 35–40, IEEE, 2012

25. Conradt J, Cook M, Berner R, Lichtsteiner P, Douglas RJ, Delbruck T (2009) A pencil balancing robot using a pair of aer dynamic vision sensors. In: 2009 IEEE international symposium on circuits and systems, pp 781–784, IEEE

26. Blum H, Dietmüller A, Milde M, Conradt J, Indiveri G, Sandamirskaya Y (2017) A neuromorphic controller for a robotic vehicle equipped with a dynamic vision sensor. In: Robotics science and systems, RSS 2017

27. El Gamal A, Eltoukhy H (2005) Cmos image sensors. IEEE Circuits Devices Mag 21(3):6–20

28. Euler T, Haverkamp S, Schubert T, Baden T (2014) Retinal bipolar cells: elementary building blocks of vision. Nat Rev Neurosci 15(8):507–519

29. Almatrafi M, Hirakawa K (2019) Davis camera optical flow. IEEE Trans Comput Imag 6:396–407

30. Yin Z, Kaiser MA-A, Camara LO, Camarena M, Parsa M, Jacob A, Schwartz G, Jaiswal A (2022) Iris: Integrated retinal functionality in image sensors, pp 2022–2028

31. Zheng Y, Ghosh S, Das S (2024) A butterfly-inspired multisensory neuromorphic platform for integration of visual and chemical cues. Adv Mater 36(13):2307380

32. Guo P, Jia M, Guo D, Wang ZL, Zhai J (2023) Retina-inspired in-sensor broadband image preprocessing for accurate recognition via the flexophototronic effect. Matter 6(2):537–553

33. Gallego G, Delbrück T, Orchard G, Bartolozzi C, Taba B, Censi A, Leutenegger S, Davison AJ, Conradt J, Daniilidis K, Scaramuzza D (2022) Event-based vision: a survey. IEEE Trans Pattern Anal Mach Intell 44(1):154–180

34. Li N, Chang M, Raychowdhury A (2024) E-gaze: gaze estimation with event camera. IEEE Trans Pattern Anal Mach Intell

35. Li Z, Su H, Li B, Luan H, Gu M, Fang X (2024) Event-based diffractive neural network chip for dynamic action recognition. Optics & Laser Technol 169:110136

36. Tan ZP, Hsu K, Tan J-M (2024) asyncelf: development of event-based 3d imaging for hypersonic fsi measurements. In: AIAA SCITECH 2024 Forum, p 2492

37. Qu X, Ma C, Hu W, Den H, Yang S (2024) A robust autofocusing method for microscopic imaging based on an event camera. Opt Lasers Eng 175:108025

38. Ralph NO, Marcireau A, Afshar S, Tothill N, Van Schaik A, Cohen G (2023) Astrometric calibration and source characterisation of the latest generation neuromorphic event-based cameras for space imaging. Astrodynamics 7(4):415–443

39. Davies M, Srinivasa N, Lin T-H, Chinya G, Cao Y, Choday SH, Dimou G, Joshi P, Imam N, Jain S et al (2018) Loihi: a neuromorphic manycore processor with on-chip learning. IEEE Micro 38(1):82–99

40. Benjamin BV, Gao P, McQuinn E, Choudhary S, Chandrasekaran AR, Bussat J-M, Alvarez-Icaza R, Arthur JV, Merolla PA, Boahen K (2014) Neurogrid: a mixed-analog-digital multichip system for large-scale neural simulations. Proc IEEE 102(5):699–716

41. Furber SB, Galluppi F, Temple S, Plana LA (2014) The spinnaker project. Proc IEEE 102(5):652–665
42. Pehle C, Billaudelle S, Cramer B, Kaiser J, Schreiber K, Stradmann Y, Weis J, Leibfried A, Müller E, Schemmel J (2022) The brainscales-2 accelerated neuromorphic system with hybrid plasticity. Front Neurosci 16:795876
43. Roser M, Ritchie H, Mathieu E (2024) What is moore's law? Our world in data
44. Chakraborty I, Jaiswal A, Saha A, Gupta S, Roy K (2020) Pathways to efficient neuromorphic computing with non-volatile memory technologies. Appl Phys Rev 7(2)
45. Zahoor F, Azni Zulkifli TZ, Khanday FA (2020) Resistive random access memory (rram): an overview of materials, switching mechanism, performance, multilevel cell (mlc) storage, modeling, and applications. Nanoscale Res Lett 15:1–26
46. Wong H-SP, Raoux S, Kim S, Liang J, Reifenberg JP, Rajendran B, Asheghi M, Goodson KE (2010) Phase change memory. Proc IEEE 98(12):2201–2227
47. Tehrani S, Slaughter J, Chen E, Durlam M, Shi J, DeHerren M (1999) Progress and outlook for mram technology. IEEE Trans Magn 35(5):2814–2819
48. Sharad M, Augustine C, Panagopoulos G, Roy K (2012) Spin-based neuron model with domain-wall magnets as synapse. IEEE Trans Nanotechnol 11(4):843–853
49. Sebastian A, Le Gallo M, Khaddam-Aljameh R, Eleftheriou E (2020) Memory devices and applications for in-memory computing. Nat Nanotechnol 15(7):529–544
50. A. Jaiswal, I. Chakraborty, A. Agrawal, and K. Roy, "8t sram cell as a multibit dot-product engine for beyond von neumann computing," *IEEE Transactions on Very Large Scale Integration (VLSI) Systems*, vol. 27, no. 11, pp. 2556–2567, 2019
51. Rathi N, Chakraborty I, Kosta A, Sengupta A, Ankit A, Panda P, Roy K (2023) Exploring neuromorphic computing based on spiking neural networks: algorithms to hardware. ACM Comput Surv 55

Towards Intelligent Active Particles

Hartmut Löwen and Benno Liebchen

Abstract In this book chapter we describe recent applications of artificial intelligence and in particular machine learning to active matter systems. Active matter is composed of agents, or particles, that are capable of propelling themselves. While biological agents like bacteria, fish or birds naturally possess a certain degree of "intelligence", synthetic active particles like colloidal microswimmers and electronic robots can be equipped with different levels of artificial intelligence, either internally (as for robots) or via a dynamic external control system. This book chapter briefly discusses existing approaches to make synthetic particles increasingly "intelligent" and then focuses on the usage of machine learning to approach navigation and communication problems of active particles. Basic questions are how to steer a single active agent through a complex environment to reach or discover a target in an optimal way and how active particles need to cooperate to efficiently collect a distribution of targets (e.g. nutrients or toxins) from their complex environment.

1 Introduction

The past few years have seen a tremendous development in combing active matter [1] with artificial intelligence (AI) [2]. Active matter typically comprises agents (or 'particles") like bacteria, algae, vibrated granulates or synthetic colloidal microswimmers that convert energy from their environment into directed mechanical motion [3, 4]. These agents are out of equilibrium, in qualitative difference to passive particles. Since more than two decades [5] it is possible to create synthetic active particles in the colloidal regime, i.e. at the micron- or the nano-scale, where they are typi-

H. Löwen (✉)
Institut für Theoretische Physik II: Weiche Materie, Heinrich-Heine Universität, Universitätsstraße 1, 40225 Düsseldorf, Germany
e-mail: hlowen@hhu.de

B. Liebchen
Institut für Physik kondensierter Materie, Technische Universität Darmstadt, Hochschulstrasse 8, 64289 Darmstadt, Germany
e-mail: benno.liebchen@pkm.tu-darmstadt.de

M. te Vrugt (ed.), *Artificial Intelligence and Intelligent Matter*, Machine Intelligence for Materials Science, https://doi.org/10.1007/978-3-032-04129-6_13

cally embedded in a surrounding liquid [3], or sometimes in a complex plasma [6]. Activity then implies that the particles are actively moving (self-propelling) relative to the surrounding medium and are, unlike passive particles, not just advected by it. Accordingly, colloidal active particles are also viewed as micro- and nano-motors that do not require any movable parts and that can transport cargo at much smaller scales than conventional motors. They are also frequently discussed for their potential to perform useful tasks in the future such as delivering drugs to cancer cells, performing microsurgery, and collecting microplastics. However, this would either require to externally steer each individual particle in a suitable way, which is not easy if they are for example deep inside the human body, or to equip the particles with some degree of artificial intelligence that allows them to navigate autonomously. To date it remains as a persistent experimental challenge to directly integrate the required hardware for artifical intelligence into active colloidal particles [7, 8] which are far too small to equip them with sensors, actuators and information processing units. Thus, other, simpler, approaches are required, some of which we discuss in the following.

Figure 1 shows particles of increasing complexity that are equipped with seven different levels of sophistication or "intelligence". Here the individual levels can be viewed as steps between an ordinary particle without any intelligence and a complete sensor-processor-actuator system that can perceive information and can perform computations to determine actions that are executed by the actuator. Concretely, panel (a) represents a passive particle with no "intelligence" at all. This particle is randomly kicked by a Newtonian solvent (such as water) leading to Brownian motion. Such particles are the subject of traditional colloid science which has started more than 100 years ago with famous contributions from Einstein, Perrin, Derjaguin, Landau, Verwey and Overbeek and has been brought to maturity by using almost monodisperse suspensions even at high particle density as ideal classical model systems for equilibrium phase transitions [9]. In (b) the particle is susceptible to an external field and can be steered by suitable variations of the field in space and time. Possible fields are an external solvent flow field, a laser-optical field (leading to optical laser-tweezers), an electric or a magnetic field [10]. Here, the particle itself also has no intelligence at all, but the setup allows a human controller to use her/his intelligence to steer the particle in a desired way. This can be done based on quick feedback effects (e.g. using optical tweezers) such that the particle motion can be changed at each point in time almost at wish. The next level, (c), is an active colloidal particle which has internal degrees of freedom and which can self-propel, i.e. it is self-actuated. Such active colloids are also called microswimmers and they are the elementary building blocks of active soft matter. The trajectory of a microswimmer typically has an initial ballistic part (apart from Brownian motion at very short times) which crosses over to long-time Brownian motion albeit with a so-called active diffusion coefficient which typically is orders of magnitude larger than the diffusion coefficient of passive colloids [3]. The intelligence of such microswimmers is rather primitive; it is embodied in the internal motor of the particles which reacts to environmental changes. For instance, it is now well known that active colloids which catalyze a certain chemical reaction

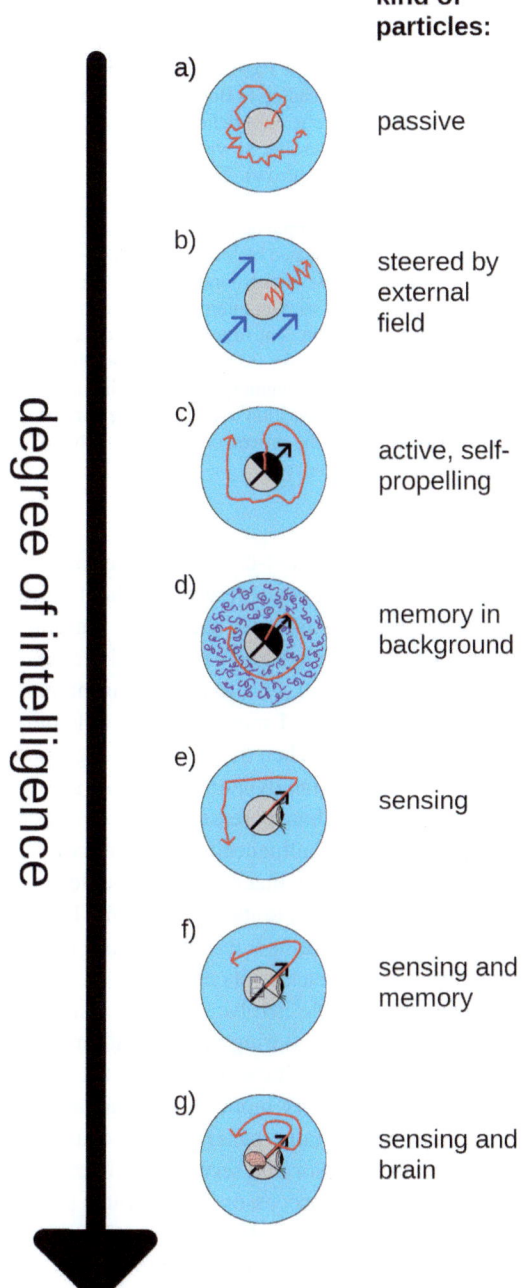

Fig. 1 (Continued)

◀ **Fig. 1** Schematic illustration of micron-sized colloidal particles (grey) in a fluid solvent (blue background) with different levels of sophistication or "intelligence" (increasing from top to bottom). Typical trajectories are shown in red color: **a** Passive particle with no "intelligence" at all; the particle is randomly kicked by the solvent leading to Brownian motion. **b** Particle susceptible to an external field (flow, optical electric, magnetic etc.) indicated by the black arrows such that it can be steered by an "intelligent" external controller at wish. **c** Active particles with an inner motor (actuator) leading to self-propulsion (black arrow). These colloidal "microswimmers" can move in a way that depends on their local environment. This essentially equips them with the ability to sense and react to their environment in a primitive way (e.g. via synthetic taxis). **d** Particles in a non-Newtonian (viscoelastic) medium (indicated by blue polymer coils) exhibit additional memory effects which are induced externally by the background. These particles react also to the history of their environment. **e** Particles equipped with an internal sensor, that allows them e.g. to see and react to other particles within a certain vision cone or to "smell" and react in a sophisticated way to chemical cues. **f** Particle equipped with a sensor (as indicated by a schematic eye) and an internal memory storage (as indicated by a schematic memory chip) allowing it to react to its environment in a way that depends on present and past sensation (experience). **g** Particle that is equipped with a full sensor-processor-actor system, where the processor serves as a "brain" that can compute and autonomously make decisions on how to move

on part of their surface only and use a self-created concentration gradient to achieve self-propulsion also react to external concentration gradients. Effectively, this reaction is similar to the reaction of chemotactic bacteria to nutrient or toxin gradients (albeit based on a very different mechanism) [11]. In this sense, active colloids are self-actuated agents that can sense and react to their environment. The situation can be made more complex (see (d)) when dispersing the particle into a non-Newtonian solvent (such as a polymer solution). Then the solvent background exhibits memory effects which leads to a non-Markovian dynamics. In some sense, compared to case (c), the "intelligence" of the system has increased as it now remembers the past (via its environment) and its dynamics is influenced by its history (or memory). In panel (e) the particle is equipped with a sensor that is directly coupled to the self-propulsion mechanism such that the particle reacts to its environment. That is the particle measures an external stimulus and transforms it instantaneously into motion. In contrast to case (c) where the particle also reacts to its environment, in (e) the sensation can be more general and can for example involve a certain vision cone [12]. If this is augmented with an internal memory storing past sensations (panel f), the particle is more "intelligent" in the sense that it can not only react to its environment in a way that depends on its present sensory input but that can also depend on the past, i.e. on the particle's experience. Finally, panel (g) shows an active particle that is equipped with a sensor and a processor that serves as an internal brain. Such a particle represents a full actuator-sensor-processor system. Such systems can be realized in the form of programmable electronic robots at centimeter if not millimeter sizes, but not yet at the microscale. However, while we are not yet able to synthetically create intelligent micro-devices with the properties indicated in panel (g), nature has already realized them in the form of bacteria and other microorganisms.

While we are far away from being able to synthetically realize a fully intelligent micro- or nanoparticle (level (g)), AI has been applied in the situations (b)–(e) in various ways. In particular when using external fields lead to actuate motion of particles and a feedback control system [12–15] "intelligent" behavior can be implemented. Here, the learning process takes place externally in a computer (or some other machine or human). Accordingly, one focus of current studies is to steer particles (actuated or/and self-propelled) with external fields that are dynamically adapted to the motion and environment of the particle. Alternatively, one can also use small robots as active particles [16]. Then the learning process takes place internally. In both cases AI can be applied either to solve learning problems for an *individual* swimmer or for an ensemble of active agents. Problems for an individual swimmer can concern e.g. the problem to find the optimal route to a food source (e.g. the fastest path) that is located within a complex environment or to find the optimal body-shape deformations of a deformable swimmer to optimize the self-propulsion speed. Here, challenges in determining e.g. optimal navigation or swimming strategies arise from the smallness of the swimmers. This is because microswimmers are subject to significant fluctuations due to Brownian motion, errors and delays in the steering protocol, changes and fluctuation in the environment; hence they cannot accurately predict the outcome of their navigational maneuvers and this hast to be taken into account for real applications. The second problem concerns *interacting active particles* that can in general communicate with each other e.g. via self-induced gradients in a phoretic field (e.g. a chemical concentration field, temperature or electric potential) or via light sources and sensors [16]. Here, artifical intelligence can be used for example to optimize the communication rules among the agents as we exemplarily discuss further below. In the future, AI could be also used to optimize communication rules to make the agents cooperate to collectively perform a certain task. For macroscopic robots this has been established to a certain extent, but if we have an army of agents at the micrometer level (as required e.g. for a drug delivery task in our human body [17–19]) this is getting enormously difficult due to fluctuations and hydrodynamic interactions. In the real living world of biological microswimmers, there are many examples of such concerted collective behavior (that is typically evolutionarily optimized). Examples are sperm cells which sense gradients in the concentration of the chemicals emitted by the egg and phagocytes hunting pathogenic bacteria [20, 21]. In contrast, while synthetic microswimmers can also sense and react to their environment, even via self-produced chemical signals [11, 22], they do not yet cooperate in a way that optimizes their collective behavior to perform useful tasks.

In this book chapter we shall focus on applications of AI on microswimmers, both single and collective. We first focus on reinforcement learning techniques to guide and steer single microswimmers in a complex environment. Then we discuss predator-prey systems and many-body problems which feature applications of AI for communicating agents. Here we emphasize recent applications and do not aim to be exhaustive. Instead, for more exhaustive discussions of applications of artificial intelligence to active matter we would like to guide the reader to the articles [7, 8] and to the bookchapter by G. Volpe, and the one by J. Jeggle and R. Wittkowski. For an introduction to reinforcement learning see the bookchapter by M. Schilling.

2   Artificial Intelligence Applied to a Single Microswimmer

An agent that has the ability to mechanically move its limbs can use different strategies to achieve self-propulsion. A necessary condition to achieve a net propulsion at
low Reynolds numbers is to break time-reversal symmetry of the mechanical motion
pattern. This was exemplified in different model swimmers such as the three-linked-
sphere microswimmer of Najafi and Golestanian [23] where three aligned linked
spheres move relatively to each other. Another example is the push-me-pull-you
swimmer that requires only two spheres that additionally change their size [24].
Such model swimmers have been generalized towards circular swimming [25] and
have also been studied in confinement [26]. The three-linked microswimmer can
use different locomotoric gaits (different shape-deformation patterns), obtained via
reinforcement learning, e.g. to optimize self-propulsion speed or to steer (navigate)
in a desired way as recently demonstrated in [27–30]. A similar idea can also be
applied to mechanical rotations of the swimmer [31]. A second type of problem for
which reinforcement learning plays a key role is to learn optimal strategies to find a
target of unknown position by chemotaxis such as e.g. a static food source [32, 33].
Here, an internal learning scheme with a genetic algorithm which leads to successfully finding the target was proposed in [34, 35]. A third typical problem for a single
microswimmer that can freely control its self-propulsion direction but not its speed
is to find the fastest path from a prescribed starting point A to another prescribed
target point B, i.e. the connecting path that leads to the shortest traveling time. Some
simple cases can be solved analytically [36, 37] but for more complex environments
reinforcement learning techniques [38] have been applied to find the optimal path,
see e.g. [39–47]. Some works have used tabular Q-learning, for self-thermophoretic
active particles [48], or for learning to navigate optimally inside a Mexican hat potential [49]. Other works have used actor-critic methods [41] and also reinforcement
learning combined with neural networks [47, 50]. In particular, in Ref. [47] a reinforcement learning-based method has been developed to determine *asymptotically
optimal trajectories* which has shown that it is possible to systematically learn the
result of an optimal control problem without having to do an explicit calculation.
As an example of such optimal navigation problems we now consider an active particle in a prescribed motility landscape, that is in a spatially varying external field
that controls the self-propulsion speed of the particle. (Such motility patterns can be
created e.g. for light-driven microswimmers in light-intensity patterns [51].) Suppose that a particle in such a motility pattern can freely choose its self-propulsion
direction, but not its speed and let us ask for the fastest route that connects a given
starting point and a target point [52]. Using reinforcement learning and training the
agent in different motility landscapes allows one to find a navigation strategy that
leads to trajectories that closely approximate the exact optimal path, although there
is typically no guarantee that the resulting path is really optimal in all cases [52].
In particular, there is the risk that the reinforcement learning approach converges to
some local optimum rather to the global one, which can be largely avoided by using
a suitable on-policy approach [47]. Exemplaric results are shown in Fig. 2 where the
learned trajectory gets closer and closer to the optimal one as training proceeds.

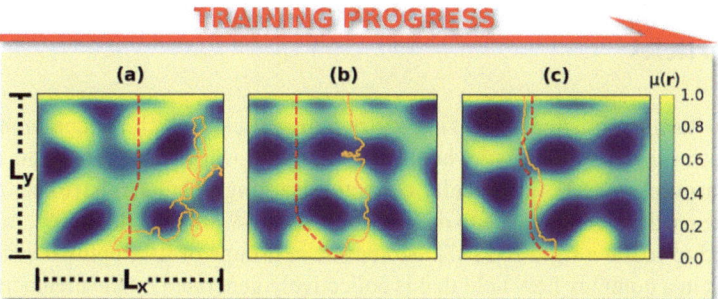

Fig. 2 Three typical trajectories (orange solid lines) in three different motility fields $\mu(\mathbf{r})$ over the course of the training procedure of a Q-learning active Brownian particle. The particle's objective is to cross the box of lateral sizes L_x and L_y from bottom to top as fast as possible. The motility fields are generated with the help of modified isotropic Gaussian random waves. The probabilities ϵ to perform random actions are **a** $\epsilon = 1.0$, **b** $\epsilon \approx 0.58$, **c** $\epsilon \approx 0$. The red dashed lines highlight the optimal trajectories obtained with Dijkstra's algorithm [53]. Motility fields are generated as modified isotropic Gaussian random waves. In the horizontal direction reflecting boundary conditions are employed. From Ref. [52]. Licensed under the terms of Creative Commons CC BY 4.0

3 Artificial Intelligence Applied to Predator-Prey-Like Systems

A somewhat more complex problem occurs if the active particles does not simply navigate towards a stationary target but if the target moves, either randomly or by reacting to the active particle. The latter situation resembles a predator-prey problem where the predator hunts the prey and where the prey tries to escape from the predator. Such problems are very common in nature at the macroscale, but there are also popular examples at the microscale such as, e.g. a macrophage chasing a pathogenic bacteria. Recently, such predator-prey-like systems have also been artificially realized in the context of micron-sized particles, see e.g. [54]. (Corresponding models mainly for chemotactic coupling have been explored theoretically [55–57].) It is interesting to consider a situation where both the predator and the prey are trained in a reinforcement learning framework and learn by self-play. This has been considered recently for simple lattice models [58] relevant for ecology. Recently, Ref. [59] has used reinforcement learning also to explore a predator-prey problem at the colloidal scale. In this study a single predator is surrounded by a gas of prey particles and the prey learns to escape the predator. Different sensing situations are discussed in this work, leading to different strategies. This model can be extended to a group of predators exploring group chasing strategies which have been considered in the context of active particles [60]. In Ref. [43] an interesting application has been proposed where a swarm of predators is trained to find a passive but Brownian cargo particle ("prey") in a maze-like environment. Also the control of particle groups by chasers or controllers (shep-dog in the macroscopic world) [61] are interesting problems where learning strategies will be fruitful.

4   Artificial Intelligence Applied to Groups of Active Particles

We now turn to situations involving a group of particles which are communicating with each other [62, 63] and where the group as a whole can learn strategies for different purposes [64]. In particular, collective learning techniques have been applied to active particles for instance to make them learn to swarm and flock [65], to cluster [66] and to rotate a rigid rod [67]. Moreover, in groups of fish, optimal swimming patterns in a complex flow field that is collectively generated by the members of the group themselves can also be learned based on deep reinforcement learning [68]. In this case the impact of fluid vortices is key to understand the collective behavior.

Let us now discuss a specific example where a group of communicating agents learns to cooperate in a way that enhances their nutrient consumption [69]. Concretely, we consider a group of active agents each of which can sense the local food gradient (typical for chemotactic microorganisms) and consumes nutrients (or remove toxins) at a certain rate in the vicinity of its momentaneous position. The different agents communicate via chemical signaling (quorum sensing); that is each agents emits certain chemicals which spread out by diffusion and other agents can sense (measure) the gradient in the chemical concentration (Fig. 3). That is, overall, each agent can either greedily follow the nutrient concentration gradient which bears the risk of ending up in some small local concentration maximum, or it can try to find

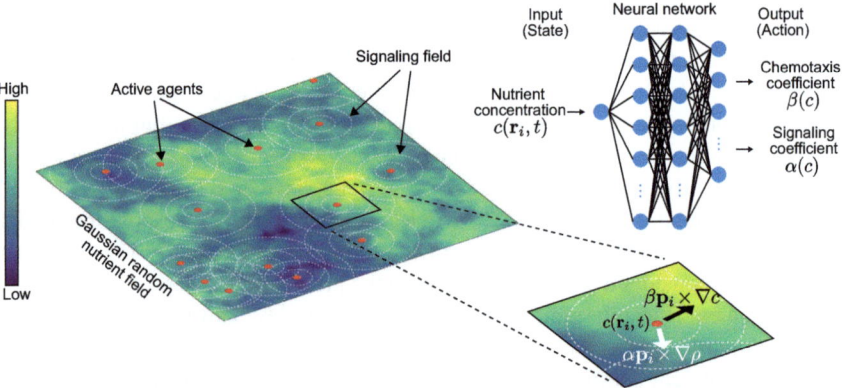

Fig. 3 Left: Schematic illustration of active agents (red dots) that move in a nutrient concentration field (background color indicates nutrient concentration) and communicate with each other to coordinate their motion. The agents communicate by producing quorum sensing molecules (white circles illustrate iso-concentration lines of the signaling molecules due to each agent). Right: Each particle feeds the nutrient concentration at its current position into a neural network. The network helps predicting coupling coefficients β and α that determine to which extent agents greedily move up nutrient concentration gradients ∇c and to which extent they follow the concentration gradient of the signaling molecules $\nabla \rho$, respectively. From Ref. [69]. Licensed under the terms of Creative Commons CC BY 4.0

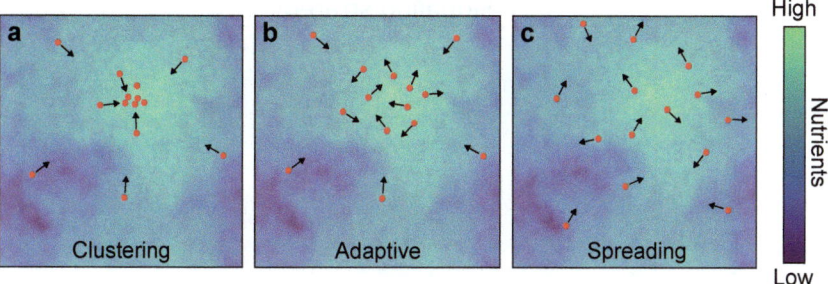

Fig. 4 Illustration of the collectively learned motion patterns. **a** clustering strategy: all agents behave cooperatively and swarm together. **b** adaptive strategy: the agents either follow others or avoid them depending on the local nutrient concentration. **c** spreading strategy: the agents avoid each other and spread out over the entire domain. From Ref. [69]. Licensed under the terms of Creative Commons CC BY 4.0

a suitable compromise between following the food concentration gradient and the gradient of signaling molecules. The latter allows the agents for example to benefit from other agents that have already found regions of higher nutrient concentration. The key question is then: what is the optimal compromise? To find a suitable compromise, one can use reinforcement learning combined with neural networks. This leads to three qualitatively different motion patterns of the agents [69], which are shown in Fig. 4. Panel (a) shows a scenario where the particles cluster in a region of high nutrient concentration. In panel (b) particles behave adaptively; they move towards locations of high particle density when they are in regions of low food concentration and they tend to avoid others when they are in regions of high food concentration. Panel (c) show a case where particles move away from each other and spread over the full system. Figure 5 shows a state diagram where one can see which of these three strategies leads to a higher average nutrient consumption depending on the consumption rate of the agents and the agent density (in suitably reduced units). This example shows that machine learning can be used to coordinate and direct collective behavior in a way that allows a group of agents to approach a common goal.

5   Further Applications

There are many other applications of machine learning to active matter physics. For brevity we only sketch a few of them here. For instance, machine learning is currently used to identify phase transitions in active matter, which is a challenging problem even in relatively simple active matter systems due to the strong fluctuations in these systems. However, recent works have exemplified the possibility to identify phase transitions across a wide range of physical systems, based on various machine learning techniques hinging on supervised learning [71], unsupervised learning [72, 73] or combinations thereof [74]. Very recently, such (or similar) ideas have been applied

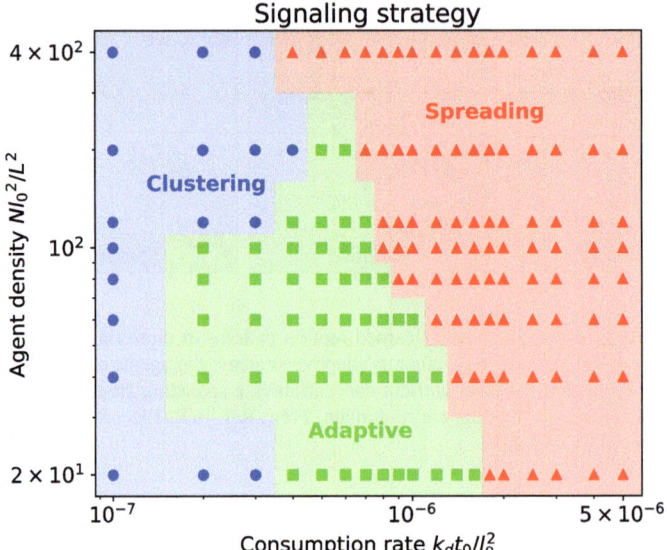

Fig. 5 State diagram of learned nutrient collection strategies. The active agents employ different strategies depending on the reduced agent density $N l_0^2 / L^2$ and the reduced nutrient consumption rate $k_0 t_0 / l_0^2$. Clustering strategy (blue dots): the agents behave cooperatively and aggregate; adaptive strategy (green dots): the agents either approach others or avoid them, depending on the local nutrient concentration; spreading strategy (red dots): the agents avoid each other at all nutrient concentrations. See [69] for details Licensed under the terms of Creative Commons CC BY 4.0

to classify active matter [75, 76]. In particular, to achieve this, Ref. [75] has applied machine learning techniques to determine in which phase an active test particle is based on the particles in the vicinity of that particle. This has been discussed in the context of motility-induced phase separation of active particles. A different application of machine learning techniques to analyze phase diagrams has been reported in Ref. [77] where learning techniques have been used to identify different sources for bacterial swarming based on experimental and simulation data [77]. For this complex system, the full space-time phase diagram of bacterial swarm expansion was analyzed and classified by machine learning techniques.

Another popular and highly relevant application of machine learning methods to active matter concerns the learning of governing equations describing active matter [70, 78]. Such methods use data (e.g. particle trajectories) from experiments or stochastic particle based simulations to predict governing equations describing the system at coarse grained scales. The enormous potential of such data-driven approaches has already been demonstrated in recent works that exploit techniques like symbolic regression, autoencoder networks, and sparse regression [70, 79–81]. Exemplaric applications to active matter have recently been proposed in Refs. [70, 78]. Such methods are highly useful to find new phenomenological hydrodynamic descriptions of fluctuating active matter and beyond. See Fig. 6 for

(a)   **(b)**   **(c)**   **(d)**

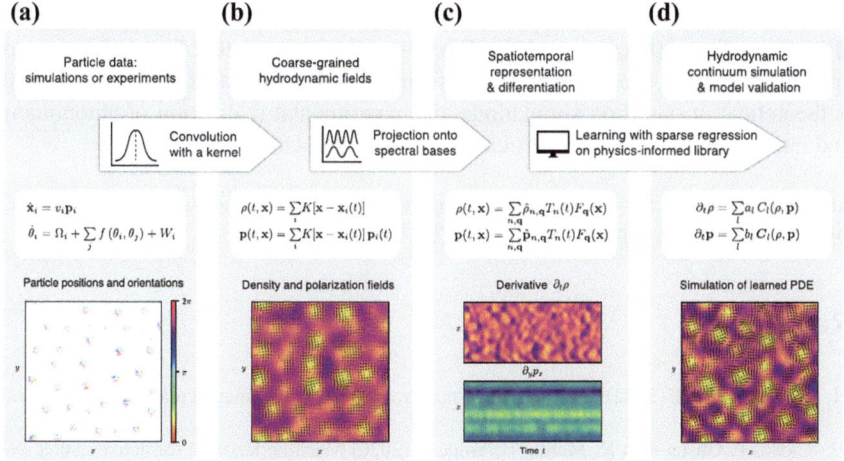

Fig. 6 Illustration of the learning approach in Ref. [70]. **a** Inputs are trajectories, e.g. of particle positions $\mathbf{x}_i(t)$ and orientation angles $\theta_i(t)$, $i = 1, 2, ..., N$, from experiments or particle based simulations. **b** Continuous hydrodynamic fields, such as density $\rho(\mathbf{x}, t)$ and polarization $\mathbf{p}(\mathbf{x}, t)$ are obtained via spatial kernel coarse-graining of the discrete microscopic variables. **c** The coarse-grained fields are sampled on a grid and projected onto spectral basis functions. Then, spectral filtering is used to smoothen the fields before calculating derivatives. **d** From these derivatives a library of candidate terms is constructed (consistent with prior knowledge about conservation laws and symmetries) and a sparse regression algorithm is used to determine phenomenological coefficients to obtain a sparse hydrodynamic model. From Ref. [70]. Licensed under the terms of Creative Commons CC BY-NC-ND 4.0

a very recent example. Here not the strategies of the many-body system are learned but the underlying governing equations.

6 Conclusions and Outlook

In conclusion, micron-sized active particles can be equipped with different levels of "intelligence". In particular it is now possible to dynamically control them with external feedback control systems which can be combined with suitable machine learning approaches. This results in "smart" active particles that can sense and react to their environment, communicate with each other and learn from past experience (although the actual learning process takes place in a computer). However, as these particles depend on their external control system and on the external learning platform they do not yet reach the autonomy of biological active particles like bacteria or other microorganisms. In fact, equipping synthetic agents with sensors, actuators and processors to make them smart and autonomous remains as a persistent challenge at the micro-scale. (At the macroscale such an approach leads to electronic robots, which are many orders of magnitude larger than active colloidal particles.)

Today we are much closer to the vision of emergent intelligence formulated by Hillis more than 30 years ago [82]. Overall a flourishing future for applications of artificial intelligence to active particles is lying ahead. While most of the current work is theoretical or based on simulations, the experimental realization of autonomous and intelligent micron-sized particles remains highly challenging.

Acknowledgements The work of HL was supported by the German Research Foundation (DFG) via project LO 418/29-1 (project number 522595197).

References

1. Sagués Mestre F (2022) Colloidal active matter: concepts, experimental realizations, and models. CRC Press
2. Cichos F, Gustavsson K, Mehlig B, Volpe G (2020) Machine learning for active matter. Nat Mach Intell 2:94
3. Bechinger C, Di Leonardo R, Löwen H, Reichhardt C, Volpe G, Volpe G (2016) Engineering sensorial delay to control phototaxis and emergent collective behaviors. Rev Mod Phys 88:045006
4. Gompper G, Winkler RG, Speck T, Solon A, Nardini C, Peruani F, Löwen H, Golestanian R, Kaupp UB, Alvarez L (2020) The 2020 motile active matter roadmap. J Phys: Condens Matter 32:193001
5. Paxton WF, Kistler KC, Olmeda CC, Sen A, Angelo SKSt, Cao Y, Mallouk TE, Lammert PE, Crespi VH (2004) Catalytic nanomotors: autonomous movement of striped nanorods. J Am Chem Soc 126:13424
6. Nosenko V, Luoni F, Kaouk A, Rubin-Zuzic M, Thomas H (2020) Active Janus particles in a complex plasma. Phys Rev Res 2:033226
7. Nasiri M, Löwen H, Liebchen B (2023) Optimal active particle navigation meets machine learning (a). EPL 142:17001
8. Cichos F, Landin SM, Pradip R (2023) Intelligent Nanotechnology. Elsevier, p 113
9. Yethiraj A (2007) Tunable colloids: control of colloidal phase transitions with tunable interactions. Soft Matter 3:1099
10. Löwen H (2013) Introduction to colloidal dispersions in external fields. Eur Phys J Spe Top 222:2727
11. Liebchen B, Löwen H (2018) Synthetic chemotaxis and collective behavior in active matter. Acc Chem Res 51:2982
12. Lavergne FA, Wendehenne H, Bäuerle T, Bechinger C (2019) Group formation and cohesion of active particles with visual perception-dependent motility. Science 364:70
13. Khadka U, Holubec V, Yang H, Cichos F (2018) Active particles bound by information flows. Nat Commun 9:3864
14. Bäuerle T, Fischer A, Speck T, Bechinger C (2018) Self-organization of active particles by quorum sensing rules. Nat Commun 9:3232
15. Sprenger AR, Fernandez-Rodriguez MA, Alvarez L, Isa L, Wittkowski R, Löwen H (2020) Active Brownian motion with orientation-dependent motility: theory and experiments. Langmuir 36:7066
16. Mijalkov M, McDaniel A, Wehr J, Volpe G (2016) Engineering Sensorial Delay to Control Phototaxis and Emergent Collective Behaviors. Phys Rev X 6:011008
17. Patra D, Sengupta S, Duan W, Zhang H, Pavlick R, Sen A (2013) Intelligent, self-powered, drug delivery systems. Nanoscale 5(4):1273
18. Alapan Y, Yasa O, Schauer O, Giltinan J, Tabak AF, Sourjik V, Sitti M (2018) Soft erythrocyte-based bacterial microswimmers for cargo delivery. Sci. Robot. 3:eaar4423

19. Gu H, Hanedan E, Boehler Q, Huang TY, Mathijssen AJ, Nelson BJ (2022) Artificial microtubules for rapid and collective transport of magnetic microcargoes. Nat Mach Intell 4:678
20. Spehr M, Gisselmann G, Poplawski A, Riffell JA, Wetzel CH, Zimmer RK, Hatt H (2003) Identification of a testicular odorant receptor mediating human sperm chemotaxis. Science 299:2054
21. Eisenbach M, Giojalas LC (2006) Sperm guidance in mammals – an unpaved road to the egg. Nat Rev Mol Cell Biol 7(4):276
22. Tsang AC, Demir E, Ding Y, Pak OS (2020) Roads to smart artificial microswimmers. Adv Intell Syst 2:1900137
23. Najafi A, Golestanian R (2004) Simple swimmer at low Reynolds number: Three linked spheres. Phys Rev E 69:062901
24. Avron JE, Kenneth O, Oaknin DH (2005) Pushmepullyou: an efficient micro-swimmer. New J Phys 7:234
25. Ledesma-Aguilar R, Löwen H, Yeomans JM (2012) A circle swimmer at low Reynolds number. Eur Phys J E 35:9746
26. Daddi-Moussa-Ider A, Lisicki M, Hoell C, Löwen H (2018) Swimming trajectories of a three-sphere microswimmer near a wall. J Chem Phys 148:134904
27. Tsang ACH, Tong PW, Nallan S, Pak OS (2020) Self-learning how to swim at low Reynolds number. Phys Rev Fluids 5:074101
28. Zou Z, Liu Y, Young YN, Pak OS, Tsang AC (2022) Gait switching and targeted navigation of microswimmers via deep reinforcement learning. Commun Phys 5:158
29. Abdi H, Pishkenari HN (2023) Self-learning swimming of a three-disk microrobot in a viscous and stochastic environment using reinforcement learning. Eng Appl Artif Intell 123:106188
30. Qin K, Zou Z, Zhu L, Pak OS (2023) Reinforcement learning of a multi-link swimmer at low Reynolds numbers. Phys Fluids 35:032003
31. Liu Y, Zou Z, Tsang ACH, Pak OS, Young YN (2021) Mechanical rotation at low Reynolds number via reinforcement learning. Phys Fluids 33:062007
32. Kaur H, Franosch T, Caraglio M (2023) Adaptive active Brownian particles searching for targets of unknown positions. Mach. Learn.: Sci. Technol. 4:035008
33. Goh S, Winkler RG, Gompper G (2022) Noisy pursuit and pattern formation of self-steering active particles. New J Phys 24:093039
34. Hartl B, Hübl M, Kahl G, Zöttl A (2021) Microswimmers learning chemotaxis with genetic algorithms. Proc Natl Acad Sci 118:e2019683118
35. Ramakrishnan RO, Friedrich BM (2023) Learning run-and-tumble chemotaxis with support vector machines. EPL 142:47001
36. Liebchen B, Löwen H (2019) Optimal navigation strategies for active particles. EPL 127:34003
37. Daddi-Moussa-Ider A, Löwen H, Liebchen B (2021) Hydrodynamics can determine the optimal route for microswimmer navigation. Commun Phys 4:15
38. Sutton RS, Barto AG (2018) Reinforcement learning: an introduction. MIT Press
39. Yang Y, Bevan MA (2018) Optimal navigation of self-propelled colloids. ACS Nano 12:10712
40. Yang Y, Bevan MA, Li B (2020) Efficient navigation of colloidal robots in an unknown environment via deep reinforcement learning. Adv Intell Syst 2:1900106
41. Biferale L, Bonaccorso F, Buzzicotti M, Clark Di Leoni P, Gustavsson K (2019) Zermelo's problem: optimal point-to-point navigation in 2D turbulent flows using reinforcement learning. Chaos 29:103138
42. Yang Y, Bevan MA, Li B (2020) Micro/nano motor navigation and localization via deep reinforcement learning. Adv Theory Simul 3:2000034
43. Xu K, Yang Y, Li B (2021) Brownian cargo capture in mazes via intelligent colloidal microrobot swarms. Adv Intell Syst 3:2100115
44. Alageshan JK, Verma AK, Bec J, Pandit R (2020) Machine learning strategies for path-planning microswimmers in turbulent flows. Phys Rev E 101(4):043110
45. Zhu Y, Pang JH (2023) A numerical simulation of target-directed swimming for a three-link bionic fish with deep reinforcement learning. Proc Inst Mech Eng C J Mech Eng Sci 237:2450

46. Zhu Y, Pang JH, Tian FB (2022) Point-to-point navigation of a fish-like swimmer in a Vortical flow with deep reinforcement learning. Front Phys 10:237
47. Nasiri M, Liebchen B (2022) Reinforcement learning of optimal active particle navigation. New J Phys 24:073042
48. Muiños-Landin S, Fischer A, Holubec V, Cichos F (2021) Reinforcement learning with artificial microswimmers. Sci Robot 6(52):eabd9285
49. Schneider E, Stark H (2019) Optimal steering of a smart active particle. EPL 127:64003
50. Gunnarson P, Mandralis I, Novati G, Koumoutsakos P, Dabiri JO (2021) Learning efficient navigation in vortical flow fields. Nat Commun 12:7143
51. Lozano C, Ten Hagen B, Löwen H, Bechinger C (2016) Phototaxis of synthetic microswimmers in optical landscapes. Nat Commun 7(1):12828
52. Monderkamp PA, Schwarzendahl FJ, Klatt MA, Löwen H (2022) Active particles using reinforcement learning to navigate in complex motility landscapes. Mach Learn: Sci Technol 3:045024
53. Dijkstra EW (1959) A note on two problems in connexion with graphs. Numer Math 1:269
54. Meredith CH, Moerman PG, Groenewold J, Chiu YJ, Kegel WK, van Blaaderen A, Zarzar LD (2020) Predator–prey interactions between droplets driven by non-reciprocal oil exchange. Nat Chem 12:1136
55. Sengupta A, Kruppa T, Löwen H (2011) Chemotactic predator-prey dynamics. Phys Rev E 83:031914
56. Schwarzendahl FJ, Löwen H (2021) Barrier-mediated predator-prey dynamics. EPL 134:48005
57. Liebchen B, Löwen H (2020) Chemical kinetics: beyond the textbook. World Scientific, p 493
58. Wang X, Cheng J, Wang L (2020) A reinforcement learning-based predator-prey model. Ecol Complex 42:100815
59. Gerhard M, Jayaram A, Fischer A, Speck T (2021) Hunting active Brownian particles: Learning optimal behavior. Phys Rev E 104:054614
60. Janosov M, Virágh C, Vásárhelyi G, Vicsek T (2017) Group chasing tactics: how to catch a faster prey. New J Phys 19:053003
61. Ranganathan A, Heyde A, Gupta A, Mahadevan L (2022) Emergent strategies for shepherding a flock. arXiv:2211.04352
62. Zampetaki AV, Liebchen B, Ivlev AV, Löwen H (2021) Collective self-optimization of communicating active particles. Proc Natl Acad Sci 118:e2111142118
63. Ziepke A, Maryshev I, Aranson IS, Frey E (2022) Multi-scale organization in communicating active matter. Nat Commun 13:6727
64. Panait L, Luke S (2005) Cooperative multi-agent learning: the state of the art. Auton Agents Multi-Agent Syst 11:387
65. Durve M, Peruani F, Celani A (2020) Learning to flock through reinforcement. Phys Rev E 102:012601
66. Speck T, Bialké J, Menzel AM, Löwen H (2014) Effective Cahn-Hilliard equation for the phase separation of active Brownian particles. Phys Rev Lett 112:218304
67. Tovey S, Zimmer D, Lohrmann C, Merkt T, Koppenhoefer S, Heuthe VL, Bechinger C, Holm C (2023) Environmental effects on emergent strategy in micro-scale multi-agent reinforcement learning. arXiv:2307.00994
68. Verma S, Novati G, Koumoutsakos P (2018) Efficient collective swimming by harnessing vortices through deep reinforcement learning. Proc Natl Acad Sci 115:5849
69. Grauer J, Löwen H, Schwarzendahl F, Liebchen B (2024) Optimizing collective behavior of communicating active particles with machine learning. Mach Learn: Sci Technol 5:015014
70. Supekar R, Song B, Hastewell A, Choi GP, Mietke A, Dunkel J (2023) Learning hydrodynamic equations for active matter from particle simulations and experiments. Proc Natl Acad Sci 120:e2206994120
71. Carrasquilla J, Melko RG (2017) Machine learning phases of matter. Nat Phys 13:431
72. Rodriguez-Nieva JF, Scheurer MS (2019) Identifying topological order through unsupervised machine learning. Nat Phys 15:790

73. Wetzel SJ (2017) Unsupervised learning of phase transitions: From principal component analysis to variational autoencoders. Phys Rev E 96:022140
74. Van Nieuwenburg EP, Liu YH, Huber SD (2017) Learning phase transitions by confusion. Nat Phys 13:435
75. Dulaney AR, Brady JF (2021) Machine learning for phase behavior in active matter systems. Soft Matter 17:6808
76. Xue T, Li X, Chen X, Chen L, Han Z (2022) Machine learning phases of active matter arXiv:2210.00161
77. Jeckel H, Jelli E, Hartmann R, Singh PK, Mok R, Totz JF, Vidakovic L, Eckhardt B, Dunkel J, Drescher K (2019) Learning the space-time phase diagram of bacterial swarm expansion. Proc Natl Acad Sci 116:1489
78. Maddu S, Vagne Q, Sbalzarini IF (2022) Learning deterministic hydrodynamic equations from stochastic active particle dynamics. arXiv:2201.08623
79. Brunton SL, Proctor JL, Kutz JN (2016) Discovering governing equations from data by sparse identification of nonlinear dynamical systems. Proc Natl Acad Sci 113:3932
80. Rudy SH, Brunton SL, Proctor JL, Kutz JN (2017) Data-driven discovery of partial differential equations. Sci Adv 3:e1602614
81. Champion K, Lusch B, Kutz JN, Brunton SL (2019) Data-driven discovery of coordinates and governing equations. Proc Natl Acad Sci 116:22445
82. Hillis WD (1988) Intelligence as an emergent behavior; or, the songs of eden. Daedalus 117:175

Intelligent Matter Consisting of Active Particles

Julian Jeggle and Raphael Wittkowski

Abstract In this book chapter, we review how systems of simple motile agents can be used as a pathway to intelligent systems. It is a well known result from nature that large groups of entities following simple rules, such as swarms of animals, can give rise to much more complex collective behavior in a display of *emergence*. This begs the question whether we can emulate this behavior in synthetic matter and drive it to a point where the collective behavior reaches the complexity level of intelligent systems. Here, we will use a formalized notion of "intelligent matter" and compare it to recent results in the field of active matter. First, we will explore the approach of *emergent computing* in which specialized active matter systems are designed to directly solve a given task through emergent behavior. This we will then contrast with the approach of *physical reservoir computing* powered by the dynamics of active particle systems. In this context, we will also describe a novel reservoir computing scheme for active particles driven ultrasonically or via light refraction.

1 Introduction

The term "intelligence" is typically attributed to biological entities as an extreme form of *emergent* behavior as the (relatively) simple biochemical and biophysical interactions between neurons give rise to dramatically more complex psychological phenomena. Many approaches of artificial intelligence thus try to create synthetic analogs of these biological structures. An example are artificial neural networks which derive from a strongly simplified view on neuron activity [1]. However, we can also find a form of intelligence in other systems with emergent behavior such

J. Jeggle · R. Wittkowski
Institute of Theoretical Physics, Center for Soft Nanoscience, University of Münster, Münster, Germany

R. Wittkowski (✉)
Department of Physics, RWTH Aachen University, Aachen, Germany
e-mail: rgwitt25@dwi.rwth-aachen.de

DWI – Leibniz Institute for Interactive Materials, Aachen, Germany

© The Author(s) 2026
M. te Vrugt (ed.), *Artificial Intelligence and Intelligent Matter*, Machine Intelligence for Materials Science, https://doi.org/10.1007/978-3-032-04129-6_14

as swarms of bacteria [2], schools of fish [3], or even groups of humans [4]. These systems exhibit collective behavior that transcends the capabilities of each individual and allows the system as a whole to adapt to its environment in an intelligent manner (e.g., to avoid a perceived threat). The observation of this "swarm intelligence" in natural systems is an inspiration for an alternative path to artificial intelligence based on systems of synthetic motile agents or, as it is more commonly known as, synthetic *active matter* [5].

As an inherently nonequilibrium system, active matter exhibits a wide range of collective effects not found in nonactive (i.e., passive) materials. Examples include motility-induced phase separation [6, 7], swarming [8], or self-assembly into complex structures [9]. It is the persistent energy flux caused by the energy dissipation of the active agents that permits these advanced behavior patterns in the first place. This correlates well with the aim to produce intelligent materials as such an energy flux is indeed necessary for adaptive (and ultimately intelligent) materials [10] and has inspired much work into bringing about artificial intelligence on the platform of active matter.

In this chapter, we will give an overview of recent advances in this topic. However, given the large size of the active matter community we will limit ourselves to active matter systems on the microscale where the individual agents do not possess any intelligence themselves nor are they controlled by an external (artificial) intelligence. In other words, we restrict ourselves to systems where intelligence emerges as a purely collective phenomenon. This is justified as current manufacturing capabilities severely limit the complexity of micro- and nanoparticles while systems with larger agents (e.g., robotic swarms) suffer from obvious scaling problems. It must be stressed that this is not a principal limitation, however, as is clearly demonstrated by the remarkable complexity of microorganisms found in nature. Readers interested in the *application* of machine learning on active matter systems are best referred to the chapter of Löwen and Liebchen as well as the chapter by Volpe in this book.

Before diving into the concrete approaches of this field, it is well advised to first establish a common understanding of what the defining properties of an intelligent system are. In this chapter, we will use the definition proposed by Kaspar et al. in Ref. [11]. There, it is argued that four key functionalities are required[1] for a class of materials to be capable of "intelligence", which is here understood as the ability of a system to adapt to and learn from its environment. It should be noted that this definition is not limited to material systems comprised of individual agents as is the case for active particle systems, but can also be applied to materials with tighter integration such as molecular systems (e.g., reaction networks [12]). In multiagent systems, the functional requirements described in the following only need to be realized by the material as a whole and not necessarily by each individual agent.

First, there needs to be a sensing component that allows the system to both receive information about its environment and harvest energy for its internal workings. Sec-

---

[1] It should be noted that the presence of these components does not immediately prove that the system is intelligent. Such a proof typically requires the system to actually solve a complex task that requires adaptivity and learning.

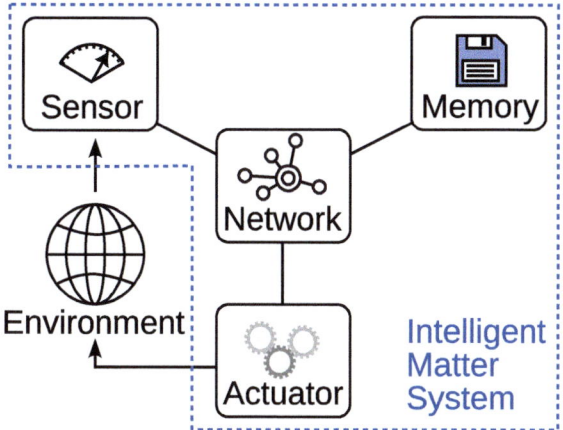

Fig. 1 Overview of the four components of intelligent matter as defined by Ref. [11] and their interaction with the environment of the system. The dashed blue outline indicates the boundary of the material system. The black lines denote signal pathways along which information is passed along. They can allow either directional, one-way information flow (direction indicated by an arrow head) or bidirectional, two-way information flow (no arrow heads). The sensor component serves as an information input to the system while the actuating component allows information output. The network component allows for information exchange within the material itself between the other three components. It should be noted that the information pathways inside the material are not directional as internal feedback between components is possible ·

ondly, there needs to be an actuation component that changes the material properties[2] and thus effectively providing information output to the environment. The third required component must be capable of information storage, i.e., memory, thus allowing the system to keep track of previous information and enabling learning. Finally, the fourth component is the information network that connects the previous three components. For learning, it is necessary that this information network also permits feedback, i.e., the actuation result of the system must also be available as an information input. A schematic overview of these components is shown in Fig. 1. Besides *intelligent* matter, Kaspar et al. also formalize the terms *responsive* matter as matter that has only the sensing and actuation components as well as *adaptive* matter as a matter system that has sensing, actuation, and network components.

When it comes to designing active matter systems with the aim of producing an intelligent material, there are two major routes that are being investigated at the moment. The first route is to design the active matter system directly such that its behavior solves a given task. To make this feasible, extensive insight into the dynamics of the specific active system in question is required as the connection between the properties of the active agents and the collective behavior is hard to

---

[2] Typically, a change in the behavior on the microscopic level will also lead to a different physical behavior of the system as a whole.

predict in general. The second route is to use the active matter system as the basis for a (physical) reservoir computing approach, which is an established computational archetype in machine learning. In this approach, the active matter system does not solve the task at hand directly, but rather serves as strongly nonlinear dynamical system that maps low-dimensional input signals to a high-dimensional space of internal states. This high-dimensional state can then be further processed by, e.g., linear classification to achieve complex tasks.

In this book chapter, we will first discuss existing advances and design frameworks for active particle systems solving concrete tasks and discuss the principal challenges of this approach that still remain. Next, we will discuss the reservoir computing approach for active particle systems and review successful implementations of this kind. Following this, we will discuss the potential of optically and acoustically driven particles in the context of reservoir computing. Finally, we will summarize the state of the field and give an outlook to future developments.

2 Emergent Computing with Active Particles

Designing active particles to solve a specific task can be seen as a special case of *emergent computing*, i.e., a decentralized computational paradigm of locally interacting agents giving rise to more abstract global behavior that can be interpreted as computation. Most results in this field use specialized cellular automata as the theoretical backbone of designing systems for specific tasks [13, 14]. While these automata are typically designed to emulate the motion of active agents (see, e.g., Refs. [15, 16]), it is not immediately clear how they can be translated to the continuous particle models typically analyzed in the field of active matter. One can also look at the problem from the other side and start with a typical active particle models such as the Vicsek model [17] and then attempt to tune the model parameters to achieve a formulated goal. However, for more complex models, such as those with spatiotemporal variations in agent activity, the dimensionality of the parameter space becomes large enough that advanced techniques for multidimensional optimization must be employed as brute-force approaches fail. In Ref. [18], Falk et al. demonstrate that machine learning techniques such as reinforcement learning can be used effectively to explore such a high-dimensional space of parameters and gain physical insights.

With regard to the four functional components of intelligent matter, the sensing and actuation components come naturally with the paradigm of active particles as they utilize (and therefore sense) energy fluxes in their environment and disturb the environment through their actuation, e.g., by inducing flows in the medium or by pushing objects in the environment. Additionally, there are typically strong interactions between active particles (e.g., if two particles collide with opposing directions) which enables signal propagation within the active particle system, e.g., in the form of acoustic waves [19]. However, the implementation of a memory component in active particle systems is more difficult. Embedding per-particle artificial memory within microparticles transcends current (mass-)manufacturing technology, so other methods of realizing memory have to be found. One solution is to exploit the inertia of each particle to store information about its past states. It was found that including inertia into consideration of active matter strongly influences the possible collective phases [20, 21], allows for properties not seen in overdamped active matter such as different modes of sound propagation [19] or a classical analogue of the tunnel effect [22], and even helps explain the motion patterns observed in motile microorganisms [23]. While using inertial memory is convenient, it is as of yet unclear if it is sufficient for the requirements of constructing intelligent matter systems. An alternative approach for *collective* memory storage is presented by Couzin et al. in Ref. [24]. Here, the authors describe how information can be stored on the group-level by exploiting hysteresis effects in the transition between collective dynamical states.

To the best of our knowledge, there is as of yet no demonstration of an intelligent system of active particles as current implementations have not yet shown the ability for autonomous learning, i.e., they rely on external training to solving a specific task. However, there are numerous promising results for active particle systems showing adaptivity. In the interest of brevity we only give some examples here and refer to the review of Li et al. in Ref. [25] for a more exhaustive overview of colloidal self-assembly. For one, there are efforts to use magnetic nanoparticles as the building blocks of adaptive microswarms. An impressive result for this kind of system is given by Yu et al. in Ref. [26] where they show a microswarm navigating through complex boundary geometries. A general overview of the field of magnetic active matter systems with functional collective behavior can be found in the account of Jin et al. in Ref. [27]. A second example are active particle systems showing phototaxis, i.e., a directed motion towards or away from a light source, a common motif found in biological systems as well. In Ref. [28], Dai et al. present a tunable microswimmer system propelled by light-induced self-electrophoresis that shows collective photo-taxis similar to green algae. Mou et al. demonstrate in Ref. [29] how a phototactic flock of TiO_2 micromotors can perform collective tasks such as transport of cargo significantly larger than the micromotors. In a yet more complex setup, Liang et al. show in Ref. [30] that phototactic microswarms of unequal individuals can self-organize into hierarchical structures which allows for implementing multiresponsiveness and thus further increases the design possibilities of this class of active particles. It should be noted that phototaxis is not the only taxis behavior possible as shown in the review by Ji et al. in Ref. [31].

To conclude this section, we also wish to point out the field of *swarm robotics* as a relevant adjacent field of research. While robots are rarely called active particles, the challenge of solving complex tasks collectively with a very limited budget for the complexity of each swarm member is remarkably similar to the field of active particles as we have discussed above. An introduction to the field of swarm robotics can be found in Ref. [32]. In Ref. [33], Savoie et al. explore how a swarm of "smarticles" (smart active particles in the form of robots following simple rules) can perform higher-level functions such as phototaxis and thus effectively act as a larger robot without a central control unit. Another recent example can be found in Ref. [34] where Saintyves et al. present "granulobots" as a bridge between soft robotics and active granular materials. These robots take the form of a self-actuating roller that can magnetically connect to its neighbors. Using only local interactions with their neighbors, they exhibit a wide variety of possible collective phases from rigid self-assembly to fluid-like states.

3   Physical Reservoir Computing with Active Matter

Reservoir computing was developed as a framework for simplifying the training process of recurrent neural networks (see the chapter by te Vrugt for a general introduction to this topic). The key idea is to split the network into a *reservoir* part that nonlinearly maps its input to a large space of internal states and a much simpler *readout layer* (often a simple linear classification) that computes the overall network output by observing the output of the reservoir. Training is then only performed on the readout layer in the hope that classification of inputs is easier in the high-dimensional state space of the reservoir. In *physical* reservoir computing, the reservoir part of the network is realized as a dynamical system.[3] There is a wide variety of examples of dynamical systems that can be used as reservoirs such as waves on a water surface [35], mechanical networks of anharmonic oscillators [36], or DNA-based reaction networks [12]. However, not just any dynamical system is also an *effective* reservoir. In Ref. [37], Konkoli et al. illuminate the challenge of designing a good reservoir and give an overview of successful implementations. In particular, they describe how a reservoir must ensure that all input states that are to be classified differently by the network must produce distinct internal states in the reservoir (*separation property*). Additionally, a physical reservoir must fulfill the *echo state property*, i.e., it must ensure that earlier inputs eventually fade from the memory of the system. Other in-depth reviews of physical reservoir computing are given by Tanaka et al. in Ref. [38] with an even more detailed overview of current realizations and by Nakajima in Ref. [39] with a focus on using soft robotics as a basis for physical reservoir computing. Unfortunately, the research area of physical reservoir computing using active matter systems as the dynamical system for their reservoir is still in its infancy. In the following, we present three promising results that highlight the potential of this direction of research. It should be noted that while these approaches do in principle contain all the functional components for an intelligent system – sensing and actuation as part of the active particle paradigm as well as a network and memory as part of the reservoir computing paradigm—there is as of yet no implementation of an actual intelligent system using this technique since all results so far use external training.

First, in Ref. [40], Lymburn et al. present an active matter reservoir computing approach using a modified Reynolds boids model with predator avoidance. The

---

[3] The term "physical reservoir computing" is somewhat misleading since in principle any time-dependent system is a candidate for physical reservoir computing, even if the dynamical system does not describe an actual physical system. However, since unphysical dynamics require some form of costly emulation (typically on conventional computing hardware), the potential of physical reservoir computing for more efficient computation is only realized when using physical dynamics.

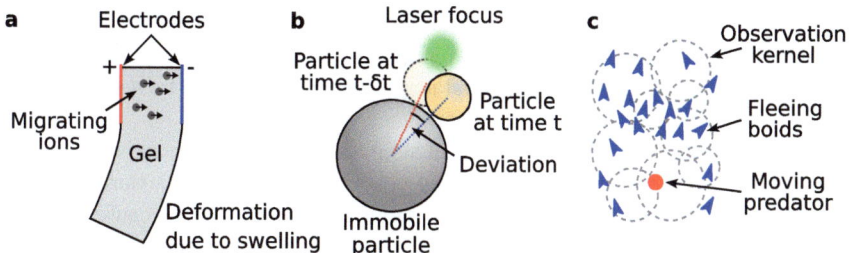

Fig. 2 Schematic setups for active particle reservoir computing platforms. **a** Setup used by Strong et al. in Ref. [42]. Here, a gel block made from an electroactive polymer is deformed due to the osmotic swelling caused by ionic migration induced by applying an electric field. The setup is submerged in a liquid tank not shown here. Data is fed into the system by modulating the applied voltage and the reservoir output is derived from the degree of deformation of the gel block. **b** Setup used by Wang et al. in Ref. [43]. The particle shown in yellow can move freely in the xy-plane and is driven by laser-induced thermophoresis towards an immobilized particle of larger size. The laser focus follows the particle trajectory with a delay δt thus causing the active particle to rotate about the immobile particle. The input information is encoded by modulating the laser focus position and information is extracted from the system by observing the time evolution of the deviation angle between the particle positions at times $t - \delta t$ and t. **c** Setup used by Lymburn et al. in Ref. [40]. Here, the system consists of a two-dimensional swarm of boids that are evading a single predator moving through the system. Information is fed into the system in the form of the predator's trajectory and extracted by observing the boid densities through a set of randomly placed Gaussian kernels

motion of the predator serves as the information input to the system and is chosen to follow the dynamics of the Lorenz system [41] in the chaotic regime. The reservoir computer is then trained to predict future positions of the predator. It was found that using the particle coordinates directly as reservoir output variables does not yield good predictions as such an output definition is inconsistent with the fact that in a system of identical particles any permutation of particles does not change the physical state of the system. To solve this, the authors employed an additional observational layer that observes in random, but fixed locations the state of nearby particles using a Gaussian kernel. With this correction, the authors were able to significantly improve the accuracy of the prediction. It was also observed that using the repulsion force as a control parameter, a phase transition between ordered flock states and disordered states can be triggered and that the reservoir performs optimally when close to this phase transition.

Another example for active matter reservoir computing is given by Wang et al. in Ref. [43]. Here, the dynamics of a single active particle is used as the foundation for constructing a reservoir computer. In Fig. 2, the particle setup is shown schematically. The motion of the active particle is governed by a nonlinear delay dynamics. Importantly, this dynamics fulfills the echo state property, i.e., the system has fading memory. To obtain a sufficiently large reservoir for complex tasks, multiple physical setups are run in parallel and time-multiplexing is used to construct multiple virtual nodes for each physical system. With a reservoir constructed from these virtual nodes it was possible to predict the chaotic Mackey-Glass series with good accuracy. The authors note that the noise naturally present in the dynamics of the particle has severely detrimental effects to the function of the reservoir computer. This was mitigated by including past reservoir states in the output layer which significantly improved the reservoir computing performance.

The final example is given by Strong et al. in Ref. [42]. Here, the active agents are hydrogen ions in an electroactive polymer gel (EAP) driven by an external electric field. The resulting ion migration causes osmotic flows which in turn cause localized swelling and deformation of the material. Notably, this deformation is subject to hysteresis effects such that each consecutive deformation decreases the magnitude of future deformations. The dynamics of EAP deformations thus exhibits inherent memory effects. Additionally, the deformations possess intrinsic variability and as such can only be modelled probabilistically. To use EAPs as computing devices, the authors developed encoding schemes for input data as discrete stimulation voltages and for gel deformations as output data. With these they were able to model the EAP behavior in terms of probablistic Moore automata, a probabilistic extension of finite state automata. This was then further extended to reservoir computing by recording the sequence of deformation states of the gel as the input voltages are applied and using this sequence as reservoir output. It is noteworthy that the role of the active particles in this approach is only to provide a nonlinear response of the electroactive polymer, so their dynamics is only implicitly relevant for the reservoir construction, in contrast to the other two examples we gave in this section.

4   Propulsion Mechanisms for Active Particle Reservoir Computing

As noted in the previous section, the topic of using active particle systems to drive reservoir computing remains largely unexplored as of today. We have seen in the few results pioneering this research field that designing reservoirs with high information bandwidth, i.e., large numbers of nodes, is difficult. Thus, implementations often resort to time-multiplexing to construct virtual nodes at the cost of serializing information processing. Furthermore, we have seen that inputting and extracting information from an active particle reservoir requires dedicated schemes for data encoding to achieve good reservoir performance. Finally, we have discussed that memory design is critical for the design of a good reservoir due to the echo state property. However, active particle systems often introduce memory effects as an unwanted side product of the propulsion mechanism. For example, chemically fueled particles will generally leave concentration gradients in their wake and will eventually deplete the fuel in the system [44, 45] while bulk systems of thermophoretic particles can incidentally introduce memory in the form of the average system temperature [46]. In this section, we will give a description of two propulsion schemes for active particles that we expect to mitigate these issues, but that have not been used for the purpose of reservoir computing to the best of our knowledge. In Fig. 3, we sketch a possible

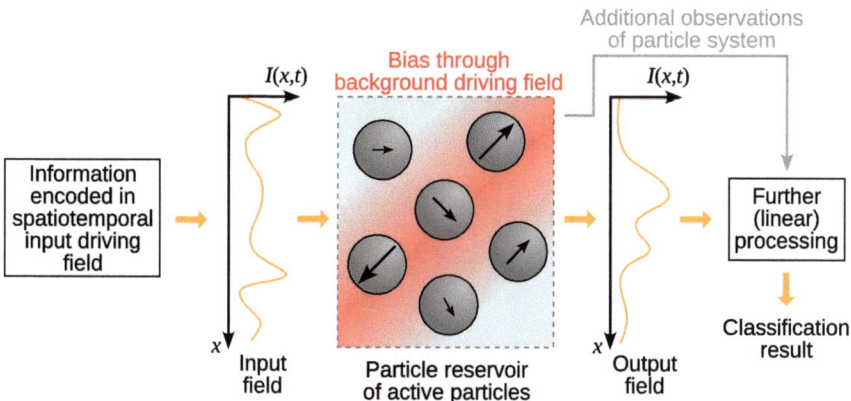

Fig. 3 A possible setup for implementing reservoir computing with bulk active particle systems driven by spatiotemporally modulated driving fields. In this example, we show information encoding in the intensity $I(x, t)$ of the input driving field, but one can also imagine information encoding in other quantities affecting the particle propulsion, e.g., polarization for optically driven particles. To achieve greater freedom in the design of the information encoding scheme for the input light field, the particles are also subject to a bias light field with fixed spatiotemporal structure. The readout layer of this reservoir computing scheme is fed with a measurement of the driving fields after they have passed through the system and optionally with other observations of the particle system (e.g., density or velocity averages of particles in specific regions of the reservoir). This scheme can be used for particle reservoirs in two or three spatial dimensions

setup for using these propulsion mechanisms in a reservoir computing context. The core idea behind this setup is to encode information in the spatiotemporal structure of the input driving field and thus achieving high information bandwidth in the reservoir. So far, there have been no realizations of such a system and as such its efficacy remains speculative.

The first propulsion mechanism useful in this context is based on acoustically driven particles [48–50]. In this type of active matter, ultrasonic pressure waves interact with microparticles which induces flows in the medium around the particle driving the particles forward. An illustration of this principle is provided in Fig. 4. Notably, the acoustic propulsion does not rely on creating secondary gradients (e.g., in concentrations or temperature) in the medium and thus does not introduce secondary memory effects. An ultrasonic field can be spatially modulated and recorded with high resolution as has recently been demonstrated by Ma et al. [51]. The design of the propulsion offers many degrees of freedom since the acoustic propulsion is determined by the shape of the particles [52] and their material which can be tuned by microfabrication techniques. Possible designs include nanorods [48–50], particles including microbubbles [53], and screw-shaped microrobots [54]. With these designs, a wide range of propulsion properties can be realized (e.g., reversal of the propulsion direction by tuning the aspect ratio [47]). In combination, these properties make acoustically propelled particle systems a promising candidate for active particle reservoir computing.

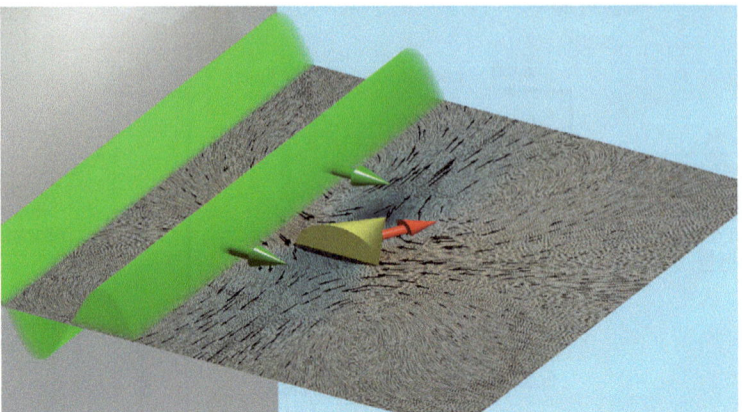

Fig. 4 Illustration of a planar ultrasound wave (shown schematically in green) interacting with a microparticle submerged in a fluid. Boundary-driven acoustic streaming creates a flow field around the particle driving it in the direction indicated by the red arrow. The wavelength of the ultrasonic field was shortened significantly for illustrative purposes and would be much larger than the particle if shown to scale. This image was created by Johannes Voß and Raphael Wittkowski, published in Ref. [47] under the Creative Commons Attribution 3.0 Unported Licence (see *creativecommons.org/licenses/by/3.0/*), and is reproduced here without changes

The second propulsion mechanism we present uses the momentum transfer that occurs on light refraction to generate directed motion in microparticles with a symmetry-broken refractive index profile (SBRIP) [55–57] as shown in Fig. 5a and **b**. These particles can be produced experimentally by microfabrication techniques such as two-photon polymerization [58]. Such a system can then be used as a building block for all-optical reservoir computing. This is particularly promising as photonic systems have been one of the most successful approaches to physical reservoir computing so far [38, 59]. One of the principal difficulties in these types of systems is the requirement for nonlinearity in the reservoir. By coupling the dynamics of an active particle system to the light field, this requirement for nonlinearity can be fulfilled without compromises like the time-multiplexing used in delay-based photonic reservoirs which necessitates serialization of the input and output data. When it comes to driving active matter systems optically, a refraction-based approach has multiple advantages. Since the light does not interact with the medium directly, side effects (e.g., concentration gradients) from the propulsion are avoided. Compared to absorptive or reflective propulsion methods, refraction allows for a much larger attenuation length of the light thus permitting active bulk systems. When using the refractive index profile of the particles to achieve the necessary breaking of symmetry for propulsion, one can effectively decouple shape-dependent hydrodynamic and steric particle interaction properties from the propulsion mechanism, thus allowing for a much greater freedom in particle design. In Fig. 5c, it is demonstrated that the extensive possibilities to modulate light fields in space and time can be used to tailor system behavior. Even the simple case of a two-dimensional system of active parti-

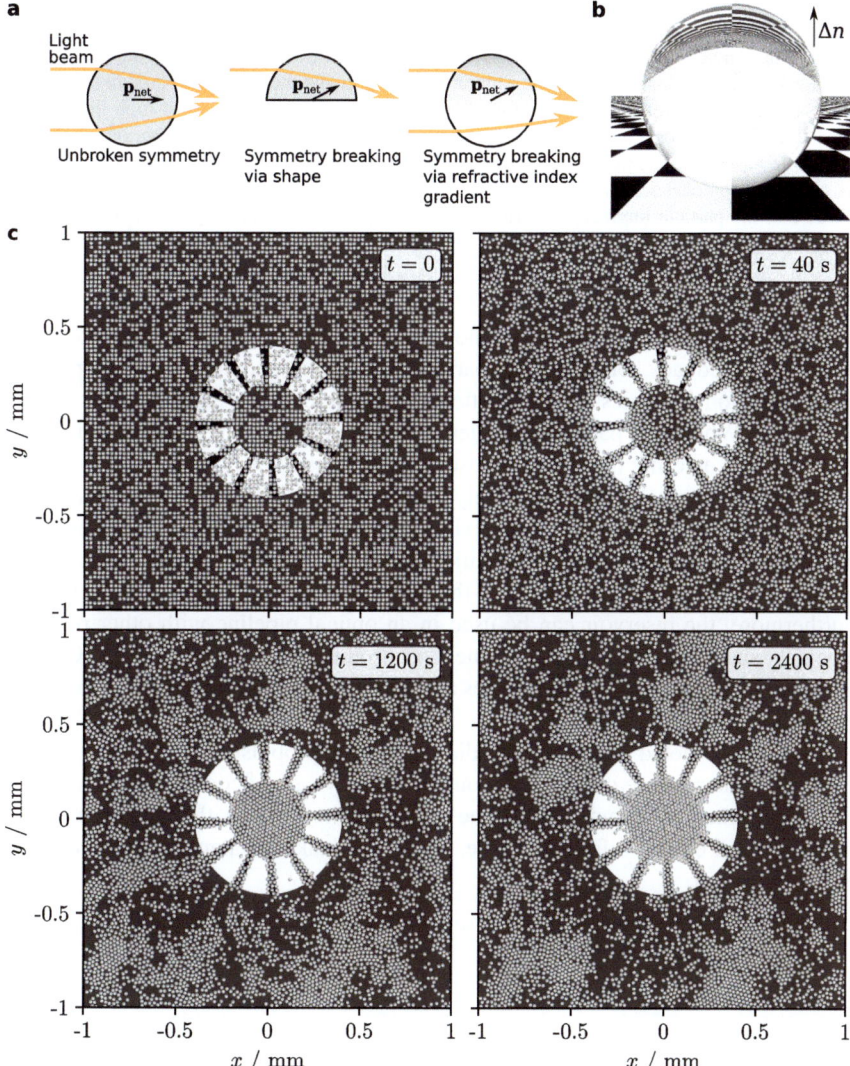

Fig. 5 Overview of symmetry-broken refractive microparticles. **a** Sketch showing the principle behind the refractive propulsion. Momentum conservation dictates that, in general, light refraction transfers momentum $\mathbf{p}_{net}$ from the light field to the refracting material. To achieve orientation-dependent propulsion, the particle needs to break symmetry either in shape or in the refractive index profile (here symbolized by a color gradient). **b** Raytracing rendering of a particle with a symmetry-breaking refractive index gradient (the direction is denoted by Δn). The center of the viewport is on the same height as the center of the spherical particle. The distorted image of the horizon thus demonstrates the breaking of symmetry. **c** Dynamic simulation of a system of refractive microparticles in two spatial dimensions illuminated perpendicular to the particle domain. The illumination intensity is shown as the background color with white denoting a high intensity

(Continued)

◀ **Fig. 5** (Continued) and black denoting a low intensity. It can be observed that the higher propulsion velocity in the bright region displaces particles into the darker regions. With the illumination structure used here, this leads to the formation of walls made of slowly moving particles that funnel incoming particles to the center of the illumination pattern and the associated growth of a crystal in the center of the domain. Notably, this crystal also grows into the areas of high illumination intensity. This behavior is an indication of *adaptivity* as the overall particle distribution is no longer the simple superposition of the reaction of each particle to the local illumination intensity, but rather we observe a nonlocal dependency of the average particle density on the illumination pattern. This shows that the particle interactions form a network capable of exchanging information between different parts of the system

cles with position-dependent propulsion speed can show adaptive system behavior by creating feedback between steric particle interactions and the particle propulsion: The propulsion dynamics leads to particle accumulation in the areas of lower propulsion speed. When high particle densities are reached in these areas, new particles are stopped from entering them. This causes the particles to drift along the edge of the dark region instead and thus effectively alters the propulsion velocity in this region. In bulk systems, the possible complexity grows further as then the light field itself becomes coupled to the particle dynamics: The light field induces changes in the particle density which in return changes the light field as it passes through the system. Furthermore, the reservoir can be used in an optical pipeline with other photonic elements such as mirrors or delay lines to introduce additional feedback. Overall, these properties make SBRIP particles a promising platform for the development of physical reservoir computing.

It should be noted that while we did outline the propulsion mechanisms in this section for the purpose of reservoir computing, they can, of course, also be used for the task of emergent computing. To the best of our knowledge, this direction has not been explored either so far. Therefore, we cannot make a statement on the efficacy of this approach.

5   Conclusions and Outlook

In summary, we have shown how active matter systems are a promising platform for constructing intelligent matter due to their inherently nonequilibrium nature that allows complex collective behavior patterns. While truly intelligent active matter systems have not been realized yet, there are results showing advanced adaptive behavior in these kinds of systems. In particular, there is a growing field of emergent computation with active particles where the individual particles are following simple rules and yet produce a swarm behavior capable of solving useful tasks. However, a large challenge that remains in this domain is the implementation of memory which is necessary for the step to intelligent systems.

We have also seen that there are efforts to implement physical reservoir computing using active particle systems as the underlying dynamical system. While this field is very much in its infancy, the first successful results seem to indicate that this approach works in principle. One of the major challenges still open in this field is to integrate the learning procedure into the system itself (e.g., by integrating a reinforcement learning scheme into the physical system), which is considered a requirement for intelligent matter systems. The success of the reservoir computing approach begs the question whether active particle systems can also be used to emulate other machine learning archetypes, e.g., to emulate neural networks by networks of active particle suspensions acting as neurons. Indeed, there remains much work to be done to discover the full potential of active matter systems for the construction of artificial intelligence.

Acknowledgements This work is funded by the Deutsche Forschungsgemeinschaft (DFG, German Research Foundation)—Project-ID 433682494–SFB 1459.

References

1. Schmidhuber J (2015) Deep learning in neural networks: an overview. Neural Netw 61:85
2. Ben-Jacob E (2009) Learning from bacteria about natural information processing. Ann N Y Acad Sci 1178(1):78
3. Pitcher TJ (1998) Shoaling and schooling behaviour in fishes. In: Greenberg G, Haraway MM (eds) in Comparative psychology: a handbook. Garland, New York, pp 748–760
4. Kerr NL, Tindale RS (2004) Group performance and decision making. Annu Rev Psychol 55(1):623
5. Ramaswamy S (2010) The mechanics and statistics of active matter. Ann Rev Condensed Matter Phys 1(1):323
6. Cates ME, Tailleur J (2015) Motility-induced phase separation. Ann Rev Conden Matter Phys 6(1):219
7. Jeggle J, Stenhammar J, Wittkowski R (2020) Pair-distribution function of active Brownian spheres in two spatial dimensions: simulation results and analytic representation. J Chem Phys 152(19):194903
8. Wensink HH, Löwen H (2012) Emergent states in dense systems of active rods: from swarming to turbulence. J Phys: Condens Matter 24(46):464130
9. Mallory SA, Valeriani C, Cacciuto A (2018) An active approach to colloidal self-assembly. Annu Rev Phys Chem 69(1):59
10. Walther A (2019) Viewpoint: From responsive to adaptive and interactive materials and materials systems: a roadmap. Adv Mater 32(20):1905111
11. Kaspar C, Ravoo BJ, van der Wiel WG, Wegner SV, Pernice WHP (2021) The rise of intelligent matter. Nature 594(7863):345–355
12. Goudarzi A, Lakin MR, Stefanovic D (2013) reservoir computing: a novel molecular computing approach. In: Soloveichik D, Yurke B (eds) DNA computing and molecular programming (DNA, 2013) Springer International Publishing. Lect Notes Comput Sci 8141:76–89
13. Crutchfield JP, Mitchell M (1995) The evolution of emergent computation. Proc Natl Acad Sci USA 92(23):10742
14. Olson RL, Sequeira RA (1995) Emergent computation and the modeling and management of ecological systems. Comput Electron Agric 12(3):183
15. Andrés Arroyo M, Cannon S, Daymude JJ, Randall D, Richa AW (2018) A stochastic approach to shortcut bridging in programmable matter. Nat Comput 17(4):723–741
16. Daymude JJ, Richa AW, Scheideler C (2023) The canonical amoebot model: Algorithms and concurrency control. Distrib Comput 36(2):159–192

17. Vicsek T, Czirók A, Ben-Jacob E, Cohen I, Shochet O (1995) Novel type of phase transition in a system of self-driven particles. Phys Rev Lett 75:1226
18. Falk MJ, Alizadehyazdi V, Jaeger H, Murugan A (2021) Learning to control active matter. Phys Rev Res 3:033291
19. te Vrugt M, Jeggle J, Wittkowski R (2021) Jerky active matter: A phase field crystal model with translational and orientational memory. New J Phys 23(6):063023
20. Caprini L, Gupta RK, Löwen H (2022) Role of rotational inertia for collective phenomena in active matter. Phys Chem Chem Phys 24(40):24910–24916
21. Nagai KH, Sumino Y, Montagne R, Aranson IS, Chaté H (2015) Collective motion of self-propelled particles with memory. Phys Rev Lett 114:168001
22. te Vrugt M, Frohoff-Hülsmann T, Heifetz E, Thiele U, Wittkowski R (2023) From a microscopic inertial active matter model to the Schrödinger equation. Nat Commun 14(1):1302
23. Mayer Martins J, Wittkowski R (2022) Inertial dynamics of an active Brownian particle. Phys Rev E 106:034616
24. Couzin ID, Krause J, James R, Ruxton GD, Franks NR (2002) Collective memory and spatial sorting in animal groups. J Theor Biol 218(1):1
25. Li Z, Fan Q, Yin Y (2022) Colloidal self-assembly approaches to smart nanostructured materials. Chem Rev 122(5):4976
26. Yu J, Wang B, Du X, Wang Q, Zhang L (2018) Ultra-extensible ribbon-like magnetic microswarm. Nat Commun 9(1):3260
27. Jin D, Zhang L (2022) Collective behaviors of magnetic active matter: Recent progress toward reconfigurable, adaptive, and multifunctional swarming micro/nanorobots. Acc Chem Res 55(1):98
28. Dai B, Wang J, Xiong Z, Zhan X, Dai W, Li C-C, Feng S-P, Tang J (2016) Programmable artificial phototactic microswimmer. Nat Nanotechnol 11(12):1087–1092
29. Mou F, Zhang J, Wu Z, Du S, Zhang Z, Xu L, Guan J (2019) Phototactic flocking of photo-chemical micromotors. iScience 19:415
30. Liang X, Mou F, Huang Z, Zhang J, You M, Xu L, Luo M, Guan J (2020) Hierarchical microswarms with leader-follower-like structures: Electrohydrodynamic self-organization and multimode collective photoresponses. Adv Func Mater 30(16):1908602
31. Ji F, Wu Y, Pumera M, Zhang L (2023) Collective behaviors of active matter learning from natural taxes across scales. Adv Mater 35(8):2203959
32. Belkacem Khaldi FC (2015) An overview of swarm robotics: swarm intelligence applied to multi-robotics. Int J Comput Appl 126(2):31
33. Savoie W, Berrueta TA, Jackson Z, Pervan A, Warkentin R, Li S, Murphey TD, Wiesenfeld K, Goldman DI (2019) A robot made of robots: Emergent transport and control of a smarticle ensemble. Sci Robot 4(34):eaax4316
34. Saintyves B, Spenko M, Jaeger HM (2024) A self-organizing robotic aggregate using solid and liquid-like collective states. Sci Robot 9(86):eadh4130
35. Fernando C, Sojakka S, Pattern recognition in a bucket. In: Banzhaf W, Ziegler J, Christaller T, Dittrich P, Kim JT (eds) Advances in artificial life (ECAL, 2003). Springer, Berlin Heidelberg. Lect Notes Comput Sci 2801:588–597
36. Coulombe JC, York MCA, Sylvestre J (2017) Computing with networks of nonlinear mechanical oscillators. PLoS ONE 12(6):e0178663
37. Konkoli Z, Nichele S, Dale M, Stepney S (2018) Reservoir computing with computational matter. In: Stepney S, Rasmussen S, Amos M (eds) Computational Matter. Springer International Publishing, Cham, pp 269–293
38. Tanaka G, Yamane T, Héroux JB, Nakane R, Kanazawa N, Takeda S, Numata H, Nakano D, Hirose A (2019) Recent advances in physical reservoir computing: a review. Neural Netw 115:100
39. Nakajima K (2020) Physical reservoir computing–an introductory perspective. Jpn J Appl Phys 59(6):060501
40. Lymburn T, Algar SD, Small M, Jüngling T (2021) Reservoir computing with swarms. Chaos: An Interdiscip J Nonlinear Sci 31(3):033121

41. Lorenz EN (1963) Deterministic nonperiodic flow. J Atmospher Sci 20(2):130
42. Strong V, Holderbaum W, Hayashi Y (2022) Electroactive polymer gels as probabilistic reservoir automata for computation. iScience 25(12):105558
43. Wang X, Cichos F (2024) Harnessing synthetic active particles for physical reservoir computing. Nat Commun 15(1):774
44. Liebchen B, Marenduzzo D, Cates ME (2017) Phoretic interactions generically induce dynamic clusters and wave patterns in active colloids. Phys Rev Lett 118:268001
45. Hokmabad BV, Dey R, Jalaal M, Mohanty D, Almukambetova M, Baldwin KA, Lohse D, Maass CC (2021) Emergence of bimodal motility in active droplets. Phys Rev X 11:011043
46. Auschra S, Bregulla A, Kroy K, Cichos F (2021) Thermotaxis of Janus particles. Eur Phys J E 44(90):123
47. Voß J, Wittkowski R (2022) Acoustically propelled nano- and microcones: Fast forward and backward motion. Nanoscale Adv 4:281
48. Li J, Mayorga-Martinez CC, Ohl C-D, Pumera M (2022) Ultrasonically propelled micro- and nanorobots. Adv Func Mater 32(5):2102265
49. Ren L, Soto F, Huang L, Wang W (2022) Ultrasound-powered micro-/nanorobots: Fundamentals and biomedical applications. In: Sun Y, Wang X, Yu J (eds) Field-Driven Micro and Nanorobots for Biology and Medicine. Springer International Publishing, Cham, pp 29–60
50. McNeill JM, Mallouk TE (2023) Acoustically powered nano- and microswimmers: From individual to collective behavior. ACS Nanoscience Au 3(6):424
51. Ma Z, Melde K, Athanassiadis AG, Schau M, Richter H, Qiu T, Fischer P (2020) Spatial ultrasound modulation by digitally controlling microbubble arrays. Nat Commun 11(1):4537
52. Voß J, Wittkowski R (2020) On the shape-dependent propulsion of nano- and microparticles by traveling ultrasound waves. Nanoscale Adv 2:3890
53. Bertin N, Spelman TA, Combriat T, Hue H, Stéphan O, Lauga E, Marmottant P (2017) Bubble-based acoustic micropropulsors: active surfaces and mixers. Lab Chip 17(8):1515–1528
54. Deng Y, Paskert A, Zhang Z, Wittkowski R, Ahmed D (2023) An acoustically controlled helical microrobot. Sci Adv 9(38):5260
55. Volpe G, Maragò OM, Rubinsztein-Dunlop H, Pesce G, Stilgoe AB, Volpe G, Tkachenko G, Truong VG, Chormaic SN, Kalantarifard F, Elahi P, Käll M, Callegari A, Marqués MI, Neves AAR, Moreira WL, Fontes A, Cesar CL, Saija R, Saidi A, Beck P, Eismann JS, Banzer P, Fernandes TFD, Pedaci F, Bowen WP, Vaippully R, Lokesh M, Roy B, Thalhammer-Thurner G, Ritsch-Marte M, Garcíía LP, Arzola AV, Castillo IP, Argun A, Muenker TM, Vos BE, Betz T, Cristiani I, Minzioni P, Reece PJ, Wang F, McGloin D, Ndukaife JC, Quidant R, Roberts RP, Laplane C, Volz T, Gordon R, Hanstorp D, Marmolejo JT, Bruce GD, Dholakia K, Li T, Brzobohatý O, Simpson SH, Zemánek P, Ritort F, Roichman Y, Bobkova V, Wittkowski R, Denz C, Kumar GVP, Foti A, Donato MG, Gucciardi PG, Gardini L, Bianchi G, Kashchuk AV, Capitanio M, Paterson L, Jones PH, Berg-Sørensen K, Barooji YF, Oddershede LB, Pouladian P, Preece D, Adiels CB, Luca ACD, Magazzù A, Ciriza DB, Iatí MA, Swartzlander GA (2023) Roadmap for optical tweezers. J Phys Photon 5(2):022501
56. Denz C, Jimenez AJ, Wittkowski R (2019) Photonic propulsion of artificial micro-swimmers (Conference Presentation). In: Sekkat Z, Omatsu T (eds) Molecular and nano machines II. International Society for Optics and Photonics, (SPIE, 2019), vol 11098, p 110980C
57. Denz C, Jurado A, Rueschenbaum M, Hallekamp J, Jeggle J, Wittkowski R (2021) Light-driven microrobots: light fuels motion. In: Galvez EJ, Rubinsztein-Dunlop H, Andrews DL (eds) Complex light and optical forces XV. International society for optics and photonics, (SPIE, 2021), vol 11701, p 1170112
58. Rüschenbaum M, Burczyk N, Jeggle J, Hallekamp J, Denz C, Wittkowski R (2021) Light propelled artificial micro-machines. In: Sekkat Z, Omatsu T (eds) Molecular and nano machines IV, International society for optics and photonics, (SPIE, 2021), vol 11812, p 1181209
59. Van der Sande G, Brunner D, Soriano MC (2017) Advances in photonic reservoir computing. Nanophotonics 6(3):561

Neural Networks Consisting of DNA

Michael te Vrugt

Abstract Neural networks based on soft and biological matter constitute an interesting potential alternative to traditional implementations based on electric circuits. DNA is a particularly promising system in this context due its natural ability to store information. In recent years, researchers have started to construct neural networks that are based on DNA. In this chapter, I provide a very basic introduction to the concept of DNA neural networks, aiming at an audience that is not familiar with biochemistry.

1 Introduction

Recent years have seen an increasing amount of work (some of which is also covered in this book) on implementing machine learning methods in physical systems, and the concept of intelligent matter [18] is closely related to this idea. While many approaches of this type employ electronic, magnetic, or photonic systems, it is in principle a relatively natural idea to use soft and biological matter as a basis for physical neural networks. After all, artificial neural networks are inspired by the brain, and the brain is a soft matter system.

DNA, the carrier of genetic information, naturally suggests itself for such approaches. It is a soft matter system that has evolved specifically for the purpose[1] of storing and processing information. Moreover, there is an established tradition of using DNA for performing artificial computational tasks in the framework of *DNA computing* [1]. Therefore, a number of authors [8, 11, 19, 27, 30] have explored the possibility of using DNA as a basis for artificial neural networks.

In this chapter, I will provide an introduction to the topic of DNA-based neural networks, in particular the approach presented in Ref. [5]. The presentation will be on

---

[1] Provided one can speak about things like "purposes" of natural objects in a scientific context, see Ref. [15] for a discussion.

M. te Vrugt (✉)
Institut für Physik, Johannes Gutenberg-Universität Mainz, Mainz, Germany
e-mail: tevrugtm@uni-mainz.de

© The Author(s) 2026
M. te Vrugt (ed.), *Artificial Intelligence and Intelligent Matter*, Machine Intelligence for Materials Science, https://doi.org/10.1007/978-3-032-04129-6_15

a relatively elementary level, without assuming prior knowledge in biochemistry (and thereby aiming at an audience coming from physics or computer science interested in such approaches). After a brief overview over the biology of DNA (Sect. 2) and DNA computing (Sect. 3), I will present the basic ingredients of a DNA neural network, namely winner-take-all networks (Sect. 4) and DNA gates (Sect. 5). Then, I will present two approaches to DNA-based artificial intelligence, namely winner-take-all networks operating with DNA (Sect. 6) and DNA reservoir computing (Sect. 7). Finally, I discuss some advantages and disadvantages of this approach (Sect. 8). I conclude in Sect. 9. In my presentation, will follow Refs. [3] (in Sect. 2), [20] (in Sect. 3), [26] (in Sect. 5), [5] (in Sects. 4 and 6), and [12] (in Sect. 7).

2 Biochemistry for Beginners

I have promised that this chapter does not require prior exposure to biochemistry, and consequently I will start with a brief introduction to what DNA is.

DNA (short for *deoxyribonucleic acid*) is a biological macromolecule that stores the genetic information. A macromolecule is (surprise!) a large molecule. Macromolecules often consist of many repeated subunits (*monomers*), in this case they are referred to as *polymers*. DNA is a polymer whose subunits are nucleotides. Each nucleotide consists of a base, deoxyribose (a sugar), and a phosphate group. The deoxyribose of one nucleotide binds to the phosphate group of the next one (this is referred to as *phosphodiester bond*).

What matters most for our purposes is the base, of which there exist five types: adenine (A), guanine (G), cytosine (C), thymine (T), and uracil (U). In DNA, one only finds adenine, guanine, cytosine, and thymine. These can bind to each other via hydrogen bonds (*base pairing*). More specifically, adenine always binds to thymine and cytosine always binds to guanine (the bonding between G and C is a bit stronger). DNA is therefore typically found in a helix structure consisting of two *strands* (chains of nucleotides). The nucleotides within a strand are bound together by the phosphodiester links, nucleotides in different strands are held together via base pairing. Since each base only binds to one specific other base, knowing the composition of one strand allows to infer the composition of the other one (the strand are *complementary*). For example, if the first strand is ACCCGAT, the second one has to be TGGGCTA.

In protein synthesis, DNA is converted into proteins (biological macromolecules that perform a variety of functions). This occurs in several steps. First (*transcription*), DNA is copied to ribonucleid acid (RNA), specifically to so-called messenger RNA (mRNA). RNA is similar to DNA, with differences being that it is usually single-stranded, that it contains ribose rather than deoxyribose, and that it uses uracil rather than thymine as the complementary base to adenine. At the *ribosome*, the mRNA is then translated to proteins (*translation*) according to the *genetic code*, where a sequence of three nucleotides corresponds to one amino acid. For example, if UUC corresponds to the amino acid phenylalanine and AGG to the amino acid arginine, then the sequence UUCAGG tells the ribosome to put phenylalanine and arginine together. (Usually these sequences are of course longer.) This flow of genetic information is summarized in the *central dogma of molecular biology* [6, 7].

3   DNA Computing

The first DNA computer was realized in 1994 by Adleman [1], who used it to solve the so-called "directed Hamiltonian path problem". Here, the aim is to find, for directed graph and a given starting and end point, a path that visits each node exactly once. This is a classical problem in computer science, since it is easy to pose, but very difficult ("NP-hard") to solve. Adleman's procedure allowed, exploiting the parallelism of DNA computing, to solve this problem with a procedure where the number of steps grew only linearly with the number of vertices. This is a notable efficiency since the number of potential solutions increases combinatorially with the number of vertices, as a consequence of which traditional computing approaches do not achieve such a scaling [2]. A discussion of the many developments and applications of DNA computing that emerged after Adleman's work can be found in review articles [9, 20, 22, 31]. Of particular interest in the context of this book is the development of intelligent systems based on DNA (see Ref. [9] for a review).

Some major advantages of DNA in the context of computing are [20].

- **Parallelism:** DNA computers can perform a large number of tasks in parallel, which can lead to extremely good performances compared to modern supercomputers. For instance, Adleman's DNA computer already had a performance of 100 Teraflops.
- **Storage:** DNA allows for extremely efficient data storage, requiring just 1 cubic nanometer for one bit of information. All the datat hat humanity has generated by 2025 could be stored in the size of a ping-pong ball, in a manner that makes the stored data easy to maintain and to copy [16].
- **Energy Efficiency:** DNA computers, being based on chemical reactions, do not require any electricity and are very energy efficient compared to traditional computers.

However, there are also significant disadvantages [20]:

- **Accuracy:** The biochemical processes involved in DNA computers are prone to errors, with the error probability increasing exponentially with the number of operations.
- **Resources:** DNA computers are fairly difficult to handle, requiring familiarity with molecular biology and biochemical experiments. Carrying out these experiments requires human interventions in most steps.

(These statements about advantages and disadvantages refer to what the technology is in principle capable of, currently it is in its infancy and is not easily available to practitioners who desire, say, high parallelism, as alternative to electronic approaches.)

4   The Winner Takes It All

The DNA-based implementation of neural networks discussed in this chapter is based on so-called *winner-take-all computation* [23]. Here, the basic idea is that neurons compete with each other. They inhibit other neurons while activating themselves. Thereby, it can be ensured that only the neuron with the largest input stays active, while all the other ones become inactive. A CMOS implementation of a winner-take-all function was proposed in Ref. [21]. How exactly this architecture can be implemented in a DNA system, and how input signals can be encoded in DNA, will be discussed in Sect. 6.

In the context of neural networks, a winner-take-all computation can connect the penultimate and final layer of a neural network, which have the same number of neurons. Let us denote the ith neuron in the penultimate layer by s_i and the ith neuron in the final layer by y_i. The connection is set up in such a way that y_i is one if s_i is the largest value in the penultimate layer and zero otherwise.

How is this achieved? Thinking of one layer of a neural network as a vector $\mathbf{x}$ with components x_i, the next layer is constructed by multiplying the vector by a matrix and then applying a nonlinear function to each component of the resulting vector. Denoting the elements of the weight matrix by w_{ij}, the jth component of the vector that results from applying the weight matrix to the penultimate layer is

$$s_i = \sum_j p_{ij} \tag{1}$$

with the products

$$p_{ij} = w_{ij} x_j. \tag{2}$$

The nonlinear function is then one that selects the largest component of the vector $\mathbf{s}$:

$$y_i = \begin{cases} 1 \text{ if } s_i > s_j \forall j \neq i, \\ 0 \text{ otherwise.} \end{cases} \tag{3}$$

Such a setup allows the network to have memory, which is encoded in the weights, and to solve pattern recognition tasks. Suppose the network has been trained to remember two patterns corresponding to the vectors $\mathbf{w}_1$ and $\mathbf{w}_2$ (for example, two different letters) with components w_{1i} and w_{2i}. Then, it receives as an input a vector $\mathbf{x}$ corresponding to some pattern, and it is supposed to figure out whether it is more similar to $\mathbf{w}_1$ or to $\mathbf{w}_2$. Then, one has to simply multiply the vector $\mathbf{x}$ by the matrix w_{ij} to get the vector $\mathbf{s}$ with components s_i given by Eq. (1). Whether or not the pattern is more similar to $\mathbf{w}_1$ or $\mathbf{w}_2$ can then be figured out by checking whether $s_1 = \mathbf{w}_1 \cdot \mathbf{x}$ or $s_2 = \mathbf{w}_2 \cdot \mathbf{x}$ is larger. Consequently, a network of this form requires nm weights to remember m n-bit patterns.

5 DNA Gates

A key technique in this context is *toehold-mediated strand displacement* [32, 33] (see Ref. [28] for a review). This process involves a single-stranded DNA (the *input*) and a double-stranded DNA, whose strands are referred to as *gate* and *output*. One strand of the double-stranded DNA (the gate) has an overhanging piece (the *toehold*) that the input can bind to. Via branch migration (a process by which one DNA strand is exchanged for another, see Ref. [14] for an introduction), the input strand then gradually starts binding to the gate strand and thereby replaces the output strand, which is thereby released. This process is more likely to start if the toehold is longer, and consequently, the toehold length can be used to control the reaction rate [25]. Specifically, the reaction rate increases exponentially with the toehold length [33]. (In practice stochastic fluctuations are of course likely to be relevant here, what exactly their influence is has not been systematically investigated.)

This process is visualized in Fig. 1. (A more complex illustration can be found in Ref. [26].) Figure 1a shows the initial configuration. The input strand consists of three segments, labelled S_1, T, and S_2. Here, S_1 and S_2 are *recognition domains*, which are relatively long (15 nucleotides), and T is the *toehold domain*, which is shorter (5 nucleotides). The other necessary ingredient is the *gate-output-complex*, which consists of two strands (gate and output) that are bound to each other. The gate has a recognition domain S_2' with a toehold domain T' before and after it, while the output has two recognition domains S_2 and S_3 separated by a toehold domain T. Here, a prime indicates a complementary base sequence (e.g., if T =AGGAT, then T'=TCCTA). Due to the DNA base-pairing rules (see Sect. 2), the segments S_2' and T' bind to the segments S_2 and T of the output, respectively.

Since the gate has two toeholds and the output has only one, the gate-output complex has a free toehold T'. This toehold is complementary to the free toehold T of the input. As a consequence, the input binds to the gate-output complex, leading to a complex consisting of three DNA strands. Now, although the recognition domain S_2' of the gate is currently bound to the output, it could by the DNA base-pairing rules equally well bind to the recognition domain S_2 of the input. Gradually, via branch migration, the recognition domain of the output ceases to bind to the S_2 recognition domain of the output and instead binds to the corresponding domain of the input (as shown in Fig. 1b). In the final state, shown in Fig. 1c, the gate is now bound to the input, forming a *gate-input complex*, while the output has been released.

To summarize, what can be achieved via this process is to convert an input signal (which is a DNA strand) into a desired output signal (which is also a DNA strand) with a certain rate. In particular, this method allows to realize so-called *seesaw gates* [25, 26]. In addition to input and output, also *fuel* (a third type of strand) is present. In the simple setup presented above, input strands release the output strands from the gates and are then no longer available for further reactions. If fuel is present, however, the fuel strand can (via toehold exchange) free the input strand and thereby make it available for further reactions. Thereby, a small amount of input can, provided enough fuel is available, in principle trigger the release of an arbitrary amount of output. The fuel therefore acts as a catalyst [26].

6 A DNA Neural Network

We have now assembled all ingredients that are necessary to understand a simple winner-take-all neural network based on DNA, namely the one realized by Cherry and Qian [5]. What it is supposed to do is to recognize handwritten numbers based on the general approach discussed in Sect. 4 (and it turns out that the DNA neural network presented in Ref. [5] is in fact able to recognize handwritten numbers – which is a standard benchmark task for artificial neural networks). The input vector (representing the image) is multiplied by a weight matrix (matrix multiplication consists of multiplication of the elements of the vector with scalar weights and summation over the results), afterwards the entries of the resulting vector are compared, the largest component is increased and the other ones are eliminated to generate a definite output.

First, we need to find a way of encoding, for example, a hand-written digit in the form of DNA strands, to allow for processing via chemical reactions. Consider as an example a nine-pixel image of the letter L, as shown in Fig. 2. We choose a set of nine distinct DNA molecules (A1, A2, A3, B1, B2, B3, C1, C2, C3) that represent the nine pixels of the image. In the case of the letter L, five of the nine pixels (A1, B1, C1, C2, C3) are black and the others white. Consequently, we can encode the letter L as an input by putting the DNA molecules A1, B1, C1, C2, and C3 into the

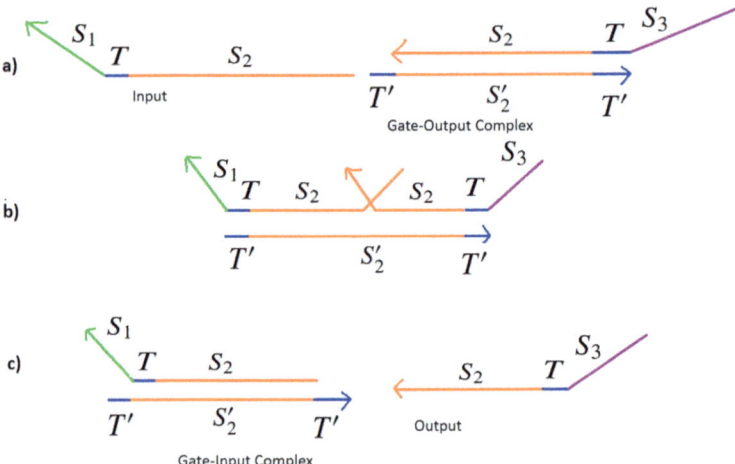

Fig. 1 Visualization of toehold exchange. (A similar illustration can be found in Ref. [26])

Fig. 2 Feeding a hand-written letter (L) as an input signal in the form of DNA into the system. The nine pixels are represented by nine distinct DNA molecules. An L is encoded by the presence of A1, B1, C1, C2, and C3. (A similar illustration can be found in Fig. 1b of Ref. [5] and in Fig. 2b of Ref. [29])

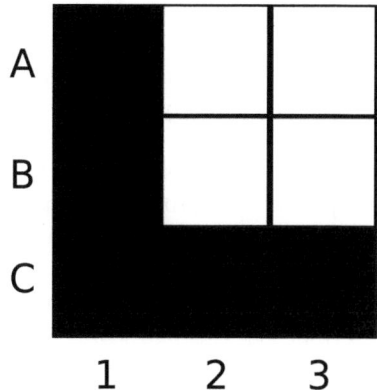

DNA computer. (In practice one would instead use 100 pixels represented by 100 distinct molecules [5], but that does not change the general idea.)

Second, we have to find a way of implementing the matrix multiplication (1) and the nonlinear function (3) via DNA. Let us start with the matrix multiplication. Importantly, one is concerned here with a binary input, i.e., all x_i are either zero or one). In the context of number recognition, where the x_i correspond to pixels, this might for example indicate whether the pixel represented by x_i is black or white. The entries of the weight matrix w_{ij}, on the other hand, are analogue numbers. What one requires here is thus a reaction that converts w_{ij} to p_{ij} if and only if x_i is present, regardless of the exact concentration of x_i. This is achieved by a catalytic cycle involving x_i, w_{ij} and fuel. The species x_i binds to a gate, corresponding to w_{ij}, that then releases a product p_{ij}. Afterwards, the fuel releases the x_i strand from the gate, such that it can trigger releases of p_{ij} elsewhere. Provided that a sufficient amount of fuel is present, this process produces a concentration of p_{ij} corresponding to the concentration of w_{ij}, but only if x_i is present.

Summation is relatively straightforward: Recall that seesaw gates allow to convert a certain input signal (a chemical substance, more specifically a DNA strand) into an output signal (another chemical substance, more specifically another DNA strand). For summation, one therefore simply requires gates that convert all species p_{ij} to the same species s_j. The concentration of s_j is then the sum of the concentrations of the p_{ij}.

The next step is to determine which of the s_i is the largest one. Given that the s_i correspond to the concentrations of chemical species, this is achieved by pairwise annihilation reactions, implemented by annihilator molecules. If two strands s_i and s_j bind to an annihilator, it splits into two waste molecules that are not able to participate in further reactions. After a while, only the species that had the largest concentration will remain. Then, a signal-restoration reaction that converts s_i to the output y_i takes place. This is again implemented catalytically via seesawing reactions – if s_i is present, in any concentration, then the fuel ensures that a sufficient amount of y_i is produced, but no production of y_i takes place if no s_i is present. Thereby, one has found a biochemical implementation of the nonlinear function 3. Finally, the output is converted to a fluorescent signal.

Given that the neural network consists of chemicals floating around in a test tube, it is not immediately clear why these calculations should take place in the desired order (as they do in a traditional neural network implemented in a computer). For most steps, this is ensured by the fact that a reaction can only take place if the reactants are present. The products of one step (one reaction) are the reactants for the next one. The exception are the annihilation and restoration reactions, where the reactants (the s_i) are the same. Here, the desired order is ensured by different reaction rates: The annihilation reaction is very fast compared to the restoration reaction. Thereby, the s_i quickly eliminate each other until only one is left, the survivor's concentration then gets very slowly increased again. Specifically, the toeholds of the annihilator have two extra nucleotides, which due to the exponential scaling of the reaction rate with the toehold length lead to a reaction rate increased by a factor of about 100.

7 DNA Reservoir Computing

Finally, I will discuss a different approach to DNA-based machine learning. This approach, proposed in a theoretical study by Goudarzi et al. [12], uses *reservoir computing* [17, 24]. See the introductory chapter on this topic in this volume for an explanation of how reservoir computing works and the chapters by Kathy Lüdge/Lina Jaurigue, Atreya Majumdar/Karin Everschor-Sitte, and Julian Jeggle/Raphael Wittkowski in this volume for other implementations.

In a nutshell, reservoir computing employs a dynamical system (the reservoir) that is driven by an input signal. The response of the system then serves as the input for a neural network with a single layer (the *readout layer*) that converts this response into the output layer. This readout layer is the only part of the system that is changed during the training process. Since the reservoir does not have to be changed, the reservoir can also be a physical system (for example one consisting of DNA). It is widely assumed[2] that it is helpful if the reservoir operates close to criticality, such that the dynamics is rich enough to allow for interesting things to be read out from it (as opposed to, say, a system where all trajectories approach a certain stationary state irrespective of the input).

Goudarzi et al. [12] suggest that such a rich transient dynamics can be found in a network consisting of coupled chemical oscillators, realized by a microfluidic chamber containing different interacting DNA species. As a starting point, they use a network proposed by Farfel and Stefanovic [10]. Substrate molecules enter a reaction chamber and bind to gate molecules, by which they are then converted to product molecules. The presence of the product molecules inhibits the reaction of substrates and gates. Moreover, the products flow out of the chamber with a certain rate. This

---

[2] It is not really clear at present to what extend this is actually the case, see the chapter on reservoir computing in this book.

gives rise to a reaction-inhibition cycle leading to sustained oscillations of the product concentrations. Denoting these by P_i, and moreover the substrate concentrations by S_i, the substrate influx rates per volume by $S_{i,\text{in}}$, the gate concentrations by G_i, the efflux rate by e, the volume of the reactor by V, the well-mixed fraction of the reactor by h, and the reaction rate by β, this behavior is described by the dynamical system [12]

$$\dot{P}_1 = h\beta S_1(G_2 - P_3) - \frac{e}{V}P_1, \tag{4}$$

$$\dot{P}_2 = h\beta S_2(G_2 - P_1) - \frac{e}{V}P_2, \tag{5}$$

$$\dot{P}_3 = h\beta S_3(G_3 - P_2) - \frac{e}{V}P_3, \tag{6}$$

$$\dot{S}_1 = S_{1,\text{in}} - h\beta S_1(G_1 - P_3) - \frac{e}{V}S_1, \tag{7}$$

$$\dot{S}_2 = S_{2,\text{in}} - h\beta S_2(G_2 - P_1) - \frac{e}{V}S_2, \tag{8}$$

$$\dot{S}_3 = S_{3,\text{in}} - h\beta S_3(G_3 - P_2) - \frac{e}{V}S_3. \tag{9}$$

As detailed in Ref. [12], a linear stability analysis of this dynamical system allows to find the parameter region in which it exhibits oscillatory behavior. A sustained oscillation exists for $h\beta(S_1 S_2 S_3)^{\frac{1}{3}}/2 - e/V = 0$.

The influx rate $S_{1,\text{in}}$ is used to drive the system, this is the input layer. The reservoir state is specified by the concentrations of the various substances. Of particular importance here are the product concentrations P_1, P_2, and P_3. It is assumed that one can read out the reservoir state using a fluorescent probe. The known values of P_1, P_2, and P_3 are then fed into the readout layer (which is trained via linear regression.). The readout layer can be implemented on a computer, or as a part of the physical system. In total, one thus has a neural network consisting of a microfluidic DNA chamber and a single-layered neural network (for example on a computer) that receives an input signal in the form of an influx rate and returns an output signal that is obtained by feeding the measured product concentrations in the DNA chamber into the readout layer.

8 Advantages and Disadvantages of This Approach

While it is certainly impressive that test tubes filled with some chemicals can be used for handwritten number recognition, it should of course be noted that this is certainly not the most efficient approach if the recognition of handwritten numbers is our primary goal. If the numbers are primarily a proof of principle, what then can these methods be useful for?

DNA neural networks require the input signal to have the form of a DNA strand. In general, this is a disadvantage since converting general input signals to DNA is quite an effort. This aspect can, however, turn into an advantage in contexts where the input signal takes the form of DNA strands (or at least that of biomolecules) anyway. This will primarily be the case in biomedical applications of neural networks. Suppose, for example, that a neural network has been trained to recognize genetic dispositions for a certain disease. If this network is implemented in DNA form, then one could just take a DNA sample from the patient, put it into a test tube, and then see a glowing test tube indicating that the gene one looks for is (or is not) present.

A further aspect to note here are the time scales involved here. Photonic neural networks (see the chapters by Kathy Lüdge/Lina Jaurigue and by Lennart Meyer/Rongyang Xu/Wolfram Pernice) have the attractive feature that they operate with the speed of light, i.e., they are extremely fast. This can certainly not be said about DNA computers. Their computational speed depends on how fast the chemical reactions take place, which can be of the order of many minutes. An advantage, in contrast, is that DNA-based approaches are well suited for parallelization. These (dis)advantages are familiar from DNA computing in general.

Moreover, there can be contexts where having a neural network that operates slower can actually be an advantage.[3] A good example would be a network that processes temporal input signals, as is required, e.g., in speech recognition. In this case, it is advantageous if the system's dynamics takes place on roughly the same time scales as the input signal rather than being significantly faster. Consider reservoir computing, where the employed physical system possesses a fading memory, as an example – if, after the second word of a sentence to be processed, all memory of the first word has already vanished, the system cannot process the sentence as a whole. A particular advantage of DNA neural networks in this context is that the speed at which the reactions take place (and thereby the speed at which the network operates) can be tuned by the experimenter, namely by changing the lengths of the toeholds (see Sect. 5).

Finally, from a conceptual or pedagogical point of view, it should be noted that (if we ignore the details of the biochemical processes involved here) the neural network presented in Sect. 6 has an extremely simple architecture. It is therefore possible to explicitly model and understand how each entry of an intermediate state vector and each weight contributes to the final output of the network. Therefore, this approach helps to teach basic concepts of artificial intelligence to high school and undergraduate university students, a teaching concept based on this idea is outlined in Ref. [29]. Moreover, studying networks of this type allows to get a better intuition for how neural networks generally come to the conclusions they come to, a goal that, in the framework of "explainable artificial intelligence" [13] has motivated several works studying simple network architectures in other contexts (including biomolecular ones [4]). For the same reason, the template of DNA neural networks can provide inspiration for other attempts to implement artificial intelligence in soft matter systems (see, for example, the chapters by Hartmut Löwen/Benno Liebchen, Julian Jeggle/Raphael Wittkowski, Giovanni Volpe, and Jannes Freiberg/Roshani Madurawala

---

[3] Thanks to Raphael Wittkowski for bringing this to my attention.

in this volume), and it is a helpful case study for philosophical discussions of physical computing (see the chapter by Luis Lopez in this volume).

Note added in proof: During the production process of this book, a new study has appeared [K. M. Cherry and L. Qian, Nature 645, 639–647 (2025)] showing that an in vitro training of DNA neural networks is possible. This removes a disadvantage that this approach previously had (the need for in silico training).

9 Summary

In this chapter, I have provided a brief introduction to neural networks consisting of DNA, using as an example the winner-take-all network proposed in Ref. [5]. The input data is provided as a DNA strand and is processed via biochemical reactions. On this basis, it is possible to recognize handwritten digits using DNA. Moreover, I have briefly discussed a proposal for DNA-based reservoir computing [12]. Such approaches constitute a promising starting point for the development of intelligent matter based on biological materials, and might also find applications in, for instance, medical contexts where input data is already present in a biochemical form.

Acknowledgements This work was funded by the Deutsche Forschungsgemeinschaft (DFG, German Research Foundation) in the framework of SFB 1551; Project No. 464588647 and SFB 1552; Project No. 465145163. The author also acknowledges funding from the Mainz Institute of Multiscale Modeling, M^3ODEL.

References

1. Adleman LM (1994) Molecular computation of solutions to combinatorial problems. Science 266(5187):1021–1024
2. Baumgardner J, Acker K, Adefuye O, Crowley ST, DeLoache W, Dickson JO, Heard L, Martens AT, Morton N, Ritter M, Shoecraft A, Treece J, Unzicker M, Valencia A, Waters M, Campbell A, Heyer LJ, Poet JL, Eckdahl TT (2009) Solving a Hamiltonian path problem with a bacterial computer. J Biol Eng 3:1–11
3. Bolsover SR, Hyams JS, Shephard EA, White HA, Wiedemann CG (2004) Cell biology: a short course, 2nd edn. Wiley, Hoboken
4. Braghetto A, Orlandini E, Baiesi M (2023) Interpretable machine learning of amino acid patterns in proteins: a statistical ensemble approach. J Chem Theory Comput 19:6011–6022
5. Cherry KM, Qian L (2018) Scaling up molecular pattern recognition with DNA-based winner-take-all neural networks. Nature 559(7714):370–376
6. Crick F (1970) Central dogma of molecular biology. Nature 227(5258):561–563
7. Crick FH (1958) On protein synthesis. In: Sanders FK (ed) Symposia of the society for experimental biology, number XII: the biological replication of macromolecules. Cambridge University Press, Cambridge, pp 138–163
8. Evans CG, O'Brien J, Winfree E, Murugan A (2024) Pattern recognition in the nucleation kinetics of non-equilibrium self-assembly. Nature 625(7995):500–507
9. Ezziane Z (2005) DNA computing: applications and challenges. Nanotechnology 17(2):R27

10. Farfel J, Stefanovic D (2006) Towards practical biomolecular computers using microfluidic deoxyribozyme logic gate networks. In: Carbone A, Pierce NA (eds) DNA computing. DNA 2005, Springer, Berlin, Heidelberg, pp 38–54

11. Genot AJ, Fujii T, Rondelez Y (2013) Scaling down DNA circuits with competitive neural networks. J R Soc Interface 10(85):20130212

12. Goudarzi A, Lakin MR, Stefanovic D (2013) DNA reservoir computing: a novel molecular computing approach. In: Soloveichik D, Yurke B (eds) DNA computing and molecular programming. DNA 2013, Springer, Cham, pp 76–89

13. Gunning D, Stefik M, Choi J, Miller T, Stumpf S, Yang GZ (2019) XAI—explainable artificial intelligence. Sci Robot 4(37):eaay7120

14. Hsieh P, Panyutin I (1995) DNA branch migration. In: Eckstein F, Lilley DM (eds) Nucleic acids and molecular biology, vol 9. Springer, Berlin, Heidelberg, pp 42–65

15. Hundertmark F (forthcoming) Selektion und Disposition: Über die Existenz und Natur biologischer Funktionen und Dysfunktionen. In: te Vrugt M (ed) Wissenschaft und Metaphysik, Nomos, Baden-Baden

16. Ionkov L, Settlemyer B (2021) DNA: the ultimate data-storage solution. Scientific American. Accessed 01 Dec 2023. https://www.scientificamerican.com/article/dna-the-ultimate-data-storage-solution/

17. Jaeger H (2001) The "echo state" approach to analysing and training recurrent neural networks. Bonn, Germany: German National Research Center for Information Technology GMD Technical Report 148

18. Kaspar C, Ravoo BJ, van der Wiel WG, Wegner SV, Pernice WHP (2021) The rise of intelligent matter. Nature 594(7863):345–355

19. Kim J, Hopfield J, Winfree E (2004) Neural network computation by in vitro transcriptional circuits. In: Saul L, Weiss Y, Bottou L (eds) Advances in neural information processing systems, MIT Press, vol 17

20. Kumar T, Namasudra S (2023) Chapter one - introduction to DNA computing. In: Namasudra S (ed) Perspective of DNA computing in computer science, Advances in computers, vol 129. Elsevier, Cambridge (MA), pp 1–38

21. Lazzaro J, Ryckebusch S, Mahowald M, Mead CA (1988) Winner-take-all networks of O(N) complexity. In: Touretzky D (ed) Advances in neural information processing systems, Morgan-Kaufmann, vol 1

22. Ma Q, Zhang C, Zhang M, Han D, Tan W (2021) DNA computing: principle, construction, and applications in intelligent diagnostics. Small Struct 2(11):2100051

23. Maass W (2000) On the computational power of winner-take-all. Neural Comput 12(11):2519–2535

24. Maass W, Natschläger T, Markram H (2002) Real-time computing without stable states: a new framework for neural computation based on perturbations. Neural Comput 14(11):2531–2560

25. Qian L, Winfree E (2011) Scaling up digital circuit computation with DNA strand displacement cascades. Science 332(6034):1196–1201

26. Qian L, Winfree E (2011) A simple DNA gate motif for synthesizing large-scale circuits. J R Soc Interface 8(62):1281–1297

27. Qian L, Winfree E, Bruck J (2011) Neural network computation with DNA strand displacement cascades. Nature 475(7356):368–372

28. Simmel FC, Yurke B, Singh HR (2019) Principles and applications of nucleic acid strand displacement reactions. Chem Rev 119(10):6326–6369

29. te Vrugt M (2024) Ein einfaches neuronales Netz und seine Anwendung in der Biotechnologie. In: Huwer J, Becker-Genschow S, Thyssen C, Thoms LJ, Finger A, von Kotzebue L, Kremser E, Meier M, Bruckermann T (eds) Kompetenzen für den Unterricht mit und über Künstliche Intelligenz. Perspektiven, Orientierungshilfen und Praxisbeispiele für die Lehramtsausbildung in den Naturwissenschaften, Waxmann, Münster, pp 82–85

30. Xiong X, Zhu T, Zhu Y, Cao M, Xiao J, Li L, Wang F, Fan C, Pei H (2022) Molecular convolutional neural networks with DNA regulatory circuits. Nat Mach Intell 4(7):625–635

31. Xu J, Tan G (2007) A review on DNA computing models. J Comput Theor Nanosci 4(7–8):1219–1230

32. Yurke B, Turberfield AJ, Mills AP Jr, Simmel FC, Neumann JL (2000) A DNA-fuelled molecular machine made of DNA. Nature 406(6796):605–608
33. Zhang DY, Winfree E (2009) Control of DNA strand displacement kinetics using toehold exchange. J Am Chem Soc 131(47):17303–17314

Intelligence Without a Brain—Perspectives from the Aneural Slime Mold as a Computing Model for Artificial Intelligence

Jannes Freiberg and Roshani Madurawala

Abstract The slime mold *Physarum polycephalum* is presented as an alternative model due to its complex behavior, learning abilities, and promising electrical properties, despite being a single-celled organism. Its unique characteristics, such as forming efficient networks and demonstrating memristive properties, make it a compelling candidate for new computing architectures. The slime mold's natural energy efficiency and problem-solving capabilities offer valuable insights for AI, network design, material science, and environmental monitoring. This chapter proposes a framework for exploring *Physarum*'s potential both as a biological material with neuromorphic qualities and as an inspiration for bio-inspired computing, suggesting that it may hold the key to advancing beyond traditional computing paradigms.

1 Introduction

When embarking into 'perspectives of the aneural slime mold as a computing model for artificial intelligence', it is essential to understand what a computer model is and the importance of a computer model for artificial intelligence (AI). Artificial intelligence, in simple terms, enables computers and machines to simulate human intelligence and problem solving. The first question that arises from this simple definition is, 'How do computers and machines perform AI?'. To answer this question, we need to look back to the evolution of computing models which are from here forth known as computer architectures. The second question of the definition is 'what is

J. Freiberg (✉)
Institut Für Psychologie, Christian-Albrechts-Universität Zu Kiel, Olshausenstrasse, Kiel, Germany
e-mail: freiberg@psychologie.uni-kiel.de

R. Madurawala
Functional Nanomaterials, Institute for Materials Science, Faculty of Engineering, Christian-Albrechts-Universität Zu Kiel, Kiel, Germany

M. te Vrugt (ed.), *Artificial Intelligence and Intelligent Matter*, Machine Intelligence for Materials Science, https://doi.org/10.1007/978-3-032-04129-6_16

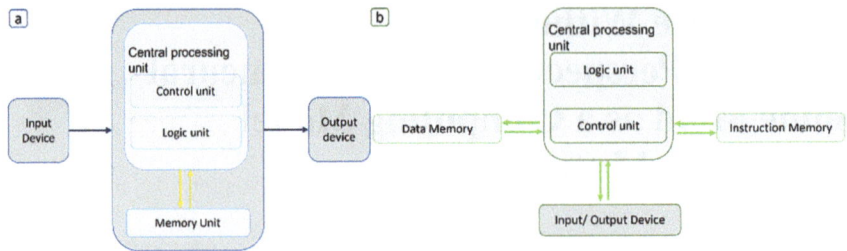

Fig. 1 Schematic of the **a** von Neumann architecture: the yellow arrow signifies the Von Neumann bottle neck. **b** Harvard Architecture. The memory—where data and instructions are stored, the control unit—which directs operations, the arithmetic/logic unit—which performs calculations, and the input/output systems—which interact with the outside world are sketched here

human intelligence and problem solving?'. To answer this question, we need to look into the definition of intelligence and cognition.

The evolution of computer architectures has been marked by significant milestones, starting with foundational models such as the Harvard (1939), von Neumann (1945) and modified Harvard (post 1945) architectures. A computer architecture organizes the components (processor, memory, input/output devices) which make up a computer system as shown in Fig. 1. The rules that contribute to the computer architecture define how the software and hardware communicate in the device. These early designs laid the groundwork for modern 'computing' but have encountered notable challenges, especially in the context of big data and artificial intelligence (AI).

The von Neumann architecture is characterized by its use of a single memory space for both instructions and data. This is very beneficial in implementing and self-modifying code. While the von Neumann architecture is easy to implement and simple to use because the instructions are executed in a linear fashion, one after the other, it suffers from a phenomenon called the 'Von Neumann bottleneck'. This drawback occurs due to the shared memory system where instruction and data use the same communication path. In other words, this system can only retrieve one item from the shared memory at a time. The bottleneck becomes a critical limitation when dealing with immense data sets required for training AI models, leading to inefficiencies and slower processing times.

In contrast, in the Harvard architecture the Von Neumann bottleneck is prevented by implementing separate memory units for storing data and instructions. Therefore, a level of parallelism is introduced since the data and instructions can be retrieved simultaneously. This significantly improves the system's overall performance regarding speed and efficiency. While the von Neumann architecture has a low implementation cost, and a simpler design that fits into a small chip, the Harvard design results in high costs due to its complex design. However, modern processes typically use modified Harvard architecture which provides a smooth balance between the ease of implementation, cost and complexity.

While artificial intelligence aspires to reach human intelligence there is a need for more energy efficient computing architectures possibly inspired by nature. "What other inspiration could there be if not for the highest inspiration of all time: the human brain?" The architecture of the human brain is strikingly distinct, featuring distributed memory and processing through every neuron, as well as parallel processing in separate regions of the brain. Consequently, the brain is not particularly adept at performing linear tasks such as calculations; however, it is highly proficient in training and pattern recognition. To illustrate, during a conversation with a friend, one's brain may simultaneously process the other person's attire, the content of their discourse, and the relevance of the topic to other similar experiences, while also monitoring the emotions and body language of the individual. To emulate these processes, contemporary AI programs frequently employ artificial neural networks (ANN) as software on conventional computing architectures. The fact that computer architectures are not yet optimized for parallel and distributed processing leads to excessive energy use. As AI and machine learning applications continue to grow, the need for more efficient computing paradigms has become apparent.

This necessity has driven the exploration of neuromorphic computing. Neuromorphic computing is an approach inspired by the architecture/structure and the functionality of the human brain. Neuromorphic computing aims to replicate the neural structure and synaptic activities of the brain to create hardware capable of learning and processing information more efficiently. Potential benefits of a neuromorphic computing architecture include three-dimensional parallel processing as well as plasticity, which lead to learning and adaptability, high performance and, most of all, to energy efficiency.

Implementing neuromorphic architectures in hardware involves designing specialized systems that can emulate the synaptic activity and neural computations of the brain. These hardware implementations are crucial for realizing the full potential of neuromorphic computing. This will enable the development of AI systems that are not only more powerful but also more sustainable in terms of energy consumption.

Emerging computing architectures such as neuromorphic computing as well as quantum computing and biocomputing are still quite immature in research. For example, the von Neumann architecture, which established a clear and structured approach to computing, separates tasks into distinct, hierarchical components. This clarity and structure enabled the development of general-purpose computers, where different types of hardware could operate under the same principles, allowing software to function uniformly regardless of specific hardware details. As a result, it is challenging to deviate from this well-established architecture.

Establishing hierarchy and structure for neuromorphic architecture and hardware is sophisticated due to our lack of understanding of the brain. The anatomy of the brain and the properties of many parts are understood, but research on the true functionality is still complicated due to the size and complexity of the system. Most complex behaviors and computations are still often dubbed an "emergent property" that occurs when a sufficient number of neurons is connected, as the exact mechanisms are still unclear. The same problem can be observed in artificial neural networks: While the number of neurons and various parameters can be optimized for certain tasks, it is unclear what happens inside an artificial neural network and how it produces its output.

To address these topics, in this book chapter we would like to propose investigating a second possible computing architecture that may go hand in hand with the neuromorphic architecture, which is the '**slime mold architecture**'. Slime molds are unicellular amoebae, and the most widely studied slime mold species is scientifically designated as *Physarum polycephalum*. It is recognized for its complex, intelligent behavior, including the ability to respond to stimuli and learn from past experiences, which is quite advanced for a single cellular organism. In addition, *Physarum polycephalum* exhibits promising electrical behavior. The electrical properties of *Physarum* such as oscillations and multiple resistances are important to understand the electrochemical mechanisms of the slime mold. These mechanisms have the potential to facilitate the incorporation of slime mold as an electrical component or the construction of artificial components exhibiting analogous properties.

From our point of view, *Physarum polycephalum* can be studied under two aspects. The first aspect would be to study the abilities and the functions of the slime mold to mimic or be inspired by its behavior giving rise to a new computing architecture. The second aspect would be to utilize the slime mold as a biological material with significant electrical properties to support the neuromorphic architecture. Both possible aspects are converged in slime mold as a new model organism for AI.

Accordingly, this book chapter will present a framework for considering the intelligence of unicellular organisms, after providing an overview of the behavioral capabilities of the slime mold that could be relevant for emulation. Subsequently, we will address the second aspect, namely the electrical properties that make it a potential foundation for neuromorphic computing.

2   Intelligence, Cognition and Aneural Cognition

Although the term "intelligent" was attributed to the slime mold in the introduction, there is a degree of controversy within the scientific community regarding the appropriateness of this linguistic designation in this particular context. When speaking of intelligence and cognition, both biological and artificial, the terms are often inherently associated with brains and nervous systems. It is universally accepted and proven that the brain serves as the centralized control unit of the human body. Sensory information is evaluated in a complex process involving correlation and integration of multisensory stimuli, memory and higher-order cognitive functions as thinking about thinking, making predictions based on past outcomes and more. The human brain is one of the most complex structures known to mankind, and the pinnacle of what the evolution of nervous systems has achieved so far. Consequently, the term cognition was first coined for describing human capabilities over half a century ago, during a movement that later became known as the cognitive revolution. It coincided with the first inception of artificial intelligence in the 1950s. At the time, psychology as a field was facing a revolution of its own. Before, behaviorism was the mainstream in psychology: the assumption that every behavior is the result of a stimulus or a learned reaction to a stimulus, while viewing the brain as a black box. Its focus on observable behavior helped psychology to become an objective field of science, but viewing the human mind as a black box denied any interpretation about underlying mechanics. At that time, it became apparent that this approach could mostly describe, but not explain complex behavior. Theories about the mental state and mind of humans were needed, and the core principles of behaviorism were slowly abandoned. These theories about information-processing in the brain made fruitful interdisciplinary work possible, where the fields of psychology, philosophy, linguistics, anthropology, neuroscience, and computer science could join forces in what became called 'cognitive sciences' in 1978 [1].

The main goals of the cognitive sciences are threefold: To provide explanations for what cognition is, what it does and how it works [2]. Since its inception fifty years ago, no unified definition of cognition was found and agreed upon. Instead, the historical approach was to wait until enough data has been accumulated, but despite an enormous mass of collected data, no definition has emerged [3]. Both cognition and intelligence can be applied in both broad and narrow contexts. Due to the strong connections to psychology, historically cognition was seen as a human trait, hence calibrated against the human experience [4]. In psychology, cognition is often described without a clear definition, instead lists of traits considered cognitive such as affect, learning, problem solving, communication, anticipation, or intentionality. Descriptions of these may read, for example, in problem solving as a listing of "rational decision making" and "abstract thinking". Another example for communication is given by Lyon: *"Verbal or written symbol systems whose units have semantic content (meaning) and their deployment is organized according to rules (syntax), both of which are conventionally established expressly for the purpose of information exchange."* [4]. In animals, communication is often not verbal and never written, while problem solving behavior may be observed, abstract thinking might only be inferred.

In human-centric sciences, describing human behavior in human terms seems natural. However, problems arise when evaluating non-human intelligence, as it is still measured against the human benchmark. If, for example, an animal displays certain traits and behaviors that are deemed cognitive in humans, it is then said to possess some degree of cognition, which typically is considered inferior to human cognitive capabilities. Lyon [3] defines this as the anthropogenic approach to cognition, while proposing another approach called biogenic cognition. Before exploring biogenic cognition, we should first investigate further into the anthropogenic approach to see why a different approach might be proposed.

In the work of Lyon [3], a few traits of the anthropogenic approach to cognition are listed. Since human cognition is based on the brain and nervous system, anthropogenic cognition assumes that cognition is a function of the brain. Often, it is not referred to particular parts of the brain, as sometimes the neural substrate is not entirely known. This prerequisite is simultaneously too broad and too narrow. Too broad in the sense that not all parts of the brain and central nervous system are involved in what is considered cognitive, many others are just regulating bodily functions. Therefore, claiming the brain as a prerequisite but not identifying which parts of the brain seems inaccurate. This is often further specified as a brain of 'sufficient complexity' without giving any indication of which level would be sufficient. The scientific consensus on that shifted drastically over time.

During the cognitive revolution, the prevailing view was that learning and cognition were exclusive to higher organisms. These higher organisms at first included mostly mammals, while every other animal was considered evolutionary lower, less sophisticated, and not capable of cognitive functions. Invertebrates of any kind were excluded on behalf of their lower number of neurons. Invertebrates were thought of as purely running on instinct, executing their functions like pre-programmed robots [5]. Despite this, in the 1950s to 1960s, researchers such as McConnell successfully showed learning and memory in planarians, a type of flatworm with less than 50,000 neurons [6]. At the time, these findings were inconsistent with the prevailing theoretical frameworks. Planarian researchers speculated that in these invertebrates memory was not stored in the brain, but biochemically in special strands of RNA [7]. The search for a "memory molecule" failed, not because it might not exist, but because planarians and other invertebrates, appear to rely on their nervous system for cognition. In the following decades, more and more evidence of learning in invertebrate organisms was investigated. These studies extended from the highly evolved octopus to insects and crustaceans [8], even to jellyfish [9] and other cnidarians [10]. Due to their different body plan, Cnidarians lack a centralized brain- instead featuring a dispersed nerve-net, in some species with localized ganglia spread around the body. With increasing amounts of data showing learning and memory in many formerly underestimated phylogenetic groups of animals, from reptiles to fish to invertebrates, the focus shifted from requiring a brain for cognition to merely having a nervous system as a prerequisite.

It must be reiterated that, while lacking a clear definition of such a big concept may seem to be a big oversight, the anthropogenic approach made sense both in a historical and practical way. As humans, it would be intuitive to start with the human brain and mind, because we have access to our own thoughts and can also communicate them. A critical examination of the historical perspective on animal cognition reveals the inherent limitations of this approach. It was assumed that animals considered most similar to humans were also the most accessible, thus facilitating the assumption that they possessed cognitive abilities. Down the evolutionary tree, where animals were more different from humans, therefore harder to interpret intuitively, were seen as less cognitively advanced. Sometimes, this view would even be reinforced by applying unsuitable experimental designs to the animals under study.

For example, snakes and other reptiles were previously considered to exhibit deficiencies in cognitive abilities because they were often tested in settings designed for mice and rats, even though their metabolic rates are significantly slower. Mice are compelled to ingest food on a near-constant basis to sustain their metabolic processes, thereby making them highly motivated by food. In experimental settings where food is used as a reward, mice have a high intrinsic motivation to explore the setting, take risks and interact in any way possible to get to food. In contrast, reptiles have the capacity to survive for extended periods without food, therefore have other options when confronted with an unfamiliar and potentially hazardous environment. This typically manifests as a tendency to observe and move as little as possible, a behavior that may be perceived as a form of risk aversion. For decades, this behavior was viewed as a sign of the animal not being able to perform the given mental task, when in fact more recent research with better suited methods contradict this assumption [11].

The top-down anthropogenic approach fails to account for other potential instances of cognition. If the measure of cognition is defined as a human-like experience of life and the presence of a brain, then non-human animals, non-neural organisms, and artificial intelligence cannot be considered cognitive. This approach to a phenomenon is unique to the field of psychology. In biology, the typical method for studying a phenomenon is not to begin with the most complex organism that exhibits the phenomenon but instead to focus on the simplest organisms where that phenomenon is present. This allows us to access and correlate observations with genetic and epigenetic data in order to get an understanding of the underlying systems. With that knowledge, it is possible to advance to more complex organisms to assess both similarities and differences, to get an understanding of the broader principles of the phenomenon [12].

The application of principles derived from biological research to the field of cognition has resulted in the emergence of the biogenic approach to cognition. The biogenic approach is less well-known and accepted within the scientific community, although it has gained some traction over the past few decades. The primary tenet of the biogenic approach is the assumption that complex cognitive abilities have an evolutionary history, evolving from more basic forms of cognition. This would permit the advent of novel cognitive traits during evolution, while maintaining incremental distinctions between organisms. Moreover, it would facilitate the pursuit of the fundamental elements and most basic forms of cognition. From a biological perspective, this approach facilitates a more comprehensive comparison of cognitive processes with other features observed in living organisms. Additionally, it extends the scope of cognitive inquiry beyond the domain of humans and their evolutionary relatives, acknowledging the existence of cognitive abilities in other species [3].

To understand the importance of including other biological entities besides animals, it might help to look at the evolutionary history of life on earth. The oldest signs of life on earth are around 3.5 billion years old. The first fossil of what might have been a eukaryote is about 1.5 billion years old. This means that life has been on earth for at least 3.5 billion years. For the first 2 billion years of that, it was prokaryotic life- meaning that life on earth consisted of cells with no nucleus such as bacteria. Subsequently, with the rise of the eukaryotic life-forms, still nearly another billion years passed until the first animals emerged, with the oldest fossils being around 560 million years old. Therefore, it can be clearly stated that life itself survived for around 3 billion years without any nervous system, and it also had time to evolve in many ways. When looking at the phylogenetic tree of eukaryotic life-forms, it becomes abundantly clear that what we know as animals, plants and fungi are just a few very

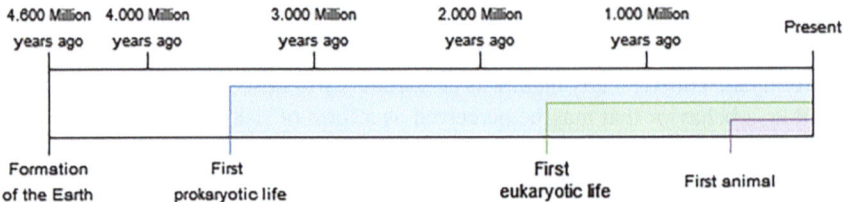

Fig. 2 The timeline of the development of life on Earth illustrates that animals, and therefore organisms with a nervous system, exist for only a small fraction of the entire existence of life itself

successful groups in a much larger mass of living beings. Most of the other groups contain mainly unicellular organisms, making up most of earth's biomass (Fig. 2).

Survival is dependent on multiple factors, of which the adaptation to changes in the environment and the avoidance of life-threatening situations play a major role. As LeDoux [13] states: *"As soon as there was life, there was danger"*. Therefore, a system to detect danger and respond to it must have evolved long before animals and nervous systems. This appears to be the case with numerous mechanisms that are involved in the functions of the nerve cell. The primary function of this mechanism, namely the generation and propagation of electric action potentials, is not exclusive to animals; it is also observed in plants [14] as well as in the unicellular paramecium [15]. Very similar systems can be observed even in bacteria: As McKinnon showed, the structure and operation of the cellular ion channels to achieve electrical signals in the bacterium *Bacillus subtilis* is very similar to that in a neuron. The bacterium even uses that electrical signaling for both long-distance communication and coordination of group behavior, not unlike neurons [16].

If organisms without neurons already have some of the tools for cognition available, how do they use them? There is already a small but growing range of evidence for interesting adaptive behavior in unicellular organisms. While many of the cited examples are rather recent, coinciding with the rise of the biogenic approach, it has to be noted that this view on unicellular organisms was not uncommon before the cognitive revolution. Darwin already thought of an evolutionary ancestor to learning and complex behavior, not excluding any form of life out hand. In 1894, Binet wrote a book called *"The psychic life of Micro-Organisms—A study in experimental psychology"*, detailing his thoughts and work on unicellular organisms. In his book he claimed: *"We could, if it were necessary, take every single one of the psychical faculties [of animals] and show that the greater part of those faculties belonged equally to Micro-organisms"* [17]. Jennings, who worked extensively on the behavior of unicellular organisms, found already in 1902 that the reactions of the ciliate *Stentor roeseli* greatly varied both between individuals and different trials on the same individual, calling their behavior complex and deeming them capable of learning [18].

Paramecium, a unicellular ciliate capable of moving freely through water, has been colloquially dubbed a "swimming neuron" in the last few years. They produce electrical action potentials as a reaction to stimuli, using them to modulate their swimming behavior. As a free-living cell, their behavior is complex. *Paramecium* are capable of detecting various stimuli and moving either towards or away from them, and even show simple social behavior. Their movement is sometimes described as "trial and error", because they do not have full control over every direction—instead, using action potentials, they can swim backwards and rotate in a new, semi-random direction, before swimming forward again. They thus can achieve variations of both the duration and frequency of the reversal. The results of multiple escape experiments, in which a *Paramecium* was trapped and remained so for as long as it continued to follow its normal patterns of behavior, demonstrated that the *Paramecium* was capable of altering its behavior in a number of ways. Even more impressive, it seems to also retain some memory of the experience. When a *Paramecium* is tested multiple times it takes much less time to switch to the new behavior in subsequent trials [15].

Perhaps an even greater feat is associative memory, in which a connection between two previously unrelated stimuli, or a behavior and a stimulus, is learned and shown through the emergence of new reactions to old stimuli. In psychology, this feat has first been described by behaviorists as associative learning, and later in the paradigms of classical and operant conditioning. This process requires not only learning and memory, but also the information processing to correlate different sensorial stimuli and change the behavior accordingly. In 2019 it was shown that multiple species of amoeba can associate weak electric fields with a normally unrelated food stimulus. After training they completely reverted their reaction to the electric field even without the presence of the food stimulus [19]. This finding of classical conditioning in amoeba shows the enormous capability of information processing and behavioral flexibility in a single cell. [20].

While classical, anthropogenic approaches to cognition fail to account for these examples of non-animal behavioral adaptivity, in the biogenic approach they point to the evolutionary ancestors of our cognition. These phenomena have been found in many instances of eukaryotic life and are currently being investigated in bacteria. Looking for shared traits among different organisms can elucidate how similar some processes are, and how some traits evolved into others. Genetic study on planarians showed that 95% of the genes related to their nervous system, both involved in brain and neural morphogenesis, are shared with other, not closely related animal species like the fruit fly, the nematode *C. elegans*, and humans. This strongly implies a common ancestor for all their nervous systems. Approximately 30% of these genes are also shared by plants and yeast. These findings show that, much like the behavior and its mechanisms, the genetic building blocks of the nervous system are also found across the tree of life [21].

An approach to categorize the processes and mechanisms that allow organisms to ensure survival by meaningfully interacting with their surroundings, even long before nervous systems evolved, is the so called 'basal cognition'. This concept, also a biogenic approach to cognition, has been developed to describe and investigate these instances of adaptive behavior that the anthropogenic approach to cognition often dismisses, in search for similarities and common mechanisms [12]. From this essence of the biogenic approach, a toolkit of basal cognitive capacities has been developed and refined, with definitions for cognitive phenomena that allow for clear testing without benchmarking on human intuition or outright excluding organisms. In this toolkit, the definition of learning is given as: *"The capacity to adapt behavior to salient stimuli according to past experience by altering the threshold for or the nature of a response"*. For the capacity of valence, which in anthropogenic approaches is often called affect, the definition is given as: *"An organism's capacity to assign a value (advantage/good, harm/bad) to a particular stimulus or the summary of information about its surroundings relative to its own current state. The "building block" of affect"*. Analysis using this toolkit allows for the search for cognitive behavior in any organism. Additionally it facilitates the comparison between behavioral capabilities and their underlying mechanisms among species [12].

This approach with its toolkit allows the search for the minimal mechanisms capable of cognition. This could not only benefit the understanding of intelligence but also the development of artificial intelligence. Instead of imitating human behavior and human brains due to the anthropogenic approach, it allows for a more open view on cognition, both in consideration of what models to use and what artificial intelligence is supposed to achieve. Under this framework, the unicellular slime mold *Physarum polycephalum* can be deemed intelligent and capable of cognition, despite lacking a nervous system. Therefore, in the next section, we will elaborate on what *Physarum* is capable of and why it can be an interesting model for artificial intelligence.

3   The Slime Mold Physarum Polycephalum

Due to their unconventional biology, slime molds have been puzzling biologists for a long time. As the name implies, initially they were categorized as a special type of fungus. This was due to the end of their life cycle, in which they produce fruiting bodies called sporangia full of spores, visually similar to those of fungi. However, with further observations, it became clear that they are not related to fungi but are instead a special type of amoeba. In the case of acellular slime molds like *Physarum*, their spores hatch into haploid amoeboid cells. These cells combine pairwise with each other to form a diploid amoeba cell. Unlike other amoeba, this cell does not normally undergo cell division for asexual reproduction. Instead, only its nuclei multiply by division, while the cell itself constantly grows, forming a giant, multi-nucleate cell called a 'plasmodium'. This macroscopic amoeboid cell is one of the biggest unicellular organisms in the world, being visible to the naked eye reaching a few centimeters in diameter. The plasmodial stage is the main part of the slime mold's life cycle. The plasmodium forages for nutrition, until it has matured. After maturation, an external stimulus triggers the transformation into the spore filled fruiting bodies once more, thereby ending the life of the current plasmodium [22]. It must be noted that other types of amoeboid slime molds with other life cycles exist in nature. One of these is the well-studied cellular slime mold *Dictyostelium discoideum*, which is often used as a model organism. This organism has unique modes of organization and reproduction, but for this book chapter we will focus only on the plasmodial *Physarum polycephalum*.

In nature, *Physarum polycephalum* is found in temperate regions across the globe, predominantly in forests and woodland areas. Despite being widespread and featuring big, bright yellow plasmodia, *Physarum polycephalum* often goes unnoticed. The plasmodia frequently remain hidden under the leaf litter or within decomposing wood, where they feed on both the decomposing plant matter and the bacteria, fungi and other organisms that inhabit it. Plasmodia are mostly similar to other amoebae, except for their size. They move about by extending pseudopodia, though in larger specimens these pseudopodia may fuse together to form wave-like structures. Like other amoebae, they feed by phagocytosis, engulfing their prey in vacuoles, and transporting it inside the cell for digestion. They are only active under humid conditions, as they depend on humidity to survive; however, this does not mean they are exclusive to damp environments. Under sub-optimal circumstances, a plasmodium may undergo a process known as sclerotization. The removal of water from the system results in the formation of a dry, hardened mass, which serves as a storage for cellular components and nutrients while maintaining a state of complete dormancy and inertia. This sclerotium has the capacity to survive for extended periods, up to several months or even years. When conditions become more favorable, the sclerotium absorbs water and revives into a healthy plasmodium. Consequently, Physarum polycephalum is also found in more arid regions of the world, with its activity constrained to the period following precipitation [22] (Fig. 3).

The plasmodial stage exhibits complex physiological and behavioral features, easily observable due to their macroscopic size. The plasmodial stage is long-lived and can be maintained in a laboratory setting for months, and subsequently stored for months as sclerotia. The plasmodium's amoeboid nature enables it to divide into multiple independent parts, or it can be divided manually without affecting the health of either part. The combination of these characteristics allows for rapid propagation of plasmodia. Maintenance is also relatively straightforward, as oat flakes have been established as a main food source. This has led to numerous researchers adopting Physarum as a model organism, providing us with a wealth of information about this organism, from its complete genome [24] to its biology and behavior.

What methods can be employed to analyze the behavior of a plasmodium?

By observing the organism's reactions to stimuli, it can be determined whether the organism is able to detect them. In the plasmodium, two main movement parameters are considered: The rhythmic movement of the cytoplasm within the cell and the movement of the plasmodium itself are of particular interest. The former is a slow-moving, oscillating transport mechanism of cytoplasm throughout the cell, reversing the direction of the flow every 60–120 s. This phenomenon, known as shuttle transport, is vital for the distribution of nutrients and chemicals throughout the cell. It reacts to stimuli, becoming faster or slower depending on the environmental conditions.

The second way to analyze the behavior of the slime mold is by observing its movement across its substrate. The movement of an organism in response to a stimulus is called taxis. Positive taxis is a guided movement towards the stimulus and negative taxis is a movement away from it. Submitting a plasmodium to a controlled environment with a gradient of a specific stimulus allows to determine if it perceives the stimulus. If it moves repeatedly and systematically towards or away from the

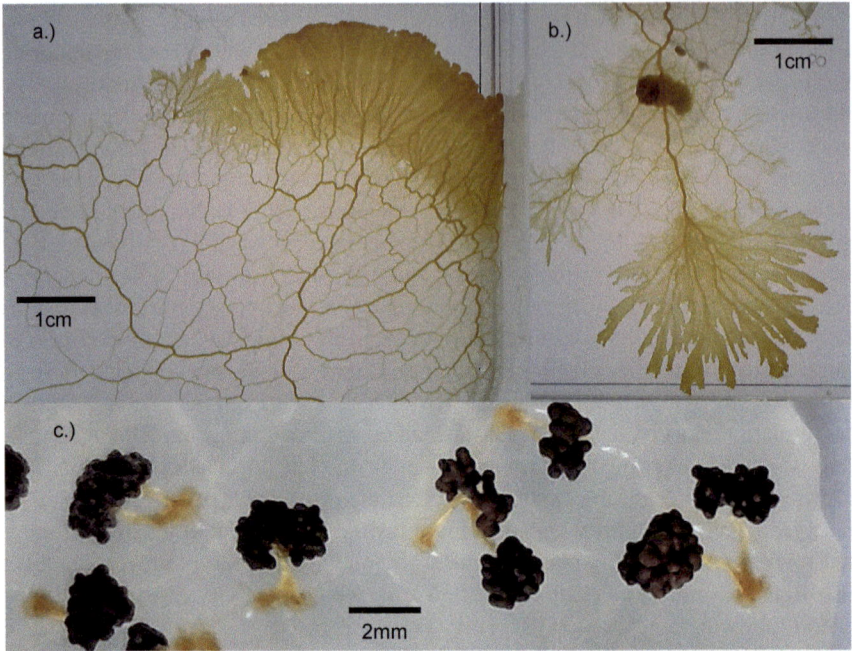

Fig. 3 a Plasmodium of *Physarum polycephalum*, moving as a wave of combined pseudopodia and leaving behind a network of tubuli to connect with other parts of the plasmodium. **b** Plasmodial movement with a spread of pseudopodia not connected at the tips. **c** Mature sporangia of *P. polycephalum* filled with reproductive spores, the end of the plasmodial lifecycle

stimulus, it is reasonable to assume that it perceives that stimulus, and gauging the direction allows for speculation if the organism assigns positive or negative valence to the stimulus. In contrast, if the organism does not move systematically but in random patterns, it is possible to assume that it does either not perceive the stimulus or does not assign any value to it. The following sections will present an overview of the known stimuli that elicit a response from *Physarum*.

Chemical sensing—The world of microorganisms is mainly a chemical world, and this is true for *Physarum* as well. It uses chemical senses both to find nourishment and avoid harmful stimuli. Physarum responds with positive taxis to chemicals supporting the growth of the plasmodium, such as some carbohydrates like glucose, maltose or peptone. Others, however, like sucrose, fructose or ribose were ignored, implying that they are either not relevant and/or cannot be sensed by the slime mold [23]. The same applies to amino acids. Some evoke positive taxis, others negative, and some do not influence movement at all [24]. Inorganic salts like chloride salts of potassium or sodium and some others work as repellents [25]. These reactions are not only dependent on the presence, but also on the chemical concentration of the chemicals. High concentrations of carbohydrates are harmful to plasmodia, and it has been shown that they are actively avoided. When foraging, *Physarum* evaluates nutrient ratios, choosing not the highest concentration but the best ratio for its sustenance [26]. In Boisseau et al. acidic media have been used as repellents as well, indicating that *Physarum* is capable of sensing acidity in a medium [27]. In recent years it was also found that slime mold reacts positively to plants with sedative properties, strongest of all to Valerian root. While the reason is unclear, many animals also show positive chemotaxis towards Valerian [28].

A special case of repellent is the extracellular slime consisting mainly of a glycoprotein that is constantly exuded by moving plasmodia. This slime is avoided by the plasmodium on contact and therefore serves *Physarum* as an external spatial memory, marking the cell's path and guiding foraging to new, unexplored areas. This reaction however is also not absolute: When a plasmodium is completely surrounded by extracellular slime, it stops avoiding it, so it cannot be trapped by its own slime [29].

Light sensing—*Physarum* has no complex optical sensors to perceive the world around it in shapes, but it is capable of sensing light and differentiating between different wavelengths. As a single cell, a plasmodium is vulnerable to UV-light. Prolonged exposure to UV light initially causes a stress reaction, then irregular streaming and disruption of the plasmodium until death. In many experiments it could be shown that a plasmodium moves away from the UV light whenever possible [30]. This negative phototaxis could be shown for many wavelengths from UVC to UVA and blue light. Negative phototaxis could also be found on the other end of the light spectrum. *Physarum* exhibits an avoidance reaction to far red light close to infrared. At first this is at odds with the fact that the spore-filled fruiting bodies are always found in well-lit, raised places. To explain this it has been speculated that a well-fed, healthy plasmodium ready to produce spores does not change its negative phototaxis to positive. Instead, it inhibits its photoavoidance, while relying on other systems like gravitaxis to find higher places for sporulation [31]. Furthermore, it could be shown that prolonged irradiation with blue light and UV-light, as well as far-red light, could induce a reaction called photofragmentation. After prolonged irradiation, the plasmodium breaks up into equally sized spherules, forming small individual plasmodia. It is not known however if this process is a stress-reaction, trying to break up into smaller individuals to have more chances of escape, or instead part of the sporulation process. The photofragmentation-process due to far-red light could be inhibited by presenting red light with shorter wavelengths. This hints at similar processes to those found in plants, suggesting that light-sensitivity is mediated by phytochrome-like proteins [32]. The differentiation between red and far-red light is also present in the behavior: *Physarum* moves towards red light, as long as it is not too bright, being the only positive phototaxis the plasmodium expresses [33]. An experiment measuring the oscillations of *Physarum* under various illuminations showed that it not only reacted to the presence of red, blue, green, and white light, but also had distinct oscillation patterns that could only be seen under red light and blue light respectively, showing that it is capable of sensing visible light and especially differentiating between blue and red light [34]. Calcium-inhibitors have been shown to inhibit the phototactic reactions of *Physarum*, indicating that calcium ions and in extension electrical signals play an important role in light sensing [35]. Quick pulses

of illumination alternating with darkness are capable of reversing the negative phototaxis, showing a possible link between the internal oscillations of the slime mold and
its behavior [36].

Temperature sensing—*Physarum polycephalum* is found in a wide geographic
range, and therefore seems to withstand a broad range of temperatures. However,
it has been shown that the internal plasma transport is dependent on the outside
temperature, indicating that some temperatures are more favored than others [37]. It
is also known that cold temperatures close to freezing lead to sclerotization and
therefore seem to be unfavorable to *Physarum* [38]. When given a temperature
gradient, *Physarum* shows both positive and negative thermotaxis—moving away
from both extremes. While authors differ on what the optimal temperature is, with
some claiming 28–30 °C [39] or rather 24–26 °C [40], both studies show that the
plasmodium seeks out an optimum. A single plasmodium is simultaneously capable
of both positive and negative thermotaxis, withdrawing from the colder and warmer
sides to arrange itself in the optimal zone [40]. An experiment with oscillating temperatures as stimuli showed that timing the oscillation of the temperature in relation to
the shuttle-transport may reverse the thermotaxis. This suggests that the internal
protoplasm-oscillation may play an important role in the thermotaxis [41].

Gravity sensing—Few studies exist on the effect of gravity on *Physarum*, but the
main consensus is that it prefers to wander towards gravity. This has been shown by
BLOCK ET AL., where plasmodia were put on vertical plates. This could further be
shown in a glass cylinder that was rotated in regular intervals, in which the plasmodia
immediately changed movement direction after each rotation [42]. However, it must
be mentioned that gravitaxis may change during the lifecycle of a plasmodium,
as mature plasmodia might actively seek out elevated places in the light for better
spore dispersal. So far it is not known if this process is guided by gravitaxis. In
an experiment where slime mold was brought into zero gravity conditions during a
space-shuttle mission, it was observed that the oscillation of the cytoplasm slowly
increased after gravity reached zero, showing a response to missing gravity [43].

Electrical current sensing—Galvanotaxis, the orientation and movement along
electrical fields, can be found in many unicellular organisms. Therefore, it is not
surprising that it has been shown in *Physarum* as well. In a weak electric field,
Physarum generally tends to move towards the cathode. Instead of a guided movement, the electric field seems to inhibit movement towards the anode, so rather than
exhibiting a specific taxis the electric field seems to interfere with the slime molds
coordination of movement [44]. A further study has shown that the electric field
mainly influences the distribution of potassium throughout the plasmodium, causing
an overall loss of potassium in the cell [45]. While an electric field can be used to
guide the movement of *Physarum* [46], galvanotaxis does not seem like an intentional
behavior.

Magnetic field sensing—A recent discovery is that Physarum also exhibits
magnetotaxis, indicating a response to magnetic fields. Given its demonstrated
capacity for galvanotaxis, researchers postulated that it might also react to magnetic
fields. When maintained in a magnetic field, the plasmodium consistently demonstrates a tendency to move in alignment with the magnetic field lines. The direction

of movement appears to be influenced by multiple factors, including the size of the plasmodium and the strength of the magnetic field. Magnetotaxis is also found in other organisms like bacteria and migratory birds, but both groups of organisms use magnetosomes, magnetite crystals responding to geomagnetism. In *Physarum*, these crystals have not yet been reported. Furthermore, unlike in *Physarum*, magnetotactic bacteria are not able to change their magnetotaxis like *Physarum* does in relation to its size, implying that a more complex mechanism is at work [47]. Magnetotaxis might also explain the findings of another experiment in which plasmodia were put into T-labyrinths showed an unexpected preference for the right arm. Lateralization as we know it from bilaterally symmetrical animals at first seems baffling, since the slime mold is amorphous without permanent structures. But in the discussion the authors also noted that the right arm of the labyrinths was facing north, so a different explanation might be that the slime molds instead followed the magnetic field of the earth itself [48].

Considering a plasmodium is just a single cell, it can gather a plethora of information from its surroundings. To coordinate unified reactions throughout the plasmodium, it becomes necessary to process and integrate the information from the various senses, especially considering that, due to the size of a plasmodium, different parts of the cell might be in contact with different stimuli. The Skin-Brain-Hypothesis proposes that early multicellular animals evolved a nervous system because they faced a similar challenge, with both information processing and movement coordination equally important for survival of larger cell colonies [51]. While the mechanisms might be different, slime mold shows many behaviors not unlike those of animals with nervous systems. The following section will examine the evidence for complex behavior and assess its potential to be classified as intelligent.

One of the most widely known findings is that *Physarum* is capable of reliably finding the shortest path through a maze. In an experiment, the plasmodium was allowed to spread out through a maze, while a food-source was placed on each of the two ends of the maze. After exploring, the plasmodium retracted until only the most efficient connection between the food sources remained—the shortest path [49]. In a series of studies, it could be shown that in various contexts the plasmodia were able to find the shortest paths between multiple food sources, showing complex structures obeying mathematical rules [50]. This in turn inspired new mathematical models for pathfinding in mazes and roadmaps [51]. In complex situations with multiple food sources *Physarum* does not always use the shortest paths, but instead seems to optimize between robustness and shortest-path-efficiency, including fault tolerance for spontaneous disconnection [52]. This alone inspired many mathematical models, for example new solutions to the traveling salesman-problem.

Physarum is capable of decision-making and evaluating different food sources. In experiments with two choices [53] and three choices, the plasmodia took more time to decide for one of the food sources if the overall nutritional value was low, showing a more careful behavior in a low resource situation. Similar behavior is observed in multiple species of higher animals, including humans [54]. Furthermore, individual plasmodia from the same species show different foraging types, with some making consistently faster decisions then others [53]. *Physarum* has a preferred ratio of carbohydrates and protein in their food. As mentioned above, when presented with multiple choices, all featuring different ratios but without presenting an ideal one, *Physarum* solves the problem by spreading out: They choose a few of the different ratios available, so that the sum of the choices made is also the optimal ratio, demonstrating the integration of various sources of information. [26].

Slime molds are also capable of integrating multiple types of stimuli. When given a choice between low-quality food in the dark and higher quality food in a well-lit area that the plasmodium would normally avoid, it chooses the safer option in the dark—unless the difference in caloric value between the food sources is high enough, causing a shift in preferences regardless of the light. This shows that the slime mold is capable of comparison between food sources and of changing its foraging strategy accordingly. In another version of the same experiment, the nutritional values of the food sources were balanced, so that the *Physarum* had no strong preference for one or the other. Then, an even lower valued food source in the dark was introduced as a third, least desirable option. In previously starved plasmodia it had no effect, but in well-fed plasmodia a significant shift in their preferences away from the food source in the light could be found. While no plasmodium took the third option, it still influenced the overall outcome. This irrational behavior, commonly seen in various animal and human trials, shows that despite its efficient seeming nature *Physarum* does evaluate its choices on comparative values [55]. The decision making in *Physarum* has been further tested using the two-armed bandit problem from gambling. In the experiment, the plasmodium has to choose between two paths with varying food contents by testing both paths and using the information gained to ultimately make a decision. Various scenarios were tested, using both predictable and non-predictable configurations. *Physarum* was able to find good solutions even in unpredictable environments, performing at the level of vertebrate animals [56].

These findings demonstrate that *Physarum* is capable of interacting in various ways with its world, often using complex behavior to solve complex problems. But is it also capable of adapting to change, by forming memory and learning?

One of the first findings concerning memory was the anticipation of recurring events. In the experiment, the plasmodia were subjected to aversive stimuli at regular intervals, causing them to slow down their movement during the application of the stimulus. After only three applications, the slime mold would respond after another interval had passed—even in the absence of the stimulus. While the reaction faded in the continued absence of a stimulus, it could also be renewed with a single new application, leading to another anticipatory reaction after the interval had elapsed [57].

One of the simplest forms of learning is habituation: The decrease of a response to a stimulus after repeated or prolonged exposure. In an experimental setup showing habituation, plasmodia had to cross a path of agar enriched with the repellent quinine to reach a food source. At first, the plasmodia would stop at the quinine for a while, and when they extended along the bridge, the connection formed was narrow to avoid exposure to the repellent. Over the course of 9 repetitions, the time it took to cross the quinine-agar became shorter, the surface of the plasmodium on the quinine agar greater. At the end of the experiment, the plasmodia behaved no different than the control slime molds that had to cross normal agar without quinine. Moreover, if the repellent was withheld for two days, the aversive reaction to it spontaneously returned. Additionally, if a slime mold habituated to quinine was confronted with another repellent (caffeine), an aversive reaction occurred, showing specificity in habituation. This shows that the slime mold is capable of habituating to aversive stimuli while at the same time differentiating between them [27]. In a follow-up this experiment was repeated with sodium chloride as a repellent, and it could even be shown that the memory of habituation can be transferred to other plasmodia. When a habituated plasmodium is brought into contact with another non-habituated plasmodium, they typically form connections and exchange cytoplasm. After severing the connection, the non-habituated plasmodium shows the habituation-response as well [58].

So far, only one experiment on associative learning has been reported. In an environment with a temperature gradient without other stimuli the plasmodia would move towards the warmer side due to their positive thermotaxis. Food, however, was only placed on the colder side. During one week of experiments, the plasmodia acquired the tendency to move towards the colder side, reverting to the original thermotaxis. Even the authors concede that this might either show associative learning or just habituation to the cold, giving not enough evidence to completely assume associative learning has happened [59]. It is, however, likely that *Physarum* might show true associative learning in further studies, as it has already been shown in distantly related amoeba species [20].

While only few experimental observations of slime mold behavior exist compared to animals, they show that it is an organism capable of complex information processing, decision-making and behavioral flexibility, despite being one cell instead of a whole nervous system. To understand the cellular mechanisms leading to this basal cognition might be a promising path to understanding both cognition in more complex organisms, as well as emulating it to create artificial intelligence- not on the level of human cognition, but as a more energy efficient autonomous problem-solving machine, as many tasks for robots do not require human level intelligence. Therefore, *Physarum* is already a promising behavioral model for artificial intelligence. The behavior, however, is not the only interesting part—investigations into the electrical properties revealed that *Physarum* itself might be an interesting biomaterial and substrate.

4   A Brief History of the Electrical Experiments on the Slime Molds

Electrical experiments are performed to understand basic electrical principles such as current (the flow of electric charge), voltage (the potential difference that drives the current) and resistance (opposition to the current flow). Electrical experiments are also carried out to study how different materials conduct electricity. This is important in identifying suitable materials for various applications such as conductors as electrical wires and insulators as protection for the wires and semiconductors for computer chips. Many electrical experiments are performed to innovate and develop new technologies, such as more efficient solar cells and batteries and even new computing architectures.

The plasmodium state of *Physarum polycephalum* is a unicellular organism with multiple nuclei. Such unicellular organisms have always been attractive in investigating their biological processes and the driving force of these processes. Electrical experiments on the slime mold *Physarum polycephalum* were first published in 1950 by Kamiya et al. [60]. They observed that the protoplasmic flow, which is the back-and-forth movement of the slime mold's protoplasm, generates a potential difference within the organism. The authors introduced a novel method to measure this electromotive force, which is the electric potential of the slime mold. This method involves two plasmodia of *Physarum polycephalum* connected by a protoplasm strand exhibiting lively movement. They found that the electromotive force changes rhythmically, corresponding to the cyclic back-and-forth streaming, with amplitude variations within 10 mV under constant conditions. Their findings include:

- The rhythmic potential changes occur regardless of whether the flow is free or obstructed by local blocking or balancing pressure.
- Local differences in bio-electric potential are not directly caused by the protoplasmic flow or streaming potential and do not explain the flow as electroendosmosis, which is the movement of water or solution in response to an electric field.
- Besides slow cyclic variations, there are quick, transient potential variations occurring spontaneously or induced by weak mechanical shocks (Fig. 4).

The effect of mechanical shocks on *Physarum polycephalum* was explored in a subsequent study by Tasaki et al. [61]. They observed that the resistance of the plasmodial strand between electrodes was generally about 5.0 Megohm, varying rhythmically up to 20 percent. Furthermore, they identified key differences between the responses of plasmodia and nerve cells.

- The response amplitude in the slime mold increases with the strength of the stimulus within certain limits, whereas in nerve cells, the response follows an all-or-none pattern.
- The slime mold shows summation when two stimuli are applied shortly apart, unlike nerve cells which have an absolute refractory period until activity ceases. The duration of the negative variation in the plasmodium is about 2000 times longer than the chronaxie,[1] whereas in nerve cells, it is only several times longer.

---

[1] Chronaxie is a standardized method used in the field of biology to characterize the electrical excitability of muscle cells and neurons. This allows for comparison between experiments and types of cells. Excitation occurs at a faster rate in relation to the strength of the electrical stimulus. For the characterization of tissue, the minimum stimulus (rheobase) and the time required for a reaction to occur in response to it are recorded and compared to the so-called chronaxie, which

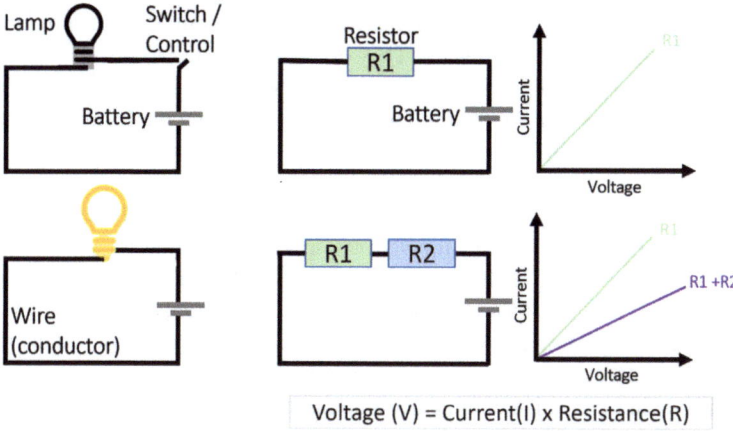

Fig. 4 Some basic parameters helpful in understanding electrical properties. In the two left images, a battery is connected to a lamp via a conducting wire. A battery supplies a voltage difference which drives a current through the circuit. An electric current is just the flow of charged particles through the wire. When the circuit is open, as shown in the top right image at the switch, the current flow is interrupted. Therefore, the lamp is switched OFF. However, in the bottom left image, the circuit is closed therefore the lamp is ON denoted by yellow. This shows that the chargers leaving the battery should go through the closed circuit and come back to the battery for the lamp to be lit. In the two schematics to the right, the lamp is replaced by a resistor which is an electrical component that implements electrical resistance. According to Ohm's law ($V = IR$) the resistance can be calculated by the ratio of the applied voltage and the current output. The higher the resistance, lower the current that passes through a circuit for a given voltage. When the two resistors are added in series the total resistance is the summation of the two. This is shown in the Current–Voltage plot (I-V plot) on the bottom right image

The study concluded that a stimulating current flowing outward through the protoplasm surface is more effective than one flowing inward. Furthermore, the underlying process of electrical responses is consistent regardless of the stimulation method, and protoplasmic flow continues unimpeded when electrical responses are induced.

Following these findings, Burr posed a key question: '*Does the change in direction of protoplasmic movement result in changed polarity, or does changing polarity cause the reversal flow?*' [62]. At the time, *Physarum polycephalum* was studied to understand the vertebrate brain's behavior. Burr's experiments confirmed that electrical changes affect the growth patterns of the slime mold, similar to findings by Kamiya and Abe. However, it remained unclear whether the streaming produces the potential or vice versa. Burr observed protoplasmic movement without characteristic potential changes and polarity changes without movement.

Burr also confirmed *Physarum*'s response to mechanical stimuli, similar to observations by Tasaki et al. He noted that *Physarum* not only responds to stimuli but also transmits them through its protoplasm, akin to nerve fibers. However, later Burr et al.

---

represents the time required for cells to react to a stimulus with double the strength of the rheobase. In this case, it is used to highlight the difference in excitability between plasmodia and neurons.

dismissed the hypothesis that the protoplasmic thread is similar to a nerve fiber due to *Physarum*'s inconsistent responses. They noted that while *Physarum* can transmit stimuli, it differs from other tissues. They referenced Tuac et al. where no action potential was observed in *Physarum*, possibly due to membrane formation over the electrode. Despite this, they concluded that a primitive life form like *Physarum* yields an action potential when stimulated electrically or mechanically [63].

In 2007, Adamatzky introduced the concept of "*Physarum* machines" to compute minimum spanning trees (MST) for various applications such as network design [64]. In network design, finding an MST helps to determine the most efficient way to connect all locations with the minimum total connection cost, avoiding redundant paths. By using *Physarum polycephalum*, which naturally forms efficient networks to explore its environment, Adamatzky et al. aimed to demonstrate how biological systems can solve complex computational problems like finding the MST in a graph. Although *Physarum* did not compute an exact MST, it demonstrated stable and adaptive computation. This research established *Physarum* as a model for studying behavioral intelligence, leading to several mathematical models explaining its behavior.

The exploration of *Physarum*'s electrical characteristics continued in 2014 when Gale, Adamatzky and De Lacy Costello questioned if *Physarum* could function as a 'memristor' [65].

5 *Physarum* as Memristor

A memristor, an abbreviation of "memory resistor," is a specific type of electrical component that regulates the flow of electrical current in a circuit. It possesses a distinctive property, namely the capacity to retain the amount of charge that has previously flowed through it. This memory capability results in a change in the resistance of a memristor based on the history of the current that has passed through it. This renders it a valuable component for incorporation into memory storage, signal processing, and neural networks. Memristors were theorized by Chua in 1971 [66] and were later physically realized by researchers at HP labs in 2008 [67]. Memristors are considered the fourth fundamental passive circuit element alongside resistors, capacitors and inductors.[2]

Gale et al. showed that *Physarum* exhibits memristive properties and hysteresis. Hysteresis occurs when the response of a material to an external force depends on the system's history. Furthermore, the memristive response is attributed to the living protoplasm since it disappeared when *Physarum* moved and abandoned the measurement area. The study suggested that *Physarum*'s memristive behavior is similar to neurological components, indicating a possible evolutionary use of memristance in

---

[2] A passive circuit element is a component that does not generate energy but can store or dissipate it. These components do not have the capability to amplify or create electrical signals, instead manage the flow of electrical energy within a circuit.

learning systems. This study further suggests that future electronic designs could incorporate slime mold networks for high-density, low power computing elements. The slime mold memristors in this study had a lifespan of 3–5 days. However, the authors suggest that enhancements with functional materials could extend this and provide varied electronic properties. Furthermore, growing electronic circuits from *Physarum* could reduce water use and toxic waste in the electronics industry [65].

Further advancing the field, Adamatzki's 2015 publication 'A would be nervous system made from slime mold', proposed the concept of a *Physarum* neuron. This study explored creating a living neural system with *Physarum*, comparing their protoplasmic tubes to neurons axons and dendrites. The slime mold's ability to propagate calcium waves and exhibit memristive properties suggested it could function as a biological neural network. Proposed potential applications include biohybrid sensors and hybrid wetware-hardware devices for detecting non-lethal substances for living cells. The slime mold's self-growing, low-maintenance nature makes it a promising candidate for integration into bio-inspired robots and fluidic controllers [68].

In an experiment by Braund and Miranda, *Physarum polycephalum* was used as a biological computing substrate in the field of unconventional computation [69]. Inspired by Miranda et al. [70], the extracellular membrane potential of *Physarum* was recorded to create a bionic musical instrument. In this setup, *Physarum* serves as a cost effective and resource-efficient biological computing substrate. The bionic instrument generates sounds from the electrical activity of the slime mold, including oscillations in the protoplasmic tubes and the electricity produced by the plasmodium as it moves over a substrate.

6 Growing Bio-Memristors from Slime Mold

An ideal memristor, hypothesized by Chua, is a passive circuit element that does not store energy, with its I-V curve passing through zero [66]. However, the slime mold showed an imperfect memristive curve (deviation from ideality) with an offset in the hysteresis curve, which the authors related to the oscillating cytoplasmic movement in the protoplasmic tubes. This movement coupled with ion flow, generates a current. This phenomenon echoes the observations of Kamiya et al. in the 1950s, who noted changes in resistance in the protoplasmic tubes, although they did not correlate this to memristive behavior [60]. Miranda et al. modeled *Physarum polycephalum* memristors as a 'two port black box that contains a battery and a memristor' [71] (Fig. 5).

Miranda et al. further elaborated on a method for growing bio-memristors using *Physarum polycephalum*. They used 3D printing to create receptacles for the slime mold. High-impact polystyrene was used to print the chambers, lids and base, while conductive polylactic acid (PLA) was used for the electrodes. The receptacles presented in this work showed promising results for implementing *Physarum polycephalum* memristors, demonstrating the potential of this approach for practical bio-computing applications.

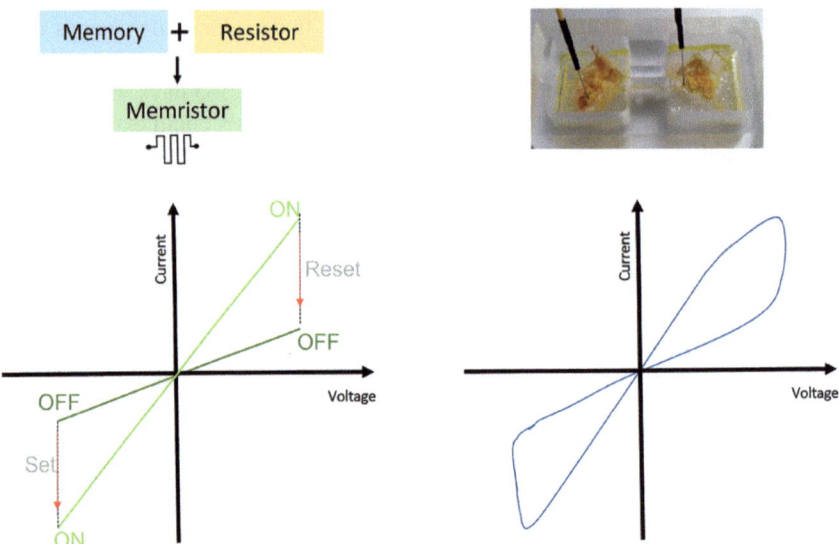

Fig. 5 A memristor is a potential electrical component that has the capacity to mimic synaptic functions, thereby enabling brain-inspired neuromorphic computing. An array containing memristors has the potential to provide synaptic characteristics, which, when combined with a neural network, could be used for data processing in a neuromorphic computer. The term "memristor" is derived from the combination of "memory" and "resistor." In essence, it signifies the ability to retain the amount of current flowing through it over time. An ideal I-V curve for a memristor would be represented by the plot on the left. The low resistance state is indicated by "on," while the high resistance state is indicated by "off." The transition from the high resistance state (off) to the low resistance state (on) is referred to as a "set," while the transition from the low resistance state (on) to the high resistance state(off) is referred to as a "reset." The plot on the right illustrates a schematic of how such an ideal I-V curve for the slime mold would appear

At a future time point, with an increasing understanding of the brain and proper hierarchy and structure for neuromorphic computing architecture is established, the electrical properties of the slime mold such as the memristance, oscillations, response to electrical and mechanical stimuli and embedded potential battery characteristics could be harnessed from *Physarum polycephalum*. Furthermore, due to these properties slime mold itself could be a biomaterial for such an architecture.

7 Slime Mold Computing—Future Perspectives

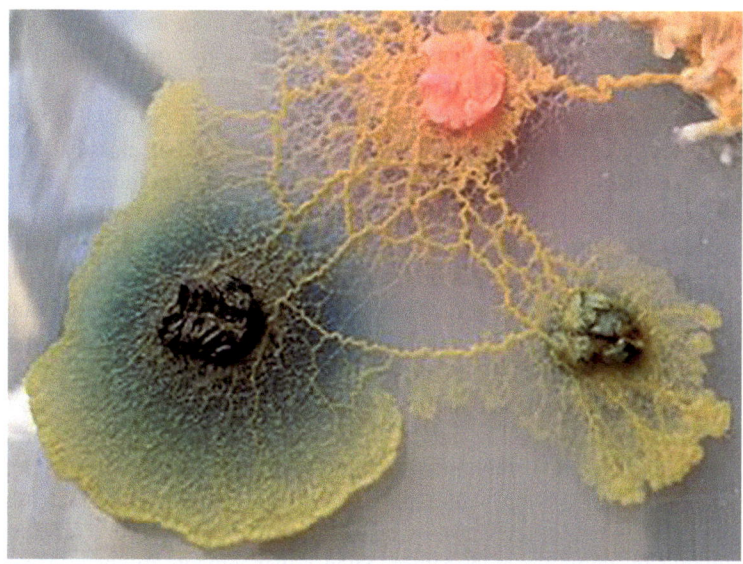

The study of slime mold has witnessed a gradual yet consistent advancement over time, uncovering a multitude of fascinating characteristics and frequently giving rise to new inquiries. This ongoing discovery process suggests that slime molds, particularly *Physarum polycephalum*, might hold the key to solving various complex problems that span over multiple fields, such as:

- **Computing and network design**: *Physarum*'s ability to form efficient networks can inspire algorithms for optimizing network paths and resource distribution.
- **Material science and engineering**: *Physarum*'s growth patterns and adaptive behaviors can inspire the development of new materials and structures that self-organize and self-repair.
- **Environmental monitoring and management**: *Physarum*'s sensitivity to environmental changes can be investigated to create biosensors for detecting pollutants or monitoring ecosystem health.
- **Artificial intelligence and robotics**: *Physarum*'s simple yet effective problem-solving capabilities can contribute to the development of bio-inspired (*Physarum*-inspired) algorithms and robots that mimic biological processes.

These unicellular organisms demonstrate remarkable complexity and sophisticated behaviors, challenging traditional notions of intelligence and supporting the biogenic approach to cognition proving that intelligence does not necessarily require a brain or even a nervous system.

Physarum polycephalum, in particular, is a prime example of what a single cell can achieve. Its behavior exhibits a degree of complexity and adaptability that challenges multi-cellular organisms. However, its unicellular nature makes it easier to study compared to a network of cells. The slime mold's electric properties are linked to its behavior in a complex manner. These properties show both similarities and differences to neuronal activity, suggesting that while *Physarum*'s behavior can be partially understood through the lens of neuroscience, it also offers unique insights that differ from traditional neural models.

Neural networks have often inspired the development of Artificial Intelligence (AI). However, simulating artificial neural networks (ANNs) on classical computer architectures is highly energy insensitive and inefficient. This inefficiency has driven a shift towards new computing hardware architectures based on memristors which are electronic components that mimic the synaptic behavior of neurons. While this shift aims to emulate the biological structure of nervous systems, *Physarum polycephalum* already naturally exhibits networked structures with oscillations and memristive properties. Moreover, it achieves this with remarkable energy efficiency, operating solely on simple nutrients like oat flakes.

Given its natural efficiency and computational capabilities, *Physarum polycephalum* presents a compelling model for investigation and inspiration. It aligns closely with the principles of memristive computing and demonstrates the potential for complex problem solving. Although a single slime mold cannot match human intelligence, understanding and harnessing its capabilities could lead to revolutionary advancements in computer architecture. Such advancements might transcend the limitations of traditional computing and redefine our concepts of intelligence.

In summary, slime molds, particularly *Physarum polycephalum*, challenge our understanding of intelligence and computation. Their unique properties and behaviors offer valuable insights that could inform the development of more efficient and effective computing systems. As research continues, the potential applications of a slime mold-based computing architecture could transform technology and expand our understanding of cognitive processes.

Acknowledgements We would like to acknowledge and thank very specially Prof. Rainer Adelung for creating the projects and bringing forth a collaboration between materials science and psychology through the slime mold. The optimism and the creativity of Prof. Adelung has sparked a genuine curiosity and ignited into further studies on the slime mold. We would also thank the Deutsche Forschungsgemeinschaft (DFG, German research foundation) for the funding under the project SFB 1461, Project-ID 434434223.

References

1. Miller GA (2003) The cognitive revolution: a historical perspective. Trends Cogn Sci 7:141–144. https://doi.org/10.1016/S1364-6613(03)00029-9
2. Bechtel W, Abrahamsen A, Graham G (2017) The Life of Cognitive Science. In: A Companion to Cognitive Science. John Wiley & Sons, Ltd, pp 1–104
3. Lyon P (2006) The biogenic approach to cognition. Cogn Process 7:11–29. https://doi.org/10.1007/s10339-005-0016-8

4. Lyon P (2015) The cognitive cell: bacterial behavior reconsidered. Front Microbiol 6:. https://doi.org/10.3389/fmicb.2015.00264
5. Rilling M (1996) The mystery of the vanished citations: James McConnell's forgotten 1960s quest for planarian learning, a biochemical engram, and celebrity. Am Psychol 51:589–598. https://doi.org/10.1037/0003-066X.51.6.589
6. Romero R, Baguñà J (1991) Quantitative cellular analysis of growth and reproduction in freshwater planarians (Turbellaria; Tricladida). I. A cellular description of the intact organism. Invertebr Reprod Dev 19:157–165. https://doi.org/10.1080/07924259.1991.9672170
7. Block RA, Mcconnell JV (1967) Classically Conditioned Discrimination in the Planarian, Dugesia dorotocephala. Nature 215:1465–1466. https://doi.org/10.1038/2151465a0
8. Perry CJ, Barron AB, Cheng K (2013) Invertebrate learning and cognition: relating phenomena to neural substrate. WIREs Cognit Sci 4:561–582. https://doi.org/10.1002/wcs.1248
9. Bielecki J, Dam Nielsen SK, Nachman G, Garm A (2023) Associative learning in the box jellyfish *Tripedalia cystophora*. Curr Biol 33:4150-4159.e5. https://doi.org/10.1016/j.cub.2023.08.056
10. Botton-Amiot G, Martinez P, Sprecher SG (2023) Associative learning in the cnidarian Nematostella vectensis. Proc Natl Acad Sci 120:e2220685120. https://doi.org/10.1073/pnas.2220685120
11. Wilkinson A, Huber L (2012) Cold-Blooded Cognition: Reptilian Cognitive Abilities. In: Shackelford TK, Vonk J (eds) The Oxford Handbook of Comparative Evolutionary Psychology. Oxford University Press, p 0
12. Lyon P, Keijzer F, Arendt D, Levin M (2021) Reframing cognition: getting down to biological basics. Philosophical Transactions of the Royal Society B: Biological Sciences 376:20190750. https://doi.org/10.1098/rstb.2019.0750
13. LeDoux JE (2021) As soon as there was life, there was danger: the deep history of survival behaviours and the shallower history of consciousness. Philosophical Transactions of the Royal Society B: Biological Sciences 377:20210292. https://doi.org/10.1098/rstb.2021.0292
14. Oda K, Abe T (1972) Action potential and rapid movement in the main pulvinus ofMimosa pudica. Bot Mag Tokyo 85:135–145. https://doi.org/10.1007/BF02489510
15. Brette R (2021) Integrative Neuroscience of Paramecium, a "Swimming Neuron." eNeuro 8:. https://doi.org/10.1523/ENEURO.0018-21.2021
16. Prindle A, Liu J, Asally M et al (2015) Ion channels enable electrical communication in bacterial communities. Nature 527:59–63. https://doi.org/10.1038/nature15709
17. Binet A (1889) The Psychic life of micro-organisms. Open court Publishing Company
18. Jennings HS (1902) Studies on reactions to stimuli in unicellular organisms. ix.—on the behavior of fixed infusoria (stentor and vorticella), with special reference to the modifiability of protozoan reactions. American Journal of Physiology-Legacy Content 8:23–60. https://doi.org/10.1152/ajplegacy.1902.8.1.23
19. De la Fuente IM, Bringas C, Malaina I et al (2019) Evidence of conditioned behavior in amoebae. Nat Commun 10:3690. https://doi.org/10.1038/s41467-019-11677-w
20. Carrasco-Pujante J, Bringas C, Malaina I, et al (2021) Associative Conditioning Is a Robust Systemic Behavior in Unicellular Organisms: An Interspecies Comparison. Front Microbiol 12. https://doi.org/10.3389/fmicb.2021.707086
21. Mineta K, Nakazawa M, Cebrià F et al (2003) Origin and evolutionary process of the CNS elucidated by comparative genomics analysis of planarian ESTs. Proc Natl Acad Sci 100:7666–7671. https://doi.org/10.1073/pnas.1332513100
22. Rojas C, Stephenson SL (2021) Myxomycetes - Biology, Systematics, Biogeography and Ecology, 2nd edn. Academic Press
23. Carlile MJ (1970) Nutrition and Chemotaxis in the Myxomycete Physarum polycephalum: the Effect of Carbohydrates on the Plasmodium. Microbiology 63:221–226. https://doi.org/10.1099/00221287-63-2-221
24. McClory A, Coote JG (1985) The chemotactic response of the myxomycete Physarum polycephalum to amino acids, cyclic nucleotides and folic acid. FEMS Microbiol Lett 26:195–200. https://doi.org/10.1111/j.1574-6968.1985.tb01590.x

25. Adamatzky A (2010) Routing Physarum with repellents. Eur Phys J E 31:403–410. https://doi. org/10.1140/epje/i2010-10589-y
26. Dussutour A, Latty T, Beekman M, Simpson SJ (2010) Amoeboid organism solves complex nutritional challenges. Proc Natl Acad Sci 107:4607–4611. https://doi.org/10.1073/pnas.091 2198107
27. Boisseau RP, Vogel D, Dussutour A (2016) Habituation in non-neural organisms: evidence from slime moulds. Proceedings of the Royal Society B: Biological Sciences 283:20160446. https://doi.org/10.1098/rspb.2016.0446
28. Adamatzky A (2011) On attraction of slime mould Physarum polycephalum to plants with sedative properties. Nat Prec 1–1. https://doi.org/10.1038/npre.2011.5985.1
29. Reid CR, Latty T, Dussutour A, Beekman M (2012) Slime mold uses an externalized spatial "memory" to navigate in complex environments. Proc Natl Acad Sci 109:17490–17494. https:// doi.org/10.1073/pnas.1215037109
30. Shebaby W, Shraideh Z (1997) Effect of UV-irradiation on protoplasmic streaming, behavior and structural organization in the slime mold Physarum polycephalum. Cytobios 89:199–207
31. Nakagaki T, Umemura S, Kakiuchi Y, Ueda T (1996) Action Spectrum for Sporulation and Photoavoidance in the Plasmodium of Physarum polycephalum, as Modified Differentially by Temperature and Starvation. Photochem Photobiol 64:859–862. https://doi.org/10.1111/j. 1751-1097.1996.tb01847.x
32. Kakiuchi Y, Takahashi T, Murakami A, Ueda T (2001) Light Irradiation Induces Fragmen- tation of the Plasmodium, a Novel Photomorphogenesis in the True Slime Mold Physarum polycephalum: Action Spectra and Evidence for Involvement of the Phytochrome¶. Photochem Photobiol 73:324–329. https://doi.org/10.1562/0031-8655(2001)0730324LIIFOT2.0.CO2
33. Hato M, Ueda T, Kurihara K, Kobatake Y (1976) Phototaxis in True Slime Mold Physarum polycephalum. Cell Struct Funct 1:269–278. https://doi.org/10.1247/csf.1.269
34. Adamatzky A (2013) Towards slime mould colour sensor: Recognition of colours by Physarum polycephalum. Org Electron 14:3355–3361. https://doi.org/10.1016/j.orgel.2013.10.004
35. Häder D (1985) Role of Calcium in Phototaxis of Physarum polycephalum. Plant Cell Physiol 26:1411–1417. https://doi.org/10.1093/oxfordjournals.pcp.a077041
36. Nakagaki T, Yamada H, Ueda T (1999) Modulation of cellular rhythm and photoavoidance by oscillatory irradiation in the Physarum plasmodium. Biophys Chem 82:23–28. https://doi.org/ 10.1016/S0301-4622(99)00099-X
37. Halvorsrud R, Laane MM, Giaever I (1995) A novel electrical method to study plasmodial contractions in Physarum. Synchrony and temperature dependence. Biol Rhythm Res 26:316– 330. https://doi.org/10.1080/09291019509360345
38. Jump JA (1954) Studies on Sclerotization in Physarum polycephalum. Am J Bot 41:561–567. https://doi.org/10.2307/2438716
39. Tso W-W, Mansour TE (1975) Thermotaxis in a slime mold, Physarum polycephalum. Behav Biol 14:499–504. https://doi.org/10.1016/S0091-6773(75)90672-0
40. Wolf R, Niemuth J, Sauer H (1997) Thermotaxis and protoplasmic oscillations inPhysarum plasmodia analysed in a novel device generating stable linear temperature gradients. Proto- plasma 197:121–131. https://doi.org/10.1007/BF01279890
41. Matsumoto K, Ueda T, Kobatake Y (1988) Reversal of thermotaxis with oscillatory stimulation in the plasmodium of Physarum polycephalum. J Theor Biol 131:175–182. https://doi.org/10. 1016/S0022-5193(88)80235-2
42. Wolke A, Niemeyer F, Achenbach F (1987) Geotactic behavior of the acellular myxomycete Physarum polycephalum. Cell Biol Int Rep 11:525–528. https://doi.org/10.1016/0309-165 1(87)90014-2
43. Block I, Wolke A, Briegleb W et al (1992) Graviresponse of Physarum — Investigations in actual weightlessness. Cell Biol Int Rep 16:1097–1102. https://doi.org/10.1016/S0309-165 1(05)80035-9
44. Anderson JD (1951) GALVANOTAXIS OF SLIME MOLD. J Gen Physiol 35:1–16. https:// doi.org/10.1085/jgp.35.1.1

45. Anderson JD (1962) Potassium Loss during Galvanotaxis of Slime Mold. J Gen Physiol 45:567–574. https://doi.org/10.1085/jgp.45.3.567

46. Ma Q, Johansson A, Tero A et al (2013) Current-reinforced random walks for constructing transport networks. J R Soc Interface 10:20120864. https://doi.org/10.1098/rsif.2012.0864

47. Shirakawa T, Konagano R, Inoue K (2012) Novel taxis of the Physarum plasmodium and a taxis-based simulation of Physarum swarm. The 6th International Conference on Soft Computing and Intelligent Systems, and The 13th International Symposium on Advanced Intelligence Systems. IEEE, Kobe, Japan, pp 296–300

48. Dimonte A, Adamatzky A, Erokhin V, Levin M (2016) On chirality of slime mould. Biosystems 140:23–27. https://doi.org/10.1016/j.biosystems.2015.12.008

49. Nakagaki T, Yamada H, Tóth Á (2000) Maze-solving by an amoeboid organism. Nature 407:470–470. https://doi.org/10.1038/35035159

50. Nakagaki T, Kobayashi R, Nishiura Y, Ueda T (2004) Obtaining multiple separate food sources: behavioural intelligence in the Physarum plasmodium. Proc R Soc Lond B 271:2305–2310. https://doi.org/10.1098/rspb.2004.2856

51. Tero A, Kobayashi R, Nakagaki T (2006) *Physarum* solver: A biologically inspired method of road-network navigation. Physica A 363:115–119. https://doi.org/10.1016/j.physa.2006.01.053

52. Tero A, Nakagaki T, Toyabe K et al (2010) A method inspired by Physarum for solving the Steiner problem. Int J Unconv Comput 6:109–123

53. Dussutour A, Ma Q, Sumpter D (2019) Phenotypic variability predicts decision accuracy in unicellular organisms. Proceedings of the Royal Society B: Biological Sciences 286:20182825. https://doi.org/10.1098/rspb.2018.2825

54. Marshall JAR, Reina A, Hay C et al (2022) Magnitude-sensitive reaction times reveal non-linear time costs in multi-alternative decision-making. PLoS Comput Biol 18:e1010523. https://doi.org/10.1371/journal.pcbi.1010523

55. Latty T, Beekman M (2010) Irrational decision-making in an amoeboid organism: transitivity and context-dependent preferences. Proceedings of the Royal Society B: Biological Sciences 278:307–312. https://doi.org/10.1098/rspb.2010.1045

56. Reid CR, MacDonald H, Mann RP et al (2016) Decision-making without a brain: how an amoeboid organism solves the two-armed bandit. J R Soc Interface 13:20160030. https://doi.org/10.1098/rsif.2016.0030

57. Saigusa T, Tero A, Nakagaki T, Kuramoto Y (2008) Amoebae Anticipate Periodic Events. Phys Rev Lett 100:018101. https://doi.org/10.1103/PhysRevLett.100.018101

58. Vogel D, Dussutour A (2016) Direct transfer of learned behaviour via cell fusion in non-neural organisms. Proceedings of the Royal Society B: Biological Sciences 283:20162382. https://doi.org/10.1098/rspb.2016.2382

59. Shirakawa T, Gunji Y-P, Miyake Y (2011) An associative learning experiment using the plasmodium of *Physarum polycephalum*. Nano Commun Networks 2:99–105. https://doi.org/10.1016/j.nancom.2011.05.002

60. Kamiya N, Abe S (1950) Bioelectric phenomena in the myxomycete plasmodium and their relation to protoplasmic flow. J Colloid Sci 5:149–163. https://doi.org/10.1016/0095-8522(50)90016-X

61. Tasaki I, Kamiya N (1950) Electrical response of a slime mold to mechanical and electrical stimuli. Protoplasma 39:333–343. https://doi.org/10.1007/BF01249160

62. Burr HS (1955) Certain electrical properties of the slime mold. J Exp Zool 129:327–341. https://doi.org/10.1002/jez.1401290207

63. Burr HS, Seifriz W (1955) Response of the Slime Mold to Electric Stimulus. Science 122:1020–1021. https://doi.org/10.1126/science.122.3178.1020

64. Adamatzky A (2007) Physarum machines: encapsulating reaction–diffusion to compute spanning tree. Naturwissenschaften 94:975–980. https://doi.org/10.1007/s00114-007-0276-5

65. Gale E, Adamatzky A, de Lacy CB (2015) Slime Mould Memristors. BioNanoSci 5:1–8. https://doi.org/10.1007/s12668-014-0156-3

66. Chua L (1971) Memristor-The missing circuit element. IEEE Transactions on Circuit Theory 18:507–519. https://doi.org/10.1109/TCT.1971.1083337
67. Strukov DB, Snider GS, Stewart DR, Williams RS (2008) The missing memristor found. Nature 453:80–83. https://doi.org/10.1038/nature06932
68. Adamatzky A (2015) A Would-Be Nervous System Made from a Slime Mold. Artif Life 21:73–91. https://doi.org/10.1162/ARTL_a_00153
69. Braund E, Miranda ER (2017) On Building Practical Biocomputers for Real-world Applications: Receptacles for Culturing Slime Mould Memristors and Component Standardisation. J Bionic Eng 14:151–162. https://doi.org/10.1016/S1672-6529(16)60386-4
70. Miranda ER, Adamatzky A, Jones J (2011) Sounds Synthesis with Slime Mould of *Physarum Polycephalum*. J Bionic Eng 8:107–113. https://doi.org/10.1016/S1672-6529(11)60016-4
71. Miranda ER, Braund E (2017) A Method for Growing Bio-memristors from Slime Mold. J Vis Exp 56076. https://doi.org/10.3791/56076

Integrated Photonics for Neuromorphic Computing

Lennart Meyer, Rongyang Xu, and Wolfram Pernice

Abstract In this chapter, we will explore the potential of integrated photonic computing using neuromorphic circuit design to meet the increasing demand for computing power. We will discuss the limitations of electronic devices and how we can leverage the inherent properties of light to overcome them. We will also examine the latest developments in all-optical neural networks and hardware accelerators that use integrated photonics, as well as analyze the underlying physics.

1 Advantages of the Optical Domain and Neuromorphic Computing

The development of transistor-based computational devices has led to the realisation of previously unimaginable advances across scientific and commercial domains. Most of these devices are based on the von Neumann Architecture, which separates the memory and computational units. This architecture, however, faces challenges with the increasing demand for computational power, particularly due to advances in Artificial Intelligence (AI) and other areas. In large systems with significant data movement bandwidth and power consumption become critical limiting factors [1]. To tackle these limitations, new architectures are explored which reduce power consumption and optimize the process of data transmission. Hardware accelerators in the form of Field-Programmable Gate Arrays (FPGAs) [2], Graphics Processing Units (GPUs) [3], and Tensor Processing Units (TPUs) [4] have been demonstrated to increase the data throughput by orders of magnitude compared to classical Central

L. Meyer (✉) · R. Xu · W. Pernice
Kirchhoff-Institute for Physics, Heidelberg, Germany
e-mail: lennart.meyer@kip.uni-heidelberg.de

R. Xu
e-mail: rongyang.xu@kip.uni-heidelberg.de

W. Pernice
e-mail: wolfram.pernice@kip.uni-heidelberg.de

© The Author(s) 2026
M. te Vrugt (ed.), *Artificial Intelligence and Intelligent Matter*, Machine Intelligence for Materials Science, https://doi.org/10.1007/978-3-032-04129-6_17

331

Processing Units (CPUs) [5, 6]. For 'neural' applications, this increase in performance stems mainly from the high parallelization within those systems [6, 7], often optimized for computational heavy tasks like Matrix-Vector Multiplication (MVM).

One of the electronic alternatives to mitigate the issues of data transmission is analog, in-memory computing. This can be accomplished through crossbar structures with a memory cell at each junction, which can be considered analogous to a physical representation of a matrix [8–10]. The weights of such a system are represented by the conductance of each cell and can be tuned continuously. Multiplication and summation can be performed through Ohm's law and Kirchhoff's current law when measured at the outputs of the system [11, 12]. One of the key differences compared to working in the digital domain is that all connections in these structures are represented by physical traces. For specific applications e.g. image processing or convolution filtering, these devices can achieve low latency and low energy consumption [13, 14]. However, besides the challenge of finding the optimal implementation for the in-memory cells [15–17], they still exhibit some of the inherent properties of electronics such as parasitic capacitance and electromagnetic interference (EMI) [18].

Even with the latest advances in electronic computing, the switch to the optical domain offers what appears to be many great advantages. One of the key features of optics is the drastically improved bandwidth, on the order of terahertz, as well as the inherent possibility for parallelization. Different properties of light can be used for multiplexing, e.g. Wavelength Division Multiplexing (WDM), Polarization Division Multiplexing (PDM), or Mode Division Multiplexing (MDM) [19]. These schemes can strongly increase the throughput of a single device without generally increasing its footprint and fewer devices performing higher amounts of operations greatly reduce latency. Although photonic devices are typically larger compared to similar electronic circuits, they can be comparable or even advantageous in terms of compute density [20]. Further advantages are the energy savings coming from data movement compared to electronic devices since no interconnects have to be charged or discharged, as well as the lack of inter-channel crosstalk [21, 22].

Compared to analog and digital computing as well as optical communication, *neuromorphic* (in the context of computational circuits) is a rather new term. It was coined in 1990 by Carver Mead to describe large-scale analog electronic systems inspired by the biological information processing [23]. Nowadays, the term is used more broadly, ranging from the strict process to emulate synaptic processes based on neuroscientific principles to a very broad meaning of highly parallelized, brain-inspired basic processes [24]. In all cases, the new architectures aim to achieve high efficiency in terms of energy consumption, data throughput, and the inherent error correction mechanism.

In this chapter, we focus on systems that combine the advantages of the optical domain with new neuromorphic architectures. In the first part of the chapter, we discuss integrated photonics, providing an intuitive understanding of the fundamental techniques used to manipulate light on a chip. We introduce the basic building blocks and also briefly go over the different available materials platforms. In the second part, we give an overview of a few state-of-the-art implementations of on-chip Optical

Neural Networks (ONN) as well as the most promising optical architectures for MVM.

2 An Introduction to Integrated Photonics

2.1 An Intuitive Approach to Integrated Waveguides

Guiding light inside a waveguide of any kind is based on the principle of total internal reflection. We can observe this in the simple geometry of a so-called slab waveguide using the ray picture of light.

Figure 1 shows a slab on the left side, comprised of three dielectric materials with different refractive indices. From top to bottom, we call the layers cladding, core, and substrate with the corresponding refractive indices n_{cl}, n_{core}, and n_{sub}. When light is injected into this slab, reflections will occur on the core-cladding and the core-substrate interface. From Snell's law of refraction, the conditions for total internal reflection at the core-substrate or core-cladding interface are:

$$\theta_{tot} \geq \arcsin\left(\frac{n_x}{n_{core}}\right), n_x = n_{cl}, n_{sub} \tag{1}$$

$$n_{core} > n_{cl}, n_{sub} \tag{2}$$

If both conditions are fulfilled, light can propagate inside the slab as seen in Fig. 2. For the light to be guided, the wavefronts of different parts of the ray have to interfere constructively, and only a discrete amount of angles can fulfil this condition denoted as waveguide modes. We can distinguish between transverse-electric (TE) and transverse-magnetic (TM) modes, characterized by the orientation of the electromagnetic field vectors. Whether a guided mode inside the waveguide can exist depends on the wavelength of the light, the refractive indices of the materials, and the geometry. To be more precise, the mode-specific refractive index, called effective

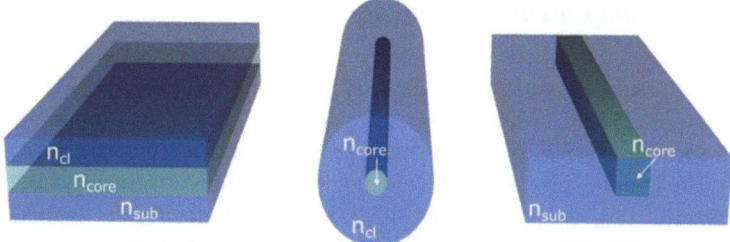

Fig. 1 Different waveguide geometries, all based on the same fundamental concept of total internal reflection, from left to right: slab-, fibre-, strip-waveguide

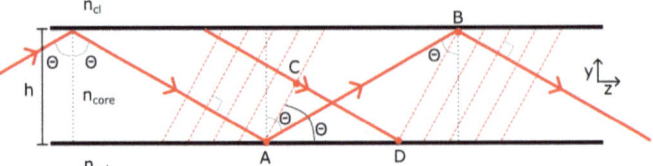

Fig. 2 Light propagating inside a slab structure. Solid lines represent plane waves. Dashed lines represent the corresponding wavefronts. Points A and C share a common wavefront, as well as B and D. The ray propagating from A to B is reflected twice while the one from C to D is not reflected at all. The phase difference accumulated along the longer ray must equal an integer multiple of 2π

refractive index n_{eff}, has to lie within the ones of the core and the surrounding materials $n_{core} > n_{eff} > n_{sub, cl}$. If only one guided TE and TM mode exists, the waveguide is called a single mode waveguide.

Although the ray picture provides an intuitive understanding of the behaviour, it lacks some details compared to a more rigorous treatment of the problem. One important aspect to note is that even for guided modes the electric fields extend into the cladding and substrate. The reason for this is that for angles greater than the critical angle, the propagation constant of the wave in the cladding region becomes purely imaginary. This leads to an exponential decay in that area instead of a sharp stop in the field distribution.

In Fig 1 two further waveguide geometries are shown, respectively the strip and fiber waveguide, which confine light in more dimensions than the slab we looked at before. Usually, the waveguide's shape is partially determined by the properties of the material used. The detailed analysis of these waveguides involves further approximations, but the underlying principle remains the same [25].

Figure 3 shows the intensity distribution of the fundamental TE mode in a typical single mode strip waveguide for a wavelength of 1550 nm. It is important to notice that the mode is not fully confined inside the waveguide but extends into the neighbouring layers. In the following, we discuss how this extension into the cladding enables different building blocks to control the light on-chip.

2.2 Basic Components of Optical Circuits

Photonic integrated circuits (PICs) compactly incorporate a range of photonic components onto a single chip, enabling diverse applications, such as data communications [26–28], sensing [29, 30], quantum computing [31], and neuromorphic computing [32–35]. The building blocks of PICs encompass active devices, such as lasers [36], modulators [37–39], and photodiodes [40, 41], as well as passive devices like directional couplers [42, 43], ring resonators [44], and Mach-Zehnder interferometers [45]. This section will briefly describe the passive devices mentioned above.

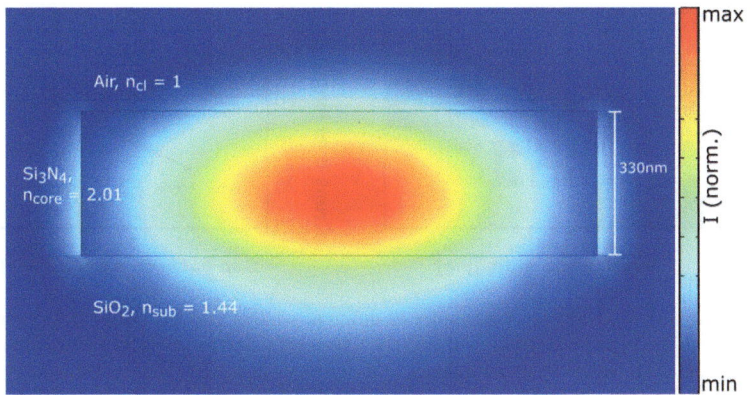

Fig. 3 Simulated intensity distribution of the fundamental TE mode inside a strip waveguide. The waveguide core is a 1.2 μm wide and 330 nm high Si_3N_4 strip on a SiO_2 substrate covered by air. The wavelength of the light is 1550 nm

2.2.1 Directional Couplers

A directional coupler (DC) is one of the most commonly used light-splitting elements in PICs and a key element used in photonic hardware accelerators such as crossbar arrays [42]. A DC usually consists of two parallel waveguides of the same width, as shown in Fig. 4a. For two neighbouring waveguides, light propagating in one waveguide can be gradually coupled to the other when the neighbouring waveguide is within the range of the evanescent field. If the two waveguides are long enough, light can reversibly couple back and forth between the two waveguides, as shown in the inset of Fig. 4a. Figure 4b shows that by choosing the length of the waveguides and the distance between the two waveguides, the percentage of light coupled to the second waveguide can be precisely engineered.

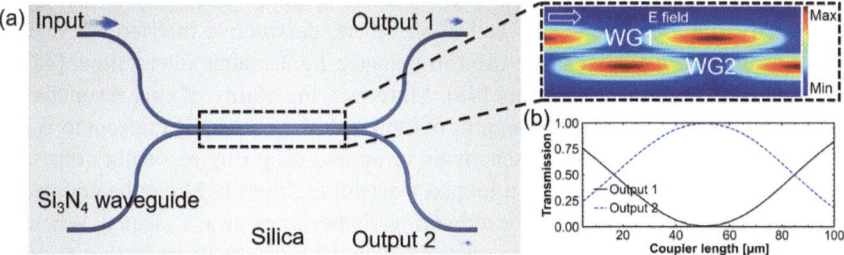

Fig. 4 a Schematic of a Si_3N_4 directional coupler on a silica substrate. The inset shows that light can travel back and forth between two parallel waveguides of the same dimensions. **b** For a given waveguide height, width, and gap, the light splitting ratio can be controlled by varying the coupler length. The width and height of the waveguide are 1.2 μm and 335 nm, respectively, and the gap between the waveguides is 300 nm

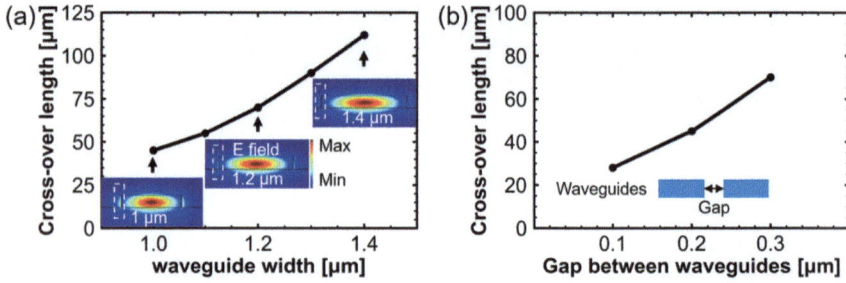

Fig. 5 a Cross-over lengths of waveguides of different widths. The gap between 335 nm-thick Si_3N_4 waveguides on a silica substrate is 300 nm. The inset shows the electric field distribution of the fundamental TE mode in the waveguides of different widths. The field confinement improves with increasing waveguide width. **b** Cross-over length for different waveguide gaps. The thickness and width of the waveguides are 335 nm and 1.2 μm, respectively

The length of the waveguide required to fully couple light from the first waveguide to the second waveguide is called the cross-over length. Taking the fundamental TE mode as an example, the cross-over length is mainly affected by the waveguide width and the gap between the two waveguides. Figure 5a shows that the cross-over length of the fundamental TE mode increases with the increase of waveguide width for a fixed gap size of 0.3μm. This is because a wider waveguide has better field confinements on the fundamental TE mode, thus weakening the coupling between neighbouring waveguides. Similarly, for a fixed waveguide width, the wider the gap between the waveguides, the weaker the coupling and hence the longer the cross-over length, as shown in Fig. 5b.

2.2.2 Ring Resonators

Ring resonators allow the resonant wavelength of light to be circulated through the resonator multiple times [46], wherein at the resonant frequency of the device constructive interference occurs and at all others, destructive interference occurs. Ring resonators in PICs can be used to enhance light-matter interactions [47] or as filters with high quality factors [48]. Moreover, the ability of ring resonators to selectively couple specific wavelengths of light makes them highly relevant to WDM technology. Figure 6a shows a schematic of an add-drop ring resonator consisting of two straight waveguides and a looped waveguide. Input light can be coupled to the looped resonator, which can be either ring-shaped or racetrack-shaped. When the round-trip phase of coupled light accumulated in the loop equals an integer multiple of 2π, the light interferes constructively and the ring resonator is in resonance. Equation 3 shows the round-trip length L_{rt} of the resonator dependent on the ring radius r and the straight waveguide length of racetrack-shaped resonators L_w [49]. For ring-shaped resonators, L_w is equal to 0. The light under constructive interference will finally be coupled to the second waveguide and leave from the drop port. The resonant

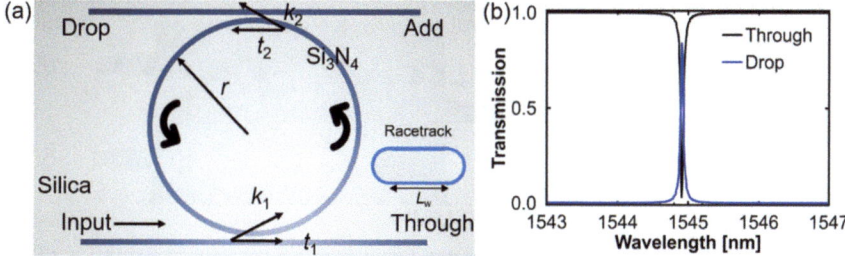

Fig. 6 **a** Schematic of an add-drop ring resonator. If light of a specific wavelength can constructively interfere in the ring resonator, it will eventually couple to the second waveguide and exit through the "Drop" port. Light of the remaining wavelength components will exit through the "Through" port. **b** Example transmission spectrum of an add-drop ring resonator

wavelength λ_{res} can be adjusted by the roundtrip length of the looped waveguides and effective refractive indices of the waveguide mode (affected by waveguide materials, geometries of the waveguide, and surrounding medium of waveguides), as shown in Eq. 4 [46].

$$L_{rt} = 2\pi r + 2L_w \qquad (3)$$

$$\lambda_{res} = \frac{n_{eff} L_{rt}}{m}, m = 1, 2, 3.. \qquad (4)$$

2.2.3 Multimode Interferometers

Multimode interferometers (MMI) are commonly used as optical power splitters to equally distribute an input light (or multiple input lights) to multiple output ports. Figure 7a shows a schematic of a 1×2 MMI consisting of a narrow input waveguide, a wide multimode section, and two narrow outputs that can be used as a 50:50 splitter. The splitting ratio is achieved by adjusting the width and length of the multimode waveguide and the separation distance between the two output waveguides. MMIs operate based on the self-imaging principle [50], where the input optical field profile undergoes repetitive reconstruction at periodic intervals along the direction of light propagation in the multimode waveguide. The interference of multiple modes in the multimode waveguide produces a specific output pattern. As shown in Fig 7b, placing output waveguides near the location of the interference pattern allows light to leave the structure in an equal manner. MMI-based splitters are not limited to 1×2 but can be flexibly designed as $N \times N$ MMI splitters [51]. In addition, MMIs can also be used for optical switching and mode conversion [52].

Fig. 7 **a** Schematic and
b electric field distribution of
a 1×2 MMI splitter

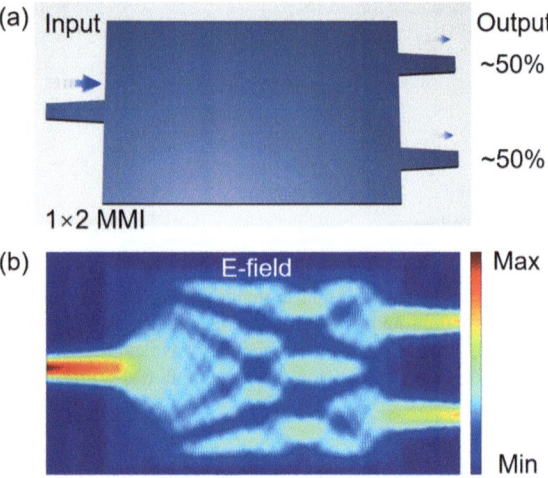

2.2.4 Mach-Zehnder Interferometers

Mach-Zehnder Interferometers (MZIs) are indispensable components of PICs and are widely used to implement optical switches and modulators. Figure 8 shows a schematic of a balanced MZI, which consists of a 50:50 splitter, two waveguides of equal length L, and a 50:50 combiner. Notably, the exact type of splitter is not relevant for this use case.

The first splitter splits the incoming light I_i equally into two paths ($I_i/2$), afterwards, phase modulation is applied to one or both paths. The propagation constants of light in the two waveguides, β_1 and β_2, respectively, are related to the effective refractive indices $n_{1,2}$ of the modes in the waveguide (Eq. 5 [49]), and in turn, are dependent on the refractive index of the waveguide materials. By changing the refractive index,

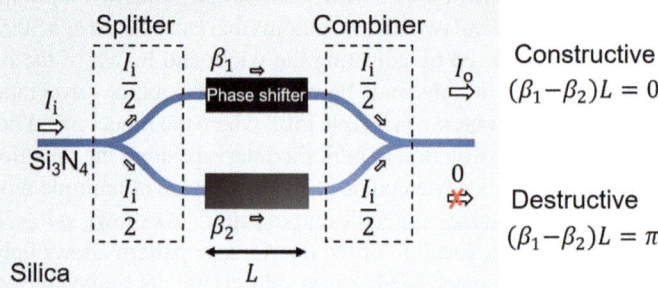

Fig. 8 Schematic of a balanced MZI. Incoming light is divided equally into two branches by a splitter. Phase shifters can be applied to each branch independently to modulate the phase of light. If there is no phase difference between the two branches of light, light will constructively interfere with the combiner, resulting in high transmission. With a phase difference of π, two branches of light interfere destructively at the combiner, resulting in negligible transmission

the phase difference between each arm can be tuned, and then the recombined light at the output of the combiner I_o can be precisely manipulated.

Equation 6 shows the output signal of a balanced MZI [49]. For simplicity, insertion loss and propagation loss are neglected. Because the waveguides of both arms have the same length, only the difference in phase constants ($\Delta\beta = \beta_1 - \beta_2$) needs to be considered. If there is no phase difference between the two arms of light ($\Delta\beta L = 0$) arriving at the combiner, the light from both arms will interfere constructively ($I_o = I_i$). If there is a π phase shift ($\Delta\beta L = \pi$), destructive interference results in output suppression ($I_o = 0$). The refractive index of materials can be adjusted based on thermo-optic effects [53, 54] and electro-optical effects [55], and thereby produce a tunable amplitude in response to external stimuli.

$$\beta_{1,2} = \frac{2\pi n_{1,2}}{\lambda}, \tag{5}$$

$$I_o = \frac{I_i}{2}[1 + \cos(\Delta\beta L)], \tag{6}$$

2.2.5 Phase-Change-Material

Combining integrated photonics with the properties of Phase-Change-Materials (PCM) enables many possibilities to manipulate light in a reconfigurable, non-volatile way [56, 57]. Most people have encountered PCMs in the form of rewritable data storage as in CDs, DVDs, and Blu-rays for many years [58]. The interesting property of PCMs is their ability to switch between their amorphous and often multiple crystalline states rapidly and reversibly [58]. Through changing their state these materials can also change their electrical and optical properties drastically [59, 60], and when used in combination with a waveguide it offers a way to manipulate the light inside passive components.

The ternary phase diagram of germanium (Ge)—antimony (Sb)—tellurium (Te) includes multiple well-studied PCM alloys that have been used in state-of-the-art devices [56, 59, 61]. In the upcoming parts, we will focus on the common alloy: $Ge_2Sb_2Te_5$ (GST).

Figure 9a shows a sketch of the GST lattice in two different states. In the amorphous state (a-GST) no long-range order is present, unlike the crystalline state (c-GST) where atoms are aligned in an energetically favourable lattice. Switching the PCM between the amorphous and the crystalline state is done by heating. For a material to adopt a crystalline state, it must be heated above the glass transition temperature to provide enough energy for the atoms to rearrange in a crystalline lattice. For an amorphous state, the material must be heated above the melting temperature, thereby allowing the system to flow, and then cool rapidly enough to prevent the atom's rearrangement [62]. In Fig. 9b a qualitative curve for the required temperatures can be seen.

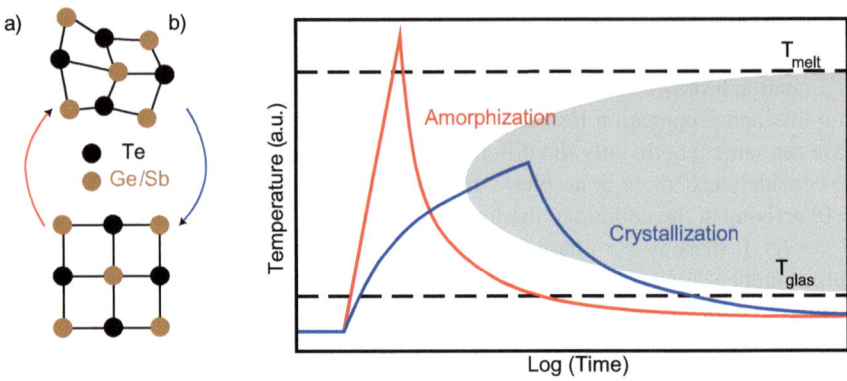

Fig. 9 **a** Schematic of an amorphous and a crystalline state of $Ge_2Sb_2Te_5$. **b** Temperature dependence of the switching dynamics of PCM, heating the PCM above the melting point will result in an amorphous phase if the cool downtime does not allow for recrystallization. Heating only above the glass temperature will allow for the rearrangement of the atoms and a crystalline phase will form

For the application of PCM in PICs, the contrast in refractive index between the amorphous and crystalline states is of interest. For GST the refractive indices at 1550 nm telecommunication wavelength are [63]:

$$n_{a-GST} = 3.94 + i0.045; \quad n_{c-GST} = 6.11 + i0.83 \qquad (7)$$

A simple form of application is to deposit a small patch of GST on top of an optical waveguide. After the deposition of GST, a second protective layer, such as tin oxide (ITO), is applied to prevent oxidation while not significantly affecting the optical properties [64].

In Fig. 10a and b, the electrical field inside a Si_3N_4 single mode waveguide with a 10 nm thick GST patch on top is shown for the amorphous and crystalline state of GST. For the amorphous state, the influence of the GST is rather weak compared to the bare waveguide. Switching to the crystalline state, the mode gets drawn towards the GST patch because of the high refractive index of the GST compared to the core and cladding, which can be qualitatively observed in Fig. 10c. Both the real and imaginary parts of the refractive index change, thereby impacting the absorption of the material. We can therefore manipulate the intensity of the light traveling through the waveguide by simply changing the state of GST on top. The required power to heat the material can be provided by optical pulses inside a waveguide, since a-GST is still slightly absorbing or external heaters and electrical pulses can be used as well. The different techniques offer different advantages in terms of accessibility and energy consumption [65].

This simple structure has been used successfully as an in-memory computing building block for larger photonic systems. Multi-level memories [57], an all-optical abacus [66], on-chip neural networks [64], and hardware accelerators for MVM [42, 67] have been demonstrated.

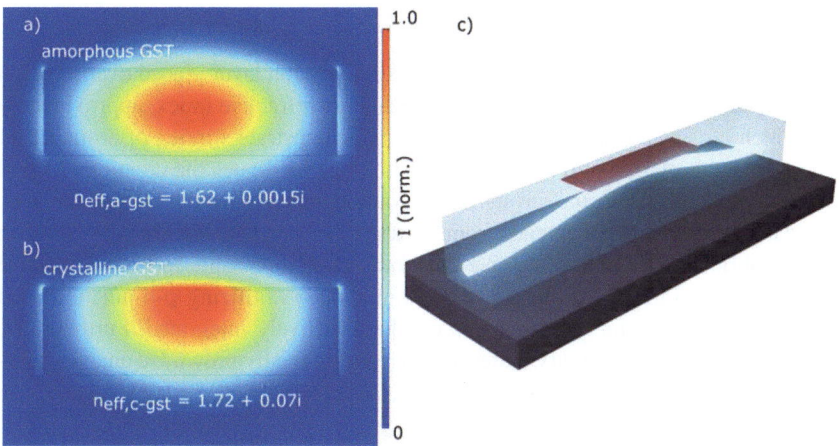

Fig. 10 Simulated intensity distribution of the fundamental TE mode in a Si_3N_4 waveguide with a 10 nm thick GST patch on top, **a** in the amorphous phase and **b** in the crystalline phase **c** PCM-cell consisting out of a waveguide with a crystalline GST patch on top, switching the GST in its state changes the propagation path of the light inside the waveguide

2.3 Available Material Platforms for Integrated Photonics

The properties of materials used in creating photonic integrated circuits are crucial in defining light behaviour and manipulation capabilities within these systems. Key characteristics of these materials include refractive index, transparency window, manufacturability, propagation losses, and nonlinear effects. In the following sections, we'll explore various platforms available for photonic integration and highlight their unique advantages.

2.3.1 Silicon Photonics

Silicon has emerged as a highly favoured material in the photonics world of technology, largely due to its unique photonic properties of the material stack and its compatibility with the established Complementary metal-oxide-semiconductor technology (CMOS) [68, 69]. This compatibility is particularly beneficial as it allows the transfer of well-developed CMOS processes to PICs, thereby greatly expanding the possibilities in fabrication.

The two primary forms of Silicon Photonics are Silicon-on-Insulator (SOI) and Silicon Nitride (Si_3N_4 or SiN). The wide availability of both SOI and SiN processes from various foundries highlights the accessibility and adaptability of Silicon Photonics, confirming its pivotal role in advancing various technological applications. A further upcoming representative of Silicon Photonics is Silicon Carbide (SiC). Also well known from the usage in electronic devices it is still rather new in the photonic realm.

SOI

SOI features a structure that includes a crystalline silicon layer placed over a silicon oxide layer. The crystalline silicon possesses a high refractive index, approximately 3.5, which creates a stark contrast with the oxide substrate's index of around 1.44 at a wavelength of 1550 nm [70]. This high contrast facilitates the creation of small footprint devices due to the strong mode confinement within the waveguide. The transparency range of SOI extends from 1.1 μm to 3.7 μm [71, 72]. Besides allowing for the fabrication of typical passive components, it is possible to integrate active components into the platform. High-speed modulators utilizing p-n-junctions and slower speed modulators with heaters can be effectively integrated. Furthermore, the incorporation of secondary materials such as epitaxially grown Ge enables high-speed on-chip detection and modulation [73, 74].

Si_3N_4

On the other hand, Silicon Nitride (Si_3N_4 or SiN) has a refractive index of approximately 2.0 at 1550 nm [75, 76], which is sufficiently high to provide the necessary index contrast for standard passive components. Since SOI absorbs wavelengths below 1.1 μm, SiN serves as an excellent alternative in these wavelength regimes, with its transparency window reaching down to 500 nm [77]. SiN is predominantly used for passive components and offers a major advantage over SOI in terms of lower propagation loss, making it an attractive option even for longer wavelengths. In terms of propagation loss, a comparable waveguide geometry at the same wavelength in SOI typically has losses around 1–2 dB/cm, while SiN waveguides have significantly lower losses, in the range of 0.1–1 dB/cm [72, 77, 78].

Silicon Carbide

SiC is a type of third-generation semiconductor that is gaining attention due to its unique optical properties and compatibility with CMOS technology. It has a high refractive index of around 2.6 and a wide transparency range from 0.34 μm to 5.6 μm [79]. These properties make it a promising candidate for photonic applications, coupled with high second and third order nonlinear coefficients [80]. Due to the small lattice mismatch between SiC and GaN, the combination of both materials has the potential to be used as a monolithic platform for UV optoelectronic devices, including emitters [81]. SiC offers a significant advantage due to its optically active defects that can function as quantum emitters with a controllable electronic spin [82, 83]. This feature makes it a promising option for fully integrated quantum PICs, as it provides the potential to combine single photon sources and nonlinear processing on a single platform.

2.3.2   Lithium Niobate (LN)

LN is an artificial material and is considered one of the most versatile and promising platforms in the photonics community [84]. The transparency window ranges from 400 nm to 5 μm and the large electro-optic (Pockels) effect enables incredibly high speed on-chip modulation. Strong second- and third-order nonlinear coefficients enable various applications, which are not possible on other platforms. The refractive index is $n_{LN} \approx 2.2$ and therefore enough for high performing passive components. Different fabrication methods strongly influence the propagation losses in LN [85]. Losses well below 1dB/cm have been achieved, even down to 0.027dB/cm [86]. Besides these advantageous properties, difficult fabrication processes alongside challenges in the integration of light sources remain [87].

2.3.3 III-V Semiconductors

The whole group of III-V Semiconductors e.g. AlN, GaAs, AlGaAs, and InP plays a crucial role in photonics. They are used to realize e.g. high quality light sources [88–91], detectors [92, 93], amplifiers [94, 95] and modulators [96], but can be utilized as complete photonic platforms as well. Integrating these materials into other photonic platforms for active on-chip components is an ongoing field of research [97–99].

Gallium Arsenide (GaAs)

GaAs is a well understood platform with a refractive index of 3.4 at a telecom wavelength of 1550 nm and a wide transparency window from 870 nm to 19 μm [100]. The second order nonlinear effects (Pockels effect) can be used efficiently for fast modulation compatible with LN devices [96, 101]. A common configuration of GaAs waveguides is in an Aluminum Gallium Arsenide (AlGaAs) heterostructure with a GaAs core. The index contrast from core to cladding depends on the percentage of Al and lies in the order of 0.2 for $Al_{0.3}Ga_{0.7}As$ cladding. One promising application of GaAs is in the field of quantum technology. It is possible to achieve single photon emission by integrating single quantum dots into photonic structures [94, 95]. Additionally, single photon detection is also possible [96], which opens up the possibility of fully integrated quantum circuits [93].

Indium Phosphide (InP)

InP has been of particular interest since it allows for the monolithic integration of multiple active components like lasers, amplifiers, modulators and detectors with passive optical structures [100, 102]. It also features high nonlinear coefficients e.g. Pockels and Kerr coefficients at the telecom wavelength. Multiple foundries have been established and continue to improve their possibilities in fabrication [102]. Drawbacks of a fully integrated InP chip compared to Silicon Photonics are high material costs, lower yield and smaller wafer sizes [68]. Additionally, InP platforms suffer from weak mode confinement because of the small index contrast. This might improve through the upcoming InP membrane technology, which could drastically decrease the footprints of the devices [103–105].

Aluminum Nitride (AlN)

AlN is a CMOS-compatible material with the largest band gap among the semiconductors of 6.2 eV. The transparency window spans from the UV to mid-infrared, from 0.2 μm up to 13.6 μm [106] with a refractive index of 2.12 at 1550 nm [107]. Because of a strong piezoelectric effect, it is often used in micro-electro-mechanical systems [108], but can also realize optomechanical devices [109]. As opposed to other CMOS-compatible materials, second and third order nonlinear effects can be utilized for various applications [110, 111]. In combination with GaN, AlN can enable UV light emitters and photodetectors [112, 113]. The AlN platform's combination of electro-optics, piezo-opto-mechanics, nonlinear processes, and low-loss passive devices makes it an attractive option for future applications.

2.3.4 Further Platforms and Conclusions

Besides the above listed material, many other possible platforms exist with unique properties. Among them is Gallium Nitride (GaN) which has been mentioned in combination with some of the platforms above and also offers attractive optical properties on its own in terms of a wide band gap and strong nonlinear coefficients [114]. Also, diamond with exceptional mechanical and thermal properties is an upcoming material for integrated photonics. It features room-temperature single photon emission, a wide band gap and low loss waveguides, to list just a few characteristics [115].

Overall, the variety of possible materials is enormous. While silicon based photonics had a head start related to its CMOS-compatibility, other platforms establish themselves in the photonic community by further developing their fabrication possibilities. Finding a suitable platform for integration is application-specific and often difficult to achieve in a single chip. Therefore packaging and interconnecting multiple PICs from different platforms is a common solution and challenge.

3 State of the Art Approaches to Optical Neuromorphic Computing

3.1 On-chip Integrated Optical-Neural Networks

To develop ONNs, various methods can be employed to replicate the entire neural network structure and synaptic dynamics at the hardware level. Our discussion will initially focus on the circuit setup and the functions of individual components. We will examine two distinct approaches: one involving the use of PCM cells and another based on an optoelectronic Ring Modulator. Following this exploration of the circuit configurations and their operational mechanisms, we will then compare the advantages and challenges inherent to each approach.

3.1.1   An All-Optical On-Chip Neural Network with Self-learning Capabilities

In their research, Feldmann *et al.* [116] developed an on-chip ONN comprising a single-layer system with four neurons, each with 15 synapses. This setup was capable of both supervised and unsupervised training, showcasing the practical application of all-optical neurons in constructing layered neural network structures.

Figure 11a and b illustrate the fundamental components and operational principles of the neuron. Figure 11c shows three individual neurons, each with four synaptic connections. Each neuron is composed of four input devices, linked to a ring resonator via a PCM-integrated waveguide. The ring resonators from all inputs converge into a single common bus waveguide. This bus waveguide, in turn, connects to a larger ring resonator with an additional PCM patch atop. Finally, this larger ring is connected to the output waveguide.

In this setup, light is coupled onto the chip through grating couplers (depicted as black triangle features in Fig. 11). These couplers serve as the entry points for

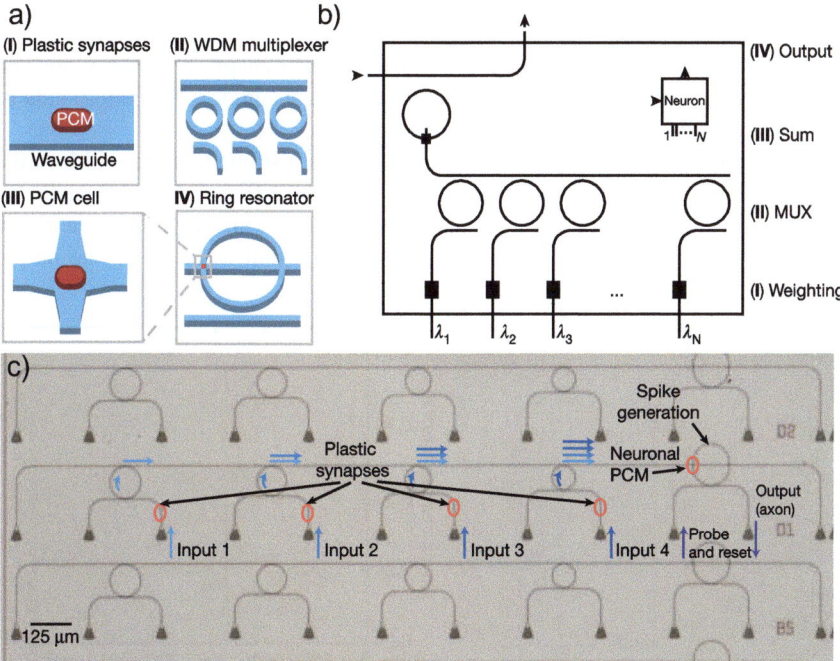

Fig. 11 **a** Schematics of the individual building blocks used in the all-optical neuron. **b** Working principle of the neuron: Inputs are encoded to different wavelengths, PCM-waveguides as input weights, rings for multiplexing all weighted input signals into a common waveguide, and PCM on a ring as the physical implementation of the nonlinear activation function. **c** Optical micrograph of the neuron on a SiN platform. Reprinted with permission from [116], Springer Nature. All Rights Reserved

the light and in a single-layer network, they function directly as network inputs. However, in a multi-layer neural network, these inputs are substituted by waveguide connections that link to the neurons in the preceding layer. This part of the circuit effectively represents the pre-synaptic neurons. In our scenario, we assume each input possesses a unique wavelength.

Positioned on top of the waveguide that connects the input to the ring is a patch of PCM, illustrated as red ovals. The PCM's state can be altered - it increases transmission through the waveguide when amorphous and decreases it when crystalline, representing the synaptic strength. These PCM-integrated waveguides form the weighted connections between the pre-synaptic and post-synaptic neurons (Dendrites) and each connection is then linked to a ring resonator. The opposite side of every ring is connected to a common bus waveguide. By leveraging the different resonance frequencies of the rings through WDM, all weighted input signals can be summed into the bus waveguide. The power within this waveguide represents the weighted sum for the post-synaptic neuron.

Continuing with the simplified comparison to a biological neuron, this weighted sum must then interact with a nonlinear activation system to produce an output.

In this research, the nonlinear activation function is realized using another ring resonator. Figure 12 (*left*) provides a detailed view of the larger ring resonator, as referenced in Fig. 11. The summed up signals of the post-synaptic dendrites propagate towards the ring resonator inside the common bus waveguide. The bottom side of the resonator is coupled to an output waveguide (axon).

Within this output waveguide, a probe pulse is transmitted. If the PCM on the resonator is in its crystalline state, the ring's resonance frequency is aligned with the probe pulse moving through the output waveguide. This alignment causes the pulse to couple into the ring, resulting in minimal signal detection in the output

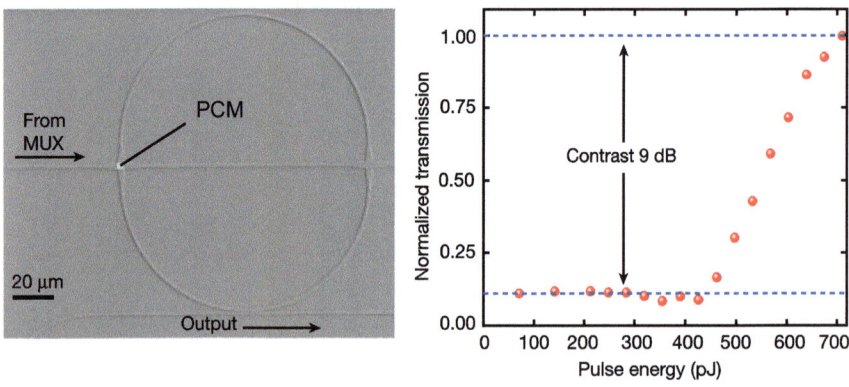

Fig. 12 Close-up of "neural" ring resonator from Fig. 11 (*left*) and the corresponding transmission function (*right*). Coupling of the multiplexed signals to the PCM patch on top of the ring can detune the ring resonance. A pulse travelling in the output waveguide is used to probe the resonance. If the weighted sum exceeds a certain threshold, the absorption of the PCM cell is reduced and the neuron starts to fire. Reprinted with permission from [116], Springer Nature. All Rights Reserved

waveguide. However, when the weighted sum from the bus waveguide reaches the neural resonator, it heats the PCM patch atop it. If this heat is sufficient to raise the PCM's temperature above its melting point, the material partially transitions to an amorphous state. This alteration in the PCM's state detunes the resonator's frequency, reducing its interaction with the output waveguide and consequently triggering an output spike. After each firing event, the PCM patch has to be reset into the ground state by a high power writing pulse.

Figure 12 (*right*) illustrates the relationship between the transmission through the output port and the power of the weighted sum. Until the threshold power level is reached, the ring's resonance remains unchanged, leading to low transmission in the output waveguide. This mechanism is analogous to the operation of the rectified linear unit (ReLU) function.

The various components of the circuit mirror the behaviour of their biological equivalents, enabling them to effectively emulate the dynamics of a single neuron.

The neurons in this model can be trained in a supervised manner by fine-tuning the PCM cells on top of the waveguides. This adjustment effectively tailors the weights to match desired outcomes. Additionally, an unsupervised training approach is feasible through the incorporation of a feedback loop splitting the output waveguide of the neuron and connecting one part back to all the PCM weights. This setup allows the output and the input pulse to overlap, given that the pulse duration is sufficiently long. In the synapses with a strong input pulse, this overlap will be sufficient to set the PCM into an amorphous state and increase that connection strength. In reverse, the synaptic connections with low input pulses will have less power generated through the overlap and therefore further crystallize, dampening synaptic strength. In this manner, a neuron can be trained to recognize a certain input pattern.

In their foundational experiment, the researchers describe a single-layer network with four neurons, each equipped with 15 synapses. This basic network successfully differentiates and accurately classifies four distinct 15-pixel images. Although simplistic, this example demonstrates the network's core image processing and recognition capabilities, setting the stage for more complex architectures.

Expanding beyond this single-layer model, the creation of multi-layer networks requires forming connections between neurons across different layers, achieved physically through waveguides on the hardware. This structure allows for efficient inter-layer communication, crucial for building advanced neural network architectures.

3.1.2   An Adaptive Approach to ONNs Through Encoding Structural Information in the Light's Properties

In another study, done by Brückerhoff-Plückelmann *et al.* [44], we explore a more adaptable approach to ONN structures, also based on PCM. As a practical application, the system was trained to differentiate between German and English text samples. This design proposes a solution for creating an event-driven ONN that is both scalable and flexible. It utilizes a photonic circuit with a non-fixed structure, which enables the

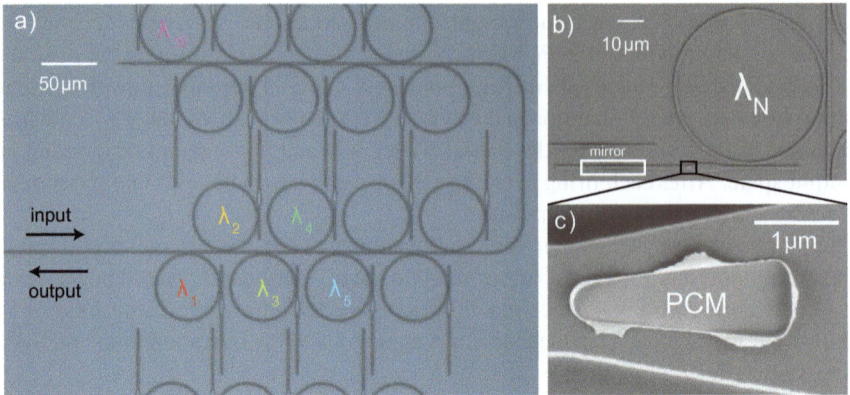

Fig. 13 Pictures of one of the neuron subsets used in [44] built on a SiN platform, **a** 16 neurons with individual resonant frequencies of the ring resonators are connected to the same waveguide. PCM on the drop ports of the rings is used as an activation function, **b** a close-up of a neuron consisting of a wavelength filter, a PCM cell, and a mirror. **c** PCM patch on top of the MMI structure. Figures adapted from [44]. Reprinted with permission from AAAS. All Rights Reserved

integration of over 11,000 on-chip neurons and versatile connectivity options. This flexibility allows for dynamic reconfiguration, which can emulate certain aspects of both long-term memory and short-term adaptation.

Figure 13b and c illustrate one of the system configurations used in this study. In this setup, the input waveguide is connected to a ring resonator, which in turn is linked to another waveguide. This subsequent waveguide features an MMI topped with a PCM patch and concludes with a Bragg mirror. The design ensures that light enters the input waveguide couples into the ring and then into the PCM waveguide. The level of light absorption here is determined by the PCM's crystallinity. After interacting with the PCM, the light reflects, recoupling into the ring and returning through the input waveguide, but in the opposite direction.

This configuration, which uses the same ring for both de-multiplexing and multiplexing the signal, addresses the challenge of fabricating two separate rings with identical resonance frequencies, while also reducing the footprint. The inclusion of the MMI structure in the waveguide improves the coupling efficiency of light to the PCM by enlarging the area of interaction. Additionally, the double-pass arrangement, achieved with the Bragg mirror, further enhances the potential contrast effect of the PCM. This design effectively maximizes the interaction between the light and PCM, crucial for the system's functionality.

Figure 13a illustrates a configuration where multiple neurons are connected to a shared bus waveguide, a setup we will refer to as a 'subset' henceforth. Due to the distinct resonant frequencies of the rings, WDM can be employed to selectively target specific neurons through the same waveguide, allowing for the simultaneous addressing of multiple neurons. In this structure, the neurons function primarily as (de)multiplexing units and as part of the activation function. Notably, there are no weighted synapses or interlayer connections between neurons implemented in this design.

A key distinction in the activation function of this study, as compared to previous work, lies in the initial state of the PCM. Unlike starting in a fully crystalline state, the PCM patches here are partially amorphous at baseline. This modification results in a voltage response that slightly deviates from the expectations set by the previous study. In contrast to the ReLU function depicted in Fig. 12b, the current activation function introduces an inhibitory phase preceding the excitatory phase. As the power of the weighted sum increases, it initially reduces the transmission through the neuron by promoting crystallization. When the power is sufficient to heat the patch above its melting point, it starts to amorphize, leading to an observed increase in transmission. This mechanism allows for the attenuation of a neuron's activity, even when only positive synaptic weights are present.

This PIC architecture concept leverages the characteristics of input light to encode synaptic information. The synaptic connections between neurons are implemented using different wavelengths and a bus waveguide in the subset, while the pulse power of the light encodes the weight of these connections.

Figure 14 illustrates the application of this structure in simulating a neural network. Consider a scenario where we aim to connect a small unit consisting of three pre-synaptic neurons and one post-synaptic neuron. This connection is established by selecting specific wavelengths for the pre-synaptic neurons and encoding the strength of their connections to the post-synaptic neuron in the power of the light pulses. These pulses travel along the bus waveguide, are demultiplexed by the ring resonators, and interact with the neurons' activation functions (PCM). The signals, now weighted and multiplied, are multiplexed back into the same waveguide but travel in the reverse direction. The combined signal is then detected by a photodiode, whose output represents the weighted sum of the synaptic inputs.

To set the post-synaptic neuron, an optical write pulse, proportional to the weighted sum and matched in wavelength, is sent into the system to alter the state of the post-synaptic neuron's PCM cell. The neuron's state can subsequently be determined using a low-power read pulse, completing the process of synaptic trans-mission and neural activation in this photonic neural network model, this way the layer structure can be built iteratively.

This system's design allows for the flexible integration of multiple neurons, enabling the creation of complex structures suited for various tasks. In a demon-stration, six neurons were utilized to differentiate between German and English text samples. For this, a neural network was trained directly on the chip, with five neurons encoding the vowel distribution and the sixth neuron's transmission distinguishing between the two languages. This demonstrates the system's adaptability, with PCM cells in each neuron allowing for the encoding of diverse information, and the on-chip training enhancing efficiency for practical applications.

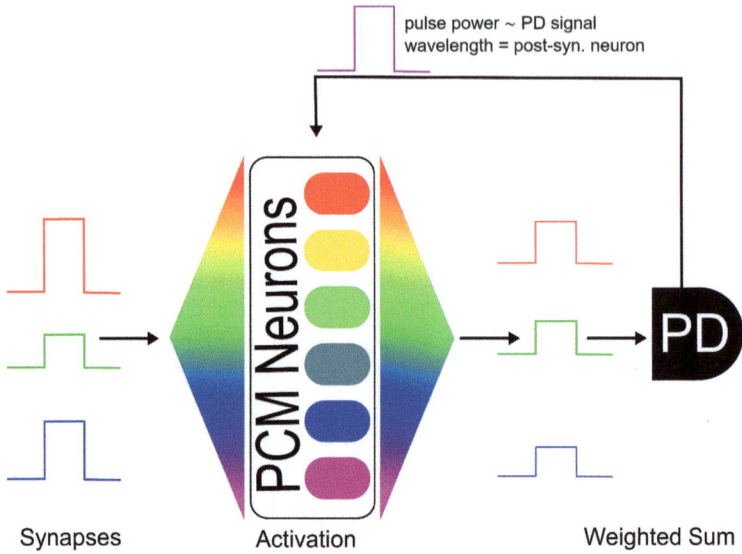

Fig. 14 Schematic of the operation of the PIC developed in [44]. Input pulses contain synaptic information (power and wavelength) and are multiplied with the corresponding activation functions selected through wavelength demultiplexing by the ring resonators. After the activation, all signals are multiplexed again. The accumulated power of all pre-synaptic neurons is summed up at the photodiode. A write pulse proportional to the weighted sum is sent to a beforehand selected post-synaptic neuron. Figures adapted from [44]. Reprinted with permission from AAAS. All Rights Reserved

3.1.3 An Integrated End-to-End ONN for Image Classification With High Speed Computing Capabilities

In this publication by Ashtiani *et al.* on on-chip ONNs [117], we explore an alternative method to replicate synaptic dynamics used in constructing a fully integrated photonic deep neural network (PDNN).

Figure 15 outlines the neuron design employed in this approach, featuring N possible connections. A key difference from earlier methods is the use of electro-optic attenuators and ring modulators instead of PCM components for implementing weights and activation functions. The synaptic connections are realized through waveguides with optical attenuators, based on PIN junctions. These attenuators control the light intensity by adjusting the applied current. The output from each of the N attenuators is detected by a photodiode, and the collective current from these diodes represents the weighted sum.

The nonlinear activation function is realized using a ring resonator, similar to previous studies. This ring is coupled to an output waveguide that transmits a probe pulse. When the ring's resonance is aligned with the probe pulse, most light is coupled into the ring, resulting in low transmission. The modulation of the ring's resonance, and thereby the output pulse, is achieved by embedding the ring in a PN junction.

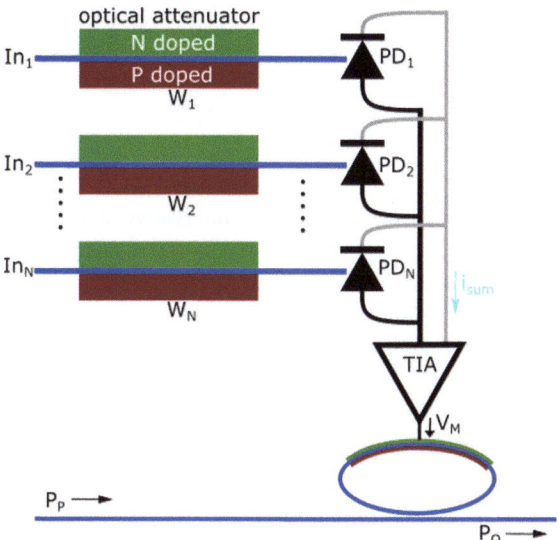

Fig. 15 Schematic of the neuron used in [117]. The synapses are based on waveguides and optical attenuators. The weighted inputs are measured by photodiodes and summed up by accumulating the current of all diodes. A trans-impedance amplifier creates a voltage proportional to the current. This voltage modulates the resonance of a ring modulator. Detuning the resonance decouples the ring from the output port enabling the transmission of a read pulse. Figures adapted and reprinted with permission from [117], Springer Nature. All Rights Reserved

Adjusting the voltage applied to this junction alters the ring's refractive index and resonance, allowing control over the output waveguide's transmission. The voltage-dependent transmission through the output waveguide mirrors a nonlinear activation function, specifically a ReLU function. To relate this voltage to the weighted sum, the photodiodes' accumulated currents are amplified and converted to a voltage using off-chip Trans-Impedance Amplifiers (TIAs).

This neuron design presents a highly controllable and scalable solution for ONN architectures. It has been effectively utilised in a PDNN for image classification. The process begins with 30 input pixels being directed onto the classification chip. These pixels are then divided into four overlapping sub-images, each comprising 12 pixels. Each sub-image is then fed into one of four neurons, each with 12 inputs, creating a convolutional layer. The architecture of this PDNN includes a second layer, which is fully connected to the first layer and consists of three neurons, each with four inputs. This second layer, in turn, is fully connected to the output layer, which comprises two neurons with three inputs each. The interlayer connections in this neural network are facilitated through off-chip TIAs.

This specific PDNN demonstrates its capability to perform both two-class and four-class image classification tasks. Remarkably, it achieves end-to-end classification times of just 570 picoseconds, a speed comparable to a single clock cycle of a conventional digital computing platform. This efficiency illustrates the potential of photonic neural networks in high-speed image processing and classification tasks, marking a significant advancement in the field of neuromorphic computing.

3.1.4   Conclusion

The three studies of ONNs each introduce innovative approaches with distinct advantages and challenges.

Feldmann *et al.*'s research centres on an all-optical component setup using Phase-Change Material (PCM) patches for efficient in-memory computing. This design supports unsupervised learning and processes network layers in a single time step, independent of neuron count, highlighting the system's scalability and potential for biological neuron behaviour emulation.

The study by Brückerhoff-Plückelmann *et al.* adopts a versatile approach, also utilising PCM for in-memory computing but avoiding the complexities of hardware-level synapse implementation. This method leverages light's intrinsic properties for structural information encoding and features an improved activation function that allows neuron inhibition with only positive synaptic weights.

Comparing the two, we see the diverse applications of PCM in photonic computing. Feldmann *et al.* delves into the possibilities of multi-level networks. While feasible, larger systems encounter increasing complexity, especially with hardware-implemented connections. Additionally, fabricating multiple ring resonators with identical frequencies remains a challenge. On the other hand, Brückerhoff-Plückelmann *et al.*'s approach addresses these complexity issues in larger systems but at the cost of reduced computational speed. Both approaches, however, face the challenge of requiring multiple light sources, a significant factor when considering the integration of the entire system on a chip. Together, these studies showcase the potential of PCM in ONN architectures,

The study by Ashtiani *et al.* presents a fully integrated system that utilizes easily controllable electro-optic devices to achieve remarkably fast end-to-end computing speeds. This rapid processing capability is primarily limited by the bandwidth of the electro-optic devices. While this represents a significant leap forward in terms of computational speed, the system faces challenges in scalability, particularly due to the complexity of its layout when scaled up to larger structures.

Notably, this approach does not exploit the potential multiplexing capabilities inherent in optical systems. However, it compensates for this by requiring only a single light source, which simplifies the supply of light for all layers. This aspect significantly eases the challenges associated with system integration. On the down-side, the need to interface all active electro-optic components within the system introduces an additional layer of complexity.

Overall, each study contributes significantly to the evolving landscape of ONNs, showcasing different methods to surpass traditional electronic neural networks' limitations. While the first study excels in multi-layer solutions and biological neuron emulation, the second study leverages the scalability and versatility of PCM, and the third study highlights the potential of a high-speed, fully integrated photonic deep neural network. These explorations indicate a promising future for ONNs in various applications, though challenges in complexity, speed, and integration remain to be addressed.

We discussed three possible realizations of ONNs, but there are numerous innovative approaches for implementing partial or entire networks in optical computing, utilizing technologies like integrated and free-space optical interconnections. These methods include diffraction, interference, scattering, microlens arrays, holographic elements, saturated absorbers, and nonlinear optical effects [118–121].

3.2 Optical Hardware Accelerators

Crossbar arrays, ring resonator arrays, MMIs, and MZIs are common structures found in optical hardware accelerators for neuromorphic computing. By combining multiple of the previously introduced building blocks, these optical hardware accelerators can be built. The parameters of the selected photonic structures need to be optimized according to the specific requirements (e.g. operating wavelengths and sizes of the accelerator). Another important factor is the properties of the light source. Depending on the accelerator either incoherent (e.g. for crossbar arrays) [64, 122] or coherent (e.g. for MZIs) [123] light can be used as input. Therefore, we need to have a general understanding of the basic working principles of optical hardware accelerators based on these four different structures, which are described below.

3.2.1 Crossbar arrays

Figure 16a and b show an example of how MVM can be achieved using a crossbar array. The input vector, x_1 to x_5, corresponds to 5 light pulses of different wavelengths in the crossbar array. The amplitude of these pulses, associated with the values of the input vectors, can be modulated by optical modulators.

These pulses then enter 5 horizontal waveguides of the crossbar array and each pulse is equally distributed to 5 DCs of the same colour in Fig. 16b. The PCMs on the waveguides of the DCs partially absorb the light as it passes through serving as multiplication. Finally, the pulses from the DCs are coupled to vertical output waveguides. Addition operation is achieved when 5 pulses of different wavelengths are coupled to the same vertical waveguides and arrive at the same photodetector. In the case of the M×N crossbar arrays using DCs, only 1/MN (where MN is the number of rows M × number of columns N) of the input power can reach the output waveguide due to losses caused by the DC's coupling and the number of columns.

The values of the matrix, represented by the absorption of the PCMs, can be programmed using optical (on-chip [124] and off-chip [125]) and electrical (on-chip [126]) pulses. In certain applications, such as edge computing [42], where the kernel (filter matrix) is rarely changed, energy consumption is as low as 17 fJ per multiply-accumulate (MAC) operation [42].

(a)
$$\begin{pmatrix} a_{11} & a_{21} & a_{31} & a_{41} & a_{51} \\ a_{12} & a_{22} & a_{32} & a_{42} & a_{52} \\ a_{13} & a_{23} & a_{33} & a_{43} & a_{53} \\ a_{14} & a_{24} & a_{34} & a_{44} & a_{54} \\ a_{15} & a_{25} & a_{35} & a_{45} & a_{55} \end{pmatrix}^T \times \begin{pmatrix} x_1 \\ x_2 \\ x_3 \\ x_4 \\ x_5 \end{pmatrix} = \begin{pmatrix} y_1 \\ y_2 \\ y_3 \\ y_4 \\ y_5 \end{pmatrix}$$

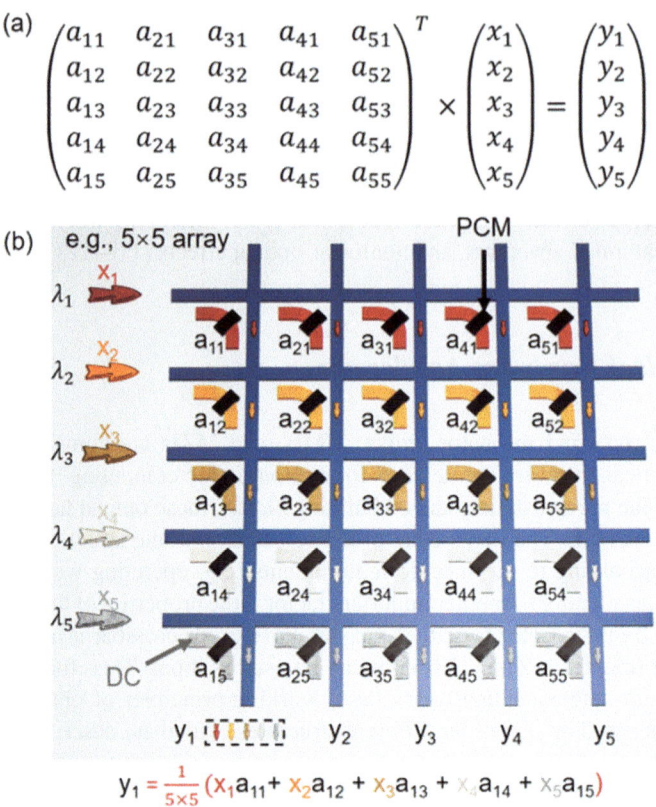

$$y_1 = \frac{1}{5 \times 5}(x_1 a_{11} + x_2 a_{12} + x_3 a_{13} + x_4 a_{14} + x_5 a_{15})$$

Fig. 16 **a** Matrix-Vector Multiplication to be calculated in the photonic crossbar structure. **b** Schematic of a 5×5 crossbar array. Light with five different wavelengths enters the five waveguides of the crossbar array and is then coupled equally to 5 DCs. After passing through the PCMs on the DCs (multiplication), the light is coupled to the vertical waveguides (addition). At each output port, light of 5 wavelengths can be detected

Transferring a MVM to the crossbar array is not entirely intuitive. Figure 17 shows a possible application of using the crossbar array for image processing. The pixel data (capital letters, yellow frame) and the kernels themselves (small letters, blue and orange frame) are shown on the left. To perform a convolution of the picture and the kernels element-wise products have to be calculated and summed up. To map this operation onto the crossbar structure the matrices are flattened. Afterwards, the pixel data is modulated onto the intensity of the light and the different wavelengths are sent into the corresponding input waveguide. The kernel data is encoded into the PCM cells. One column represents one kernel, shown on the right of Fig. 17. This way to perform a convolution of a 3×3 filter matrix a 9×1 crossbar structure is needed. With multiple columns, multiple convolutions can be processed in parallel e.g. running four edge detection operations on the same picture. Since the kernels are static the pixel data is continuously modulated to move the filter matrix over the

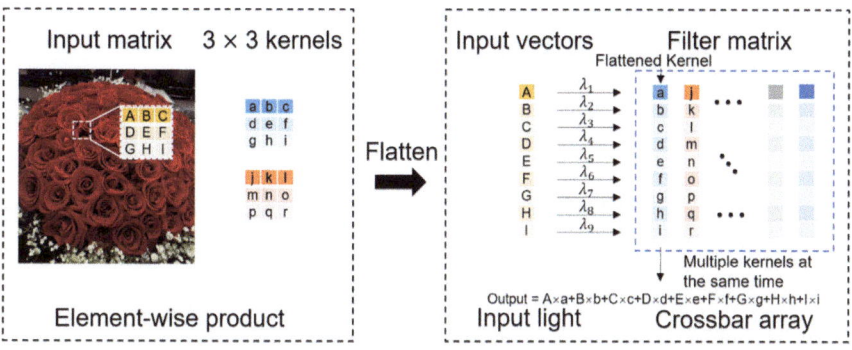

Fig. 17 Schematic of a convolution operation on an image using a crossbar array. The matrices representing the pixel data (capital letters, yellow frame) and the filter matrices (small letters, blue and orange frame) are shown on the left. After flattening the matrices the pixel data is modulated onto the light and sent into the crossbar. In each column of the crossbar, one kernel is encoded. The element-wise product and summation are performed on the light by propagating through the structure and detection at the output ports

picture. While propagating through the structure the filter values are multiplied onto the pixel data through absorption in the PCM cells. The summation is performed during detection at the different outputs of the crossbar.

The N^2 scaling of operations for a $N \times N$ crossbar is one of the biggest advantages of the structure, while also making it difficult to operate in larger systems. The inherent loss due to the small splitting ratios and the increasing amount of electro-optic devices bring up challenges in terms of fabrication and electronic interfacing.

3.2.2 Ring resonator arrays

Figure 18 shows a schematic of a photonic accelerator based on ring resonators. Similar to photonic accelerators based on crossbar arrays, the implementation uses WDM technology to perform MVM with different wavelengths of light. In a photonic accelerator based on ring resonators [127], light pulses of different wavelengths enter the waveguide and are demultiplexed by a series of ring resonators. The diameters of these ring resonators are carefully tuned to ensure that a particular wavelength of light couples to the ring resonators and then from the ring resonators to the vertical waveguides. As light passes through the PCM, partial absorption (multiplication operation) occurs. Then, ring resonators of the same size (indicated by the same colour in Fig. 18) couple the attenuated light to the output waveguide. This setup ensures that different wavelengths of light converge at the output port to perform the addition operation. By carefully tuning the ring resonators, precise wavelength control and manipulation can be achieved. This concept of configurable weight banks can be used as a photonic accelerator in and of itself. Similar structures based on ring resonators in conjunction with balanced detectors were also proposed for broadcast and weight

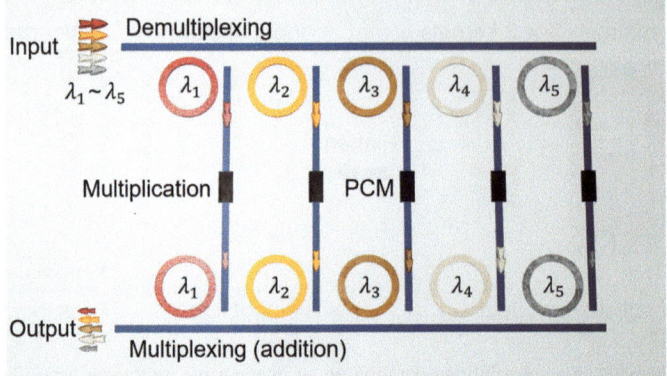

Fig. 18 Schematic of a photonic accelerator based on ring resonators. Input light of different wavelengths is demultiplexed by different ring resonators. The demultiplexed light is partially absorbed by PCMs (multiplication operation) and finally multiplexed again by the same ring resonators (addition operation)

systems, where they serve as spectral filter banks for network interconnects between different spiking neurons [128].

Despite their simplicity and versatility, ring arrays pose a fabrication challenge due to the need for precise parameter control to achieve the correct resonances.

3.2.3 Multimode Interferometer

MMIs are also promising for performing MVMs. Figure 19 shows a schematic of an MMI-based photonic accelerator. Input data is encoded as the amplitude of light at four wavelengths. The light passes through two MMIs and phase shifters and reaches square-law photodetectors. The transfer matrices of the MMIs (M), phase shifters (ϕ), and the full transfer matrix (T) of the overall structure in Fig. 19 describe how the intensities are converted while propagating through the system. Each row of the full transfer matrix can be viewed as a kernel. Negative values can be achieved by selecting one vector as a reference and subtracting it from the rest of the inputs. This is also possible for the previously discussed approaches. Thus, the device can perform three groups of convolution operations (2×2 correlated kernels) simultaneously, resulting in a high computational density photonic accelerator of 12.74 T MACs/s/mm^2 [129] (Crossbar array: 0.6 T MACs/s/mm^2 [42] and Ring resonators: 2.89 T MACs/s/mm^2 [130]). Note that the kernels of the MMI-based photonic accelerator are programmed by phase shifters based on the thermo-optic effect and thus can require higher energy per operation than ring resonator and crossbar array-based photonic accelerators using non-volatile PCMs.

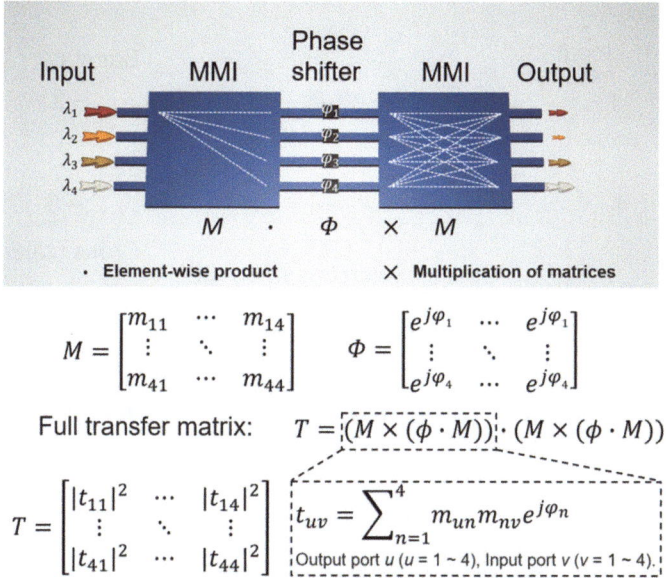

$$M = \begin{bmatrix} m_{11} & \cdots & m_{14} \\ \vdots & \ddots & \vdots \\ m_{41} & \cdots & m_{44} \end{bmatrix} \qquad \Phi = \begin{bmatrix} e^{j\varphi_1} & \cdots & e^{j\varphi_1} \\ \vdots & \ddots & \vdots \\ e^{j\varphi_4} & \cdots & e^{j\varphi_4} \end{bmatrix}$$

Full transfer matrix: $T = (M \times (\phi \cdot M)) \cdot (M \times (\phi \cdot M))$

$$T = \begin{bmatrix} |t_{11}|^2 & \cdots & |t_{14}|^2 \\ \vdots & \ddots & \vdots \\ |t_{41}|^2 & \cdots & |t_{44}|^2 \end{bmatrix} \qquad t_{uv} = \sum_{n=1}^{4} m_{un} m_{nv} e^{j\varphi_n}$$

Output port u ($u = 1 \sim 4$), Input port v ($v = 1 \sim 4$).

Fig. 19 Schematic of an MMI-based photonic accelerator featuring two 4×4 MMIs and a phase shifter array. Input vectors are encoded into different wavelengths of light. Four phase shifters are associated with convolution kernels. The transfer matrices of the MMIs, phase shifters, and the overall structure are taken from [129]

3.2.4 Mach-Zehnder Interferometers

The MZI constitutes the fundamental structure of coherent additive photonic accelerators for neuromorphic computing [123]. Figure 20a shows a schematic of a 2×2 MZI using a DC to achieve a 50:50 beam splitter (BS). The first DC evenly divides incident light between the two arms. Without a phase shift introduced by the phase shifter, light entering through port 1 will exit through port 3, representing the cross-state. This is because the second DC can be viewed as a continuation of the first DC. Similarly, light entering through port 4 will leave through port 2.

Changing the phase of the light from the upper arm causes the light from both arms to reach the second DC with a 180° phase difference, resulting in destructive interference. In this case, light entering through port 1 will exit through port 2, i.e., the bar state. Similarly, light entering from port 4 will now exit through port 3.

The effect of MZIs on light can be described mathematically using unitary matrices, as shown in Fig. 20b. Moreover, the MZIs can be placed in a cascade, depicted in Fig. 21 to realise arbitrary unitary matrix multiplication and complex valued NNs [123, 132]. Adjusting the mesh architecture with various matrix decomposition methods provides a way to improve the performance and capabilities of these devices. While MZI Meshes open up a way to programable general purpose PICs, there are a few challenges that have to be improved for larger systems, like

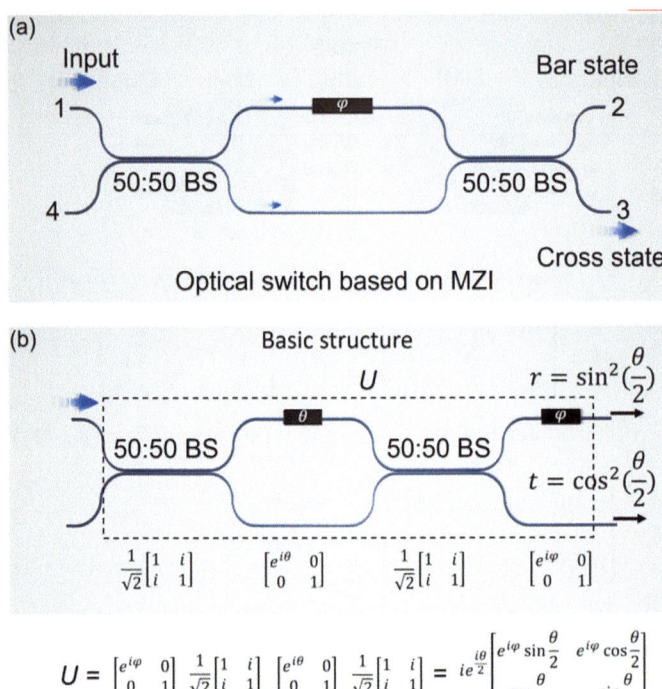

Fig. 20 **a** Schematic of an MZI-based optical switch. **b** Schematic of an MZI realising an elementary 2×2 unitary matrix operation. The matrices are taken from [131]

maintaining an exact phase control through all optical pathways and decreasing the complexity of the electronic interface [131, 133–136].

In summary, this section provides an overview of several different approaches to increase computing power by performing MVM in the optical domain. Hardware accelerators with crossbar arrays, ring resonator arrays, and MMIs can utilize several multiplexing techniques such as WDM to achieve exceptionally high parallelism in computing, which leads to high data throughput. On the other hand, accelerators based on MZIs require single-wavelength light sources, simplifying integration requirements. However, the need for phase error compensation can complicate the operation of these devices.

4 Outlook and Challenges

Transitioning to the optical domain in computing promises unmatched potential for high data throughput and computational speed due to the intrinsic properties of light as evidenced in ONNs and optical Hardware Accelerators. Unlike typical electronic

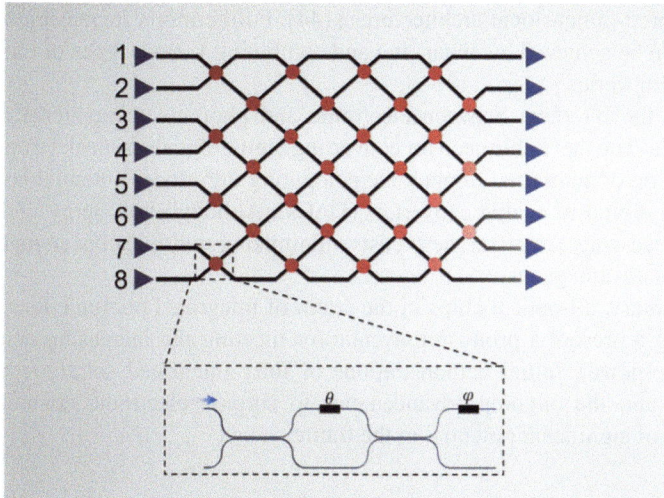

Fig. 21 A schematic of a cascaded 8×8 MZI Mesh. Each crossing (red dots) corresponds to an MZI performing a 2×2 unitary operation [131]

architectures, the bottlenecks of data transfer and processing are significantly mitigated by in-memory computing enabled by PCM or the integration of electro-optic devices. We have observed diverse implementations of neural network structures on optical chips, and specific optical circuits designed to enhance computationally intensive tasks like Matrix-Vector Multiplication. However, matching the performance and availability of commercial electronic systems presents several challenges.

A critical aspect is integrating all necessary devices on-chip and ensuring system-wide integration. Many presented solutions are still primarily confined to laboratory environments. While the main functionality is on-chip, much of the required periphery remains off-chip. Integrating active components like light sources, modulators, and photodetectors poses fabrication challenges, often necessitating the coupling of different platforms [99, 137–139]. However, the maturation of fabrication processes and the increasing availability of foundries capable of handling these challenges will significantly influence the development of fully integrated circuits [102, 139–142]. Additionally, packaging and interconnecting different chips is a research area in and of itself, with advances in 3D printing of photonic wire bonds showing promise for improving chip-to-chip connections [143].

To compete with state-of-the-art electronics, the computing density of these systems must be comparable. Although PICs have limitations in reducing footprint, they match up in computing density [20]. The limitations in footprint stem from the minimum waveguide width and bending radii required to guide light at a certain wavelength. For example, in a SiN platform, the typical width is around 1.2 μm with bending radii of 50 μm [57, 64]. However, higher index materials like SOI can reduce this to approximately 0.5 μm [117] with radii of about 10 μm at the telecom wavelength of 1550 nm. An intriguing approach to reducing the footprint of PICs is to develop three-dimensional architectures [144]. Furthermore, increasing computing density can be achieved by enhancing and combining various types of multiplexing available for optics.

Lastly, the interface between electronic and photonic components remains a crucial area. The inevitable need for converting digital signals from electronic circuits to the analog domain used in photonic computing introduces potential bottlenecks, notably in digital-to-analog converters (DACs). Although the energy savings from optical processing can offset these costs, minimizing multiple conversions between domains is advantageous.

In summary, all-optical chips in the realm of integrated photonic neuromorphic computing represent a promising avenue for meeting the increasing demands for computing power. Initial demonstrations of fully integrated solutions have been achieved, and the ongoing advancements to surpass electronic counterparts are indicative of significant potential in the future.

References

1. Backus J (1978) Can programming be liberated from the von Neumann style? A functional style and its algebra of programs. Communications of the ACM 21:613–641. https://doi.org/10.1145/359576.359579
2. Zhang C, Li P, Sun G, Guan Y, Xiao B, Cong J (2015) Optimizing FPGA-based accelerator design for deep convolutional neural networks. In: Proceedings of the 2015 ACM/SIGDA international symposium on field-programmable gate arrays. ACM, Monterey California USA, pp 161–170
3. Baji T (2018) Evolution of the GPU Device widely used in AI and Massive Parallel Processing. 2018 IEEE 2nd Electron Devices Technology and Manufacturing Conference (EDTM). IEEE, Kobe, pp 7–9
4. Jouppi NP, Cliff Young, Patil N, Patterson D, Agrawal G, Bajwa R, Bates S, Bhatia S, Boden N, Borchers A, Boyle R, Cantin P, Chao C, Clark C, Coriell J, Daley M, Dau M, Dean J, Gelb B, Ghaemmaghami TV, Gottipati R, Gulland W, Hagmann R, Ho CR, Hogberg D, Hu J, Hundt R, Hurt D, Ibarz J, Jaffey A, Jaworski A, Kaplan A, Khaitan H, Killebrew D, Koch A, Kumar N, Lacy S, Laudon J, Law J, Le D, Leary C, Liu Z, Lucke K, Lundin A, MacKean G, Maggiore A, Mahony M, Miller K, Nagarajan R, Narayanaswami R, Ni R, Nix K, Norrie T, Omernick M, Penukonda N, Phelps A, Ross J, Ross M, Salek A, Samadiani E, Severn C, Sizikov G, Snelham M, Souter J, Steinberg D, Swing A, Tan M, Thorson G, Tian B, Toma H, Tuttle E, Vasudevan V, Walter R, Wang W, Wilcox E, Yoon DH (2017) In-Datacenter Performance Analysis of a Tensor Processing Unit. In: Proceedings of the 44th annual international symposium on computer architecture. ACM, Toronto ON Canada, pp 1–12
5. Sipola T, Alatalo J, Kokkonen T, Rantonen M (2022) Artificial Intelligence in the IoT Era: A review of edge AI hardware and software. 2022 31st Conference of Open Innovations Association (FRUCT). IEEE, Helsinki, Finland, pp 320–331
6. Talib MA, Majzoub S, Nasir Q, Jamal D (2021) A systematic literature review on hardware implementation of artificial intelligence algorithms. J Supercomput 77:1897–1938. https://doi.org/10.1007/s11227-020-03325-8
7. Misra J, Saha I (2010) Artificial neural networks in hardware: A survey of two decades of progress. Neurocomputing 74:239–255. https://doi.org/10.1016/j.neucom.2010.03.021

8. Ielmini D, Wong H-SP (2018) In-memory computing with resistive switching devices. Nat Electron 1:333–343. https://doi.org/10.1038/s41928-018-0092-2

9. Alibart F, Zamanidoost E, Strukov DB (2013) Pattern classification by memristive crossbar circuits using ex situ and in situ training. Nat Commun 4:2072. https://doi.org/10.1038/ncomms3072

10. Boybat I, Le Gallo M, Nandakumar SR, Moraitis T, Parnell T, Tuma T, Rajendran B, Leblebici Y, Sebastian A, Eleftheriou E (2018) Neuromorphic computing with multi-memristive synapses. Nat Commun 9:2514. https://doi.org/10.1038/s41467-018-04933-y

11. Hu M, Li H, Chen Y, Wu Q, Rose GS, Linderman RW (2014) Memristor crossbar-based neuromorphic computing system: A case study. IEEE Trans Neural Netw Learning Syst 25:1864–1878. https://doi.org/10.1109/TNNLS.2013.2296777

12. Borghetti J, Snider GS, Kuekes PJ, Yang JJ, Stewart DR, Williams RS (2010) 'Memristive' switches enable 'stateful' logic operations via material implication. Nature 464:873–876. https://doi.org/10.1038/nature08940

13. Li C, Hu M, Li Y, Jiang H, Ge N, Montgomery E, Zhang J, Song W, Dávila N, Graves CE, Li Z, Strachan JP, Lin P, Wang Z, Barnell M, Wu Q, Williams RS, Yang JJ, Xia Q (2017) Analogue signal and image processing with large memristor crossbars. Nat Electron 1:52–59. https://doi.org/10.1038/s41928-017-0002-z

14. Li Y, Ang K-W (2021) Hardware implementation of neuromorphic computing using large-scale memristor crossbar arrays. Adv Intell Syst 3:2000137. https://doi.org/10.1002/aisy.202000137

15. Le Gallo M, Khaddam-Aljameh R, Stanisavljevic M, Vasilopoulos A, Kersting B, Dazzi M, Karunaratne G, Brändli M, Singh A, Müller SM, Büchel J, Timoneda X, Joshi V, Rasch MJ, Egger U, Garofalo A, Petropoulos A, Antonakopoulos T, Brew K, Choi S, Ok I, Philip T, Chan V, Silvestre C, Ahsan I, Saulnier N, Narayanan V, Francese PA, Eleftheriou E, Sebastian A (2023) A 64-core mixed-signal in-memory compute chip based on phase-change memory for deep neural network inference. Nat Electron 6:680–693. https://doi.org/10.1038/s41928-023-01010-1

16. Gong N, Idé T, Kim S, Boybat I, Sebastian A, Narayanan V, Ando T (2018) Signal and noise extraction from analog memory elements for neuromorphic computing. Nat Commun 9:2102. https://doi.org/10.1038/s41467-018-04485-1

17. Sebastian A, Le Gallo M, Khaddam-Aljameh R, Eleftheriou E (2020) Memory devices and applications for in-memory computing. Nat Nanotechnol 15:529–544. https://doi.org/10.1038/s41565-020-0655-z

18. Wang S, Lee FC (2010) Analysis and applications of parasitic capacitance cancellation techniques for EMI suppression. IEEE Trans Ind Electron 57:3109–3117. https://doi.org/10.1109/TIE.2009.2038333

19. Khonina SN, Kazanskiy NL, Butt MA, Karpeev SV (2022) IPSI RAS-Branch of the FSRC "Crystallography and Photonics" RAS, Samara 443001, Russia, Samara National Research University, Samara 443086, Russia, Warsaw University of Technology, Institute of Microelectronics and Optoelectronics, Koszykowa 75, Warszawa 00-662, Poland. Optical multiplexing techniques and their marriage for on-chip and optical fiber communication: a review. OEA 5:210127–210127. https://doi.org/10.29026/oea.2022.210127

20. Nahmias MA, De Lima TF, Tait AN, Peng H-T, Shastri BJ, Prucnal PR (2020) Photonic multiply-accumulate operations for neural networks. IEEE J Select Topics Quantum Electron 26:1–18. https://doi.org/10.1109/JSTQE.2019.2941485

21. Miller DAB (2017) Attojoule optoelectronics for low-energy information processing and communications. J Lightwave Technol 35:346–396. https://doi.org/10.1109/JLT.2017.2647779

22. Alduino A, Paniccia M (2007) Wiring electronics with light. Nature Photon 1:153–155. https://doi.org/10.1038/nphoton.2007.17

23. Mead C (1990) Neuromorphic electronic systems. Proc IEEE 78:1629–1636. https://doi.org/10.1109/5.58356

24. Christensen DV, Dittmann R, Linares-Barranco B, Sebastian A, Le Gallo M, Redaelli A, Slesazeck S, Mikolajick T, Spiga S, Menzel S, Valov I, Milano G, Ricciardi C, Liang S-J, Miao F, Lanza M, Quill TJ, Keene ST, Salleo A, Grollier J, Marković D, Mizrahi A, Yao P, Yang JJ, Indiveri G, Strachan JP, Datta S, Vianello E, Valentian A, Feldmann J, Li X, Pernice WHP, Bhaskaran H, Furber S, Neftci E, Scherr F, Maass W, Ramaswamy S, Tapson J, Panda P, Kim Y, Tanaka G, Thorpe S, Bartolozzi C, Cleland TA, Posch C, Liu S, Panuccio G, Mahmud M, Mazumder AN, Hosseini M, Mohsenin T, Donati E, Tolu S, Galeazzi R, Christensen ME, Holm S, Ielmini D, Pryds N (2022) 2022 roadmap on neuromorphic computing and engineering. Neuromorph Comput Eng 2:022501. https://doi.org/10.1088/2634-4386/ac4a83
25. Saleh BEA, Teich MC (1991) Fundamentals of Photonics, 1st edn. Wiley
26. Wang K, Nirmalathas A, Lim C, Wong E, Alameh K, Li H, Skafidas E (2018) High-speed indoor optical wireless communication system employing a silicon integrated photonic circuit. Opt Lett 43:3132. https://doi.org/10.1364/OL.43.003132
27. Lee W, Han S, Moon S-R, Park J, Yoo S, Park H, Lee JK, Yu K, Cho SH (2022) Coherent terahertz wireless communication using dual-parallel MZM-based silicon photonic integrated circuits. Opt Express 30:2547. https://doi.org/10.1364/OE.446516
28. Wang K, Yuan Z, Wong E, Alameh K, Li H, Sithamparanathan K, Skafidas E (2019) Experimental demonstration of indoor infrared optical wireless communications with a silicon photonic integrated circuit. J Lightwave Technol 37:619–626. https://doi.org/10.1109/JLT.2018.2889252
29. Chang J, Gao J, Esmaeil Zadeh I, Elshaari AW, Zwiller V (2023) Nanowire-based integrated photonics for quantum information and quantum sensing. Nanophotonics 12:339–358. https://doi.org/10.1515/nanoph-2022-0652
30. Kazanskiy NL, Khonina SN, Butt MA (2022) Advancement in silicon integrated photonics technologies for sensing applications in near-infrared and mid-infrared region: A review. Photonics 9:331. https://doi.org/10.3390/photonics9050331
31. Elshaari AW, Pernice W, Srinivasan K, Benson O, Zwiller V (2020) Hybrid integrated quantum photonic circuits. Nat Photonics 14:285–298. https://doi.org/10.1038/s41566-020-0609-x
32. Xu R, Taheriniya S, Ovvyan AP, Bankwitz JR, McRae L, Jung E, Brückerhoff-Plückelmann F, Bente I, Lenzini F, Bhaskaran H, Pernice WHP (2023) Hybrid photonic integrated circuits for neuromorphic computing [Invited]. Opt Mater Express 13:3553. https://doi.org/10.1364/OME.502179
33. Huang Y, Yue H, Ma W, Zhang Y, Xiao Y, Tang Y, Tang H, Chu T (2023) Parallel photonic acceleration processor for matrix–matrix multiplication. Opt Lett 48:3231. https://doi.org/10.1364/OL.488464
34. Varri A, Taheriniya S, Brückerhoff-Plückelmann F, Bente I, Farmakidis N, Bernhardt D, Rösner H, Kruth M, Nadzeyka A, Richter T, Wright CD, Bhaskaran H, Wilde G, Pernice WHP (2023) Scalable non-volatile tuning of photonic computational memories by automated silicon ion implantation. Adv Mat 2310596. https://doi.org/10.1002/adma.202310596
35. Dabos G, Bellas DV, Stabile R, Moralis-Pegios M, Giamougiannis G, Tsakyridis A, Totovic A, Lidorikis E, Pleros N (2022) Neuromorphic photonic technologies and architectures: scaling opportunities and performance frontiers [Invited]. Opt Mater Express 12:2343. https://doi.org/10.1364/OME.452138
36. Aoyama K, Kobayashi S, Wada M, Yokota N, Kita T, Yasaka H (2018) Compact narrow-linewidth optical negative feedback laser with Si optical filter. Appl Phys Express 11:112703. https://doi.org/10.7567/APEX.11.112703
37. Wang J, Xiong N, Zou W (2023) Loss compensation of an ultra-wideband electro-optic modulator in heterogeneous silicon/erbium-doped lithium niobate. Opt Lett 48:3399. https://doi.org/10.1364/OL.489988
38. Shen J, Zhang Y, Feng C, Xu Z, Zhang L, Su Y (2023) Hybrid lithium tantalite-silicon integrated photonics platform for electro-optic modulation. Opt Lett 48:6176. https://doi.org/10.1364/OL.502492
39. Fujikata J, Noguchi M, Katamawari R, Inaba K, Ono H, Shimura D, Onawa Y, Yaegashi H, Ishikawa Y (2023) High-performance Ge/Si electro-absorption optical modulator up to 85°C

and its highly efficient photodetector operation. Opt Express 31:10732. https://doi.org/10.1364/OE.484380

40. Ye H, Han Q, Wang S, Chu Y, Zheng Y, Geng L (2024) High speed evanescent waveguide photodetector with a 100 GHz bandwidth. Appl Phys Express 17:012001. https://doi.org/10.35848/1882-0786/ad0e90

41. Marzen S, Postelnicu E, Michel J, Wada K, Kimerling LC (2023) High performance germanium on silicon photodiodes for back-end-of-line photonic integration. Appl Phys Lett 123:111105. https://doi.org/10.1063/5.0153651

42. Feldmann J, Youngblood N, Karpov M, Gehring H, Li X, Stappers M, Le Gallo M, Fu X, Lukashchuk A, Raja AS, Liu J, Wright CD, Sebastian A, Kippenberg TJ, Pernice WHP, Bhaskaran H (2021) Parallel convolutional processing using an integrated photonic tensor core. Nature 589:52–58. https://doi.org/10.1038/s41586-020-03070-1

43. Nikbakht H, Khoshmehr MT, Van Someren B, Teichrib D, Hammer M, Förstner J, Akca BI (2023) Asymmetric, non-uniform 3-dB directional coupler with 300-nm bandwidth and a small footprint. Opt Lett 48:207. https://doi.org/10.1364/OL.476537

44. Brückerhoff-Plückelmann F, Bente I, Becker M, Vollmar N, Farmakidis N, Lomonte E, Lenzini F, Wright CD, Bhaskaran H, Salinga M, Risse B, Pernice WHP (2023) Event-driven adaptive optical neural network. Sci Adv 9:eadi9127. https://doi.org/10.1126/sciadv.adi9127

45. Bankwitz JR, Dijkstra J, Pradip R, McRae L, Lomonte E, Lenzini F, Pernice WHP (2023) Towards "smart transceivers" in FPGA-controlled lithium-niobate-on-insulator integrated circuits for edge computing applications [Invited]. Opt Mater Express 13:3667. https://doi.org/10.1364/OME.503340

46. Bogaerts W, De Heyn P, Van Vaerenbergh T, De Vos K, Kumar Selvaraja S, Claes T, Dumon P, Bienstman P, Van Thourhout D, Baets R (2012) Silicon microring resonators. Laser & Photonics Rev 6:47–73. https://doi.org/10.1002/lpor.201100017

47. Lee BS, Kim B, Freitas AP, Mohanty A, Zhu Y, Bhatt GR, Hone J, Lipson M (2020) High-performance integrated graphene electro-optic modulator at cryogenic temperature. Nanophotonics 10:99–104. https://doi.org/10.1515/nanoph-2020-0363

48. Feldmann J, Youngblood N, Li X, Wright CD, Bhaskaran H, Pernice WHP (2020) Integrated 256 cell photonic phase-change memory with 512-bit capacity. IEEE J Select Topics Quantum Electron 26:1–7. https://doi.org/10.1109/JSTQE.2019.2956871

49. Chrostowski L, Hochberg M (2015) Silicon Photonics Design. Cambridge University Press, Cambridge

50. Soldano LB, Pennings ECM (1995) Optical multi-mode interference devices based on self-imaging: principles and applications. J Lightwave Technol 13:615–627. https://doi.org/10.1109/50.372474

51. Cooney K, Peters FH (2016) Analysis of multimode interferometers. Opt Express 24:22481. https://doi.org/10.1364/OE.24.022481

52. Priti RB, Pishvai Bazargani H, Xiong Y, Liboiron-Ladouceur O (2017) Mode selecting switch using multimode interference for on-chip optical interconnects. Opt Lett 42:4131. https://doi.org/10.1364/OL.42.004131

53. Nishida K, Sasai K, Xu R, Yen T-H, Tang Y-L, Takahara J, Chu S-W (2023) All-optical scattering control in an all-dielectric quasi-perfect absorbing Huygens' metasurface. Nanophotonics 12:139–146. https://doi.org/10.1515/nanoph-2022-0597

54. Šik J, Hora J, Humlíček J (1998) Optical functions of silicon at high temperatures. J Appl Phys 84:6291–6298. https://doi.org/10.1063/1.368951

55. Soref R, Bennett B (1987) Electrooptical effects in silicon. IEEE J Quantum Electron 23:123–129. https://doi.org/10.1109/JQE.1987.1073206

56. Wuttig M, Bhaskaran H, Taubner T (2017) Phase-change materials for non-volatile photonic applications. Nature Photon 11:465–476. https://doi.org/10.1038/nphoton.2017.126

57. Ríos C, Stegmaier M, Hosseini P, Wang D, Scherer T, Wright CD, Bhaskaran H, Pernice WHP (2015) Integrated all-photonic non-volatile multi-level memory. Nature Photon 9:725–732. https://doi.org/10.1038/nphoton.2015.182

58. Wuttig M, Yamada N (2007) Phase-change materials for rewriteable data storage. Nature Mater 6:824–832. https://doi.org/10.1038/nmat2009
59. Raoux S, Xiong F, Wuttig M, Pop E (2014) Phase change materials and phase change memory. MRS Bull 39:703–710. https://doi.org/10.1557/mrs.2014.139
60. Raoux S, Burr GW, Breitwisch MJ, Rettner CT, Chen Y-C, Shelby RM, Salinga M, Krebs D, Chen S-H, Lung H-L, Lam CH (2008) Phase-change random access memory: A scalable technology. IBM J Res & Dev 52:465–479. https://doi.org/10.1147/rd.524.0465
61. Abdollahramezani S, Hemmatyar O, Taghinejad H, Krasnok A, Kiarashinejad Y, Zandehshahvar M, Alu A, Adibi A (2020) Tunable nanophotonics enabled by chalcogenide phase-change materials. Nanophotonics 9:1189–1241. https://doi.org/10.1515/nanoph-2020-0039
62. Berziņš J, Indrišiūnas S, Fasold S, Steinert M, Žukovskaja O, Cialla-May D, Gečys P, Bäumer SMB, Pertsch T, Setzpfandt F (2020) Laser-induced spatially-selective tailoring of high-index dielectric metasurfaces. Opt Express 28:1539. https://doi.org/10.1364/OE.380383
63. Stegmaier M, Ríos C, Bhaskaran H, Pernice WHP (2016) Thermo-optical effect in phase-change nanophotonics. ACS Photonics 3:828–835. https://doi.org/10.1021/acsphotonics.6b00032
64. Brückerhoff-Plückelmann F, Feldmann J, Wright CD, Bhaskaran H, Pernice WHP (2021) Chalcogenide phase-change devices for neuromorphic photonic computing. J Appl Phys 129:151103. https://doi.org/10.1063/5.0042549
65. Zhou W, Farmakidis N, Feldmann J, Li X, Tan J, He Y, Wright CD, Pernice WHP, Bhaskaran H (2022) Phase-change materials for energy-efficient photonic memory and computing. MRS Bulletin 47:502–510. https://doi.org/10.1557/s43577-022-00358-7
66. Feldmann J, Stegmaier M, Gruhler N, Ríos C, Bhaskaran H, Wright CD, Pernice WHP (2017) Calculating with light using a chip-scale all-optical abacus. Nat Commun 8:1256. https://doi.org/10.1038/s41467-017-01506-3
67. Dong B, Brückerhoff-Plückelmann F, Meyer L, Dijkstra J, Bente I, Wendland D, Varri A, Aggarwal S, Farmakidis N, Wang M, Yang G, Lee JS, He Y, Gooskens E, Kwong D-L, Bienstman P, Pernice WHP, Bhaskaran H (2024) Partial coherence enhances parallelized photonic computing. Nature 632:55–62. https://doi.org/10.1038/s41586-024-07590-y
68. Doerr CR (2015) Silicon photonic integration in telecommunications. Front Phys 3. https://doi.org/10.3389/fphy.2015.00037
69. Romero-García S, Merget F, Zhong F, Finkelstein H, Witzens J (2013) Silicon nitride CMOS-compatible platform for integrated photonics applications at visible wavelengths. Opt Express 21:14036. https://doi.org/10.1364/OE.21.014036
70. Li HH (1980) Refractive index of silicon and germanium and its wavelength and temperature derivatives. J Phys Chem Ref Data 9:561–658. https://doi.org/10.1063/1.555624
71. Soref R (2010) Mid-infrared photonics in silicon and germanium. Nature Photon 4:495–497. https://doi.org/10.1038/nphoton.2010.171
72. Siew SY, Li B, Gao F, Zheng HY, Zhang W, Guo P, Xie SW, Song A, Dong B, Luo LW, Li C, Luo X, Lo G-Q (2021) Review of silicon photonics technology and platform development. J Lightwave Technol 39:4374–4389. https://doi.org/10.1109/JLT.2021.3066203
73. Absil P, Croes K, Lesniewska A, De Heyn P, Ban Y, Snyder B, De Coster J, Fodor F, Simons V, Balakrishnan S, Lepage G, Golshani N, Lardenois S, Srinivasan SA, Chen H, Vanherle W, Loo R, Boufadil R, Detalle M, Miller A, Verheyen P, Pantouvaki M, Van Campenhout J (2017) Reliable 50Gb/s silicon photonics platform for next-generation data center optical interconnects. In: 2017 IEEE International Electron Devices Meeting (IEDM). IEEE, San Francisco, CA, p 34.2.1–34.2.4
74. Pantouvaki M, De Heyn P, Rakowski M, Verheyen P, Snyder B, Srinivasan SA, Chen H, De Coster J, Lepage G, Absil P, Van Campenhout J (2016) 50Gb/s Silicon photonics platform for short-reach optical interconnects. In: Optical fiber communication conference. OSA, Anaheim, California, p Th4H.4
75. Bååk T (1982) Silicon oxynitride; a material for GRIN optics. Appl Opt 21:1069. https://doi.org/10.1364/AO.21.001069

76. Sacher WD, Mikkelsen JC, Huang Y, Mak JCC, Yong Z, Luo X, Li Y, Dumais P, Jiang J, Goodwill D, Bernier E, Lo PG-Q, Poon JKS (2018) Monolithically integrated multilayer silicon nitride-on-silicon waveguide platforms for 3-D photonic circuits and devices. Proc IEEE 106:2232–2245. https://doi.org/10.1109/JPROC.2018.2860994
77. Baets R, Subramanian AZ, Clemmen S, Kuyken B, Bienstman P, Le Thomas N, Roelkens G, Van Thourhout D, Helin P, Severi S (2016) Silicon Photonics: silicon nitride versus silicon-on-insulator. In: Optical fiber communication conference. OSA, Anaheim, California, p Th3J.1
78. Xie W, Zhu Y, Aubert T, Verstuyft S, Hens Z, Van Thourhout D (2015) Low-loss silicon nitride waveguide hybridly integrated with colloidal quantum dots. Opt Express 23:12152. https://doi.org/10.1364/OE.23.012152
79. Wang S, Zhan M, Wang G, Xuan H, Zhang W, Liu C, Xu C, Liu Y, Wei Z, Chen X (2013) 4H-SiC: a new nonlinear material for midinfrared lasers. Laser & Photonics Rev 7:831–838. https://doi.org/10.1002/lpor.201300068
80. Yi A, Wang C, Zhou L, Zhu Y, Zhang S, You T, Zhang J, Ou X (2022) Silicon carbide for integrated photonics. Appl Phys Rev 9:031302. https://doi.org/10.1063/5.0079649
81. Huang C, Zhang H, Sun H (2020) Ultraviolet optoelectronic devices based on AlGaN-SiC platform: Towards monolithic photonics integration system. Nano Energy 77:105149. https://doi.org/10.1016/j.nanoen.2020.105149
82. Castelletto S, Peruzzo A, Bonato C, Johnson BC, Radulaski M, Ou H, Kaiser F, Wrachtrup J (2022) Silicon carbide photonics bridging quantum technology. ACS Photonics 9:1434–1457. https://doi.org/10.1021/acsphotonics.1c01775
83. Babin C, Stöhr R, Morioka N, Linkewitz T, Steidl T, Wörnle R, Liu D, Hesselmeier E, Vorobyov V, Denisenko A, Hentschel M, Gobert C, Berwian P, Astakhov GV, Knolle W, Majety S, Saha P, Radulaski M, Son NT, Ul-Hassan J, Kaiser F, Wrachtrup J (2022) Fabrication and nanophotonic waveguide integration of silicon carbide colour centres with preserved spin-optical coherence. Nat Mater 21:67–73. https://doi.org/10.1038/s41563-021-01148-3
84. Kösters M, Sturman B, Werheit P, Haertle D, Buse K (2009) Optical cleaning of congruent lithium niobate crystals. Nature Photon 3:510–513. https://doi.org/10.1038/nphoton.2009.142
85. Chen G, Li N, Ng JD, Lin H-L, Zhou Y, Fu YH, Lee LYT, Yu Y, Liu A-Q, Danner AJ (2022) Advances in lithium niobate photonics: Development status and perspectives. Adv Photon 4. https://doi.org/10.1117/1.AP.4.3.034003
86. Wang M, Wu R, Lin J, Zhang J, Fang Z, Chai Z, Cheng Y (2019) Chemo-mechanical polish lithography: A pathway to low loss large-scale photonic integration on lithium niobate on insulator. Quantum Eng 1:e9. https://doi.org/10.1002/que2.9
87. Qi Y, Li Y (2020) Integrated lithium niobate photonics. Nanophotonics 9:1287–1320. https://doi.org/10.1515/nanoph-2020-0013
88. Huffaker DL, Park G, Zou Z, Shchekin OB, Deppe DG (1998) 1.3 μm room-temperature GaAs-based quantum-dot laser. Appl Phys Lett 73:2564–2566. https://doi.org/10.1063/1.122534
89. Liu HY, Childs DT, Badcock TJ, Groom KM, Sellers IR, Hopkinson M, Hogg RA, Robbins DJ, Mowbray DJ, Skolnick MS (2005) High-performance three-layer 1.3-/spl mu/m InAs-GaAs quantum-dot lasers with very low continuous-wave room-temperature threshold currents. IEEE Photon Technol Lett 17:1139–1141. https://doi.org/10.1109/LPT.2005.846948
90. Gready D, Eisenstein G, Ivanov V, Gilfert C, Schnabel F, Rippien A, Reithmaier JP, Bornholdt C (2014) High speed 1.55 μm InAs/InGaAlAs/InP quantum dot lasers. IEEE Photon Technol Lett 26:11–13. https://doi.org/10.1109/LPT.2013.2287502
91. Tang M, Park J-S, Wang Z, Chen S, Jurczak P, Seeds A, Liu H (2019) Integration of III-V lasers on Si for Si photonics. Prog Quantum Electron 66:1–18. https://doi.org/10.1016/j.pquantelec.2019.05.002
92. Sun J, Han M, Gu Y, Yang Z, Zeng H (2018) Recent advances in group III–V nanowire infrared detectors. Adv Opt Mater 6:1800256. https://doi.org/10.1002/adom.201800256
93. Cao F, Liu L, Li L (2023) Short-wave infrared photodetector. Materials Today 62:327–349. https://doi.org/10.1016/j.mattod.2022.11.003

94. Haq B, Kumari S, Van Gasse K, Zhang J, Gocalinska A, Pelucchi E, Corbett B, Roelkens G (2020) Micro-transfer-printed III-V-on-silicon C-band semiconductor optical amplifiers. Laser & Photonics Rev 14:1900364. https://doi.org/10.1002/lpor.201900364
95. Davenport ML, Skendzic S, Volet N, Hulme JC, Heck MJR, Bowers JE (2016) Heterogeneous Silicon/III–V semiconductor optical amplifiers. IEEE J Sel Top Quantum Electron 22:78–88. https://doi.org/10.1109/JSTQE.2016.2593103
96. Sinatkas G, Christopoulos T, Tsilipakos O, Kriezis EE (2021) Electro-optic modulation in integrated photonics. J Appl Phys 130:010901. https://doi.org/10.1063/5.0048712
97. Justice J, Bower C, Meitl M, Mooney MB, Gubbins MA, Corbett B (2012) Wafer-scale integration of group III–V lasers on silicon using transfer printing of epitaxial layers. Nature Photon 6:610–614. https://doi.org/10.1038/nphoton.2012.204
98. Yang W, Li Y, Meng F, Yu H, Wang M, Wang P, Luo G, Zhou X, Pan J (2019) III–V compound materials and lasers on silicon. J Semicond 40:101305. https://doi.org/10.1088/1674-4926/40/10/101305
99. Liang D, Bowers JE (2021) Recent Progress in Heterogeneous III-V-on-Silicon Photonic Integration. Light: Advanced Manufacturing 2:59. https://doi.org/10.37188/lam.2021.005
100. Vyas K, Espinosa DHG, Hutama D, Jain SK, Mahjoub R, Mobini E, Awan KM, Lundeen J, Dolgaleva K (2022) Group III-V semiconductors as promising nonlinear integrated photonic platforms. Adv Phys: X 7:2097020. https://doi.org/10.1080/23746149.2022.2097020
101. Walker R (1987) High-speed electrooptic modulation in GaAs/GaAlAs waveguide devices. J Lightwave Technol 5:1444–1453. https://doi.org/10.1109/JLT.1987.1075432
102. Smit M, Williams K, Van Der Tol J (2019) Past, present, and future of InP-based photonic integration. APL Photonics 4:050901. https://doi.org/10.1063/1.5087862
103. Jiao Y, Van Der Tol J, Pogoretskii V, Van Engelen J, Kashi AA, Reniers S, Wang Y, Zhao X, Yao W, Liu T, Pagliano F, Fiore A, Zhang X, Cao Z, Kumar RR, Tsang HK, Van Veld-hoven R, De Vries T, Geluk E-J, Bolk J, Ambrosius H, Smit M, Williams K (2020) Indium Phosphide Membrane nanophotonic integrated circuits on silicon. Physica Status Solidi (A) 217:1900606. https://doi.org/10.1002/pssa.201900606
104. Jiao Y, Nishiyama N, Van Der Tol J, Van Engelen J, Pogoretskiy V, Reniers S, Kashi AA, Wang Y, Calzadilla VD, Spiegelberg M, Cao Z, Williams K, Amemiya T, Arai S (2021) InP membrane integrated photonics research. Semicond Sci Technol 36:013001. https://doi.org/10.1088/1361-6641/abcadd
105. Van Der Tol JJGM, Jiao Y, Van Engelen JP, Pogoretskiy V, Kashi AA, Williams K (2020) InP Membrane on Silicon (IMOS) Photonics. IEEE J Quantum Electron 56:1–7. https://doi.org/10.1109/JQE.2019.2953296
106. Majkić A, Puc U, Franke A, Kirste R, Collazo R, Sitar Z, Zgonik M (2015) Optical properties of aluminum nitride single crystals in the THz region. Opt Mater Express 5:2106. https://doi.org/10.1364/OME.5.002106
107. Gaeta AL, Lipson M, Kippenberg TJ (2019) Photonic-chip-based frequency combs. Nature Photon 13:158–169. https://doi.org/10.1038/s41566-019-0358-x
108. Karabalin RB, Matheny MH, Feng XL, Defaÿ E, Le Rhun G, Marcoux C, Hentz S, Andreucci P, Roukes ML (2009) Piezoelectric nanoelectromechanical resonators based on aluminum nitride thin films. Applied Physics Letters 95:103111. https://doi.org/10.1063/1.3216586
109. Fan L, Sun X, Xiong C, Schuck C, Tang HX (2013) Aluminum nitride piezo-acousto-photonic crystal nanocavity with high quality factors. Appl Phys Lett 102:153507. https://doi.org/10.1063/1.4802250
110. Li N, Ho CP, Zhu S, Fu YH, Zhu Y, Lee LYT (2021) Aluminium nitride integrated photonics: a review. Nanophotonics 10:2347–2387. https://doi.org/10.1515/nanoph-2021-0130
111. Xiong C, Pernice WHP, Sun X, Schuck C, Fong KY, Tang HX (2012) Aluminum nitride as a new material for chip-scale optomechanics and nonlinear optics. New J Phys 14:095014. https://doi.org/10.1088/1367-2630/14/9/095014
112. Zhao S, Connie AT, Dastjerdi MHT, Kong XH, Wang Q, Djavid M, Sadaf S, Liu XD, Shih I, Guo H, Mi Z (2015) Aluminum nitride nanowire light emitting diodes: Breaking the funda-mental bottleneck of deep ultraviolet light sources. Sci Rep 5:8332. https://doi.org/10.1038/srep08332

113. Spies M, Hertog MID, Hille P, Schörmann J, Gayral B, Eickhoff M Monroy E Bias-controlled spectral response in GaN/AlN single-nanowire ultraviolet photodetectors

114. Zheng Y, Sun C, Xiong B, Wang L, Hao Z, Wang J, Han Y, Li H, Yu J, Luo Y (2022) Integrated gallium nitride nonlinear photonics. Laser & Photonics Rev 16:2100071. https://doi.org/10.1002/lpor.202100071

115. Mi S, Kiss M, Graziosi T, Quack N (2020) Integrated photonic devices in single crystal diamond. J Phys: Photonics 2:042001. https://doi.org/10.1088/2515-7647/aba171

116. Feldmann J, Youngblood N, Wright CD, Bhaskaran H, Pernice WHP (2019) All-optical spiking neurosynaptic networks with self-learning capabilities. Nature 569:208–214. https://doi.org/10.1038/s41586-019-1157-8

117. Ashtiani F, Geers AJ, Aflatouni F (2022) An on-chip photonic deep neural network for image classification. Nature 606:501–506. https://doi.org/10.1038/s41586-022-04714-0

118. Zhou H, Dong J, Cheng J, Dong W, Huang C, Shen Y, Zhang Q, Gu M, Qian C, Chen H, Ruan Z, Zhang X (2022) Photonic matrix multiplication lights up photonic accelerator and beyond. Light Sci Appl 11:30. https://doi.org/10.1038/s41377-022-00717-8

119. Liu J, Wu Q, Sui X, Chen Q, Gu G, Wang L, Li S (2021) Research progress in optical neural networks: theory, applications and developments. PhotoniX 2:5. https://doi.org/10.1186/s43074-021-00026-0

120. Sui X, Wu Q, Liu J, Chen Q, Gu G (2020) A Review of optical neural networks. IEEE Access 8:70773–70783. https://doi.org/10.1109/ACCESS.2020.2987333

121. Xu R, Lv P, Xu F, Shi Y (2021) A survey of approaches for implementing optical neural networks. Opt & Laser Technol 136:106787. https://doi.org/10.1016/j.optlastec.2020.106787

122. Brückerhoff-Plückelmann F, Feldmann J, Gehring H, Zhou W, Wright CD, Bhaskaran H, Pernice W (2022) Broadband photonic tensor core with integrated ultra-low crosstalk wavelength multiplexers. Nanophotonics 11:4063–4072. https://doi.org/10.1515/nanoph-2021-0752

123. Shen Y, Harris NC, Skirlo S, Prabhu M, Baehr-Jones T, Hochberg M, Sun X, Zhao S, Larochelle H, Englund D, Soljačić M (2017) Deep learning with coherent nanophotonic circuits. Nature Photon 11:441–446. https://doi.org/10.1038/nphoton.2017.93

124. Aggarwal S, Milne T, Farmakidis N, Feldmann J, Li X, Shu Y, Cheng Z, Salinga M, Pernice WH, Bhaskaran H (2022) Antimony as a programmable element in integrated nanophotonics. Nano Lett 22:3532–3538. https://doi.org/10.1021/acs.nanolett.1c04286

125. Tanaka D, Shoji Y, Kuwahara M, Wang X, Kintaka K, Kawashima H, Toyosaki T, Ikuma Y, Tsuda H (2012) Ultra-small, self-holding, optical gate switch using Ge_2Sb_2Te_5 with a multi-mode Si waveguide. Opt Express 20:10283. https://doi.org/10.1364/OE.20.010283

126. Zhou W, Dong B, Farmakidis N, Li X, Youngblood N, Huang K, He Y, David Wright C, Pernice WHP, Bhaskaran H (2023) In-memory photonic dot-product engine with electrically programmable weight banks. Nat Commun 14:2887. https://doi.org/10.1038/s41467-023-38473-x

127. Ghazi Sarwat S, Brückerhoff-Plückelmann F, Carrillo SG-C, Gemo E, Feldmann J, Bhaskaran H, Wright CD, Pernice WHP, Sebastian A (2022) An integrated photonics engine for unsupervised correlation detection. Sci Adv 8:eabn3243. https://doi.org/10.1126/sciadv.abn3243

128. Tait AN, Nahmias MA, Shastri BJ, Prucnal PR (2014) Broadcast and weight: An integrated network for scalable photonic spike processing. J Lightwave Technol 32:4029–4041. https://doi.org/10.1109/JLT.2014.2345652

129. Meng X, Zhang G, Shi N, Li G, Azaña J, Capmany J, Yao J, Shen Y, Li W, Zhu N, Li M (2023) Compact optical convolution processing unit based on multimode interference. Nat Commun 14:3000. https://doi.org/10.1038/s41467-023-38786-x

130. Filipovich MJ, Guo Z, Al-Qadasi M, Marquez BA, Morison HD, Sorger VJ, Prucnal PR, Shekhar S, Shastri BJ (2022) Silicon photonic architecture for training deep neural networks with direct feedback alignment. Optica 9:1323. https://doi.org/10.1364/OPTICA.475493

131. Pai S, Bartlett B, Solgaard O, Miller DAB (2019) Matrix optimization on universal unitary photonic devices. Phys Rev Applied 11:064044. https://doi.org/10.1103/PhysRevApplied.11.064044

132. Zhang H, Gu M, Jiang XD, Thompson J, Cai H, Paesani S, Santagati R, Laing A, Zhang Y, Yung MH, Shi YZ, Muhammad FK, Lo GQ, Luo XS, Dong B, Kwong DL, Kwek LC, Liu AQ (2021) An optical neural chip for implementing complex-valued neural network. Nat Commun 12:457. https://doi.org/10.1038/s41467-020-20719-7

133. Du Y, Su K, Yuan X, Li T, Liu K, Man H, Zou X (2023) Implementation of optical neural network based on Mach-Zehnder interferometer array. IET Optoelectronics 17:1–11. https://doi.org/10.1049/ote2.12086

134. Tsakyridis A, Giamougiannis G, Totovic A, Pleros N (2022) Fidelity restorable universal linear optics. Adv Photonics Res 3:2200001. https://doi.org/10.1002/adpr.202200001

135. Bogaerts W, Pérez D, Capmany J, Miller DAB, Poon J, Englund D, Morichetti F, Melloni A (2020) Programmable photonic circuits. Nature 586:207–216. https://doi.org/10.1038/s41586-020-2764-0

136. Dong M, Zimmermann M, Heim D, Choi H, Clark G, Leenheer AJ, Palm KJ, Witte A, Dominguez D, Gilbert G, Eichenfield M, Englund D (2023) Programmable photonic integrated meshes for modular generation of optical entanglement links. npj Quantum Inf 9:42. https://doi.org/10.1038/s41534-023-00708-6

137. Crosnier G, Sanchez D, Bouchoule S, Monnier P, Beaudoin G, Sagnes I, Raj R, Raineri F (2017) Hybrid indium phosphide-on-silicon nanolaser diode. Nature Photon 11:297–300. https://doi.org/10.1038/nphoton.2017.56

138. Miller D (2009) Device Requirements for Optical Interconnects to Silicon Chips. Proc IEEE 97:1166–1185. https://doi.org/10.1109/JPROC.2009.2014298

139. Bogaerts W, Chrostowski L (2018) Silicon photonics circuit design: methods, tools and challenges. Laser & Photonics Rev 12:1700237. https://doi.org/10.1002/lpor.201700237

140. Giewont K, Hu S, Peng B, Rakowski M, Rauch S, Rosenberg JC, Sahin A, Stobert I, Stricker A, Nummy K, Anderson FA, Ayala J, Barwicz T, Bian Y, Dezfulian KK, Gill DM, Houghton T (2019) 300-mm monolithic silicon photonics foundry technology. IEEE J Select Topics Quantum Electron 25:1–11. https://doi.org/10.1109/JSTQE.2019.2908790

141. Rahim A, Goyvaerts J, Szelag B, Fedeli J-M, Absil P, Aalto T, Harjanne M, Littlejohns C, Reed G, Winzer G, Lischke S, Zimmermann L, Knoll D, Geuzebroek D, Leinse A, Geiselmann M, Zervas M, Jans H, Stassen A, Dominguez C, Munoz P, Domenech D, Giesecke AL, Lemme MC, Baets R (2019) Open-access silicon photonics platforms in Europe. IEEE J Select Topics Quantum Electron 25:1–18. https://doi.org/10.1109/JSTQE.2019.2915949

142. Yan Z, Han Y, Lin L, Xue Y, Ma C, Ng WK, Wong KS, Lau KM (2021) A monolithic InP/SOI platform for integrated photonics. Light Sci Appl 10:200. https://doi.org/10.1038/s41377-021-00636-0

143. Lindenmann N, Balthasar G, Hillerkuss D, Schmogrow R, Jordan M, Leuthold J, Freude W, Koos C (2012) Photonic wire bonding: a novel concept for chip-scale interconnects. Opt Express 20:17667. https://doi.org/10.1364/OE.20.017667

144. Moughames J, Porte X, Thiel M, Ulliac G, Larger L, Jacquot M, Kadic M, Brunner D (2020) Three-dimensional waveguide interconnects for scalable integration of photonic neural networks. Optica 7:640. https://doi.org/10.1364/OPTICA.388205

Time-multiplexed Reservoir Computing with Semiconductor Laser Systems

Kathy Lüdge and Lina Jaurigue

Abstract This chapter will focus on a specific variant of reservoir computing (RC) which is based on time-multiplexing the input and readout layers. It is often referred to as delay-based or time-delay RC, as it can easily be realized with a single physical systems subject to self-feedback and does not require a network of numerous coupled nonlinear units. In this time-multiplexed scheme, the nonlinear transformation and mixing of the input data is realized via the internal dynamics of the nonlinear physical system. Strongly nonlinear optical elements, for example semiconductor lasers subject to optical feedback, are promising candidates and offer fast data processing rates combined with the possibility of on-chip hardware realization. The time-delay of the feedback loop has a twofold purpose; it induces a complex response and provides short term memory. While the intrinsic dynamical nonlinearity can be exploited for the computing, it can also lead to intrinsic instabilities that prevent stable computing operation. We will discuss this in detail and elaborate how the bifurcation structure, i.e. the phase diagram in parameter space, can help to find good operation conditions for delay-based RC. Please also see the chapter by Stephan Wong, Doris E. Reiter and Sang Soon Oh for a machine learning based method to determine phase diagrams. We will also focus on the fact that the specific memory requirements for a given reservoir computing task, for example the prediction of a complex time series, are not unique and vary a lot between tasks, leading to different prerequisites for the dynamical system in use.

K. Lüdge (✉) · L. Jaurigue
Theoretische Physik II, Technische Universität Ilmenau, Ilmenau, Germany
e-mail: kathy.luedge@tu-ilmenau.de

L. Jaurigue
e-mail: lina.jaurigue@tu-ilmenau.de

© The Author(s) 2026
M. te Vrugt (ed.), *Artificial Intelligence and Intelligent Matter*, Machine Intelligence for Materials Science, https://doi.org/10.1007/978-3-032-04129-6_18

1 Introduction

Reservoir computing (RC), as explained in the chapter by Michael te Vrugt in Part I of this book, is a machine learning scheme where only the output layer is trained via one linear regression step, while the dynamical network itself is not adapted during the training phase. The RC scheme omits the complex and time consuming process of adjusting the internal connections of coupled networks, that is for example realised via error back-propagation in deep neural networks. (Please see the chapter by Tobias Wand in this book for more details on general methods for machine learning.) Instead it uses large systems with random but fixed connections. In other words, the RC approach trades the sophisticated but energy and time consuming training step with a larger but randomly connected system that hopefully provides the nonlinearities and memory needed for a particular task. The big advantage of the idea is that the random network can be any complex physical system. The requirement is that it provides high dimensional and consistent nonlinear responses to external perturbations. Thus, it does not need to be specifically designed. In recent years, there has been tremendous activity towards finding optimally suited physical reservoirs with high dimensional and nonlinear responses that can be exploited in the RC scheme. To name just a few, the physical substrates used for RC range from passive optical cavities [1–4], skyrmirons [5] and nanomagnetic devices [6], to active laser systems [7–10]. A review about physical implementations can be found in [11] and with a photonic perspective in [12]. Please also see the chapter by Akhilesh Jaiswal, Md Abdullah-Al-Kaiser and Maryam Parsa of this book for hardware implementations.

In general, the performances that can be reached for certain tasks vary a lot between the systems. This is mostly due to the different nonlinear dynamics and internal physical timescales that are involved. In this chapter we want to discuss this point in more detail and investigate the interplay between the bifurcation structure and internal timescales of the reservoir, task specific memory requirements and prediction performance. We will focus on the time-multiplexed RC scheme originally introduced in [13], which can easily be realized with a nonlinear oscillator subjected to self-feedback. In this scheme, which has been experimentally realized with various systems (e.g. [14–17]), the delay line provides short term memory and induces a complex response to the input data [4, 18]. In contrast to the spatially multiplexed RC schemes described in the introductory chapter by Michael te Vrugt of this book, the data are not injected into different network nodes but into the same node at different points in time. In analogy to the input layer with randomly chosen weights used in the conventional RC scheme, the time-multiplexed scheme uses a randomly chosen mask that is multiplied to every input data point in time. The choice of masks can also influence the performance, as has been investigated in [19]. For the numerical investigations presented here, the training itself is very fast and only involves one matrix inversion. The numerical integration of the delayed differential equations for the system dynamics is much more time consuming. For the real world applications envisioned for the RC scheme, e.g. edge computing devices, the complex internal dynamics of a laser system can be directly exploited while the training step is carried out offline.

2 Time-multiplexed Reservoir Computing

The delay-based reservoir computing setups investigated here are comprised of a semiconductor laser and an optical self-feedback line. It is an example of a dynamical RC system that can be operated within the optical domain. This can be suitable for directly processing optical data such as optical spectra or communication data. Nevertheless, also other dynamical systems with feedback (e.g. acoustically excited oscillators [20] or electro-optical oscillators [21]) will show delay induced instabilities and strong dependencies upon the respective system timescales and the discussion below can also be transferred to those systems.

The input layer of a photonic time-multiplexed RC is realized by driving the laser via either a time varying pump current or a time varying optical signal. The output layer is realized via sampling the output intensity $I(t)$ of the driven laser system at different points in time at a sampling rate of $1/\theta$, where θ is the virtual node distance. It is usually realized via a photo detector. The readout is synchronized to the input such that N_V data points of the response are measured during the time of one input data interval $T = N_V\theta$ (see Eq. (1)). An example of such a setup is depicted in Fig. 1. The nonlinear transformation and mixing of the data that enter at different points in time (time-multiplexing) takes place within the optical cavity due to the finite response times of the photons and electrons, i.e. a high electron density induced by the electrical pump current will need some time before it is seen in the photon

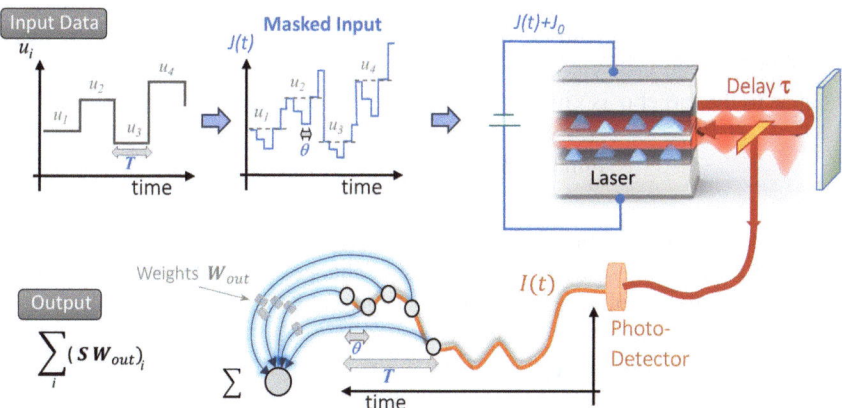

Fig. 1 Sketch of the time-muliplexed reservoir computing scheme. The input data stream u_i is transformed into a time-continuous signal via a sample and hold mechanism, multiplied by a random mask signal $m(t)$ and injected into the dynamic system as a driving force (pump current $J(t)$ as defined in Eq. (8)). The response of the system is measured via the intensity $I(t)$ as a function of time. The data points are measured N_v-times within one clock cycle T, with a temporal distance of θ (virtual node distance). The results are written into the state matrix $\mathbf{S}$ where one row contains the data measured within one clock cycle T. The weights W_{out} are determined via linear regression such that the output is closest to the desired target

density due to the light matter interaction. For the training procedure the intensity values that are measured as a function of time at the output are collected into a state matrix **S**:

$$\mathbf{S} = \begin{bmatrix} I(\theta) & I(2\theta) \ldots & I(N_V\theta) & \Big| 1 \\ I(\theta+T) & \ldots \ldots & I(N_V\theta+T) & \Big| 1 \\ \vdots & & \vdots & \Big| \vdots \\ I(\theta+N_IT) & \ldots & \ldots I(N_V\theta+N_IT) & \Big| 1 \end{bmatrix}. \tag{1}$$

The intensity response $I(t)$ of the laser is measured N_V times within the input duration T of each piece of input data, i.e. these measured values form one row of the **S** matrix. The number of rows of the state matrix is given by the size of the training data set N_I. A bias term is added to **S** in the last column to allow to have a distribution offset from the origin during the linear regression. The temporal distance between the readouts (the virtual node distance θ also indicated in Fig. 1) has to be adapted to the characteristic response time of the laser [13, 22] (or which ever physical system is used as the reservoir) and determines the data processing rate. If for example, $N_v = T/\theta = 30$ with a virtual node spacing of $\theta = 33$ ps, this results in a clock-cycle of $T = 1$ ns and thus a data processing rate of 1 GHz.

2.1 Training

The trained weights $\mathbf{W}_{\text{out}}$ for the weighted readout sum at the output layer are determined for a given task via linear regression. The linear regression minimizes the distance to the target **o** as given in Eq. (2), where $\|\cdot\|_2$ refers to the L2-norm.

$$\min_{\mathbf{W}_{\text{out}}} \left(\|\mathbf{S}\mathbf{W}_{\text{out}} - \mathbf{o}\|_2^2 + \lambda \|\mathbf{W}_{\text{out}}\|_2^2 \right). \tag{2}$$

The Tikhonov regularization parameter λ is used to avoid overfitting and has to be optimized for every system to take into account changes in the absolute values of the state matrix entries. The solution of Eq. (2) is given by

$$\mathbf{W}_{\text{out}} = \left(\mathbf{S}^T\mathbf{S} + \lambda\mathbf{I} \right)^{-1} \mathbf{S}^T\mathbf{o}, \tag{3}$$

where **I** is the identity matrix. After training and during RC operation, the output of the RC system is then computed via $\hat{\mathbf{o}} = \mathbf{S}\mathbf{W}_{\text{out}}$. To quantify the error in predicting the target **o**, we compute the normalized root mean square error ($NRMSE$)

$$NRMSE = \sqrt{\frac{1}{N_I} \sum_{k=1}^{N_I} \frac{(\hat{o}(k) - o(k))^2}{\text{var}(\mathbf{o})}}, \tag{4}$$

which incorporates the normalization by the variance of the target var($\mathbf{o}$). N_I is the number of training or testing steps. For our numerically performed prediction tasks the reservoir was trained with 10000 inputs and tested with 5000 data points. We always average over multiple randomly chosen masks in order to eliminate the impact of the mask and gain reliable insights into the parameter dependence. For an actual RC application only one mask needs to be used.

Time Series Prediction task

Reservoir computing is particularly suited for chaotic timeseries prediction due to the recurrent mixing of the input data which provides short term memory. This memory is needed for anticipating the next step in a complex time evolution. To demonstrate the interrelation between the bifurcation structure and the prediction performance of the RC system, we will mainly focus on tasks using the chaotic Lorenz 63 system [23]. We will also briefly discuss different tasks, e.g. Mackey-Glass [24] and NARMA10 [25]. The Lorenz 63 system [23] is described by the three coordinates:

$$\dot{x}(t) = \sigma(y - x); \quad \dot{y}(t) = x(\rho - z) - y; \quad \dot{z}(t) = xy - \beta z. \tag{5}$$

With the parameters set to $\sigma = 10$, $\rho = 28$, and $\beta = 8/3$, this system yields chaotic dynamics. To generate the training and testing data we integrate numerically with a Runge-Kutta 4th-order solver and a step size of $\Delta t_{sim} = 0.01$. We then sample $x(t)$ with a time step $\Delta_d = 0.1$ to yield the input vector $\mathbf{x}(k)$ (the length of the vector equals the number of training or testing steps, i.e. $k \in 1 \cdots N_I$, plus a suitably long buffer (washout) for initialising the reservoir). For a one-step-ahead x-to-x prediction task the vector $\mathbf{x}(k)$ is the input and $\mathbf{o}(k) = \mathbf{x}(k + 1)$ is the target. We note that the sampling step Δ_d has a nontrivial impact both on the difficulty of the task and on its memory requirements [26–28], i.e. a smaller sampling step Δ_d does not necessarily lead to simpler tasks.

Another benchmark task used later in the chapter is the NARMA10 task which is a nonlinear autoregressive-moving average task [25]. It is given by the recursive formula

$$A_{k+1} = 0.3A_k + 0.05A_k \left(\sum_{j=0}^{9} A_{k-j} \right) + 1.5u_{k-9}u_k + 0.1, \tag{6}$$

where A_k is the iterative NARMA10 time series and u_k are i.i.d. random numbers in [0, 0.5] serving as input. The task is to predict $\mathbf{o}(k) = \mathbf{A}(k + 1)$ from the input sequence $\mathbf{u}(k)$.

Lastly, the Mackey-Glass equation is used as a benchmarking task [29–31]. The Mackey-Glass delay-differential equation is given by [32]

$$\frac{dx}{dt} = \beta \frac{x\,(t - \tau_M)}{1 + x\,(t - \tau_M)^n} - \gamma x. \tag{7}$$

Using the standard parameters, $\tau_M = 17$, $n = 10$, $\beta = 0.2$ and $\gamma = 0.1$, this system exhibits chaotic dynamics. To numerically integrate this delay-differential equation we use a Runge-Kutta 4th-order solver with a step size of $\Delta t_{sim} = 0.01$ and cubic Hermite interpolation for the delayed mid-steps. As for the Lorenz case the dynamic output is sampled to get the input vector. We choose $\Delta_d = 1$, and the target is $\mathbf{o}(k) = \mathbf{x}(k + 10)$ when $\mathbf{x}(k)$ is the input, i.e. we perform a 10-step ahead prediction.

For the photonic computing we simulate a time continuous system as our reservoir. Therefore, the input vector $\mathbf{x}(k)$ for a specific task is first transformed into a continuous signal $X(t)$ via the sample and hold method (hold constant during the clock cycle T) and then multiplied with a mask signal $m(t)$ that repeats every T (see Fig. 1). The mask values take the role of the random input weights in conventional RC setups. In our case they are drawn from an independent and identical distribution between -1 and 1. For the case that the reservoir is driven via input to the pump current $J(t)$ (as illustrated in Fig. 1), the resulting time dependent signal is given by

$$J(t) = \eta \cdot m(t)X(t), \tag{8}$$

with an input scaling factor η that can be used for optimization.

3 Bifurcation Structure of the Dynamical System

To demonstrate the dynamics arising in semiconductor lasers with optical feedback (see sketch in Fig. 1), we use a quantum dot (QD) laser modeled via microscopically based semi-classical rate-equations, as introduced in [33]. Other laser system can be used as active elements and in the subsequent sections we show reservoir computing results for a variety of semiconductor lasers. For all systems the response time of the laser plays a critical role and the timescales of the data induced perturbations have to be adjusted accordingly. Common to all semiconductor laser setups that will be discussed is the feedback delay line, which generates a high dimensional and complex input response. Due to the mathematical properties of delay differential equations, the phase space dimension is infinite and close to bifurcations (tipping points in the dynamic response), subsets of this phase space can be exploited via sampling the transient response.

The differential equation system used for modelling the QD laser with feedback is given by Eqs. (9)–(13). We also include the possibility for external optical injection (with a normalized strength k_{inj} and an optical frequency that is detuned by Δv from the solitary laser frequency) in order to have an additional tuning parameter to enable a control of the laser dynamics. The electric pump current $J(t)$ represents the interface to the RC setup as this is where the input data of a certain task enter the dynamical system while the laser is driven at a constant value J_c of twice the threshold current.

$$\dot{E} = (g - i\delta\omega - \kappa)E + k_{\text{inj}}|E_0|2\kappa e^{-i2\pi\Delta\nu t} + 2\kappa K e^{iC} E(t - \tau) \tag{9}$$

$$\dot{\rho}_{GS,b}^{\text{act}} = -\frac{g}{f^{\text{act}} a_L N_{QD}}|E|^2 - R_{\text{sp,GS}}^{\text{act}} + S_{GS,b}^{\text{cap,act}} + S_b^{\text{rel,act}} \tag{10}$$

$$\dot{\rho}_{GS,b}^{\text{inact}} = -R_{\text{sp,GS}}^{\text{inact}} + S_{GS,b}^{\text{cap,inact}} + S_b^{\text{rel,inact}} \tag{11}$$

$$\dot{\rho}_{ES,b} = -R_{\text{sp,ES}}^{\text{act}} + S_{ES,b}^{\text{cap}} - \frac{1}{2}\left(f^{\text{act}} S_b^{\text{rel,act}} + f^{\text{inact}} S_b^{\text{rel,inact}}\right) \tag{12}$$

$$\dot{\omega}_b = -2N_{QD}\left(f^{\text{act}} S_{GS,b}^{\text{cap,act}} + f^{\text{inact}} S_{GS,b}^{\text{cap,inact}}\right) - 4N_{QD} S_{ES,b}^{\text{cap}} - R_{\text{loss}} + J_c + J(t) \tag{13}$$

The light inside the cavity is described by the slowly varying amplitude of the complex electric field E which yields the output intensity $I(t) = |E|^2$. In the electric field equation (Eq. (9)), the factor k_{inj} measures the optical injection strength normalized to the inverse photon lifetime 2κ and to the amplitude output $|E_0|$ of the solitary QD laser. $e^{-i2\pi\Delta\nu t}$ describes the phase shift due to the frequency detuning $\Delta\nu$ between the driven and the driving laser, $2\kappa K e^{iC} E(t - \tau)$ denotes delayed optical self-feedback with the feedback strength K, feedback delay time τ and feedback phase shift C.

The gain is generated by the occupation probabilities of the resonant QDs in the ground state $\rho_{GS,b}^{\text{act}}$. For the internal charge carrier dynamics also the off-resonant QDs in the ground state $\rho_{GS,b}^{\text{inact}}$, the excited state $\rho_{ES,b}$, and the 2D charge carrier densities ω_b are important dynamic variables ($b = \{e, h\}$ labels electrons and holes, respectively). More details on the QD laser modelling are given in [34, 35]. The complex linear gain is given by an amplitude gain g and the change of the instantaneous frequency $\delta\omega$. Both are described by linear equations that model their dependencies on the charge carrier occupation probabilities in the QD and the charge carrier densities in the surrounding well. The gain of the laser is given by $g = g_{GS}(\rho_{GS,e}^{\text{act}} + \rho_{GS,h}^{\text{act}} + 1)$ with the differential gain g_{GS}, while the instantaneous frequency shift $\delta\omega = \delta\omega_{ES}\left(\rho_{ES,e} + \rho_{ES,h}\right) + \delta\omega_e \cdot \omega_e + \delta\omega_h \cdot \omega_h$ models an amplitude phase coupling that dynamically depends on the laser state. The latter results in a non-constant α-factor that depends on the individual shifts ($\delta\omega_{ES}$, $\delta\omega_e$, and $\delta\omega_h$) of the carrier distributions within the involved non-resonant states which can be derived from a fully microscopic modelling approach [36]. $R_{sp}^{\text{(in)act}}$ and R_{loss} describe the spontaneous recombination losses of the charge carriers as a phenomenological rate. The microscopically-based Boltzmann scattering rates $S_{m,b}^{\text{cap,(in)act}}$ and $S_b^{\text{rel,(in)act}}$ between the different energy states contain all in- and out-scattering processes between the charge carriers and can be found in [33, 34].

3.1 Dynamics with Optical Feedback

To get an overview of the dynamical behaviour of semiconductor laser systems with
optical feedback and without the time dependent input of the RC setup or additional
optical injection, results of numerical bifurcation analysis for the QD system Eqs. (9)–
(13) are presented in Fig. 2. The observed dynamics in the feedback parameter space
of feedback phase C and feedback strength K are shown for six different delay times
τ. For every parameter set we solved the differential equation system, let the system
converge to its equilibrium dynamics and then analysed the time series $I(t)$ with
respect to its shape in time. The results are plotted in Fig. 2 as a color code. Blue
shading indicate regions where stable intensity (constant wave operation) is found,
while yellow to red colors mark delay induced oscillations of increasing complexity
(the number of extrema found in one period (or quasi-period) are counted). This
behaviour is ubiquitous for lasers with optical feedback, where the delay induces new
solutions (external cavity modes) that are born in saddle-node bifurcations [37–40].
Those new solutions exist in multiple layers within this parameter plane. Jumps from
one steady state solution to another can be seen by the non-smooth changes in the
blue shading in Fig. 2 for delay times larger than $\tau = 0.5$ ns. The longer the delay
the more frequent the jumps appear for increasing K. The periodic oscillations are
born in Hopf bifurcations of the first external cavity mode (transition from blue
to yellow) [40]. The other border of the oscillating region can be a homoclinic or
an interior crisis (see e.g. [41, 42]). The longer the delay the higher the degree of

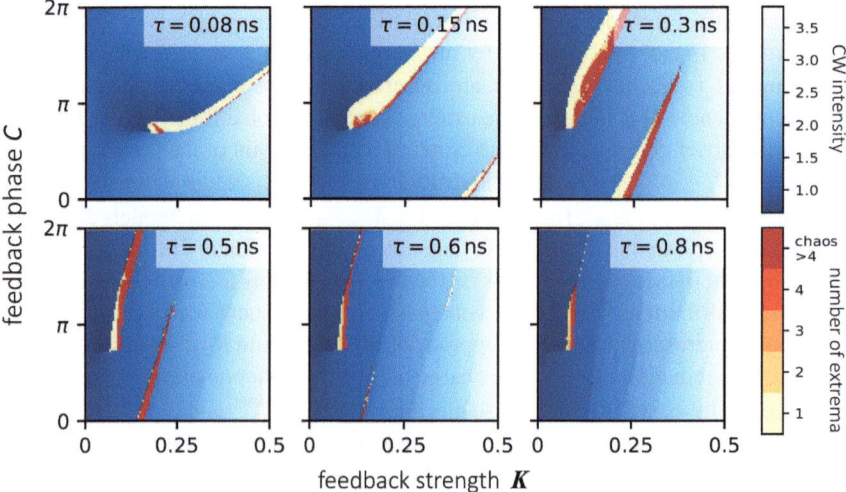

Fig. 2 Dynamics observed from a quantum-dot laser subjected to optical feedback. The panels
show parameter scans in feedback strength K and phase C for 6 different feedback delay times τ.
Blues colors indicate the constant wave intensity (stable emission) and yellow-red colors indicate
the number of maxima found within one period of the pulsating laser intensity

multistability [43] (this can also be seen by the increased number of jumps in the blue shaded region in Fig. 2 for $\tau = 0.8$ ns). This is a fact that must be kept in mind when using such a setup as a reservoir computer. Also, the feedback strength where the first instability happens decreases with the delay. The regions of oscillatory dynamics are not beneficial for RC and a feedback strength lower than the critical value at which the first lasing instabilities arise should be chosen.

3.2 Effect of Additional Optical Injection

For the case that additional optical injection is added to the laser (for example the k_{inj} term in Eq. (9)), the dynamic response changes and leads to large regions where complex dynamic response is found. Examples of optical injection dependent dynamics are shown in Fig. 3 using the same color code as in Fig. 2. The regions of complex dynamics (yellows and reds) are not useful for reservoir computing because the system does not respond consistently to the input, as will be demonstrated for a spin-VCSEL system in the next section. Stable intensity operation (blue) is achieved when the laser locks to the external injection. Since there are multiple delay induced solutions that can exist in the resonator, the locking region shows multiple tongues and the locking border is jagged (one tongue for every external cavity mode). The higher the feedback strength of the delay loop (Fig. 3b, c), the smaller are the locking cones, which leaves most of the injection parameter space with periodic (yellow) or chaotic (red) dynamics. Thus, if injection and feedback are combined a small feedback strength is better for ensuring stable intensity operation.

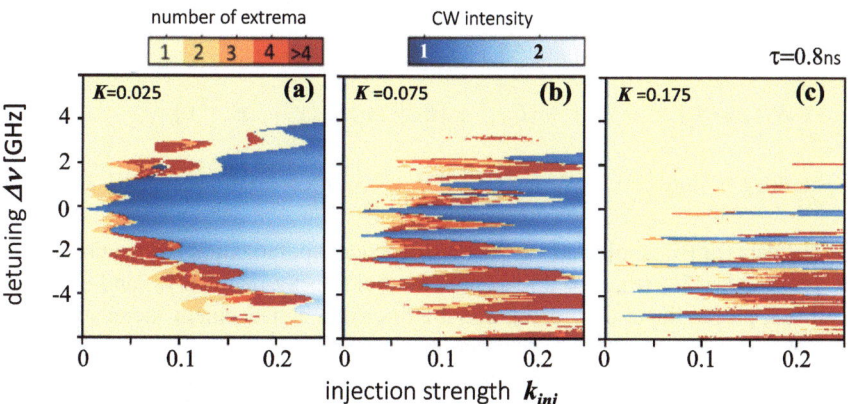

Fig. 3 Dynamics observed from an optically injected quantum-dot laser subjected to optical feedback with $\tau = 0.8$ ns and $C = 0$. Panels **a**–**c** show results for 3 different feedback strength K. Blue colors indicate constant wave emission and yellow-red colors indicate the number of maxima found in one period of the pulsating laser intensity

The delay time in Fig. 3 was chosen to be $\tau = 0.8\,\text{ns}$, which is a relatively large value. For smaller delay times the degree of multistability is smaller which then also leads to less complex locking borders [34]. We remark that spike-based reservoir computing encoding schemes can also be realized with this setup, as the system is excitable at the edge of the locking borders [34, 44, 45]. Choosing the injection parameters $\Delta\nu$ and k_{inj} close to the injection induced saddle-node bifurcation, but within the stable blue region, allows the excitation of pulses from small optical or electrical perturbations [46, 47].

4   Performance in Parameter Space

In the following section we discuss how the time series prediction performance after training the RC system, which is build by a laser with feedback delay, changes in the parameter plane of an additionally applied optical injection. We demonstrate the behaviour using a spin-VCSEL system [48, 49]. Experimental results for these optically injected lasers with feedback and optical data injection have been published e.g. in [17, 50]. In our case we inject the data via an electric polarized pump current. The dynamic equations for the spin-VCSEL are rate equations similar to the QD-laser case, however, here spin-up and spin-down carriers are modelled separately with an additional timescale describing the spin relaxation $\gamma_s = 10\,ns^{-1}$. The electric field inside the cavity is polarized either right circular (E_+) or left circular (E_-). The total sum of the charge carrier sub-populations is given by N, while m describes their difference:

$$\frac{d}{dt}E(t)_{\pm} = \kappa\,(1+i\alpha)(N \pm m - 1)E_{\pm} - (\gamma_a + i\gamma_p)E_{\mp} + 2K\kappa\,E_{\pm}(t-\tau)e^{-iC}$$
$$+ 2\kappa k_{inj}|E_0|\,e^{2\pi i\Delta\nu t} \tag{14}$$

$$\frac{d}{dt}N(t) = \gamma\,[\,J_c - N(1 + |E_+|^2 + |E_-|^2)m(|E_+|^2 - |E_-|^2)\,] \tag{15}$$

$$\frac{d}{dt}m(t) = \gamma\,J_cJ(t) - [\gamma_s + \gamma(|E_+|^2 + |E_-|^2)]m - \gamma(|E_+|^2 - |E_-|^2)N. \tag{16}$$

The data for RC is injected via the time dependent polarization pump current $J(t)$ as described in Eq. (8). $\kappa = 230\,ns^{-1}$ is the field decay rate, C, γ_a, and α are the feedback phase, linear dichrosim, and the linewidth enhancement factor which are zero in our case. $\gamma_p = 30\,ns^{-1}$ describes birefringency, $\gamma = 1\,ns^{-1}$ the carrier decay rate, and $J_c = 1.5$ the total pump current normalized to threshold. More details on the model can be found in [51].

Using the time-multiplexed reservoir computing scheme introduced in Sect. 2 we compute the normalized root mean square error (NRMSE) when predicting the x-variable of the Lorenz system one step into the future (with a time step of 0.1 Lorenz times). Figure 4 depicts the NRMSE as a color code in the injection parameter

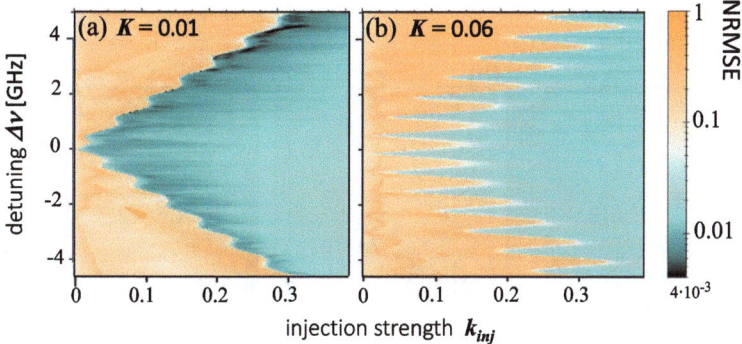

Fig. 4 Performance for the Lorenz x-to-x one step ahead time series prediction task performed with a spin-VCSEL RC setup with optical injection and optical delay τ as given in Eqs. (14)–(16). The NRMSE is plotted color coded in the injection parameter space as a function of the injection strength k_{inj} and the detuning frequency $\Delta \nu$. The optical self feedback delay is set to $\tau = 1.4$ ns, feedback phase $C = 0$, $N_V = 30$ nodes, $\theta = 33$

plane. It can be seen that good performance is only found within the locking region (dark green colors) while the regions with the pulsating dynamics do not allow for a reliable prediction (orange colors). More details are given in [51] the panels (a) and (b) in Fig. 4 depict two different feedback strength. As also observed before in the dynamics discussion, the stronger feedback strength leads to a much more jagged border. This correlates with worse performance for the larger feedback strength. The best performance is found at the border of the locking region where the transient response of the laser is complex due to the close proximity to the bifurcation, but the underlying dynamics of the un-driven system are still stable intensity emission (continuous wave emission).

4.1 Resonances and Internal Response Time

The dynamic time-scales governing the complete RC setup are particularly important for the performance. The time that is needed to feed in one masked data point is the clock-cycle T. If this time is chosen equivalent to the feedback delay time τ, a resonance effect which is detrimental to the prediction performance [52, 53] can be observed. Figure 5 shows this by scanning the time series prediction performance within the (τ-T) plane for a QD laser system with optical feedback at varying delay times (x-axis). No additional optical injection was used here. Along the diagonal, a dip in the performance is seen. A second dip is found for the resonance where $2T = \tau$ (see while arrows in Fig. 5). The reason for the performance drop is a reduction in the complexity of the coupling that occurs at such resonances [53, 54, 56]. For example, when $T = \tau$ each sampled reservoir response is only coupled via the feedback with

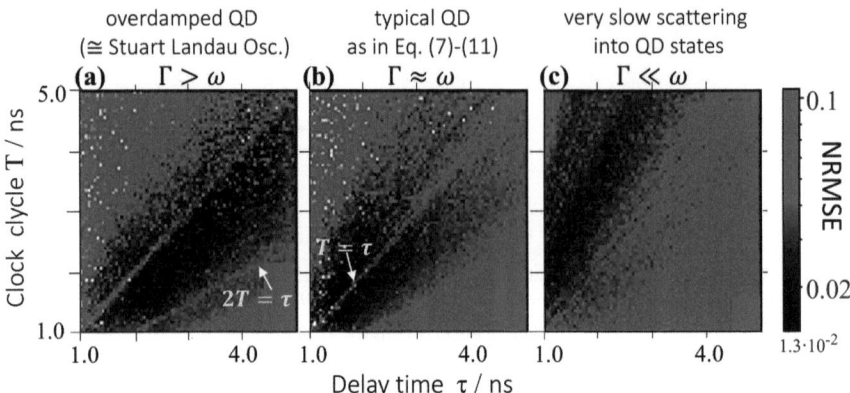

Fig. 5 Scan of the time series prediction performance (Lorenz x-to-x task one step ahead) within the RC parameter space of time delay τ and clock cycle T. The NRMSE is color coded with black being the best values. Three laser systems with different turn on damping Γ and relaxation oscillation frequency ω are chosen: **a** overdamped QD laser, **b** strongly damped like a typical QD laser modelled by Eqs. (9)–(13), and **c** a strongly spiking QD laser with very slow scattering rates

inputs associated with the same mask steps, leading to a type of self-coupling between responses separated by T. Whereas, for the commonly used desynchronized scheme with $T + \theta = \tau$ [53, 55], responses to inputs from neighbouring mask steps are coupled.

The three panels in Fig. 5 depict the results for laser systems in three different dynamical regimes. The left panel depicts an example of an overdamped laser (class-A laser) where the dynamics of the charge carriers can be adiabatically eliminated, leaving only the dynamics described by the electric field equation [56]. Such a simple nonlinear oscillator [52], is described by the so-called Stuart Landau oscillator equation. This system shows an damped response to perturbations with the characteristic damping time Γ. For this systems a ratio of $T/\tau \approx 0.7$ is the best choice for an optimal prediction performance [57] given a task that has no specifically long memory requirements (dark regions with best performance in Fig. 5 where the Lorenz x-to-x prediction task was chosen). In those cases an optimal delay is always slightly larger than the clock cycle. The value for the optimal ratio for the same task shifts for the other laser systems in Fig. 5. Figure 5b shows the results in τ-T-parameter space for a typical QD laser as described in Sect. 3. Here, a ratio of about 1 between τ and T is most beneficial [22]. The corresponding QD laser turn-on transients are in between class-A and class-B like behaviour [22, 58] with a relaxation oscillation frequency ω that is on the same order as the damping Γ. Instead, the QD laser in Fig. 5c has tremendously slow scattering rates which leads to pronounced spikes in the turn-on dynamics ($\omega << \Gamma$). There the optimal ratio is shifted to higher values ($T/\tau = 1.67$) which refers to a much slower data injection rate due to the much slower system response [22]. More details on the effect of the QD laser scattering rates on the turn-on dynamics and modulation response of QD lasers are given in

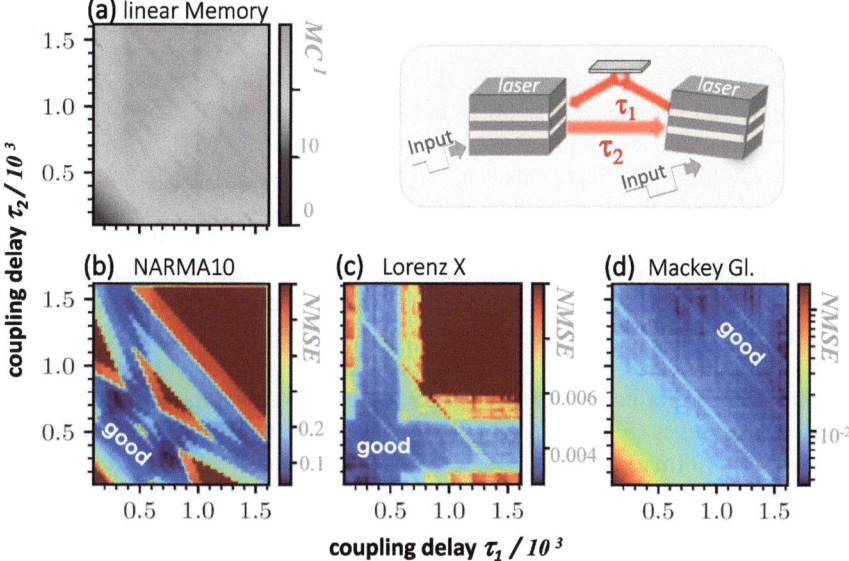

Fig. 6 Normalised mean square error (NMSE) plotted color coded as a function of the two coupling delay times τ_1 and τ_2 of two mutually coupled lasers given by Eq. (17), with times given in normalized time units [63]. **c–d** show the three different prediction tasks (NARMA10, Lorenz X and Mackey Glass) introduced in Sect. 2. **a** shows the linear memory capacity MC^1 Eq.(18). Clock cycle $T = 200$, $N_V = 100$

[59, 60], details on their effect regarding QD lasers with optical injection in [61] and regarding optical feedback in [39].

In our RC setup, one reason for the shift of the optimal τ/T-ratio for the three lasers in Fig. 5 lies in the changed system response. The response is dominated by the damping in the overdamped case (Fig. 5a) and by the relaxation oscillation frequency in the slow scattering case (Fig. 5c). If the nonlinear properties of the physical system are to be used in an optimal way, the sampling rate $(1/\theta)$ of the time-multiplexed output has to match the internal relaxation time-scale of the dynamical system [22, 57, 62].

5 Task Dependence of the Performance

To explore the task dependence of the delay-based RC scheme, we now look at a slightly different setup, two delay-coupled lasers described by Stuart-Landau oscillators. The latter is a normal form describing the behaviour near a Hopf bifurcation. The dynamic equations for the complex amplitude Z_j (which represents the electric field in case of a Class A laser as detailed in [56]) are

$$\dot{Z}_j = \left(J_j^0 + J_j(t) + (\gamma_j + i\alpha_j)|Z_j|^2\right) Z_j + K_j \, e^{iC_j} Z_{j+1} \left(t - \tau_j\right). \tag{17}$$

The index j labels the two oscillators, $J_j^0 = 0.01$ denotes the pump rate, $\gamma_j = -0.1$ the nonlinearity parameter, $K_j = 0.18$ the feedback strength, C_j and α_j represent feedback phase and amplitude phase coupling which are chosen to be zero here. τ_j are the delay-times. The input data is injected via $J_j(t)$ as described before in Eq. (8) with the difference that here each laser has a different mask.

In this setup two delay times exist (see setup sketch in Fig. 6) which can be tuned very efficiently [53]. We will investigate the performance in this parameter plane spanned by τ_1 and τ_2. To evaluate how much the achievable prediction results after training the RC system depend on the task, three different tasks (NARMA10, Lorenz-X and Mackey-Glass) were chosen (see [63] for more details). For all three the prediction performance one step into the future was determined as a function of the two delay-times (Fig. 6 panels b, c, d). Within the blue regions in Fig. 6 the RC computing performance is best. The regions of good performance are interspersed with regions of poor performance. The structure of these regions depend strongly on the delay-times and the task. While the Lorenz task is performed best with small delay lines (dark region in the lower left corner of Fig. 6c), the 10-step ahead Mackey Glass task needs rather long delay values (Fig. 6d) for a decent performance. The reason are the different memory requirements; The Mackey Glass time series results from an underlying delay system and needs a long memory for predicting the next step. This memory is provided by the delay in the RC setup. Instead, the Lorenz x-to-x prediction can be performed with very little memory [28] and small values of τ suffice.

Figure 6a presents the memory properties of the mutually coupled RC setup. The linear memory capacity can be measured by computing how well past inputs can be remembered [56, 64]. The lighter grey regions correspond to larger linear memory capacity and they appear for larger values of τ_1 and τ_2 in Fig. 6a. It is computed by first calculating the linear recall capability $C_i = \mathbf{o}^T \hat{\mathbf{o}}/||\mathbf{o}||^2$ where the target $\mathbf{o}$ is an input sequence shifted i steps back in time while $\hat{\mathbf{o}}$ is the output trained for this task with the original input. The input for this task are i.i.d. distributed random numbers drawn from $[-1, 1]$. C_i equals 1 if the reservoir remembers the ith input perfectly. The linear memory capacity is then defined as the sum over the recall capabilities of all possible linear recall tasks

$$MC^1 = \sum_{i=0}^{\infty} C_i. \tag{18}$$

It is noted that the linear MC^1 can be analytically derived from the linear response of the RC system as long as the input is a small perturbation [62]. In our case we numerically determined the value by training the reservoir to remember an input from i-steps into the past and then summing the values of the performance over i. It is striking that the correlation with the performance on a specific time-series prediction task is not straightforward, as can be seen by comparing the memory capacity in Fig. 6a with the performance of the RC setup after training for the three different time-series prediction tasks in Fig. 6b–d. The reason is that every task has specific requirements for the nonlinearity and the memory and the best operation region for one task can be very bad for another task.

Some insights into the requirements of a time series prediction task can be derived from the auto-correlation function [28]. Here the zero crossing of the AC indicates the temporal distance of uncorrelated points in time. Another approach to analyse the task and its requirements is by performing a series expansion of the time series with e.g. Legendre Polynomials as base functions [63]. This method could identify the memory needed, but the analytic relation only hold true for uncorrelated input which is not fulfilled for the case of a time series that results from a physical process.

6 Conclusion

Time-multiplexed reservoir computing is a powerful machine learning concept that can be easily realized with physical systems, for example optically with lasers and a self-feedback delay line. Due to the physical delay-based realizations there are only a few parameters that need to be optimized. A bifurcation analysis of the underlying physical system can help to identify the interesting regions that are close to complex dynamical regions but still within stable intensity regimes. For different types of semiconductor lasers, there are differences in the internal time scales which must be taken into account when setting up the details of the reservoir computer, e.g. data input rate, internal delay time and output sampling rate. However, there are also universal behaviours which strongly influence the reservoir performance and that do not depend on the specific dynamic system chosen as reservoir but are more dependent on the dynamics induced by the optical self-feedback and the data injection.

Acknowledgements We thank all former and present members of the group for contributing to the research on QD laser modelling and photonic reservoir computing. In detail we want to thank Andrej Krimlowski (QD lasers with injection and feedback), Huifang Dong (time-multiplexed RC with QD lasers), Lukas Mühlnickel (photonic RC with optically injected spin-VCSEL) and Tobias Hülser (RC with mutually coupled lasers).

References

1. Lugnan A, Katumba A, Laporte F, Freiberger M, Sackesyn S, Ma C, Gooskens E, Dambre J, Bienstman P (2020) APL Photon 5(2):020901. https://doi.org/10.1063/1.5129762
2. Bauwens I, Harkhoe K, Bienstman P, Verschaffelt G, Van der Sande G (2022) Opt Express 30(8):13434. https://doi.org/10.1364/oe.449508
3. Lupo A, Picco E, Zajnulina M, Massar S (2023) Optica 10(11):1478. https://doi.org/10.1364/optica.489501
4. Brunner D, Soriano MC, Van der Sande G (2019) Photonic reservoir computing, optical recurrent neural networks. De Gruyter, Berlin, Boston. https://doi.org/10.1515/9783110583496

5. Pinna D, Bourianoff G, Everschor-Sitte K (2020) Phys Rev Appl 14:054020. https://doi.org/ 10.1103/physrevapplied.14.054020
6. Allwood DA, Ellis MOA, Griffin D, Hayward TJ, Manneschi L, Musameh MFK, O'Keefe S, Stepney S, Swindells C, Trefzer MA, Vasilaki E, Venkat G, Vidamour I, Wringe C (2023) Appl Phys Lett 122(4):040501. https://doi.org/10.1063/5.0119040
7. Porte X, Skalli A, Haghighi N, Reitzenstein S, Lott JA, Brunner D (2021) J Phys Photonics 3(2):024017. https://doi.org/10.1088/2515-7647/abf6bd
8. Boikov IK, Brunner D, De Rossi A (2023) New J Phys 25(9):093056. https://doi.org/10.1088/ 1367-2630/acfba6
9. Tselios C, Georgiou P, Politi C, Hurtado A, Alexandropoulos D (2023) IEEE J Quantum Electron 59(5):2400308. https://doi.org/10.1109/jqe.2023.3296732
10. Owen-Newns D, Robertson J, Hejda M, Hurtado A (2023) Intell Comput 2:0031. https://doi. org/10.34133/icomputing.0031. https://spj.science.org/doi/abs/10.34133/icomputing.0031
11. Nakajima K (2020) Jpn J Appl Phys 59(6):060501. https://doi.org/10.35848/1347-4065/ ab8d4f
12. Abreu S, Boikov I, Goldmann M, Jonuzi T, Lupo A, Masaad S, Nguyen L, Picco E, Pourcel G, Skalli A, Talandier L, Vettelschoss B, Vlieg EA, Argyris A, Bienstman P, Brunner D, Dambre J, Daudet L, Domeneśch JD, Fischer I, Horst F, Massar S, Mirasso CR, Offrein BJ, Rossi A, Soriano MC, Sygletos S, Turitsyn SK (2024) Rev Phys 12:100093. https://doi.org/10.1016/j. revip.2024.100093
13. Appeltant L, Soriano MC, Van der Sande G, Danckaert J, Massar S, Dambre J, Schrauwen B, Mirasso CR, Fischer I (2011) Nat Commun 2:468. https://doi.org/10.1038/ncomms1476
14. Brunner D, Soriano MC, Mirasso CR, Fischer I (2013) Nat Commun 4:1364. https://doi.org/ 10.1038/ncomms2368
15. Nakajima M, Tanaka K, Hashimoto T (2021) Commun Phys 4(20):2399. https://doi.org/10. 1038/s42005-021-00519-1
16. Tang JY, Lin BD, Shen YW, Li RQ, Yu J, He X, Wang C (2023) Opt Express 31(2):2456. https://doi.org/10.1364/oe.478728
17. Oliverio L, Rontani D, Sciamanna M (2023) Opt Lett 48(10):2716. https://doi.org/10.1364/ol. 486383
18. Brunner D, Penkovsky B, Marquez BA, Jacquot M, Fischer I, Larger L (2018) J Appl Phys 124(15):152004. https://doi.org/10.1063/1.5042342
19. Kuriki Y, Nakayama J, Takano K, Uchida A (2018) Opt Express 26(5):5777. https://doi.org/ 10.1364/oe.26.005777
20. Lenk C, Hövel P, Ved K, Durstewitz S, Meurer T, Fritsch T, Männchen A, Küller J, Beer D, Ivanov T, Ziegler M (2023) Nat Electron 6(5):370. https://doi.org/10.1038/s41928-023-00957-5
21. Picco E, Antonik P, Massar S (2023) Neural Netw 165:662. https://doi.org/10.1016/j.neunet. 2023.06.014
22. Dong H, Jaurigue LC, Lüdge K (2025) Phys Status Solidi RRL 2025, 2400433. http://dx.doi. org/10.1002/pssr.202400433
23. Lorenz EN (1963) J Atmos Sci 20:130. https://doi.org/10.1175/1520-0469(1963)020 %3C0130:DNF%3E2.0.CO;2
24. Macho-Ortiz A, Pérez-López D, Azaña J, Capmany J (2022) Laser Photonics Rev. 2023, 17, 2200360. https://doi.org/10.1002/lpor.202200360
25. Atiya AF, Parlos AG (2000) IEEE Trans Neural Netw 11(3):697. https://doi.org/10.1109/72. 846741
26. Storm L, Gustavsson K, Mehlig B (2022) Mach Learn Sci Technol 3(4):045021. https://doi. org/10.1088/2632-2153/aca1f6
27. Tsuchiyama K, Röhm A, Mihana T, Horisaki R, Naruse M (2023) Chaos 33(6):063145. https:// doi.org/10.1063/5.0143846
28. Jaurigue L, Lüdge K (2024) Neuromorph. Comput Eng 4(1):014001. https://doi.org/10.1088/ 2634-4386/ad1d32

29. Shi Z, Han M (2007) IEEE Trans Neural Netw Learn Syst 18(2):359. https://doi.org/10.1109/tnn.2006.885113
30. Ortín S, Soriano MC, Pesquera L, Brunner D, San-Martín D, Fischer I, Mirasso CR, Gutierrez JM (2015) Sci Rep 5(1):14945. https://doi.org/10.1038/srep14945
31. Goldmann M, Köster F, Lüdge K, Yanchuk S (2020) Chaos 30(9):093124. https://doi.org/10.1063/5.0017974
32. Mackey MC, Glass L (1977) Science 197(4300):287. https://doi.org/10.1126/science.267326
33. Lingnau B, Chow WW, Lüdge K (2014) Opt Express 22(5):4867. https://doi.org/10.1364/oe.22.004867
34. Köster F, Lingnau B, Krimlowski A, Hövel P, Lüdge K (2021) Phys Status Solidi B 2021:2100345. https://doi.org/10.1002/pssb.202100345
35. Lüdge K, In: Nonlinear laser dynamics - from quantum dots to cryptography [58], chap 1, pp 3–34
36. Lingnau B, Lüdge K, Chow WW, Schöll E (2012) Phys Rev E 86(6):065201(R). https://doi.org/10.1103/physreve.86.065201
37. Mørk J, Tromborg B, Mark J (1992) IEEE J Quantum Electron 28:93. https://doi.org/10.1109/3.119502
38. Rottschäfer V, Krauskopf B (2007) Int J Bifurc Chaos 17(5):1575. https://doi.org/10.1142/s0218127407017914
39. Globisch B, Otto C, Schöll E, Lüdge K (2012) Phys Rev E 86:046201. https://doi.org/10.1103/physreve.86.046201
40. Lüdge K, Lingnau B (2020) Laser dynamics and delayed feedback, ISBN 978-1-07-160420-5. Springer Nature, New York, NY, 2020, 1st edn, pp. Synergetic–18. Encyclopedia of Complexity and Systems Science. https://doi.org/10.1007/978-3-642-27737-5_729-1. https://link.springer.com/referenceworkentry/10.1007/978-3-642-27737-5_729-1
41. Heil T, Fischer I, Elsäßer W, Krauskopf B, Green K, Gavrielides A (2003) Phys Rev E 67:066214. https://doi.org/10.1103/physreve.67.066214
42. Otto C, Globisch B, Lüdge K, Schöll E, Erneux T (2012) Int J Bifurc Chaos 22(10):1250246. https://doi.org/10.1142/s021812741250246x
43. Yanchuk S, Perlikowski P (2009) Phys Rev E 79(4):046221. https://doi.org/10.1103/physreve.79.046221
44. Ziemann D, Aust R, Lingnau B, Schöll E, Lüdge K (2013) Europhys Lett 103:14002. https://doi.org/10.1209/0295-5075/103/14002
45. Dillane M, Robertson J, Peters M, Hurtado A, Kelleher B (2019) Eur Phys J B 92(9):197. https://doi.org/10.1140/epjb/e2019-90733-6
46. Dillane M, Viktorov EA, Kelleher B (2023) Opt Lett 48(1):21. https://doi.org/10.1364/ol.475805
47. Robertson J, Deng T, Javaloyes J, Hurtado A (2017) Opt Lett 42(8):1560. https://doi.org/10.1364/ol.42.001560
48. Miguel MS, Feng Q, Moloney JV (1995) Phys Rev A 52(2):1728. https://doi.org/10.1103/PhysRevA.52.1728
49. Martin-Regalado J, Prati F, San Miguel M, Abraham NB (1997) IEEE J Quant Electron 33(5):765. https://doi.org/10.1109/3.572151
50. Estébanez I, Schwind J, Fischer I, Argyris A (2020) Nanophotonics 9(13):4163. https://doi.org/10.1515/nanoph-2020-0184
51. Mühlnickel L, Jaurigue JA, Jaurigue LC, Lüdge K (2024) Commun Phys 7:370. https://doi.org/10.1038/s42005-024-01858-5
52. Röhm A, Jaurigue L, Lüdge K, (2019) IEEE J Sel Top Quantum Electron 26(1):7700108. https://doi.org/10.1109/jstqe.2019.2927578
53. Hülser T, Köster F, Jaurigue L, Lüdge K (2022) Opt Mater Express 12(3):1214. https://doi.org/10.1364/ome.451016
54. Stelzer F, Röhm A, Lüdge K, Yanchuk S (2020) Neural Netw 124:158. https://doi.org/10.1016/j.neunet.2020.01.010

55. Paquot Y, Duport F, Smerieri A, Dambre J, Schrauwen B, Haelterman M, Massar S (2012) Sci Rep 2:287. https://doi.org/10.1038/srep00287
56. Köster F, Ehlert D, Lüdge K (2020) Cogn Comput 2020. https://doi.org/10.1007/s12559-020-09733-5. Springer. ISSN 1866-9956
57. Köster F, Yanchuk S, Lüdge K (2021) J Phys Photon 3(2):024011. https://doi.org/10.1088/2515-7647/abf237
58. Lüdge K (2012) Nonlinear laser dynamics - from quantum dots to cryptography. Wiley-VCH, Weinheim
59. Lingnau B, Lüdge K, Chow WW, Schöll E (2012) Appl Phys Lett 101(13):131107. https://doi.org/10.1063/1.4754588
60. Lüdge K, Lingnau B, Otto C, Schöll E (2013) Nonlinear Phenom Complex Syst 15(4):350
61. Pausch J, Otto C, Tylaite E, Majer N, Schöll E, Lüdge K (2012) New J Phys 14:053018. https://doi.org/10.1088/1367-2630/14/5/053018
62. Köster F, Yanchuk S, Lüdge K (2024) IEEE Trans Neural Netw Learn Syst 35(6), 7712. https://doi.org/10.1109/tnnls.2022.3220532
63. Hülser T, Köster F, Lüdge K, Jaurigue L (2023) Nanophotonics 12(5):937. https://doi.org/10.1515/nanoph-2022-0415
64. Dambre J, Verstraeten D, Schrauwen B, Massar S (2012) Sci Rep 2:514. https://doi.org/10.1038/srep00514

Neuromorphic Spintronics

Atreya Majumdar and Karin Everschor-Sitte

Abstract Neuromorphic spintronics combines two advanced fields in technology, neuromorphic computing, and spintronics, to create brain-inspired, efficient computing systems that leverage the unique properties of the electron's spin. In this book chapter, we first introduce both fields—neuromorphic computing and spintronics—and then make a case for neuromorphic spintronics. We discuss concrete examples of neuromorphic spintronics, including computing based on fluctuations, artificial neural networks, and reservoir computing, highlighting their potential to revolutionize computational efficiency and functionality.

1 Introduction

The invention of computers has been instrumental in ushering in a new era where we can quickly perform large computations in the blink of an eye and store vast amounts of data. This advancement can be directly attributed to the development of electronics, which has enabled efficient miniaturization and high processing speeds. The development of electronic technology has been so rapid that the number of transistors in integrated circuits has doubled every two years since 1970. This phenomenon has been coined the name Moore's law [1]. However, transistor miniaturization has hit a plateau in recent years due to technological limitations [2, 3]. Furthermore, advances in deep learning have heightened the demands for computation, memory, and energy. To address these issues, there is a need for new computing paradigms based on novel materials [4].

Throughout history, new epochs of civilization have been shaped by the combined innovations of ideas and the methods used to express them. For instance, Johannes Gutenberg's invention of the printing press in 1436 enabled the rapid dissemination of

A. Majumdar · K. Everschor-Sitte (✉)
University of Duisburg-Essen, Duisburg, Germany
e-mail: karin.everschor-sitte@uni-due.de

A. Majumdar
e-mail: atreya.majumdar@uni-due.de

M. te Vrugt (ed.), *Artificial Intelligence and Intelligent Matter*, Machine Intelligence for Materials Science, https://doi.org/10.1007/978-3-032-04129-6_19

knowledge through books, sparking a transformative global revolution. As computing methods and materials reach saturation, an urgent necessity arises for yet another revolutionary breakthrough. This revolution must encompass advancements in both computation as well as the materials that perform it.

The animal world, having undergone millions of years of evolutionary optimization, is a significant inspiration for many modern marvels, such as the airplane. The human brain, in particular, exhibits remarkable pattern recognition abilities while also being energy-efficient. In the late 1980s, Carver Mead introduced the concept of neuromorphic computing, a brain-inspired approach designed to mimic how the brain processes information for computational tasks [5, 6]. While conventional computing developed in tandem with silicon and other semiconductor technologies, adopting a new computational paradigm would require innovative material technologies that naturally fulfill its unique needs. Spintronics-based materials, which utilize the spin degree of freedom, constitute one such class of materials showing promise as candidates for implementing neuromorphic computing. This chapter aims to motivate the potential of neuromorphic spintronics in addressing numerous challenges that plague contemporary computing technologies.

We begin this chapter by presenting the promise that spintronics-based materials hold as a platform for neuromorphic computing. This will be followed by exploring various application domains for spintronics-based materials. First, we consider the noisy yet error-tolerant aspect of our brain as an inspiration, and in that light, we discuss computing based on fluctuations using spintronics-based materials. Next, we discuss the implementation of neural networks, which are directly inspired by how neurons and synapses in our brain are configured to perform pattern recognition tasks. The subsequent section deals with another neuro-inspired algorithm called reservoir computing, and here we emphasize the usefulness of spintronics systems as suitable *reservoirs*. In the penultimate section, we compare and contrast spintronics-based memories with other unconventional memory technologies. Lastly, we explore the future trajectory of neuromorphic spintronics, anticipating advancements in computational capabilities driven by new discoveries in the physical properties of spintronics materials.

1.1 The Promise of Spintronics

The field of electronics is based on using and manipulating the electron's charge. Like the charge, spin is another intrinsic property of the electron. Spintronics, short for "spin-based electronics", is a field in physics and engineering that focuses on this degree of freedom for developing technology [7, 8]. Spin is a quantum mechanical property of any fundamental particle. It is associated with an intrinsic angular momentum of the particle, which allows for manipulation through various means, such as electric current, electric field, magnetic field, and light, among others.

The conceptual roots of spintronics can be traced back to the early 20th century with the discovery of the electron spin by Samuel Goudsmit and George Uhlenbeck in 1925 [9]. However, it was not until several decades later that the potential for practical applications began to emerge; a significant breakthrough came in 1988 with the independent discovery of the giant magnetoresistance (GMR) effect by physicists Albert Fert and Peter Grünberg [10, 11]. This discovery revealed how small changes in applied magnetic fields could lead to significant changes in electrical resistance of certain materials. In the 1990s, the application of GMR in read heads for hard disk drives revolutionized computer data storage, dramatically increasing storage capacity and efficiency and ultimately earning Fert and Grünberg the Nobel Prize in Physics in 2007. At the end of the 20th century, further advances led to the development of the tunneling magnetoresistance (TMR) effect [12–14]. This led to the creation of Magnetoresistive Random-Access Memory (MRAM), a type of non-volatile memory that is based on Magnetic Tunnel Junctions (MTJs). The core physics behind TMR (in MTJs) and the GMR effect involves a three-layer magnetic junction, where a non-magnetic layer is sandwiched between two ferromagnetic layers. The non-magnetic layer provides an insulating tunnel effect and conduction pathway, respectively, for the MTJs and the GMR devices. One of the ferromagnetic layers is a fixed reference layer, while the other is a free layer whose magnetic orientation can be altered. Data storage in MRAM is based on the principle of magnetoresistance – the change in electrical resistance depending on the magnetic alignment of these two layers. When the magnetic moments of the free and fixed layers are parallel, the device's resistance is low, representing a binary '1'. Conversely, the resistance is high when the magnetic moments are antiparallel, representing a binary '0'.

Another important milestone was the race-track memory [15], where the motion of magnetic domain walls along nanowires is utilized. Here, the data is stored as magnetic bits along nanowire *tracks*. These magnetic bits can be moved along the track by applying current pulses, enabling the reading and writing of data. The race-track memory technology, which is yet to be commercially available, aims to achieve a high storage density by packing magnetic bits close together on tracks, which can also be arranged three-dimensionally. In addition, racetrack memories provide the potential for faster and more energy-efficient memory systems compared to traditional technologies like Dynamic Random Access Memory (DRAM) and Flash. In the last decade, other magnetic textures, such as skyrmions, have been explored as magnetic bits [16–19], while antiferromagnets have demonstrated the potential for even greater transmission speeds [20].

Unlike traditional electronics, which rely solely on the charge of electrons, spintronics also harnesses the spin of electrons to encode and process information. One key benefit is reduced energy consumption, as manipulating spin typically requires less energy than moving or storing charges. Additionally, spintronics devices can achieve faster processing speeds due to the intrinsically fast dynamics of spin states, enabling quicker data transmission and processing. This technology also holds potential for greater scalability, particularly in the context of miniaturizing devices, and offers enhanced stability and reliability, as spin currents or magnetized states are less susceptible to external perturbations such as electromagnetic interference. Furthermore, it provides many spin-dependent physical effects like TMR, spin-transfer torque (STT), spin-orbit torque (SOT), etc., which offer new opportunities for developing the next generation of storage and computing devices [21, 22].

1.2 A Case for Neuromorphic Spintronics

Conventional computers are built with silicon-based architectures and hardware, with algorithms tailored to the technological requirements and available resources. Contrary to that, neuromorphic computing is an approach that mimics the neural structure and functioning of the human brain to enhance computational efficiency and processing power. The inspiration could manifest at various levels: the architecture, the computing hardware (including the substrate), or even the software strategies and algorithms.

Spintronics materials hold significant promise for neuromorphic computing due to their inherent characteristics, which are ideally suited for this novel computational paradigm. These characteristics can be broadly classified into four categories, as shown in Fig. 1.

1. **Memory and adaptability:** In the context of computing, memory refers to the ability to store values. In spin-based systems, a device can retain the magnetic state to which it is programmed, thus providing memory. An example is MRAM, where magnetic states are represented by parallel or anti-parallel alignments, thereby serving as memory. Compared to traditional charge-based memory technologies such as static random-access memory (SRAM) and DRAM,

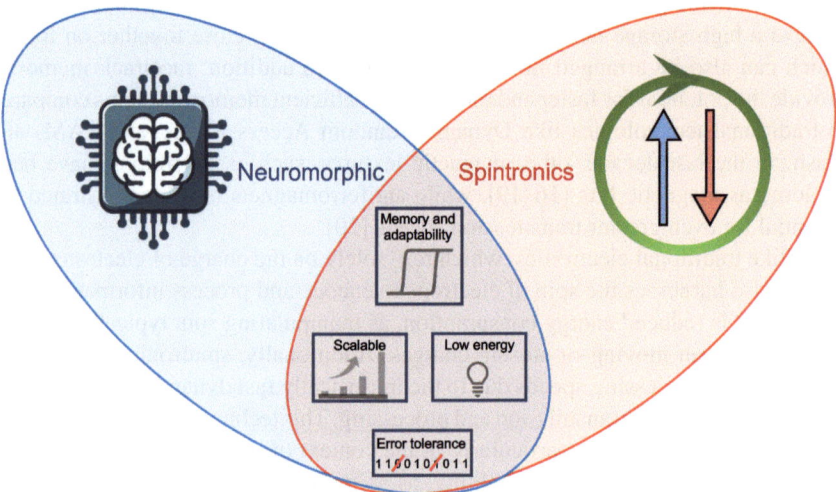

Fig. 1 Neuromorphic computing with spintronics-based devices offers many characteristics such as memory and adaptability, scalability, low-energy computation, and error tolerance

spintronics-based memories are non-volatile, meaning they retain stored values even when powered off [23, 24]. Non-volatility reduces energy consumption, as memory units do not require constant access in many practical applications. Furthermore, a single device can implement both memory and non-linear behavior, unlike other memory systems that often require multiple transistors to emulate these features [25].

In the human brain, both the abilities to memorize and to update memory are crucial for its functionality, highlighting the importance of adaptability. The brain can readjust its internal state by making its synaptic connections stronger or weaker and learning new information or modifying old memories. For spin-based systems, the magnetic state can be reversibly and controllably changed by expending a small amount of energy. The MRAM example uses a current (STT or SOT) to change the magnetic state from parallel to anti-parallel (or vice-versa). As we shall see in Sect. 3, this ability to adapt, also referred to as *plasticity*, is important for implementing neural networks with spintronics-based devices [26].

2. **Low energy:** To perform tasks such as pattern recognition, conventional computers consume orders of magnitude more power than the human brain, which uses around 20 Watts [27]. Our brain's efficiency stems from several factors, with its architecture being one of the primary reasons. Neurons (the basic computing/logic element) and the synapse (the basic memory element) are located physically together in the brain. In contrast, in conventional computers, the logic and memory units are physically separated, and a lot of energy is spent in shuttling information back and forth [28]. This bottleneck of energy consumption in conventional computing paradigms termed the *von Neumann bottleneck*, can be alleviated using spintronics-based memory elements such as MRAMs [24, 29]. Such novel memory technologies can be integrated with transistors that perform logic and arithmetic operations, thus facilitating in-memory or near-memory computing.

Another factor contributing to low energy consumption in the human brain is its method of representing information. Neurons propagate information in the form of electrical voltage spikes. The presence of spikes, rather than a continuous signal, makes the representation of information sparse. This sparsity of activation can also be emulated using spintronics-based devices, which can implement leaky-integrate and fire neurons. Such examples will be described in more detail in Sect. 3.

3. **Scalability:** The human brain has a volume of just about 1,200 cm^3 but contains 10^{11} neurons and 10^{14} synaptic connections [30–32]. The computational efficiency of our brain is partially attributed to this massive connectivity. Although achieving such high connectivity in hardware remains challenging due to limitations in 2D and 3D device fabrication techniques, spintronics devices, with their miniaturization capabilities, can enable significant advances in device integration. The underlying mechanisms of spintronics devices maintain functionality even when device dimensions are reduced to a few nanometers [33]. In contrast,

charge-based effects in traditional electronics become unreliable at such small scales due to quantum tunneling and electron leakage [34].

The huge connectivity in the brain is also responsible for its massively parallel computation. The ability to process multiple information streams in parallel is critical for the brain's adaptability and functionality, underpinning everything from basic survival instincts to higher cognitive functions. The scalability offered by neuromorphic spintronics would allow such parallelization to be possible in hardware as well, where tasks can be divided across a vast array of processing units, enabling the system to handle complex, dynamic processes efficiently.

Realistically, replicating the extensive connectivity and dynamic reorganization of our brain in hardware is not feasible. Instead, adopting a modular approach allows specific hardware components to fulfill distinct roles regulated through software. This strategy effectively addresses the challenges of connectivity and the dynamic reconfiguration of synaptic connections.

Spintronics devices easily integrate with existing CMOS technology, aligning well with ongoing advancements in silicon-based systems [33]. This integration combines the best features of both spintronics and silicon-based technologies, advancing overall technological development.

4. **Error tolerance:** While the brain is capable of pattern recognition at a low energy budget, it is also quite noisy [35, 36]. The sources behind this noise can range from fluctuations in ion channel activities and synaptic transmission variabilities to spontaneous neuronal firings. Ion channels, for instance, might open and close randomly, leading to variability in the membrane potential of neurons [35, 37].

Globally, the brain represents a highly complex, non-linear network where small perturbations can result in large-scale changes. Despite these fluctuations, the brain excels at robust pattern recognition. This capability largely stems from the substantial redundancy within the brain; multiple neurons and pathways often represent the same or similar information, allowing other pathways to compensate when one is affected by noise. The vast scalability potential of spintronic devices enables them to mimic this redundancy, enhancing their capacity for reliable information processing and pattern recognition.

Furthermore, charge-based memories, particularly those with miniaturized components, are more sensitive to environmental factors such as radiation exposure, leading to higher error rates. On the contrary, spintronics-based devices are often more robust to radiation, which can have significant advantages for extra-terrestrial applications [38].

While the complete understanding of brain function remains elusive, it is suggested that the inherent stochasticity within the brain could play a crucial role in enhancing its computational efficiency [39, 40]. Considering this, we delve into discussions around computing with fluctuations and its spintronics implementations.

2 Computing Based on Fluctuations

Usually, in the realm of computation and other practical applications, stochasticity and fluctuations are a major nuisance, necessitating extensive algorithmic and instrumental machinery to minimize their impact. The sources of noise are often intrinsic to the system, either due to the inhomogeneities present or thermal fluctuations. This inherent noise can be harnessed to perform computations in various ways, which will be presented in this section. Each of these computational paradigms has specialized use cases based on their properties.

2.1 Stochastic Computing

Originally introduced in the 1960s, stochastic computing emerged as a cost-effective computing scheme that is perfectly suited for computations where the accuracy could be traded in favor of the speed [41, 42]. A common example in everyday life is a vacuum cleaning robot designed to quickly clean the floor, where millimeter-level precision is not a concern. The core concept of this method is to compute with probability values, i.e., numbers between 0 and 1, which can be approximated by random uncorrelated streams of bits, where the proportion of ones approximates the numerical value. Computational operations are performed directly on the bitstreams. For instance, multiplying two numbers is achieved by feeding their respective bitstreams into an AND gate, as exemplified in Fig. 2. Other arithmetic calculations can also be done using gates and multiplexers [42].

The primary advantage of this method is that it circumvents the need for the complex and energy-intensive circuits typically used for binary arithmetic calculations. Instead, simple gates are employed, offering a solution with substantially lower energy consumption and reduced space requirements. Furthermore, this approach is

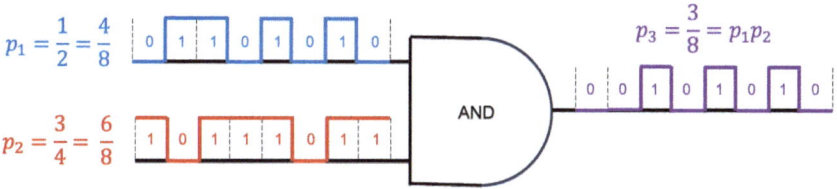

Fig. 2 Illustration of the operating principle of stochastic computing using the example of multiplication of two numbers p_1 and p_2. The (in general approximate) result of the multiplication p_3 is obtained by connecting uncorrelated random bitstream representations of p_1 and p_2 using an AND gate

more error-tolerant since single erroneous bit flips in the bitstreams only mildly affect the results compared to ordinary bit-based calculations.

Despite its advantages, such as inherent error tolerance, simplicity of design, and energy efficiency, this computing paradigm suffers from the trade-off between accuracy and calculation time. This is because the mathematical basis of stochastic computing is the law of large numbers, which states that the sample mean of a random variable converges to the true mean in the limit of a large number of samples [43]. Thus, the law of large numbers guarantees that the longer the bitstreams, the more accurate the representation of the desired probability value is on average [44].

Moreover, a key requirement for computing with bitstreams is that they need to be uncorrelated. The easiest way to understand this is by using the example of multiplication of two of the same numbers. For two identical input bitstreams representing this number (i.e. when they are maximally correlated), the output of the AND gate will be exactly the same as the input streams. Thus, the calculation result will not be the product of the input's probability values but simply the probability itself.

Spintronic devices have been proposed and demonstrated as effective generators of uncorrelated bitstreams, functioning as random number generators. For example, MTJs with a low energy barrier between the parallel and anti-parallel states, also called superparamagnetic tunnel junctions, have been proposed as random number generators [45, 46]. Here, thermal fluctuations cause random switchings between the two magnetic states, which can be interpreted as the bits 0 and 1.

Similarly, magnetic textures such as skyrmions reshufflers can be leveraged for generating uncorrelated bitstreams [47, 48]. This is achieved by representing the bitstreams as skyrmions moving under a constant current into two separate chambers. The value of each bit (0 or 1) determines which chamber receives a skyrmion. Once inside a chamber, the skyrmions undergo random thermal diffusion, effectively causing them to *reshuffle*. The output is read similarly, where receiving a skyrmion from the first chamber indicates a 0, and from the second chamber indicates a 1. This method ensures that while the probability value encoded in the bitstreams remains unchanged, the correlation among the bitstreams is decoupled.

2.2   Inverse Computing

Typically, computing involves producing the correct output from a given input through systematic processing. Identifying the inverse relationship—tracing back from the output to the original input—can also be particularly useful in applications such as integer factorization and invertible logic [49]. Inverse computing is inherently challenging as mathematically, often no unique inverse exists, i.e. multiple inputs often correspond to a single output. For instance, as illustrated in Fig. 3a, output 0 of the AND gate corresponds to three different input states.

Thus, the entire input space must be searched to find all inputs that produce the desired output, making this process typically far more resource-intensive than

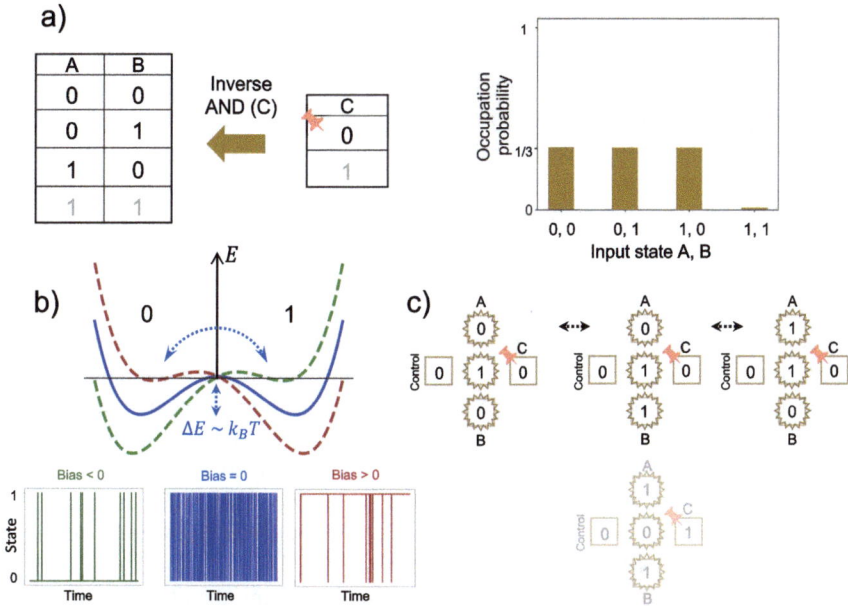

Fig. 3 Illustration of inverse computing exploiting p-bits. **a** Schematic of the general idea behind inverse computing showing an example of an AND gate. The inverse for the output "0" is not unique. Three different inputs (00, 01, 10) correspond to the output "0". An ideal inverse computing scheme utilizing fluctuations yields all three input possibilities with equal probability. **b** Key characteristics of a p-bit: the bistable system has two energy (E) minima at states representing 0 and 1 separated by an energy barrier (ΔE), which is of the order of $k_B T$, where T is the ambient temperature. Under zero bias (blue), the p-bit fluctuates equally between the two states. Under positive/negative bias (brown/green), the energy landscape is tilted, and the 1/0 state is favored, thus changing the represented probability value. **c** Operation of a p-bit with nanomagnets: a majority gate arrangement could be used to implement the AND gate with p-bits. When the output bit C is fixed (indicated by the light red pin) to the 0 state, the A and B bits fluctuate between the three possible states. The bits represented by jagged lines indicate that they can fluctuate between the two states, 0 and 1

standard forward computing. Gates that can function in reverse would be ideal; however, an additional challenge is that the input space often also grows exponentially with the number of inputs.

One possibility is to leverage fluctuations to efficiently explore the entire space of possible solutions in parallel. For the example of inverting the output "0" of the AND gate it means finding all three corresponding input states, ideally with equal probability. Generally speaking, fluctuations are utilized to efficiently explore the entire space for possible solutions. The fluctuating elements are arranged in a way that favors the correct solution over others, making the system statistically more likely to settle into the states corresponding to the desired solution, as explained below using a specific example.

A promising building block for the basic fluctuating element is the so-called p-bit. A p-bit or a probabilistic bit fluctuates between 0 and 1 [50, 51] in contrast to a classical bit, which takes a definite value, i.e. either 0 or 1. The p-bit represents the probability value as the fraction of time it spends in the state labeled as 1 relative to the entire operational period. In principle, any bistable physical system where the switching probabilities can be tuned by an external bias can be used to implement p-bits, as shown in Fig. 3b. Here, the energy needed to switch between the two minima must be in the order of the thermal energy, such that the p-bit naturally fluctuates between the two states at ambient temperature. In the figure, the 0 and 1 p-bit states are represented by two distinct physical states of the system. For a spintronics-based system, the magnetization order parameter, for example, can represent the state of the system. Ideally, under unbiased conditions, the p-bit stays in each of the two states with an equal probability and frequently switches between them. Applying an external bias alters the energy landscape, making the system more likely to reside in one of the two states. A possible spintronic implementation of p-bits utilizes superparamagnetic tunnel junctions. Here, the low energy barrier between the parallel and antiparallel alignment of the magnetic layers can be overcome by thermal fluctuations, allowing to switch between the two states [52]. The biasing of the states can be achieved through various methods, such as an external magnetic field, bias voltage, STT currents, or SOT currents [53, 54].

In Fig. 3c, we show a specific example of implementing the aforementioned p-bit-based inverted AND gate. This form of logic can be implemented using the idea of majority gates with nanomagnetic islands, which are anti-ferromagnetically coupled [55, 56]. The control bit determines the logic operation, which is set to 0 for the AND operation. For the inverse AND gate operation, the output C is fixed (indicated by a red pin in Fig. 3c), and the three p-bits in the middle column are allowed to fluctuate (indicated by the zigzag lines). If the output C is fixed to state "0", the center island will energetically prefer the "1" state due to the antiferromagnetic coupling between nearest neighbors. If the next-nearest neighbor coupling is also antiferromagnetic, the states that the p-bits A and B will assume are no longer unique and will fluctuate between the three different input states corresponding to output "0" of the AND gate (first row of Fig. 3c).

This basic example also reveals some challenges of this majority gate-based implementation of inverse computing: first, the energy landscape/interactions need to be carefully engineered to secure the correct gate operations. Second, it will be very difficult to achieve inverted gate operations that are balanced, i.e., where all correct input states are equally likely to occur. Besides further engineering the interaction and geometry of the structure, the individual p-bits can also be biased to construct balanced gates.

In summary, the primary technological advantages of inverse computing with p-bits are the parallel exploration of the input space and ultra-low energy consumption, driven by the inherent stochasticity introduced by thermal fluctuations. Another possibility for efficiently solving inverse problems is proposed in the context of memcomputing. In this alternative computing paradigm, memory elements are used directly for processing by self-organizing logic gates [57, 58]. There have also been proposals to implement memcomputing using spintronics-based devices, enabling the execution of inverse computations [33, 59].

2.3   Token-Based Brownian Computing

In magnetic p-bits, the stochasticity originates from the fluctuations in the magnetic states. The thermal Brownian motion of nanomagnetic solitons can also be harnessed for computation in a paradigm called token-based Brownian computing [60]. Whereas for stochastic computing and p-bits, the stochasticity resulted in generating bits, in token-based Brownian computing, the thermal fluctuations are utilized to randomly propagate tokens, which can physically explore the search space of a task. In this context, a token refers to a solitary discrete object whose presence or absence is interpreted as the signal. The computational task or logical operation is mapped into a circuit where the tokens undergo Brownian motion along certain paths guided by a few basic, resource-friendly elements [61, 62].

A minimal example set of basic elements for performing logical operations consists of a Hub and a C-join [61], as shown in Fig. 4a. The Hub element is a trijunction, i.e., it consists of three bidirectional wires enabling the token to move randomly along any of the connected wires. The C-join is a four-way intersection that acts as a signal synchronizer: when two tokens enter the element from different directions, the C-join releases them along the other two directions.

Figure 4b shows a possible realization of a half-adder circuit exploiting only the Hub and the C-join. This basic example illustrates when token-based Brownian computations are useful; the calculation relies only on the thermal motion of the tokens, and almost no external power is needed (besides the ultra-low power required to activate the C-joins.) The disadvantage, however, is the potentially very long time required to obtain the calculation result, as there is no guaranteed time until the tokens reach the output lines. Adding ratchets, another type of basic element, at suitable locations in the circuit can accelerate the convergence to the solution. Ratchets limit the token's movement along a certain direction, i.e., it cannot go backward after a token has passed a ratchet. In the half-adder shown in Fig. 4b, they can, for example, be placed after the C-joins.

Using spintronics systems, an energy-efficient implementation of the tokens and circuit primitives can be achieved. Skyrmions, or other topologically protected textures that undergo such Brownian motion, can play the role of the tokens [60]. The Hub has been demonstrated using circuit geometry in magnetic materials [63], the C-join through voltage-controlled magnetic anisotropy (VCMA) effects [60], and the ratchet by employing either a VCMA gradient or specially designed wire geometries [64, 65]. An additional advantage of spintronics systems is that they allow to control and enhance the diffusion of magnetic solitons and, in particular, skyrmions via different mechanisms [66].

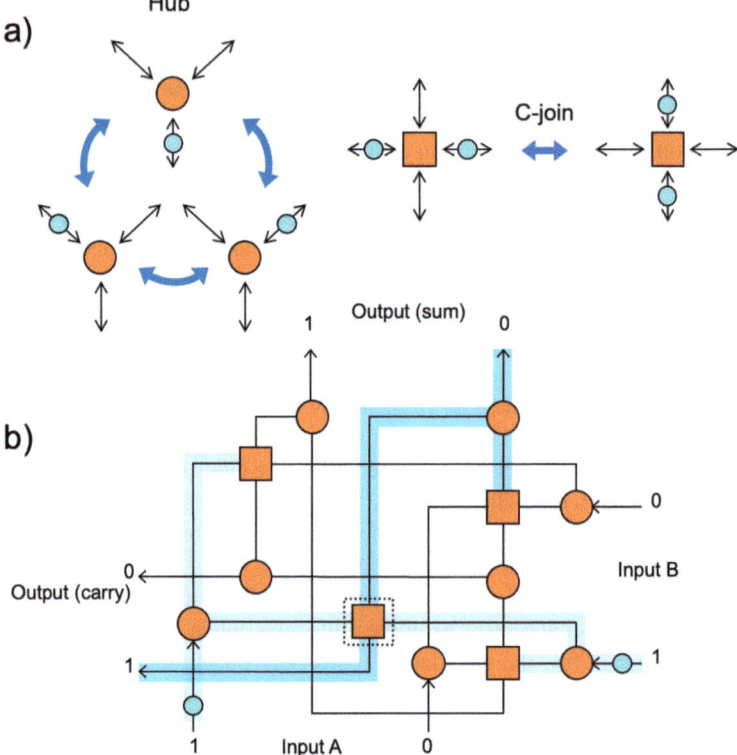

Fig. 4 Illustration of the basic principles of token-based Brownian computing. The tokens (cyan-colored circles) perform a random motion along the wires (black) and are influenced by the circuit elements (orange) according to their functions. **a** One basic set of primitive elements consists of a Hub (circle) and a C-join (Square). The hub is a triple junction that allows the token to move freely along any of the three connecting wires. The C-join is a four-way intersection that requires two tokens to enter the element from different directions, and it releases them along the other two directions. **b** Implementation of a half-adder circuit using hubs and C-joins. In the specific example, the calculation of the addition of two bits (Input A and B) with the value 1 is shown. The wires marked in (light) blue indicate along which wires the tokens are allowed to move (before) after passing the C-join. Adding two bits with the value 1 thus leads to the desired result of carry 1 and sum 0, which is where the tokens exit the circuit and can be detected. The C-join that mainly contributes to this operation is marked with a dotted square

3   Spintronics-Based Neural Networks

Human intelligence relies not only on the capacity to perform computations but also on the ability to recognize patterns. Artificial neural networks (ANNs), inspired by the neural connections in the brain, excel at pattern recognition, driving the majority of the Artificial Intelligence (AI) advancements in the present era [67, 68]. Currently, ANNs are used across various fields, including image recognition, natural language processing, the study of protein folding [69–71] and its applications have even pervaded many aspects of our daily lives [72] (see also the chapter "Introduction to Artificial Intelligence" by Tobias Wand).

An ANN essentially serves as a function approximator with a huge number of tunable parameters (e.g. GPT-3 has 175 billion parameters [73]), which are tweaked to learn a good mapping between an input and its target label [74]. The rapid increase in network size over the recent years and the development of more complex architectures have significantly increased the energy requirements for training and running these models on conventional hardware. This energy consumption has adverse environmental consequences, prompting the search for resource-friendly substrates for implementing neural networks [75].

In this chapter, we address the hardware implementation of neural networks with spintronics. Before taking a deeper dive into the implementation of ANNs using spintronics and the advantages that it delivers over conventional hardware, we summarize its basic functionalities.

3.1 Basic Mode of Operation of ANNs

ANNs consist of two main components: the neurons (or neuronal activations) and the synapses (or synaptic weights) interconnecting the neurons. A deep neural network features multiple neuron layers that transform input into output via hidden layers. Specifically, in a supervised learning setting, an input for a layer is fed into the neurons (x_i). Then, using the synaptic weights (w_{ij}) attached to the neurons, the jth neuronal activation a_j in the next layer is computed as $a_j = f\left(\sum_i w_{ij} x_i\right)$. Here, the activation function f introduces non-linearity into the network, which is indispensable for solving complex problems.

For the physical realization of ANNs, generally, factors such as energy efficiency, speed, error tolerance, compatibility with CMOS technology, endurance, low area overhead, and fabrication cost are crucial considerations. In addition, there are specific requirements for the synaptic and neuronal elements, as explained below.

3.1.1 Requirements for Synapses

The synaptic weights serve as the parameters that characterize the neural network model [67]. They can be regarded as memory elements whose main task is to store the value of the weight in a robust, non-volatile manner for inference. To enable training with materials, the physical properties of the memory elements—typically resistance or conductance—that correspond to weights must be precisely adjusted according to updates from the training algorithm. Resistance-based memories (also called memristors) offer a natural way to implement synapses in the following way [76]: if a weight w_{ij} is encoded as the conductance g_{ij} of a device at the junction between the ith input and jth output, and if the input and output are represented respectively by voltage V_i and current I_j, then using the Ohm's law and Kirchoff's current law we can write the output current as $I_j = \sum_i g_{ij} V_i$.

Thus, the conductance of any physical system could be used to implement the basic functionality of synapses. However, for neural networks, additional properties are required that current hardware does not provide. Firstly, it is essential that synapses are non-volatile, meaning the memory elements retain their programmed state even when the network is not powered. Secondly, overcoming the *von Neumann bottleneck* necessitates the co-location of computing and memory elements. It is important to note that, in a neural network, synapses outnumber neurons, making them the primary bottleneck for in-memory implementation and, thus, a critical focus for hardware development.

3.1.2 Requirements for Neurons

The neurons are the basic nodes or computational elements of the network. Their main purpose is to implement the accumulation of the inputs from the previous layer multiplied by the connecting weights and the non-linear activation function f. Such computing neurons, which typically take continuous real values, can be implemented using a few transistors. However, the implementation with transistors usually has a high area overhead, is energy-inefficient, and suffers from device-to-device variability.

As opposed to the neurons of the simple feed-forward ANNs, the neurons in our brain are dynamic, excitable cells. They accumulate the incoming electrical signal from other neurons in the form of spikes in a leaky manner. While integrating the signal, when the membrane electric potential reaches a certain threshold, it releases an electrical voltage spike. Following this discharge, the membrane potential returns to its baseline state and enters a refractory phase. Based on this idea, Spiking Neural Networks (SNNs) have been proposed as another more bio-realistic variant of neural networks where the neurons communicate with each other via sparse, discrete spiking signals [77]. Although more difficult to train in terms of speed and accuracy, SNNs are more energy-efficient because of the sparse nature of their representation [78]. To emulate SNNs, the neurons must possess a mechanism for generating spikes. Typically, the elements producing the spikes are called the leaky integrate and fire neurons. They integrate the input signal and only fire when this integrated value reaches a certain threshold [79].

3.2 Spintronics-Based Building Blocks for Neural Networks

Spintronics-based hardware is inherently more energy-efficient than current transistor technology, and research has shown that spintronics technology is congruent with implementing the different building blocks of neural networks. In the subsequent text, we present examples of both spintronics-based synapses and neurons.

3.2.1   Spintronics-Based Synapses

Spintronics-based synapses exploit the magnetic state of a physical system to store the synaptic weight. Besides the general advantages of spintronics systems mentioned in Sect. 1.2, spintronics systems offer the reliable and efficient reading and writing of non-volatile magnetic states using effects such as anisotropic magnetoresistance, STT, and SOT. This enables a variety of concepts for spintronics-based synapses, including the following.

For example, spintronics synapses using MTJs have been proposed, and the junctions can be utilized in various ways. An instance of this is using them as binary synaptic elements [80], where the two (parallel and anti-parallel) magnetic states represent the two binary states. In addition, the stochastic switching property has been leveraged as stochastic synapses [81] and even as synapses controlled by radio frequency signals [82]. Another class of spintronic synapses is based on magnetic textures like magnetic skyrmions or domain walls. Here, for example, the number of skyrmions in a region dictates the conductance and, consequently, the synaptic weight [83, 84]. For magnetic domain walls, it has been proposed to use the position of the wall as synaptic weight, as this can also change the conductance [85, 86]. Furthermore, antiferromagnetic materials have been identified to have magnetization-switching properties that can be used for synaptic applications [87, 88].

3.2.2   Spintronics-Based Neurons

Various spintronics-based implementations of neurons have been proposed to utilize magnetic states, as discussed in this subsection. In particular, they truly shine when implementing leaky integrate and fire neurons.

Spin-torque nano-oscillators, a type of MTJ, exhibit spontaneous microwave oscillations when driven by direct current. Their memory-like oscillation amplitudes mimic neuronal leaky integration, while the non-linear response of voltage oscillation amplitudes to input current or field enables the direct implementation of activation functions [89, 90]. Another example is the superparamagnetic tunnel junction, which is an MTJ with a very low energy barrier between the parallel and antiparallel states. Here, the switching rate can be tuned by magnetic fields and spin torque effects, enabling them to be used as neurons in a scheme of computing called population coding [91, 92].

Neurons, like synapses, have also been proposed with magnetic textures. These magnetic textures, such as skyrmions and domain walls, offer the advantage of being controllable through energy-efficient methods, such as applying a low current. This capability allows them to function as carriers of information, akin to neurons, effectively. Furthermore, these textures, owing to their nanoscale dimensions, can be prone to noise either from pinning sites or from thermal fluctuations [93]. This feature holds promise in emulating the functionality of stochastic neurons. Leaky integrate-and-fire neurons based on magnetic textures have also been proposed for skyrmionic or domain wall systems. In these systems, the gradual accumulation and cumulative motion of the solitons ultimately lead to switching of the magnetic state [94–96]. This type of leaky neuron has also been proposed in synthetic antiferromagnetic systems [97].

Moving forward, we delve into reservoir computing, where we highlight the diverse functionalities of the spintronics-based approach.

4   Reservoir Computing with Spintronics

Reservoir computing exploits the inherent nonlinear complex dynamics of a system (called the reservoir) to simplify a classification or prediction task. For this, the reservoir project inputs into a higher dimensional space where the classification or prediction task reduces to an easy (e.g. linear) separation task. The first reservoirs were implemented by recurrent neural networks with random but fixed synaptic weights, such that only the last output layer was trained [98]. In recent years, it has become apparent that many physical systems are well-suited as reservoirs. The emerging field of physical reservoir computing leverages the properties of physical systems to enhance computational performance [99].

Reservoir computing has been discussed in detail in the chapter "An Introduction to Reservoir Computing" by Michael te Vrugt, but here we want to highlight some key features that a physical system has to have to function as a reservoir computer [100, 101].

- **Non-linearity:** This refers to the input undergoing a non-linear transformation due to the reservoir.
- **Complexity:** The term complexity relates to the reservoir's ability to effectively project inputs into a space of higher dimensionality, where "high dimensionality" refers to a reservoir possessing considerably more degrees of freedom than the inputs have.
- **Short-term or fading memory:** The reservoir's ability to process the input signal's temporal history is characterized by its short-term memory attribute. This attribute prioritizes recent inputs while retaining information from earlier ones, with the output being influenced by past inputs, though their impact gradually fades over time. This fading memory is vital for the system to be more responsive to recent inputs, enabling it to adapt to new input patterns or discard outdated information.
- **Reproducibility:** The reservoir should yield identical responses to identical inputs, provided it has been reset between each input. This attribute, although trivial for software implementations, is a crucial prerequisite for physical reservoirs.

The strength of physical reservoir computing lies in its flexibility, allowing a variety of systems that meet the aforementioned criteria to serve as the reservoir.

4.1 Magnetic Textures for Reservoir Computing

Magnetic systems ideally fulfill all criteria to serve as suitable reservoirs due to their inherently non-linear and complex responses to external stimuli. Also, their ability to "forget" information over time—a characteristic known as fading memory—arises from Gilbert damping and other mechanisms that dissipate energy in the magnetization dynamics. Additionally, their compatibility with CMOS technology means magnetic reservoirs can seamlessly integrate into modern electronic devices.

Physical reservoir computing has been demonstrated with common magnetic building blocks such as MTJs [90] and spin-vortex nano oscillators [102]. However, despite their utility, these systems fall short in the complexity required for effective reservoir computing, especially in terms of scaling to more difficult tasks. Enhancing their complexity, for instance, through time multiplexing, is necessary to leverage them as reservoirs [103, 104]. Conversely, magnetic systems with more complexity like dipole-coupled nanomagnets [105, 106], spin wave-based reservoirs [105], magnetic metamaterials [107], skyrmion fabrics [108–114], and other complex spin textures [101, 115, 116] have been used as physical reservoir computers. Other topological spin textures, such as anti-skyrmions, hopfions, and dislocations [117–121], are gaining attention and in the future, can potentially serve as reservoir. Such diversity in terms of options for the reservoir is shown in Fig. 5.

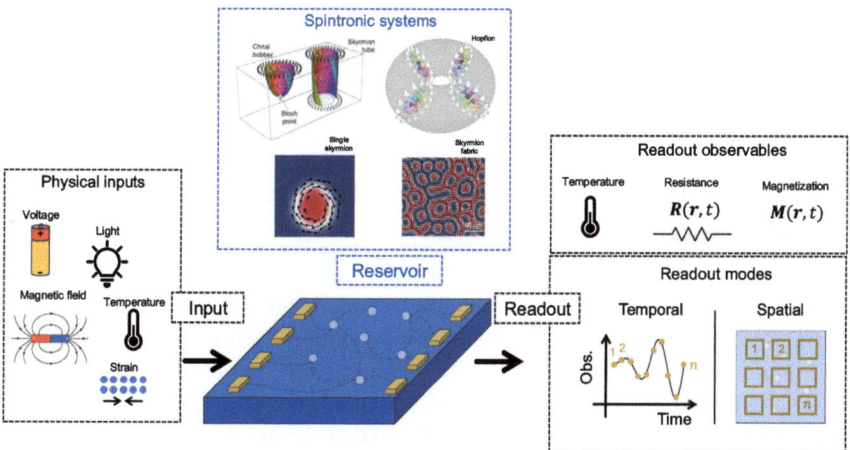

Fig. 5 Illustration depicting various physical operational and readout modes of reservoir computing. The input can be supplied to the reservoir via different physical quantities like voltage, light, magnetic field, temperature, and strain. A plethora of different spintronics systems can play the role of the reservoir. The reservoir states can be read by measuring different physical quantities like temperature, resistance, or magnetization. Furthermore, the mode of reading out the reservoir state could be spatial or temporal. (Figure adapted from figures in [100, 108, 109, 122]). All Rights Reserved. Hopfion figure created by Ross Knapman

Furthermore, the spin-textures and magnetic states can be excited by a wide range of different physical inputs like voltage, light, temperature, magnetic field, and strain, any of which could be used as the input for the reservoir [25, 101, 123]. This flexibility in inputs offers a critical advantage for magnetic reservoir computers, as it allows optimization for speed, accuracy, or power efficiency based on the specific use case, leveraging the unique benefits of each input mode. Furthermore, if a sensor outputs these physical properties, it can be directly integrated with the reservoir, eliminating the need for conversion to an intermediate quantity.

After the inputs excite the reservoir, it has to be read out for the final trained layer. This, too, can be done via different physical observables: the input drive can locally change the temperature, magnetization, and, in turn, the electrical resistance, which can be measured [100, 124]. In addition to various physical observables, the readout can be done in multiple ways. For instance, the reservoir's response can be monitored over time [110]. Alternatively, or in addition, spatial sampling of the reservoir can be employed for readouts [114].

The readout values are then utilized to train the weights of the final linear layer, which produces the output predictions. Currently, the training of these weights occurs on an external device in physical reservoir computing schemes, and further research is needed to explore how this process can be implemented directly within the material itself. Doing so could significantly reduce data transfer and the need for additional hardware, leading to faster computations, miniaturization, and reduced energy consumption. Additionally, more research needs to be done to establish a fair and reliable method for comparing the performance of different reservoirs. This includes distinguishing the impact of pre-processing and understanding the effects of artificially increasing a system's complexity.

5 Memory Technologies: Spintronic Implementations and Beyond

The current landscape of memory technologies is dominated by SRAM, DRAM, and NAND flash memories. SRAM uses bistable flip-flop circuits made of transistors to store data as long as power is supplied, making it fast and ideal for cache memory. DRAM stores data in capacitors and transistors, requiring periodic refreshing to maintain the information. It is commonly used as the main memory in computers due to its higher density and lower cost. NAND flash memory, composed of floating-gate transistors, stores data by varying charge levels in memory cells, retaining data without power. It is widely used in solid-state drives (SSDs), USB drives, and memory cards for its high storage capacity and durability. All of these rely on charge storage to define a memory state. However, a new class of memory technologies is emerging that relies on resistance instead of charge. This includes spintronics-based memories, which distinguish themselves from their conventional counterparts through their non-volatile nature, fast operation, and scalability, among other advantages. The long-term vision for this class of memories is not to fully replace existing silicon-based technology but to serve specific use cases, such as embedded applications for edge devices. In such applications, energy efficiency, speed, error tolerance, integrability with existing technology, and low area overhead are considerations of paramount importance. In terms of implementing artificial intelligence algorithms like ANNs, the switching time and endurance are important parameters. For edge learning applications where the system has to be trained locally, fast and reliable switching is necessary as the network synaptic parameters are updated very frequently. In the context of memory technologies, "endurance" refers to the durability or lifespan of a memory device, specifically measured by the number of write-erase cycles it can withstand before its performance degrades to an unacceptable level. For real tasks, the networks need to be trained for a very long time, and hence, the endurance of the devices needs to be high.

In addition to spintronic-based memories (such as MRAMs mentioned above), three other main types of memory technologies are attracting a lot of attention: oxide filamentary materials (OxRAMs), phase change memories (PCRAMs), and ferroelectric memories (FeRAMs) [125]. All of these technologies exploit the non-volatile physical properties of a material to encode information.

OxRAMs change their resistance by the formation and dissolution of a conducting filament made with oxygen vacancies within an insulating material [126, 127], whereas in PCRAM the material undergoes a phase transition that leads to differences in resistances [128–130]. Some of these technologies have matured to become commercially available [131, 132]. In FeRAMs, a relatively new form of memory, the dielectric polarization of the material changes with the application of voltage, which in turn is measured as a change in resistance using a FeFET (Ferroelectric field effect transistor) [133–135]. Despite their distinct physical properties, these RAM technologies, i.e. MRAMs, OxRAMs, PCRAMs, and FeRAMs, all store memory using resistances.

Due to their fundamentally different physical characteristics, each technology has advantages over the others. State-of-the-art MRAMs, especially utilizing the SOT mechanism, exhibit simultaneously fast switching speed ($\sim$ns) and high endurance (10^{14}) [136, 137]. Furthermore, MRAMs are relatively more robust, whereas other memory technologies, especially OxRAMs, can be prone to noise which degrades neural network performance [138, 139]. Although FeRAM technology competes with MRAM in terms of performance, particularly in energy efficiency and endurance, it remains a relatively new technology that faces challenges with scalability and device-to-device variability [140, 141].

Nevertheless, MRAM has its fair share of challenges to overcome [34]. Since the principal mechanism for resistance switching relies on the tunneling barrier's thickness, precision in the manufacturing process is important. Slight variations in thickness can result in significant performance disparities and affect the device's functionality. Other limitations of the MRAM technology include a relatively small ON/OFF ratio and the presence of only two states of resistance.

Research in spintronics-based memories is advancing towards mitigating these challenges. Utilizing effects such as SOT has already addressed some of the issues present in STT-MRAM, such as the need for high write currents and longer switching times [142]. Additionally, memory devices based on magnetic textures are being explored for their immense potential due to their unique physical characteristics. In summary, spintronics-based materials are well-suited for edge applications, and the ongoing research in this field holds promise for realizing low-power, efficient, and fast computations in the future.

6 Summary and Outlook

In this chapter, we have showcased how the interdisciplinary nature of Neuromorphic Spintronics holds immense potential to contribute positively to many facets of modern computing technologies. In Sect. 1 we highlight the suitability of spintronics materials as a substrate for implementing bio-inspired neuromorphic computing. The next Sects. 2, 3, 4, and 5 present the different specialized use cases where this idea could make an impact. Though the examples widely differ in their applications, they remain connected by their Spintronic materials-based implementation and neuromorphic concepts.

Moreover, the field of spintronics is witnessing rapid developments, greatly increasing its potential for neuromorphic applications. One exciting area of progress is the exploration of magnetic textures not only in 2D but also in 3D [123, 143], greatly expanding the range of possibilities. In the third dimension, the complexity of physical processes and responses increases due to interactions with more neighbors, offering improved functionality and tunability. Additionally, 3D systems allow for higher-density processing power and more readout nodes, leading to compact devices and increased error robustness [100].

Apart from 3D textures, substantial advancements have been made in the domain of multiferroic materials, where coexisting order parameters can significantly enhance the complexity and functionality of devices [100, 144]. By leveraging multiple order parameters, time scales, and length scales, the concept of multi-physics enhances the opportunities for parallel, in-materio, neuromorphic computing.

As computing devices become increasingly personalized, brain-computer interfaces necessitate the integration of computing elements into the human body. In this context, organic spintronics offers significant advantages due to its cost-effective fabrication, lightweight nature, biocompatibility, and biodegradability [145]. The convergence of these technologies promises to revolutionize how we interact with and utilize computing systems, paving the way for more seamless and integrated solutions. Overall, the advancements discussed throughout this chapter highlight the transformative potential of implementing intelligence with cutting-edge physical systems, paving the way for technological breakthroughs.

Acknowledgements We are grateful to Robin Msiska, Maria Azhar, Dennis Meier, Hidekazu Kurebayashi, Jack C. Gartside, and Kerem Camsari for enlightening discussions on various topics, which greatly helped us write the book chapter. We acknowledge funding from the Deutsche

Forschungsgemeinschaft (DFG, German Research Foundation) Project-ID 320163632, 403233384 (SPP Skyrmionics), 405553726 âŁ" CRC/TRR 270, project B12, 505561633 in the TOROID project co-funded by the French National Research Agency (ANR) under Contract No. ANR-22-CE92-0032, and the Emergent AI Center funded by the Carl-Zeiss-Stiftung.

References

1. Schaller RR (1997) Moore's law: past, present and future. IEEE Spectr 34(6):52–59
2. Mitchell Waldrop M (2016) The chips are down for Moore's law. Nat News 530(7589):144
3. Theis TN, Wong H-SP (2017) The end of Moore's law: a new beginning for information technology. Comput Sci & Eng 19(2):41–50
4. Editorial (2018) Big data needs a hardware revolution. Nature 554:145–146
5. Mead C (1990) Neuromorphic electronic systems. Proc IEEE 78(10):1629–1636
6. Mead C (2020) How we created neuromorphic engineering. Nat Electr 3(7):434–435
7. Žutić I, Fabian J, Sarma SD (2004) Spintronics: fundamentals and applications. Rev Modern Phys 76(2):323
8. Hirohata A, Yamada K, Nakatani Y, Prejbeanu I-L, Diény B, Pirro P, Hillebrands B (2020) Review on spintronics: principles and device applications. J Magn Magn Mater 509:166711
9. Pais A (1989) George Uhlenbeck and the discovery of electron spin. Phys Today 42(12):34–40
10. Baibich MN, Broto JM, Fert A, Dau FN, Petroff F, Etienne P, Creuzet G, Friederich A, Chazelas J (1988) Giant magnetoresistance of (001) Fe/(001) Cr magnetic superlattices. Phys Rev Lett 61(21):2472
11. Binasch G, Grünberg P, Saurenbach F, Zinn W (1989) Enhanced magnetoresistance in layered magnetic structures with antiferromagnetic interlayer exchange. Phys Rev B 39(7):4828
12. Moodera JS, Kinder LR, Wong TM, Meservey R (1995) Large magnetoresistance at room temperature in ferromagnetic thin film tunnel junctions. Phys Rev Lett 74(16):3273
13. Mathon J, Umerski A (2001) Theory of tunneling magnetoresistance of an epitaxial Fe/MgO/Fe (001) junction. Phys Rev B 63(22):220403
14. Butler WH, Zhang X-G, Schulthess TC, MacLaren JM (2001) Spin-dependent tunneling conductance of Fe/Mgo/Fe sandwiches. Phys Rev B 63(5):054416
15. Parkin SSP, Hayashi M, Thomas L (2008) Magnetic domain-wall racetrack memory. Science 320(5873):190–194
16. Fert A, Cros V, Sampaio J (2013) Skyrmions on the track. Nat Nanotechnol 8(3):152–156
17. Fert A, Reyren N, Cros V (2017) Magnetic skyrmions: advances in physics and potential applications. Nat Rev Mater 2(7):1–15
18. Tokura Y, Kanazawa N (2020) Magnetic skyrmion materials. Chem Rev 121(5):2857–2897
19. Everschor-Sitte K, Masell J, Reeve RM, Kläui M (2018) Perspective: magnetic skyrmions-overview of recent progress in an active research field. J Appl Phys 124(24):240901
20. Parkin S, Yang S-H (2015) Memory on the racetrack. Nat Nanotechnol 10(3):195–198
21. Ralph DC, Stiles MD (2008) Spin transfer torques. J Magn Magn Mate 320(7):1190–1216
22. Ramaswamy R, Lee JM, Cai K, Yang H (2018) Recent advances in spin-orbit torques: Moving towards device applications. Appl Phys Rev 5(3)
23. Upadhyay NK, Jiang H, Wang Z, Asapu S, Xia Q, Yang JJ (2019) Emerging memory devices for neuromorphic computing. Adv Mat Technol 4(4):1800589
24. Burr GW, Shelby RM, Sebastian A, Kim S, Kim S, Sidler S, Virwani K, Ishii M, Narayanan P, Fumarola A, Sanches LL, Boybat I, Gallo M, Moon K, Woo J, Hwang H, Leblebici Y (2017) Neuromorphic computing using non-volatile memory. Adv Phys X 2(1):89–124
25. Grollier J, Querlioz D, Camsari KY (2020) Karin Everschor-Sitte, Shunsuke Fukami, and Mark D Stiles. Neuromorph Spintron Nat Electr 3(7):360–370
26. Kolb B, Whishaw IQ (1998) Brain plasticity and behavior. Ann Rev Psychol 49(1):43–64

27. Attwell D, Laughlin SB (2001) An energy budget for signaling in the grey matter of the brain. J Cerebral Blood Flow & Metabol 21(10):1133–1145
28. Horowitz M (2014) Computing's energy problem (and what we can do about it). In: 2014 IEEE international solid-state circuits conference digest of technical papers (ISSCC), pp 10–14. IEEE
29. Indiveri G, Liu S-C (2015) Memory and information processing in neuromorphic systems. Proc IEEE 103(8):1379–1397
30. Cosgrove KP, Mazure CM, Staley JK (2007) Evolving knowledge of sex differences in brain structure, function, and chemistry. Biolog Psych 62(8):847–855
31. Herculano-Houzel S (2009) The human brain in numbers: a linearly scaled-up primate brain. Front Hum Neurosci 3:857
32. Pakkenberg B, Pelvig D, Marner L, Bundgaard MJ, Gundersen HJG, Nyengaard JR, Regeur L (2003) Aging and the human neocortex. Exp Gerontol 38(1–2):95–99
33. Finocchio G, Di Ventra M, Camsari KY, Everschor-Sitte K, Amiri PK, Zeng Z (2021) The promise of spintronics for unconventional computing. J Magn Magn Mater 521:167506
34. Bhatti S, Sbiaa R, Hirohata A, Ohno H, Fukami S, Piramanayagam SN (2017) Spintronics based random access memory: a review. Mater Today 20(9):530–548
35. Faisal AA, Selen LPJ, Wolpert DM (2008) Noise in the nervous system. Nat Rev Neurosci 9(4):292–303
36. Rolls ET, Deco G (2010) The noisy brain: stochastic dynamics as a principle of brain function. Oxford University Press
37. Faisal AA, White JA, Laughlin SB (2005) Ion-channel noise places limits on the miniaturization of the brain's wiring. Current Biol 15(12):1143–1149
38. Ren F, Jander A, Dhagat P, Nordman C (2012) Radiation tolerance of magnetic tunnel junctions with MgO tunnel barriers. IEEE Trans Nucl Sci 59(6):3034–3038
39. Mori T, Kai S (2002) Noise-induced entrainment and stochastic resonance in human brain waves. Phys Rev Lett 88(21):218101
40. Deco G, Rolls ET, Romo R (2009) Stochastic dynamics as a principle of brain function. Progr Neurobiol 88(1):1–16
41. Gaines BR (1969) Stochastic computing systems. Adv Inf Syst Sci 2:37–172
42. Alaghi A, Hayes JP (2013) Survey of stochastic computing. ACM Trans Embed Comput Syst (TECS) 12(2s):1–19
43. Hsu P-L, Robbins H (1947) Complete convergence and the law of large numbers. Proc Natl Acad Sci 33(2):25–31
44. Alaghi A, Qian W, Hayes JP (2017) The promise and challenge of stochastic computing. IEEE Trans Comput-Aided Des Integr Circuits Syst 37(8):1515–1531
45. Vodenicarevic D, Locatelli N, Mizrahi A, Friedman JS, Vincent AF, Romera M, Fukushima A, Yakushiji K, Kubota H, Yuasa S, Tiwari S, Grollier J, Querlioz D (2017) Low-energy truly random number generation with superparamagnetic tunnel junctions for unconventional computing. Phys Rev Appl 8:054045
46. Zhenxiao F, Tang Y, Zhao X, Kai L, Dong Y, Shukla A, Zhu Z, Yang Y (2021) An overview of spintronic true random number generator. Front Phys 9:638207
47. Pinna D, Araujo FA, Kim J-V, Cros V, Querlioz D, Bessière P, Droulez J, Grollier J (2018) Skyrmion gas manipulation for probabilistic computing. Phys Rev Appl 9(6):064018
48. Zázvorka J, Jakobs F, Heinze D, Keil N, Kromin S, Jaiswal S, Litzius K, Jakob G, Virnau P, Pinna D, Everschor-Sitte K, Rózsa L, Donges A, Nowak U, Kläui M (2019) Thermal skyrmion diffusion used in a reshuffler device. Nat Nanotechnol 14(7):658–661
49. Borders WA, Pervaiz AZ, Fukami S, Camsari KY, Ohno H, Datta S (2019) Integer factorization using stochastic magnetic tunnel junctions. Nature 573(7774):390–393
50. Camsari KY, Sutton BM, Datta S (2019) p-bits for probabilistic spin logic. Appl Phys Rev 6(1)
51. Camsari KY, Faria R, Sutton BM (2017) Datta S (2017) Stochastic p-bits for invertible logic. Phys Rev X 7(3):031014

52. Camsari KY, Salahuddin S, Datta S (2017) Implementing p-bits with embedded MTJ. IEEE Electron Dev Lett 38(12):1767–1770
53. Chowdhury S, Grimaldi A, Aadit NA, Niazi S, Mohseni M, Kanai S, Ohno H, Fukami S, Theogarajan L, Finocchio G, Datta S, Camsari KY (2023) A full-stack view of probabilistic computing with p-bits: devices, architectures, and algorithms. IEEE J Explor Solid-State Comput Dev Circ 9(1):1–11
54. Yin J, Liu Y, Zhang B, Du A, Gao T, Ma X, Dong Y, Bai Y, Lu S, Zhuo Y, Huang Y, Cai W, Zhu D, Shi K, Cao K, Zhang D, Zeng L, Zhao W (2022) Scalable ising computer based on ultra-fast field-free spin orbit torque stochastic device with extreme 1-bit quantization. In: 2022 international electron devices meeting (IEDM), pp 36.1.1–36.1.4
55. Imre A, Csaba G, Ji L, Orlov A, Bernstein GH, Porod W (2006) Majority logic gate for magnetic quantum-dot cellular automata. Science 311(5758):205–208
56. Eichwald I, Breitkreutz S, Ziemys G, Csaba G, Porod W, Becherer M (2014) Majority logic gate for 3d magnetic computing. Nanotechnology 25(33):335202
57. Ventra MD, Traversa FL (2018) Perspective: memcomputing: leveraging memory and physics to compute efficiently. J Appl Phys 123(18)
58. Traversa FL, Ventra MD (2017) Polynomial-time solution of prime factorization and NP-complete problems with digital memcomputing machines. Chaos: An Interdisip J Nonlinear Sci 27(2)
59. Finocchio G, Incorvia JAC, Friedman JS, Yang Q, Giordano A, Grollier J, Yang H, Ciubotaru F, Chumak AV, Naeemi AJ, Cotofana SD, Tomasello R, Panagopoulos C, Carpentieri M, Lin P, Pan G, Yang JJ, Todri-Sanial A, Boschetto G, Makasheva K, Sangwan VK, Trivedi AR, Hersam MC, Camsari KY, McMahon PL, Datta S, Koiller B, Aguilar GH, Temporão GP, Rodrigues DR, Sunada S, Everschor-Sitte K, Tatsumura K, Goto H, Puliafito V, Åkerman J, Takesue H, Ventra MD, Pershin YV, Mukhopadhyay S, Roy K, Wang I-T, Kang W, Zhu Y, Kaushik BK, Hasler J, Ganguly S, Ghosh AW, Levy W, Roychowdhury V, Bandyopadhyay S (2023) Roadmap for unconventional computing with nanotechnology, Nano Futures
60. Nozaki T, Jibiki Y, Goto M, Tamura E, Nozaki T, Kubota H, Fukushima A, Yuasa S, Suzuki Y (2019) Brownian motion of skyrmion bubbles and its control by voltage applications. Appl Phys Lett 114(1)
61. Peper F (2009) Exploiting noise in computation. IEICE Proc Ser 43(B0L-A1)
62. Lee J, Peper F (2010) Efficient circuit construction in brownian cellular automata based on a new building-block for delay-insensitive circuits. In: Cellular automata: 9th international conference on cellular automata for research and industry, ACRI 2010, Ascoli Piceno, Italy, September 21–24, 2010. Proceedings 9, pp 356–364. Springer
63. Jibiki Y, Goto M, Tamura E, Cho J, Miki S, Ishikawa R, Nomura H, Srivastava T, Lim W, Auffret S, Baraduc C, Bea H, Suzuki Y (2020) Skyrmion brownian circuit implemented in continuous ferromagnetic thin film. Appl Phys Lett 117(8)
64. Wang X, Gan WL, Martinez JC, Tan FN, Jalil MBA, Lew WS (2018) Efficient skyrmion transport mediated by a voltage controlled magnetic anisotropy gradient. Nanoscale 10(2):733–740
65. Bellizotti Souza JC, Vizarim NP, Reichhardt CJO, Reichhardt C, Venegas PA (2021) Skyrmion ratchet in funnel geometries. Phys Rev B 104(5):054434
66. Goto M, Nomura H, Suzuki Y (2021) Stochastic skyrmion dynamics under alternating magnetic fields. J Magn Magn Mater 536:167974
67. LeCun Y, Bengio Y, Hinton G (2015) Deep learning. Nature 521(7553):436–444
68. Goodfellow I, Bengio Y, Courville A (2016) Deep learning. MIT Press
69. Guo Y, Liu Y, Oerlemans A, Lao S, Wu S, Lew MS (2016) Deep learning for visual understanding: a review. Neurocomputing 187:27–48
70. Deng L, Liu Y (2018) Deep learning in natural language processing. Springer
71. Jumper J, Evans R, Pritzel A, Green T, Figurnov M, Ronneberger O, Tunyasuvunakool K, Bates R, Žídek A, Potapenko A, Bridgland A, Meyer C, Kohl SAA, Ballard AJ, Cowie A, Romera-Paredes B, Nikolov S, Jain R, Adler J, Back T, Petersen S, Reiman D, Clancy E, Zielinski M, Steinegger M, Pacholska M, Berghammer T, Bodenstein S, Silver D, Vinyals O, Senior AW, Kavukcuoglu K, Kohli P, Hassabis D (2021) Highly accurate protein structure prediction with alphafold. Nature 596(7873):583–589

72. Alzubaidi L, Zhang J, Humaidi AJ, Al-Dujaili A, Duan Y, Al-Shamma O, Santamaría J, Fadhel MA, Al-Amidie M, Farhan L (2021) Review of deep learning: concepts, CNN architectures, challenges, applications, future directions. J Big Data 8:1–74

73. Zhang M, Li J (2021) A commentary of GPT-3 in MIT technology review 2021. Fundam Res 1(6):831–833

74. Cybenko G (1989) Approximation by superpositions of a sigmoidal function. Math Control Signals Syst 2(4):303–314

75. Patterson D, Gonzalez J, Le Q, Liang C, Munguia L-M, Rothchild D, So D, Texier M, Dean J (2021) Carbon emissions and large neural network training. arXiv:2104.10350

76. Strukov DB, Snider GS, Stewart DR, Williams RS (2008) The missing memristor found. Nature 453(7191):80–83

77. Tavanaei A, Ghodrati M, Kheradpisheh SR, Masquelier T, Maida A (2019) Deep learning in spiking neural networks. Neural Netw 111:47–63

78. Blouw P, Choo X, Hunsberger E, Eliasmith C (2019) Benchmarking keyword spotting efficiency on neuromorphic hardware. In: Proceedings of the 7th annual neuro-inspired computational elements workshop, pp 1–8

79. Liu Y-H, Wang X-J (2001) Spike-frequency adaptation of a generalized leaky integrate-and-fire model neuron. J Comput Neurosci 10:25–45

80. Jung S, Lee H, Myung S, Kim H, Yoon SK, Kwon S-W, Ju Y, Kim M, Yi W, Han S, Kwon B, Seo B, Lee K, Koh G-H, Lee K, Song Y, Choi C, Ham D, Kim SJ (2022) A crossbar array of magnetoresistive memory devices for in-memory computing. Nature 601(7892):211–216

81. Vincent AF, Larroque J, Locatelli N, Romdhane NB, Bichler O, Gamrat C, Zhao WS, Klein J-O, Galdin-Retailleau S, Querlioz D (2015) Spin-transfer torque magnetic memory as a stochastic memristive synapse for neuromorphic systems. IEEE Trans Biomed Circuits Syst 9(2):166–174

82. Leroux N, Mizrahi A, Marković D, Sanz-Hernández D, Trastoy J, Bortolotti P, Martins L, Jenkins A, Ferreira R, Grollier J (2021) Hardware realization of the multiply and accumulate operation on radio-frequency signals with magnetic tunnel junctions. Neuromorphic Comput Eng 1(1):011001

83. Huang Y, Kang W, Zhang X, Zhou Y, Zhao W (2017) Magnetic skyrmion-based synaptic devices. Nanotechnology 28(8):08LT02

84. Song KM, Jeong J-S, Pan B, Zhang X, Xia J, Cha S, Park T-E, Kim K, Finizio S, Raabe J, Chang J, Zhou Y, Zhao W, Kang W, Ju H, Woo S (2020) Skyrmion-based artificial synapses for neuromorphic computing. Nat Electr 3(3):148–155

85. Bhowmik D, Saxena U, Dankar A, Verma A, Kaushik D, Chatterjee S, Singh U (2019) On-chip learning for domain wall synapse based fully connected neural network. J Magn Magn Mater 489:165434

86. Liu S, Xiao TP, Cui C, Incorvia JAC, Bennett CH, Marinella MJ (2021) A domain wall-magnetic tunnel junction artificial synapse with notched geometry for accurate and efficient training of deep neural networks. Appl Phy Lett 118(20)

87. Miron IM, Garello K, Gaudin G, Zermatten P-J, Costache MV, Auffret S, Bandiera S, Rodmacq B, Schuhl A, Gambardella P (2011) Perpendicular switching of a single ferromagnetic layer induced by in-plane current injection. Nature 476(7359):189–193

88. Fukami S, Zhang C, DuttaGupta S, Kurenkov A, Ohno H (2016) Magnetization switching by spin-orbit torque in an antiferromagnet-ferromagnet bilayer system. Nat Mater 15(5):535–541

89. Kiselev SI, Sankey JC, Krivorotov IN, Emley NC, Schoelkopf RJ, Buhrman RA, Ralph DC (2003) Microwave oscillations of a nanomagnet driven by a spin-polarized current. Nature 425(6956):380–383

90. Torrejon J, Riou M, Araujo FA, Tsunegi S, Khalsa G, Querlioz D, Bortolotti P, Cros V, Yakushiji K, Fukushima A, Kubota H, Yuasa S, Stiles MD, Grollier J (2017) Neuromorphic computing with nanoscale spintronic oscillators. Nature 547(7664):428–431

91. Locatelli N, Mizrahi A, Accioly A, Matsumoto R, Fukushima A, Kubota H, Yuasa S, Cros V, Pereira LG, Querlioz D, Kim J-V, Grollier J (2014) Noise-enhanced synchronization of stochastic magnetic oscillators. Phys Rev Appl 2:034009

92. Mizrahi A, Hirtzlin T, Fukushima A, Kubota H, Yuasa S, Grollier J, Querlioz D (2018) Neural-like computing with populations of superparamagnetic basis functions. Nat Commun 9(1):1533

93. Hayward TJ (2015) Intrinsic nature of stochastic domain wall pinning phenomena in magnetic nanowire devices. Sci Rep 5(1):13279

94. Sharad M, Augustine C, Panagopoulos G, Roy K (2012) Spin-based neuron model with domain-wall magnets as synapse. IEEE Trans Nanotechnol 11(4):843–853

95. Hassan N, Hu X, Jiang-Wei L, Brigner WH, Akinola OG, Garcia-Sanchez F, Pasquale M, Bennett CH, Incorvia JAC, Friedman JS (2018) Magnetic domain wall neuron with lateral inhibition. J Appl Phys 124(15)

96. Chen M-C, Sengupta A, Roy K (2018) Magnetic skyrmion as a spintronic deep learning spiking neuron processor. IEEE Trans Magn 54(8):1–7

97. Wang D, Tang R, Lin H, Liu L, Nuo X, Sun Y, Zhao X, Wang Z, Wang D, Mai Z, Zhou Y, Gao N, Song C, Zhu L, Tom W, Liu M, Xing G (2023) Spintronic leaky-integrate-fire spiking neurons with self-reset and winner-takes-all for neuromorphic computing. Nat Commun 14(1):1068

98. Lukoševičius M, Jaeger H (2009) Reservoir computing approaches to recurrent neural network training. Comput Sci Rev 3(3):127–149

99. Nakajima K (2020) Physical reservoir computing-an introductory perspective. Jpn J Appl Phys 59(6):060501

100. Everschor-Sitte K, Majumdar A, Wolk K, Meier D (2024) Topological magnetic and ferroelectric systems for reservoir computing. Nat Rev Phys 1–8

101. Lee O, Msiska R, Brems MA, Kläui M, Kurebayashi H, Everschor-Sitte K (2023) Perspective on unconventional computing using magnetic skyrmions. Appl Phys Lett 122(26):260501

102. Marković D, Leroux N, Riou M, Abreu Araujo F, Torrejon J, Querlioz D, Fukushima A, Yuasa S, Trastoy J, Bortolotti P, Grollier J (2019) Reservoir computing with the frequency, phase, and amplitude of spin-torque nano-oscillators. Appl Phys Lett 114(1):012409

103. Röhm A, Lüdge K (2018) Multiplexed networks: reservoir computing with virtual and real nodes. J Phys Commun 2(8):085007

104. Cucchi M, Abreu S, Ciccone G, Brunner D, Kleemann H (2022) Hands-on reservoir computing: a tutorial for practical implementation. Neuromorphic Comput Eng 2(3):032002

105. Nomura H, Furuta T, Tsujimoto K, Kuwabiraki Y, Peper F, Tamura E, Miwa S, Goto M, Nakatani R, Suzuki Y (2019) Reservoir computing with dipole-coupled nanomagnets. Jpn J Appl Phys 58(7):070901

106. Gartside JC, Stenning KD, Vanstone A, Holder HH, Arroo DM, Dion T, Caravelli F, Kurebayashi H, Branford WR (2022) Reconfigurable training and reservoir computing in an artificial spin-vortex ice via spin-wave fingerprinting. Nat Nanotechnol 17(5):460–469

107. Vidamour IT, Swindells C, Venkat G, Manneschi L, Fry PW, Welbourne A, Rowan-Robinson RM, Backes D, Maccherozzi F, Dhesi SS, Vasilaki E, Allwood DA, Hayward TJ (2023) Reconfigurable reservoir computing in a magnetic metamaterial. Commun Phys 6(1):230

108. Prychynenko D, Sitte M, Litzius K, Krüger B, Bourianoff G, Kläui M, Sinova J, Everschor-Sitte K (2018) Magnetic skyrmion as a nonlinear resistive element: a potential building block for reservoir computing. Phys Rev Appl 9:014034

109. Bourianoff G, Pinna D, Sitte M, Everschor-Sitte K (2018) Potential implementation of reservoir computing models based on magnetic skyrmions. AIP Adv 8(5):055602

110. Pinna D, Bourianoff G, Everschor-Sitte K (2020) Reservoir computing with random skyrmion textures. Phys Rev Appl 14(5):054020

111. Sun Y, Lin T, Lei N, Chen X, Kang W, Zhao Z, Wei D, Chen C, Pang S, Hu L, Yang L, Dong E, Zhao L, Liu L, Yuan Z, Ullrich A, Back CH, Zhang J, Pan D, Zhao J, Feng M, Fert A, Zhao W (2023) Experimental demonstration of a skyrmion-enhanced strain-mediated physical reservoir computing system. Nat Commun 14(1):3434

112. Lee M-K, Mochizuki M (2022) Reservoir computing with spin waves in a skyrmion crystal. Phys Rev Appl 18:014074

113. Lee M-K, Mochizuki M (2023) Handwritten digit recognition by spin waves in a Skyrmion reservoir. Sci Rep 13(1):19423

114. Msiska R, Love J, Mulkers J, Leliaert J (2023) Everschor-Sitte K (2023) Audio classification with skyrmion reservoirs. Adv Intell Syst 2023:2200388
115. Bechler NT, Masell J (2023) Helitronics as a potential building block for classical and unconventional computing. Neuromorph Comput Eng 3(3):034003
116. Lee O, Wei T, Stenning KD, Gartside JC, Prestwood D, Seki S, Aqeel A, Karube K, Kanazawa N, Taguchi Y, Back C, Tokura Y, Branford WR, Kurebayashi H (2024) Task-adaptive physical reservoir computing. Nat Mater 23(1):79–87
117. Azhar M, Kravchuk VP, Garst M (2022) Screw dislocations in chiral magnets. Phys Rev Lett 128(15):157204
118. Tang J, Yaodong W, Wang W, Kong L, Lv B, Wei W, Zang J, Tian M, Haifeng D (2021) Magnetic skyrmion bundles and their current-driven dynamics. Nat Nanotechnol 16(10):1086–1091
119. Wang XS, Qaiumzadeh A, Brataas A (2019) Current-driven dynamics of magnetic hopfions. Phys Rev Lett 123(14):147203
120. Kent N, Reynolds N, Raftrey D, Campbell ITG, Virasawmy S, Dhuey S, Chopdekar RV, Hierro-Rodriguez A, Sorrentino A, Pereiro E, Ferrer S, Hellman F, Sutcliffe P, Fischer P (2021) Creation and observation of Hopfions in magnetic multilayer systems. Nat Commun 12(1):1562
121. Stepanova M, Masell J, Lysne E, Schoenherr P, Köhler L, Paulsen M, Qaiumzadeh A, Kanazawa N, Rosch A, Tokura Y, Brataas A, Garst M, Meier D (2021) Detection of topological spin textures via nonlinear magnetic responses. Nano Lett 22(1):14–21
122. Zheng F, Rybakov FN, Borisov AB, Song D, Wang S, Li Z-A, Haifeng D, Kiselev NS, Caron J, Kovács A, Tian M, Zhang Y, Blügel S, Dunin-Borkowski RE (2018) Experimental observation of chiral magnetic bobbers in B20-type FeGe. Nat Nanotechnol 13(6):451–455
123. Vedmedenko EY, Kawakami RK, Sheka DD, Gambardella P, Kirilyuk A, Hirohata A, Binek C, Chubykalo-Fesenko O, Sanvito S, Kirby BJ, Grollier J, Everschor-Sitte K, Kampfrath T, You CY, Berger A (2020) The 2020 magnetism roadmap. J Phys D Appl Phys 53(45):453001
124. Tanaka G, Yamane T, Héroux JB, Nakane R, Kanazawa N, Takeda S, Numata H, Nakano D, Hirose A (2019) Recent advances in physical reservoir computing: a review. Neural Netw 115:100–123
125. Ielmini D, Philip H-S, Wong (2018) In-memory computing with resistive switching devices. Nat Electr 1(6):333–343
126. Dittmann R, Menzel S, Waser R (2021) Nanoionic memristive phenomena in metal oxides: the valence change mechanism. Adv Phys 70(2):155–349
127. Beck A, Bednorz JG, Gerber Ch, Rossel C, Widmer D (2000) Reproducible switching effect in thin oxide films for memory applications. Appl Phys Lett 77(1):139–141
128. Yamada N, Ohno E, Nishiuchi K, Akahira N, Takao M (1991) Rapid-phase transitions of GeTe-Sb2Te3 pseudobinary amorphous thin films for an optical disk memory. J Appl Phys 69(5):2849–2856
129. Wong H-SP, Raoux S, Kim SB, Liang J, Reifenberg JP, Rajendran B, Asheghi M, Goodson KE (2010) Phase change memory. Proc IEEE 98(12):2201–2227
130. Burr GW, Breitwisch MJ, Franceschini M, Garetto D, Gopalakrishnan K, Jackson B, Kurdi B, Lam C, Lastras LA, Padilla A, Rajendran B, Raoux S, Shenoy RS (2010) Phase change memory technology. J Vac Sci & Technol B 28(2):223–262
131. Zha Y, Wei Z, Li J (2017) Recent progress in RRAM technology: from compact models to applications. In: 2017 China semiconductor technology international conference (CSTIC), pp 1–4. IEEE
132. Lee YK, Song Y, Kim JC, Oh SC, Bae B-J, Lee SH, Lee JH, Pi UH, Seo B, Jung H, Lee K, Shin HC, Jung H, Pyo M, Antonyan A, Lee S, Hwang S, Jang D, Ji Y, Lee S, Lim J, Koh K-H, Hwang K, Hong H, Park K, Jeong G, Yoon JS, Jung ES (2018) Embedded STT-MRAM in 28-nm FDSOI logic process for industrial MCU/IoT application. In : 2018 IEEE symposium on VLSI technology, pp 181–182. IEEE
133. Mikolajick T, Dehm C, Hartner W, Kasko I, Kastner MJ, Nagel N, Moert M, Mazure C (2001) Feram technology for high density applications. Microelectr Reliab 41(7):947–950

134. Böscke TS, Müller J, Bräuhaus D, Schröder U, Böttger UJAPL (2011) Ferroelectricity in hafnium oxide thin films. Appl Phys Lett 99(10)
135. Trentzsch M, Flachowsky S, Richter R, Paul J, Reimer B, Utess D, Jansen S, Mulaosmanovic H, Müller S, Slesazeck S, Ocker J, Noack M, Müller J, Polakowski P, Schreiter J, Beyer S, Mikolajick T, Rice B (2016) A 28nm HKMG super low power embedded NVM technology based on ferroelectric FETs. In: 2016 IEEE international electron devices meeting (IEDM), pp 11–5. IEEE
136. Huai Y (2008) Spin-transfer torque MRAM (STT-MRAM): challenges and prospects. AAPPS Bull 18(6):33–40
137. Carboni R, Ambrogio S, Chen W, Siddik M, Harms J, Lyle A, Kula W, Sandhu G, Ielmini D (2016) Understanding cycling endurance in perpendicular spin-transfer torque (p-STT) magnetic memory. In: 2016 IEEE international electron devices meeting (IEDM), pp 21–6. IEEE
138. Majumdar A, Bocquet M, Hirtzlin T, Laborieux A, Klein J-O, Nowak E, Vianello E, Portal J-M, Querlioz D (2021) Model of the weak reset process in HfOx resistive memory for deep learning frameworks. IEEE Trans Electron Dev 68(10):4925–4932
139. Brivio S, Spiga S, Ielmini D (2022) HfO2-based resistive switching memory devices for neuromorphic computing. Neuromorphic Comput Eng 2(4):042001
140. Banerjee W (2020) Challenges and applications of emerging nonvolatile memory devices. Electronics 9(6):1029
141. Mulaosmanovic H, Breyer ET, Dünkel S, Beyer S, Mikolajick T, Slesazeck S (2021) Ferroelectric field-effect transistors based on HfO2: a review. Nanotechnology 32(50):502002
142. Saha R, Pundir YP, Pal PK (2022) Comparative analysis of STT and SOT based MRAMs for last level caches. J Magn Magn Mat 551:169161
143. Wei W-S, He Z-D, Zhe Q, Hai-Feng D (2021) Dzyaloshinsky-Moriya interaction (DMI)-induced magnetic skyrmion materials. Rare Met 40(11):3076–3090
144. Eerenstein W, Mathur ND, Scott JF (2006) Multiferroic and magnetoelectric materials. Nature 442(7104):759–765
145. Joshi VK (2016) Spintronics: a contemporary review of emerging electronics devices. Eng Sci Technol Int J 19(3):1503–1513

Machine Learning with Quantum Computers

Ivana Nikoloska

Abstract Learning compressed, latent representations of quantum states can help us understand and simulate the natural world. To this end, one needs devices that can capture and reliably process information using quantum-mechanical effects. Whilst we currently lack a device that can achieve this goal, we are well into the era of noisy, intermediate-scale quantum (NISQ) computers—quantum machines with $50-100$ noisy qubits—which have catalysed machine learning applications. This chapter provides the main ideas underpinning learning with NISQ machines. It also provides insight into the architectures and training schemes currently in use, as well as the limitations and potential futures of the field.

1 (A Very Brief) Introduction to Quantum Computing

Unlike Turing machines built on the physical implementation of the two states '0' and '1', quantum computers leverage the laws of quantum mechanics and can make use of a superposition of two quantum states $|0\rangle$ defined as and $|1\rangle$

$$|0\rangle = \begin{bmatrix} 1 \\ 0 \end{bmatrix}, \quad |1\rangle = \begin{bmatrix} 0 \\ 1 \end{bmatrix}, \tag{1}$$

which can be, for example, encoded in two distinct energy levels of an atom [1–4]. The elementary unit of quantum computation is the qubit, $|\psi\rangle = \alpha_1 |0\rangle + \alpha_2 |1\rangle$

$$|\psi\rangle = \alpha_1 \begin{bmatrix} 1 \\ 0 \end{bmatrix} + \alpha_2 \begin{bmatrix} 0 \\ 1 \end{bmatrix} \tag{2}$$

I. Nikoloska (✉)
Eindhoven University of Technology, Eindhoven, The Netherlands
e-mail: i.nikoloska@tue.nl

© The Author(s) 2026

M. te Vrugt (ed.), *Artificial Intelligence and Intelligent Matter*, Machine Intelligence for Materials Science, https://doi.org/10.1007/978-3-032-04129-6_20

where $\alpha_1, \alpha_2 \in \mathbb{C}$ are the so called complex-valued amplitudes.[1] The transformation of the qubit's state into another is carried out using unitary quantum gates, denoted by $U \in \mathbb{C}^{2 \times 2}$, since the absolute values of the squares of the complex amplitudes represents the probability to measure the qubit in the 0 or the 1 state, and quantum dynamics must maintain the property of probability conservation $|\alpha_1|^2 + |\alpha_2|^2 = 1$. More precisely,

$$U \ket{\psi} = U \begin{bmatrix} \alpha_1 \\ \alpha_2 \end{bmatrix} = \begin{bmatrix} \alpha_1' \\ \alpha_2' \end{bmatrix} = \ket{\psi'}, \tag{3}$$

where $|\alpha_1'|^2 + |\alpha_2'|^2 = 1$ must hold. For a system with N qubits, its state space is the tensor product of all its qubits and is a 2^N-dimensional Hilbert space. A state vector in this space is a complex-valued vector with 2^N elements.

Quantum gates operating on a single qubit can manipulate the basis state, amplitude or phase of a qubit, for example through the so called X-gate,

$$X = \begin{bmatrix} 0 & 1 \\ 1 & 0 \end{bmatrix} \tag{4}$$

the Z-gate

$$Z = \begin{bmatrix} 1 & 0 \\ 0 & -1 \end{bmatrix} \tag{5}$$

and the Y-gate,

$$Y = \begin{bmatrix} 0 & -i \\ i & 0 \end{bmatrix} \tag{6}$$

respectively, or put a qubit into an equal superposition $\alpha_1 = \alpha_2 = 1/\sqrt{2}$ with the Hadamard or H-gate

$$H = \frac{1}{\sqrt{2}} \begin{bmatrix} 1 & 1 \\ 1 & -1 \end{bmatrix}. \tag{7}$$

Quantum gates can also operate on multiple qubits at once and induce entanglement. These gates are often based on controlled operations that execute a single qubit operation only if another auxiliary (ancilla) or control qubit is in a certain state. Developing quantum (machine learning) algorithms makes use of such elementary gates and effects like entanglement and interference in order to create a quantum state that has a relatively high amplitude for states that represent solutions for the

---

[1] It is not clear why the amplitudes are considered complex; it is likely to draw a parallel to signal processing where complex values are used.

problem being solved. A measurement in the computational basis then collapses the quantum state and produces such a desired result with a relatively high probability.

Depending on the number of repetitions of the quantum algorithm, we can distinguish:

- Stochastic quantum computing: The quantum algorithm returns a measurement output obtained from a single run of the quantum circuit [5–7]. Due to shot noise, (i.e., to the inherent stochasticity of quantum measurements), the output is generally a random string of zeroes and ones.
- Deterministic quantum computing: The quantum algorithm returns the average of several measurement outputs (i.e., expected value of some observable) that are obtained from multiple runs of the quantum circuit [8–10].

Whilst the second flavour is much more prevalent in quantum algorithms, stochastic quantum computing is crucial for some quantum machine learning models, for example, for generative models.

2 Machine Learning with Quantum Computers

Machine learning using a quantum computer entails using data to optimise a quantum algorithm. Whilst there are many quantum equivalents of classical machine learning schemes (e.g., k-nearest neighbours [11, 12] or kernel methods [13]), we focus here on the quantum counterpart of deep learning [14]—in what follows, we will use the term quantum machine learning to describe such algorithms. In theory, unlike many famous quantum algorithms (e.g., Shor's factoring algorithm), quantum machine learning does not impose restrictions on the number and on the reliability of qubits, quantum gates, and quantum measurements. This made it a suitable application of noisy, intermediate-scale quantum (NISQ) devices.

To elucidate the main idea in quantum machine learning, note that any unitary matrix, and quantum gate by extension, can be expressed as an exponential transformation of some Hermitian generator matrix G, i.e.,

$$U = \exp\left(-iG\right). \tag{8}$$

Taking the generator as G as

$$G = \frac{\theta}{2}X \tag{9}$$

with X given by (4), one obtains the so called Pauli X rotation gate as

$$R^X(\theta) = \exp\left(-i\frac{\theta}{2}X\right)$$

$$= \cos\left(\frac{\theta}{2}\right) I - i\sin\left(\frac{\theta}{2}\right) X$$

$$= \begin{bmatrix} \cos\left(\frac{\theta}{2}\right) & -i\sin\left(\frac{\theta}{2}\right) \\ -i\sin\left(\frac{\theta}{2}\right) & \cos\left(\frac{\theta}{2}\right) \end{bmatrix}. \tag{10}$$

Similarly, using the generator G

$$G = \frac{\theta}{2} P \tag{11}$$

with $P \in \{Y, Z\}$, one obtains the Pauli Y

$$R^Y(\theta) = \begin{bmatrix} \cos\left(\frac{\theta}{2}\right) & -\sin\left(\frac{\theta}{2}\right) \\ \sin\left(\frac{\theta}{2}\right) & \cos\left(\frac{\theta}{2}\right) \end{bmatrix}. \tag{12}$$

or Pauli Z

$$R^Z(\theta) = \begin{bmatrix} \cos\left(\frac{\theta}{2}\right) - \sin\left(\frac{\theta}{2}\right) & 0 \\ 0 & \cos\left(\frac{\theta}{2}\right) + \sin\left(\frac{\theta}{2}\right) \end{bmatrix}. \tag{13}$$

rotation gates. These gates can be applied in parallel to all N qubits, yielding the parametric unitary

$$U(\boldsymbol{\theta}) = R_0(\theta) \otimes R_1(\theta) \otimes \cdots \otimes R_{N-1}(\theta), \tag{14}$$

where $\otimes$ denotes the tensor product, $R_n(\theta)$ is applied to the nth qubit and $R_n(\theta) \in \{R_n^X(\theta), R_n^Y(\theta), R_n^Z(\theta)\}$. In (14), $\boldsymbol{\theta} \in \mathbb{R}^N$, denotes the vector of variational parameters, which are also known as the rotation angles, whereby $R_n(\theta)$ uses the $\{n\}$th entry in the vector $\boldsymbol{\theta}$. Unitaries with the form (14) give rise to the so called variational quantum circuits (VQCs) [15, 16], which is the key ingredient in quantum machine learning.

Specifically, quantum machine learning models are constructed by stacking such parametric gates in layered architectures comprised of L layers, and are generally called quantum neural networks (in analogy with classical deep learning). To minimise confusion, we emphasise that these algorithms are quite distinct from the artificial neural networks used in deep learning, and not even remotely similar to biological neural networks. Quantum machine learning models are just VQCs, represented as a sequence of quantum gates. In general, a quantum neural network has the form given in Fig. 1.

The variational parameters $\boldsymbol{\theta}$ are optimised using classical optimization techniques in order to minimise some defined cost function $\mathcal{L}_{\mathcal{D}}(\boldsymbol{\theta})$, obtained using the output of the model (a random string of zeroes and ones, or an expectation value of an observable). Formally, the goal is to solve

$$\underset{\boldsymbol{\theta}}{\arg\min} \; \mathcal{L}_{\mathcal{D}}(\boldsymbol{\theta}). \tag{15}$$

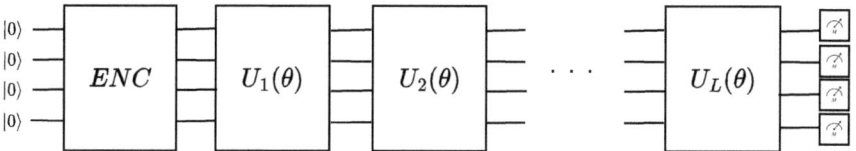

Fig. 1 A VQC comprised of L layers and $N = 4$ qubits. Similar models are often referred to as quantum neural networks

Much like in classical deep learning however, when designing a quantum machine learning algorithm, the designer needs to access a dataset $\mathcal{D}$,[2] find a suitable model, and design an effective training procedure that solves (15). Each of these is discussed in the following.

2.1 Data

The most prevalent choice of tasks for a quantum machine learning algorithm are classical whereby a quantum device is used solve tasks currently being solved using GPUs (e.g., machine vision, natural language processing [17, 18]) for which it needs access to classical data. More generally, depending on the task, the designer will use:

- Classical data: The designer has access to a classical data set $\mathcal{D} = \{(x_m, z_m)_{m=1}^M\}$. The classical data samples z_m (e.g., an image) is encoded into a quantum state representing a quantum embedding [19]. This state preparation routine is carried out by initialising all qubits in the $|0\rangle$ state, and encoding the data samples in the state amplitudes or the values of the gates' rotation angles. As a simple example, the data sample can be encoded using and encoding unitary $ENC(x_m)$, see Fig. 1, defined as the Pauli-X rotation gate as

$$ENC(\arccos(x_m)) = \exp\left(-i\frac{\arccos(x_m)}{2}X\right). \qquad (16)$$

The measurement output, possibly after averaging, encodes the model's prediction of the target variable $\tilde{z}_m$.

- Quantum data: In this case $\mathcal{D} = \{(|\psi_{x_m}\rangle, U(\theta)|\psi_{x_m}\rangle)_{m=1}^M\}$ [20], i.e., the dataset $\mathcal{D}$ is comprised of quantum states. Thereby, the quantum state $|\psi_{x_m}\rangle$ (potentially after digitisation) is post-processed by the learning model. As an example, the quantum states captured by quantum sensors could serve as an input to the quantum machine learning model in a quantum version of machine perception.

---

[2] Whilst machine learning also includes techniques such as unsupervised, and reinforcement learning, we focus here on supervised learning, whereby a designer has access to a dataset $\mathcal{D}$.

2.2 Model Architecture

The choice of the architecture of the VQC as a sequence of specific gates is akin to
the choice of the model class i.e., a neural network architecture, in classical machine
learning. The architecture of the model is referred to as the ansatz, representing an
initial estimate/assumption made of the solution to the problem. The ansatz implements the unitary transformation $U(\theta)$ which depends on a tunable (classical) real-
valued parameter vector θ. As mentioned above, the unitary is typically specified
as a sequence of one- or two-qubit gates, with each gate possibly dependent on the
parameter.

In general, choosing a good ansatz is difficult and depends on many factors. In
what follows, we review the most prevalent model classes.

2.2.1 Separable Models

An ansatz that uses only single-qubit gates is known as a separable model, or alternatively, a mean-field ansatz. The term mean-field indicates that the unitary $U_l(\theta)$
factorises as a tensor product of individual single-qubit gates. The mean field ansatz
can be comprised of L layers whereby $U(\theta)$ takes the form

$$U(\theta) = \prod_{l=0}^{L-1} R_{0,l}(\theta) \otimes R_{1,l}(\theta) \otimes \cdots \otimes R_{N-1,L-1}(\theta)$$

$$= \prod_{l=0}^{L-1} \prod_{n=0}^{N-1} R_{n,l}(\theta) \tag{17}$$

where $R_{n,l}(\theta) \in \{R^X_{n,l}(\theta), R^Y_{n,l}(\theta), R^Z_{n,l}(\theta)\}$ and where $R_{n,l}(\theta)$ is applied to the nth
qubit in the lth layer. In (17), $\theta \in \mathbb{R}^{N \times L}$ whereby $R_{n,l}(\theta)$ uses the $\{n, l\}$th entry in
θ, see Fig. 2. Since no two-qubit gates are present, a mean-field ansatz outputs a
separable state and can be easily simulated on a classical computer. Whilst simple,
the mean-field ansatz may be sufficient for some applications, or it can serve as
a (classical) benchmark to test the performance on any quantum machine learning
scheme.

Fig. 2 The unitary $U(\theta)$ for a mean-field ansatz comprised only of single-qubit gates $R_{n,l}(\theta)$, applied to the nth qubit in the lth layer. Here, $\theta \in \mathbb{R}^{N \times L}$

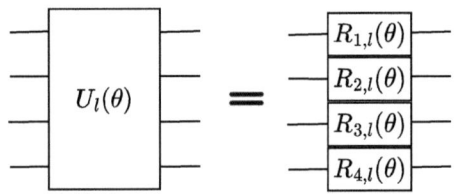

2.2.2 Hardware-Efficient Ansatz

The hardware-efficient ansatz includes both parametrized single-qubit gates and a fixed entangling unitary ENT, which is independent on the parameter vector. The entangling unitary typically is comprised of CNOT or CZ gates, whereby

$$CNOT = \begin{bmatrix} 1 & 0 & 0 & 0 \\ 0 & 1 & 0 & 0 \\ 0 & 0 & 0 & 1 \\ 0 & 0 & 1 & 0 \end{bmatrix} \tag{18}$$

and

$$CZ = \begin{bmatrix} 1 & 0 & 0 & 0 \\ 0 & 1 & 0 & 0 \\ 0 & 0 & 1 & 0 \\ 0 & 0 & 0 & -1 \end{bmatrix}. \tag{19}$$

CZ is particularly advantageous to CNOT since CZ gates commute, and as a result, they can be applied in any order without changing the result. The entangling unitary is fixed for simplicity of implementation on NISQ devices, since two-qubit gates are generally hard to implement on quantum computers [21–24].

The ansatz is comprised of general single-qubit gates applied in parallel to all qubits and an entangling circuit ENT whereby

$$U(\boldsymbol{\theta}) = \prod_{l=0}^{L-1} ENT \left(\prod_{n=0}^{N-1} R_{n,l}(\theta) \right) \tag{20}$$

where $R_{n,l}(\theta) \in \{R_{n,l}^X(\theta), R_{n,l}^Y(\theta), R_{n,l}^Z(\theta)\}$ and where $R_{n,l}(\theta)$ is applied to the nth qubit in the lth layer. Similarly to the mean-field ansatz, in (17), $\boldsymbol{\theta} \in \mathbb{R}^{N \times L}$ whereby $R_{n,l}(\theta)$ uses the $\{n, l\}$th entry in $\boldsymbol{\theta}$. The entangling unitary is typically implemented in a linear, cyclical, or complete manner whereby we distinguish:

- Linear entangling unitary: Such circuits implement two-qubit gates between successive qubits, see Fig. 3

$$ENT = \prod_{n=0}^{N-1} CZ_{n,n+1}. \tag{21}$$

The linear entangling unitary requires $N - 1$ entangling gates in each layer.
- Cyclical entangling unitary: Such circuits add to the architecture of linear entangling circuit an additional two-quit gate between the first and the last qubit, see Fig. 4. The cyclical entangling unitary requires N entangling gates in each layer.

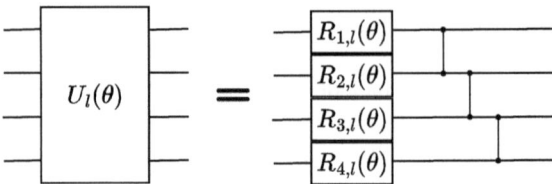

Fig. 3 The unitary $U(\theta)$ for a hardware-efficient ansatz comprised of single-qubit gates $R_{n,l}(\theta)$, applied to the nth qubit in the lth layer and a linear entangling unitary. Here, $\theta \in \mathbb{R}^{N \times L}$

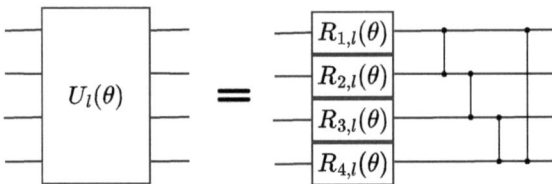

Fig. 4 The unitary $U(\theta)$ for a hardware-efficient ansatz comprised of single-qubit gates $R_{n,l}(\theta)$, applied to the nth qubit in the l-th layer and a cyclical entangling unitary. Here, $\theta \in \mathbb{R}^{N \times L}$

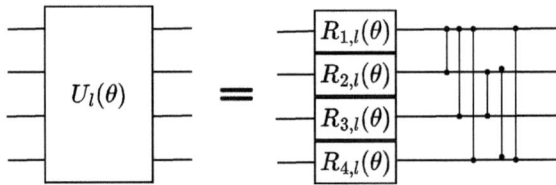

Fig. 5 The unitary $U(\theta)$ for a hardware-efficient ansatz comprised of single-qubit gates $R_{n,l}(\theta)$, applied to the nth qubit in the lth layer and a complete entangling unitary. Here, $\theta \in \mathbb{R}^{N \times L}$

- Full entangling unitary: Such circuits implement two-qubit gates between all possible pairs of qubits, see Fig. 5

$$ENT = \prod_{j<k} CZ_{k,j}. \tag{22}$$

The complete entangling unitary requires $N(N-1)/2$ entangling gates in each layer.

The hardware-efficient ansatz is problem-agnostic, and it plays a similar role to fully connected classical neural networks. It is highly expressive, meaning, it can fit many functions. On the flip side, it has been well established that the expressivity of a model is connected to barren plateaus and excessive local minima in the training landscape. In other words, the more expressive a model is, the more difficult it is to train using gradient-based schemes since the gradients vanish in an exponential manner. Problem-agnostic architectures (such as the mean-field or the hardware-efficient ansatz) are more likely to exhibit trainability issues.

2.2.3 Restricted Model Classes

One way to avoid trainability issues is to limit the expressive power of the models and encode sharp inductive biases in the choice of the architecture which is done by encoding symmetries [25–27]. On the classical side, the study of symmetries in data has lead to the blossoming field of geometric deep learning [28]. Symmetries formalize the invariance of objects under some set of operations. For example, the binding energy of a molecule does not change by permuting the order of the atoms, and a picture of a cat still depicts a cat regardless of the position of the cat within the image. Incorporating this prior knowledge into the learning architecture, as a geometric prior, gives rise to symmetry-preserving quantum geometric models which have the potential not only of mitigating barren plateau issues, but also of reducing sample complexity and improving the training efficiency.

Depending on the task, the type of data, and the symmetries present, there are several types of quantum geometric machine learning models:

- Quantum Convolutional Neural Networks (QCNNs): A QCNN is suitable for tasks benefiting from encoding invariance to translation [29–31]. As seen in Fig. 6, The model is comprised by stacking convolutional and pooling unitaries.

  - Convolutional unitary: The convolutional unitary is quasi-local and is applied to pairs of qubits. There are several choices for the type of convolutional unitaries depending on the task. As an example, one could use controlled-phase gates as well as Toffoli gates with controls in the X-basis.
  - Pooling unitary: For pooling, a fraction of qubits are measured, and their outcomes determine unitary rotations applied to nearby qubits. For example, pooling unitaries perform phase-flips on remaining qubits when the adjacent measurement yields $X = -1$

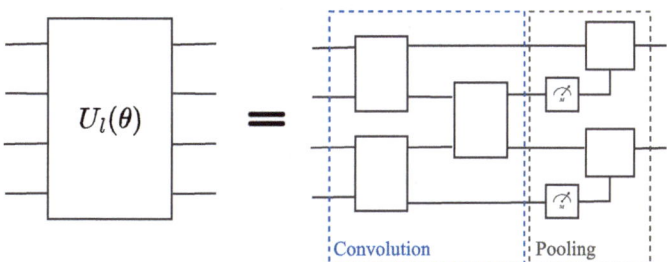

Fig. 6 The unitary $U(\theta)$ for a quantum convolutional neural network comprised of convolutional and pooling layers

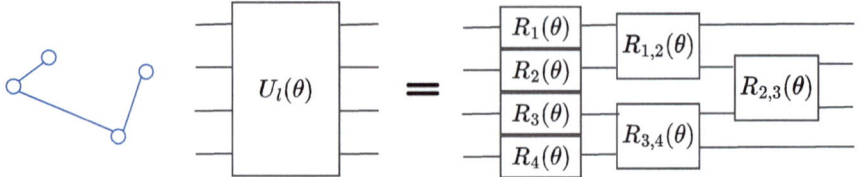

Fig. 7 The unitary $U(\theta)$ for a quantum graph neural network. Here, $\theta \in \mathbb{R}^{2 \times L}$

As a result, non-linearities in QCNNs arise from the pooling operation which reduces the number of degrees of freedom. Convolution and pooling layers are performed until the system size is sufficiently small. Finally, the outcome of the circuit is obtained by measuring a fixed number of output qubits. The QCNN has been applied to quantum phase recognition, which asks whether a given input quantum state belongs to a particular quantum phase of matter, as well as quantum error correction optimization, which asks for an optimal QEC code for a given, a priori unknown error model such as dephasing or potentially correlated depolarization in realistic experimental settings.

- Quantum Graph Neural Networks (QGNNs): QGNNs are equivarient to permutation and are tailored to represent quantum processes which have a graph structure. Let $\mathcal{A} \in R^{N \times N}$ denote the adjacency matrix of a graph. As seen in Fig. 7, QGNNs apply Hermitian operators based on the structure (i.e., the adjacency matrix) of the graph. An encoding layer can also be used to encode any node features as described above. Most generally, the QGNN unitary assumes a generator of learnable Hermitian matrices (see (8)) and takes the form

$$U(\boldsymbol{\theta}) = \prod_{l=0}^{L-1} \exp\left(-i\eta \left(\sum_{\mathcal{A}_{ij}=1} H_{i,j}(\theta^{(l)}) + \sum_{n=1}^{N} H_n(\theta^{(l)})\right)\right), \quad (23)$$

where $H_{i,j}(\theta^{(l)})$ is the edge unitary and $H_n(\theta^{(l)})$ is the node unitary [32, 33]. As an example, the edge and node unitaries can be given as

$$\sum_{\mathcal{A}_{ij}=1} H_{i,j}(\theta^{(l)}) = \sum_{\mathcal{A}_{ij}=1} \theta^{(l)} Z_i Z_j \quad (24)$$

and

$$\sum_{n=1}^{N} H_n(\theta^{(l)}) = \sum_{n=1}^{N} \theta^{(l)} X_n, \quad (25)$$

respectively. The term η represents the evolution time, which can be either a fixed or a learnable parameter. Note that, the parameters of $H_n(\theta^{(l)})$ in layer l are shared, meaning the unitary is applied with the same parameter $\theta^{(l)}$ in layer l to all qubits. The same holds for $H_{i,j}(\theta)$. Thereby, the total number of parameters is $\theta \in R^{2 \times L}$. The factor 2 comes from the fact that $H_n(\theta^{(l)})$ and $H_{i,j}(\theta^{(l)})$ do not share the same parameter θ in layer l. As a result, the model is equivariant to permutations. QGNNs can also process weighted graphs, by adding edge weights to (24) as done in [34]. They have been applied to learning Hamiltonian dynamics, graph clustering, isomorphism and graph state classification, as well as distributed processing of quantum sensor data.

The QGNN has been extended in [35] to general S_n equivariant models that have the same structure as (23) except that the placement of the entangling gates always assumes a complete graph whereby

$$U(\boldsymbol{\theta}) = \prod_{l=0}^{L-1} \exp\left(-i\eta\left(\sum_{j<i} H_{i,j}(\theta^{(l)}) + \sum_{n=1}^{N} H_n(\theta^{(l)})\right)\right), \qquad (26)$$

where the edge and node unitaries can assume the form of (24) and (25), respectively. In words, parametric two-qubit gates are placed between all pairs of qubits/nodes. This requires $N(N-1)/2$ parametric two-qubit gates in each layer.

- Quantum Recurrent Neural Networks (QRNNs): Much like reservoirs and classical RNNs discussed in Chapter 1, written by Michael te Vrugt, QRNNs are models that process temporal data [36, 37]. Formally a QRNN maps a classical sequence $x_{1:T}$ comprising T samples x_t with $t = 1, 2, \ldots, T$, to corresponding classical sequences $z_{1:T}$, comprising samples z_t with $t = 1, 2, \ldots, T$, i.e., $\mathcal{D} = \{x_{1:T}, z_{1:T}\}$. This is done by sequential application of the unitary $U(\theta)$. As such, the model is invarient to shifts in the sequence.

  Unlike the previous models, the unitary operates on a register of $N = N^A + N^B$ qubits, see Fig. 8. The unitary is applied at each time step by re-initializing the subregister of N^B qubits to the ground state $|0\rangle$. After the application of the unitary to all N qubits, the subregister of N^B qbits is measured. Since the subregister of N^A qubits is not measured, it can be used to propagate information from one time step to the next and it represents a quantum form of memory. It is noted that quantum models have been shown to describe temporal sequences with reduced memory costs [38, 39]. Therefore, the N^A-qubit subregister is referred to as the *memory subregister*, while the subsystem of N^B qubits is referred to as the *output subregister*.

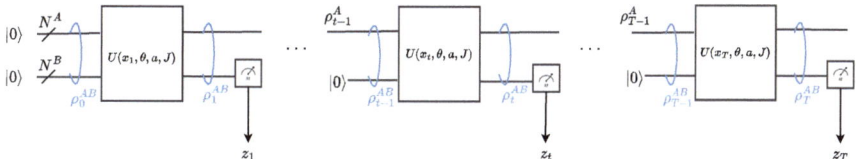

Fig. 8 The ansatz of a quantum recurrent neural network

where we have $\rho_1^B = \mathrm{Tr}_A(\rho_1^{AB})$ and $\mathrm{Tr}_X(\cdot)$ indicates the partial trace with respect to subregister $X \in \{A, B\}$. In the QRNN model, the measurement is repeated many times in order to obtain an estimate of the expectation value $\langle O^B \rangle$ of the observable O^B. After averaging over the measurement outcomes, the resulting density matrix of the memory subregister is given by $\rho_1^A = \mathrm{Tr}_B(\rho_1^{AB})$.

At any time step $t = 1, \ldots, T$, the memory subregister evolves according to the update rules

$$\rho_t^{AB} = U(x_t, \theta, a, J)\,(\rho_{t-1}^A \otimes |0\rangle \langle 0|^B)\,U(x_t, \theta, a, J)^\dagger \tag{27}$$

whereby

$$\rho_t^{AB} = |\psi_t^{AB}\rangle \langle \psi_t^{AB}| \tag{28}$$

and

$$\rho_t^A = \mathrm{Tr}_B(\rho_t^{AB}), \tag{29}$$

with initialization $\rho_0^A = |0\rangle \langle 0|^A$. The unitary $U(x_t, \theta, a, J)$ can have an arbitrary architecture, but is comprised of an encoding unitary, encoding the sample x_t and an evolution unitary, see Fig. 9. The encoding unitary is only applied to the N^B qubits and can assume a similar shape to (16) whilst the evolution unitary is applied to all qubits. It can be comprised of rotation gates as

$$U(\boldsymbol{\theta}) = \prod_{n=1}^{N-1} R_n(\theta) \tag{30}$$

whereby the rotation gates are applied with the same parameter θ to all qubits. Thereby, $\boldsymbol{\theta} \in \mathbb{R}^N$. The rotation gates are followed by a layer of fixed Hermitian matrices similar to (26). Formally,

$$U(a, J) = \exp\left(-i\eta \left(\sum_{j<i} J Z_i Z_j + \sum_{n=1}^{N} a X_n \right) \right), \tag{31}$$

where the parameters a, J are drawn randomly from a uniform distribution as $a, J \in [-1, 1]$, and they are fixed during training. The QGNN in [40] can also

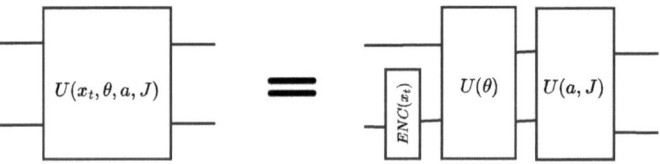

Fig. 9 The unitary $U(x_t, \theta, a, J)$ of a quantum recurrent neural network

have a recurrent form which only assumes the evolution unitary described by (23), applied with the same parameters at each layer i.e.,

$$U(\theta) = \prod_{l=0}^{L-1} \exp\left(-i\eta \left(\sum_{\mathcal{A}_{ij}=1} H_{i,j}(\theta) + \sum_{n=1}^{N} H_n(\theta)\right)\right), \tag{32}$$

where $\theta \in \mathbb{R}^2$ since $H_n(\theta)$ and $H_{i,j}(\theta)$ do not share the same parameter θ.

A QRNN model that is invariant to time-warping has been proposed in [10]. Time-warping is a peculiar type of symmetry, in the sense that it does not have a symmetry group describing the actions of a group. Rather, time-warping invariance stands for resilience to changes in the sampling rate of the sequence [28, 41]. In particular, time-series data are not naturally discrete—they are in fact obtained by sampling a continuous signal. However, in practice, the sampling rate cannot always be controlled, and moreover, it may not always be fixed. Time warping transformations capture such dynamically changing sampling rates, or basic operations such as inserting a, possibly varying, number of zeros or white spaces between the elements of an input sequence. As such, time warping can model important practical aspects such as imperfect sampling.

Formally, model class $f(\cdot, \theta)$ is said to be *time warping-invariant* with respect to the family of time-warping operations $\mathcal{C}$, if for any $z(t) = f(x(t), \theta)$ produced by the model with some model parameters θ given input $x(t)$, there exist model parameters θ' that yield the output $z(c(t)) = f(x(c(t)), \theta')$ given input $x(c(t))$ for any function $c(t)$ in $\mathcal{C}$. In words, the model class can reproduce time-warped versions of its input-output pairs. To enable this, a time-warping invariant QRNN (TWI-QRNN) has been shown to take the form

$$\rho_t^{AB} = (1 - \alpha_t)\,(\rho_{t-1}^A \otimes |0\rangle\langle 0|^B)$$
$$+ \alpha_t\, U(x_t, \theta, a, J)\,(\rho_{t-1}^A \otimes |0\rangle\langle 0|^B)\, U(x_t, \theta, a, J)^\dagger, \tag{33}$$

where $\alpha_t \in [0, 1]$ is a probability chosen equal to the time-warping derivative $dc(t)/dt$ i.e.,

$$\alpha_t = \frac{dc(t)}{dt}. \tag{34}$$

In words, the density matrix $\tilde{\rho}_{t+1}$ is produced by leaving the previous density matrix $\tilde{\rho}_t$ unchanged with probability $(1 - dc(t)/dt)$ and by applying the unitary $V_t(x_t, \theta)$ otherwise. Since the time-warping operation, and its derivative by extension, is generally unknown, the model in [10] practically replaces the unknown derivative in the model with a classical RNN that infers a suitable time-warping probability α_t from the input data sequence in a causal fashion. Specifically, the probability α_t in (34) can be written as

$$\alpha_t = \sigma(\phi_t), \tag{35}$$

where $\sigma(a) = (1 + \exp(-a))^{-1}$ denotes a logistic sigmoid with *hyperparameter vector* ϕ_t being the output of a model defined by the updates

$$\phi_t = W_x x_t + W_h h_t + b, \tag{36}$$

and

$$h_t = \tanh(W_x^h x_t + W_h^h h_{t-1} + b), \tag{37}$$

where $W = \{W_x, W_h, W_x^h, W_h^h\}$ denote learnable parameters. Stochastic counterparts of the (TWI-)QRNN have also been provided in [10].

2.3 Training

Training the model involves solving (15) using the adopted loss function. To optimise the circuit parameters, the designer can use gradient-free or gradient-based techniques.

- Gradient-free optimisation: Gradient-free solutions to (15) involve using numerical optimisers, e.g., COBLYA [42]. Such approaches are suitable for problems with a relatively small number of parameters.
- Gradient-based optimisation: Gradient-based techniques involve using gradient descent whereby the minimisation of (15) is carried out iteratively, by updating the parameters θ as

$$\theta \leftarrow \theta - \delta \frac{\partial \mathcal{L}_\mathcal{D}(\theta)}{\partial \theta}, \tag{38}$$

with δ denoting the learning rate. Since in a quantum setting one does not have access to the internal workings of the model,[3] no quantum counterpart of back-propagation exists. Hence, the computation of the gradient is an inherently more complex and less scalable operation than in classical machine learning. Still, the gradient in (38) can be approximated using numerical differentiation, or found exactly using the so-called parameter-shift rule [43–45]. In particular, each parameter θ, is shifted forward as $\theta + \frac{\pi}{2}$ and the circuit is run N_S times to estimate $\mathcal{L}_\mathcal{D}(\theta + \frac{\pi}{2})$. Then, the parameter is shifted backward as $\theta - \frac{\pi}{2}$ and the circuit is run N_S times to estimate $\mathcal{L}_\mathcal{D}(\theta - \frac{\pi}{2})$. With this the gradient can be proven to take the exact form

$$\frac{\partial \mathcal{L}_\mathcal{D}(\theta)}{\partial \theta} = \frac{\mathcal{L}_\mathcal{D}(\theta + \frac{\pi}{2}) - \mathcal{L}_\mathcal{D}(\theta - \frac{\pi}{2})}{2}. \tag{39}$$

---

[3] That is, unless one relies on a simulation of the quantum computer on a classical computer, which is only feasible for small values of N.

Therefore, the number of circuit runs scales with the number of parameters. It is important to mention again, that in sufficiently expressive architectures, the partial derivatives tend exponentially to zero in probability as N increases.

Quantum machine learning coding libraries used in practice [46, 47] offer both gradient-free and gradient-based options. It is useful to note that, using classical computers as simulators of quantum computers for machine learning tasks allows classical backpropagation; however this is only possible for a small number of qubits, which in turn, is unlikely to result in a quantum advantage. On the other hand, the number of qubits, at least in theory, is not the bottleneck of using quantum computers; however, the training procedure is not scalable.

3 Testing

As with any machine learning scheme, rigorous testing must be carried out to judge the performance of any quantum machine learning algorithm. Standard testing schemes in terms of using unobserved data apply. What is specific about quantum machine learning is that quantum machine learning schemes must be compared amongst each other, as well as against classical machine learning solving the same task. In addition, it is generally a good approach to test the model by removing all entangling gates in the ansatz. This is to ensure that the model really does require a quantum computer, or it can be easily simulated on classical hardware.

4 Outlook

Historically, scale has played a crucial role in machine learning. After all, small, shallow neural networks have been shown to overfit to data, whilst deep neural networks have enabled ChatGPT and AlphaFold2.[4] Given that we lack large-scale, reliable quantum computers, it is currently premature to make predictions on the practical utility of quantum machine learning. It is however useful to discuss the current state of field. Whilst great progress has been made in the last decade or so, the current models still need to be improved, more scalable training procedures need to be developed, and more rigorous benchmarking and testing practices have to be carried out. If we succeed in these, provided we have the hardware, we could imagine

---

[4] This is a result of the "double-descent" behaviour of neural networks: Suppose we plot the training and testing error against some measure of model complexity for a typical case. This results in curves following the same decreasing direction and the test error hovering slightly above the train error. Under the classical bias-variance trade-off, we might expect that as we move further right along the x-axis (i.e. we increase the complexity of our model), we overfit and expect the test error to increase further and further upwards. However, what we observe is quite different. Indeed the error does increase (this is expected and the model is still in the "shallow" regime), but then as we increase the complexity further, it goes back down to a new minimum (the model is in the "deep" regime.)

that quantum machine learning might not just help us unlock the secrets of the natural world, but it could even open the door to non-anthropomorphic, quantum forms of intelligence.

References

1. Shor PW (1998) Quantum computing. Doc Math 1(1000):1
2. Krantz P, Kjaergaard M, Yan F, Orlando TP, Gustavsson S, Oliver WD (2019) A quantum engineer's guide to superconducting qubits. Appl Phys Rev 6(2)
3. Schäfer V, Ballance C, Thirumalai K, Stephenson L, Ballance T, Steane A, Lucas D (2018) Fast quantum logic gates with trapped-ion qubits. Nature 555(7694):75–78
4. Kok P, Munro WJ, Nemoto K, Ralph TC, Dowling JP, Milburn GJ (2007) Linear optical quantum computing with photonic qubits. Rev Mod Phys 79(1):135
5. Coyle B, Mills D, Danos V, Kashefi E (2020) The born supremacy: quantum advantage and training of an ising born machine. npj Quant Inf 6(1):1–11
6. Nikoloska I, Simeone O (2022) Quantum-aided meta-learning for bayesian binary neural networks via born machines. In: IEEE 32nd international workshop on machine learning for signal processing (MLSP). IEEE 2022, pp 1–6
7. Liu J-G, Wang L (2018) Differentiable learning of quantum circuit Born machines. Phys Rev A 98(6):062324
8. Grover LK (1996) A fast quantum mechanical algorithm for database search. In: Proceedings of the twenty-eighth annual ACM symposium on Theory of computing, pp 212–219
9. Long G-L (2001) Grover algorithm with zero theoretical failure rate. Phys Rev A 64(2):022307
10. Nikoloska I, Simeone O, Banchi L, Veličković P (2023) Time-warping invariant quantum recurrent neural networks via quantum-classical adaptive gating. Mach Learn Sci Technol 4(4):045038
11. Schuld M, Sinayskiy I, Petruccione F (2014) Quantum computing for pattern classification. In: PRICAI 2014: trends in artificial intelligence: 13th Pacific Rim international conference on artificial intelligence, Gold Coast, QLD, Australia, December 1–5, 2014. Proceedings 13. Springer, pp 208–220
12. Basheer A, Afham A, Goyal SK (2020) Quantum k-nearest neighbors algorithm. arXiv:2003.09187
13. Mengoni R, Pierro A (2019) Kernel methods in quantum machine learning. Quant Mach Intell 1(3–4):65–71
14. Goodfellow I, Bengio Y, Courville A, Bengio Y (2016) Deep learning, vol 1, no 2. MIT Press Cambridge
15. Cerezo M, Arrasmith A, Babbush R, Benjamin SC, Endo S, Fujii K, McClean JR, Mitarai K, Yuan X, Cincio L et al (2021) Variational quantum algorithms. Nat Rev Phys 3(9):625–644
16. Chen SY-C, Yang C-HH, Qi J, Chen P-Y, Ma X, Goan H-S (2020) Variational quantum circuits for deep reinforcement learning. IEEE Access 8:141 007–141 024
17. Coecke B, de Felice G, Meichanetzidis K, Toumi A (2020) Foundations for near-term quantum natural language processing. arXiv:2012.03755
18. Meichanetzidis K, Gogioso S, De Felice G, Chiappori N, Toumi A, Coecke B (2020) Quantum natural language processing on near-term quantum computers. arXiv:2005.04147
19. Schuld M, Sweke R, Meyer JJ (2021) Effect of data encoding on the expressive power of variational quantum-machine-learning models. Phys Rev A 103(3):032430
20. Caro MC, Huang H-Y, Cerezo M, Sharma K, Sornborger A, Cincio L, Coles PJ (2022) Generalization in quantum machine learning from few training data. Nat Commun 13(1):4919
21. Leone L, Oliviero SF, Cincio L, Cerezo M (2022) On the practical usefulness of the hardware efficient ansatz. arXiv:2211.01477

22. Kandala A, Mezzacapo A, Temme K, Takita M, Brink M, Chow JM, Gambetta JM (2017) Hardware-efficient variational quantum eigensolver for small molecules and quantum magnets. Nature 549(7671):242–246
23. Tang HL, Shkolnikov V, Barron GS, Grimsley HR, Mayhall NJ, Barnes E, Economou SE (2021) qubit-adapt-vqe: an adaptive algorithm for constructing hardware-efficient ansätze on a quantum processor. PRX Quantum 2(2):020310
24. Benedetti M, Fiorentini M, Lubasch M (2021) Hardware-efficient variational quantum algorithms for time evolution. Physical Review Research 3(3):033083
25. Meyer JJ, Mularski M, Gil-Fuster E, Mele AA, Arzani F, Wilms A, Eisert J (2022) Exploiting symmetry in variational quantum machine learning. arXiv:2205.06217
26. Larocca M, Sauvage F, Sbahi FM, Verdon G, Coles PJ, Cerezo M (2022) Group-invariant quantum machine learning. arXiv:2205.02261
27. Ragone M, Braccia P, Nguyen QT, Schatzki L, Coles PJ, Sauvage F, Larocca M, Cerezo M (2022) Representation theory for geometric quantum machine learning. arXiv:2210.07980
28. Bronstein MM, Bruna J, Cohen T, Veličković P (2021) Geometric deep learning: grids, groups, graphs, geodesics, and gauges. arXiv:2104.13478
29. Cong I, Choi S, Lukin MD (2019) Quantum convolutional neural networks. Nat Phys 15(12):1273–1278
30. Oh S, Choi J, Kim J (2020) A tutorial on quantum convolutional neural networks (qcnn). In: 2020 international conference on information and communication technology convergence (ICTC). IEEE, pp 236–239
31. Wei S, Chen Y, Zhou Z, Long G (2022) A quantum convolutional neural network on nisq devices. AAPPS Bull 32(1):1–11
32. Verdon G, McCourt T, Luzhnica E, Singh V, Leichenauer S, Hidary J (2019) Quantum graph neural networks. arXiv:1909.12264
33. Mernyei P, Meichanetzidis K, Ceylan II (2022) Equivariant quantum graph circuits. In: International conference on machine learning. PMLR, pp 15 401–15 420
34. Skolik A, Cattelan M, Yarkoni S, Bäck T, Dunjko V (2023) Equivariant quantum circuits for learning on weighted graphs. npj Quant Inf 9(1):47
35. Schatzki L, Larocca M, Sauvage F, Cerezo M (2022) Theoretical guarantees for permutation-equivariant quantum neural networks. arXiv:2210.09974
36. Takaki Y, Mitarai K, Negoro M, Fujii K, Kitagawa M (2021) Learning temporal data with a variational quantum recurrent neural network. Phys Rev A 103(5):052414
37. Bausch J (2020) Recurrent quantum neural networks. Adv Neural Inf Process Syst 33:1368–1379
38. Yang C, Garner A, Liu F, Tischler N, Thompson J, Yung M-H, Gu M, Dahlsten O (2021) Provable superior accuracy in machine learned quantum models. arXiv:2105.14434
39. Elliott TJ, Yang C, Binder FC, Garner AJ, Thompson J, Gu M (2020) Extreme dimensionality reduction with quantum modeling. Phys Rev Lett 125(26):260501
40. Verdon G, Broughton M, McClean JR, Sung KJ, Babbush R, Jiang Z, Neven H, Mohseni M (2019) Learning to learn with quantum neural networks via classical neural networks. arXiv:1907.05415
41. Tallec C, Ollivier Y (2018) Can recurrent neural networks warp time? arXiv:1804.11188
42. Virtanen P, Gommers R, Oliphant TE, Haberland M, Reddy T, Cournapeau D, Burovski E, Peterson P, Weckesser W, Bright J, Walt SJ, Brett M, Wilson J, Millman KJ, Mayorov N, Nelson ARJ, Jones E, Kern R, Larson E, Carey CJ, Polat İ, Feng Y, Moore EW, VanderPlas J, Laxalde D, Perktold J, Cimrman R, Henriksen I, Quintero EA, Harris CR, Archibald AM, Ribeiro AH, Pedregosa F, Mulbregt P, SciPy 1.0 Contributors (2020) SciPy 1.0: fundamental algorithms for scientific computing in python. Nat Meth 17:261–272
43. Wierichs D, Izaac J, Wang C, Lin CY-Y (2022) General parameter-shift rules for quantum gradients. Quantum 6:677
44. Crooks GE (2019) Gradients of parameterized quantum gates using the parameter-shift rule and gate decomposition. arXiv:1905.13311

45. Banchi L, Crooks GE (2021) Measuring analytic gradients of general quantum evolution with the stochastic parameter shift rule. Quantum 5:386
46. Wille R, Van Meter R, Naveh Y (2019) Ibm's qiskit tool chain: working with and developing for real quantum computers. In: Design, automation & test in Europe conference & exhibition (DATE). IEEE 2019, pp 1234–1240
47. Bergholm V, Izaac J, Schuld M, Gogolin C, Ahmed S, Ajith V, Alam MS, Alonso-Linaje G, AkashNarayanan B, Asadi A et al (2018) Pennylane: automatic differentiation of hybrid quantum-classical computations. arXiv:1811.04968

Philosophical Aspects

Normativity and the Intelligent Matter Framework

Barnaby Crook

Abstract An emerging field at the intersection of artificial intelligence, materials science, physical chemistry, and nanotechnology targets the creation of *intelligent matter*. However, despite demonstration of interesting behaviors in molecular systems and soft materials, the production of truly intelligent matter remains elusive. To address this, Kaspar and colleagues articulate a theoretical framework purporting to guide and evaluate the progress of the field. In this chapter, I argue that the absence of a *normative criterion* by which to assess the behavior of proto-intelligent material systems prevents the framework from adequately serving its guiding and evaluative functions. Without a clear specification of what kinds of behaviors count as intelligent, researchers are unlikely to develop material systems that exhibit them. I propose to augment the IMF with a normative criterion requiring purportedly intelligent systems to implement concrete behaviors which contribute towards higher-order goals in virtue of the coordinated activity of their functional elements.

1 Introduction

The *intelligent matter* research program aims to develop physical materials which implement intelligent behavior [1–3]. To coordinate this effort, Kaspar and colleagues formulate the *Intelligent Matter Framework* (IMF), which classifies material systems on the basis of which functional elements they possess.[1] Only by possessing the right set of functional elements can a system ascend to the category of *intelligent*. However, as I will show, this framework lacks *normativity*. That is, the IMF does not proffer

---

[1] I ascribe the name *Intelligent Matter Framework* to Kaspar and colleagues' work; it is not used in their original article.

B. Crook (✉)
Department of Philosophy, University of Bayreuth, Bayreuth, Germany
e-mail: barnaby.crook@uni-bayreuth.de

© The Author(s) 2026
M. te Vrugt (ed.), *Artificial Intelligence and Intelligent Matter*, Machine Intelligence for Materials Science, https://doi.org/10.1007/978-3-032-04129-6_21

any *normative criterion* by which to judge patterns of behavior more or less intelligent.[2] Absent such a criterion, I contend, any framework for intelligence will be inadequate. For one thing, theoretical terms such as *adaptivity* become divorced from their ordinary meanings. One cannot distinguish a material system's being *adaptive* from merely variable without some normative criterion by which to judge whether patterns of stimulus-induced dispositional change are beneficial. More fundamentally, without a normative standard dictating what counts as intelligent at the level of behavior, the IMF risks losing touch with the target phenomena that motivated the research program in the first place. As such, I will argue, considerations of normativity are essential for any theoretical framework for intelligence [5–10].

I critically analyze Kaspar and colleagues' theoretical framework for the intelligent matter research program from a philosophical perspective. I suggest that the IMF serves two complementary functions: (i) an *evaluative* function, assessing progress towards the goals of the research program, and (ii) a *guiding* function, which sheds light on how such progress might be achieved. I first highlight the lack of a normative criterion in Kaspar and colleagues' presentation of the IMF, before elaborating on the claim that this omission is problematic. In particular, I argue that without a normative criterion with which to judge behavior, the framework fails to adequately serve its functions. On the one hand, its evaluations fail to track intuitive judgements regarding behavioral sophistication. On the other hand, material systems which follow its recommendations may not exhibit any intelligent behavior at all. To explore how this might be rectified, I elaborate on the concept of normativity, distinguishing between the *source* of a normative criterion and how normativity is *constituted* in a physical system. I then discuss how normativity should be integrated into the IMF. I suggest that, in light of the intelligent matter research program's commitments to interdisciplinarity, practical applicability, and decentralized information processing, the IMF should be augmented with a normative criterion requiring material systems to implement extrinsically specified concrete behaviors which serve higher-order goals in virtue of the coordinated activity of their functional elements.

The structure of the chapter is as follows. In Sect. 2, I present the IMF, describing how it divides material systems into classes based on their functional elements. In Sect. 3, I focus on the absence of normativity in the IMF, arguing that the lack of a normative criterion leaves the IMF poorly suited to fulfilling its evaluative and guiding functions. In Sect. 4, I cover normativity in more detail, clarifying the distinction between the source of a normative criterion and how it is constituted in a system. In Sect. 5, I discuss which notion of normativity is best suited for integration into the IMF. I then conclude the chapter.

---

[2] By *normative criterion* I mean any evaluative standard by which a system's behavior may be judged acceptable or unacceptable (in a dichotomous sense) or better or worse (in a continuous sense). No connotations to richer human domains of values and ethics are implied [4].

2   The Intelligent Matter Framework

In this section I introduce the IMF as described by Kaspar and colleagues in their 2021 article [1]. Before doing so, however, I briefly motivate focusing on this particular articulation of the intelligent matter research program. Since the aim of this chapter is to address conceptual issues arising in the definition and evaluation of intelligence, it is necessary to engage with a presentation explicitly delivered in conceptual terms, like the IMF. Further, research which falls under the purview of *intelligent matter* is conducted across a broad range of technical scientific disciplines, each with idiosyncratic terminology and research foci. What distinguishes the intelligent matter research program as a whole from any one discipline is its *integrative* nature. It is the challenge of combining diverse strands of technical work into a coherent research agenda that Kaspar and colleagues' IMF purports to address. And that challenge has, I will argue, particular consequences for how normativity relates to the IMF. Finally, focusing on the IMF usefully restricts the scope of the chapter.

2.1   *Introducing the Intelligent Matter Framework*

The IMF classifies material systems on the basis of their internal functional complexity. In particular, the classification is sensitive to the presence of four specific functional elements: sensors, actuators, memory, and an interaction network. These elements are defined as follows. Sensors detect external stimuli, such as heat or force, by implementing an energy transformation (e.g., converting heat into electrical potential). Actuators produce a response, such as a physical or chemical change, when external stimuli are detected. Memory is any element in a system with the ability to retain information over a temporal window such that it can be exploited to modulate behavior later. Finally, an interaction network is constituted by signal pathways through which sensor, actuator, and memory elements can interact with one another, enabling feedback and other action-relevant information to be utilized. With all four of these functional elements in place, argue Kaspar and colleagues, a material system will be capable of sophisticated information processing and learning from its environment [1, p. 347].

Kaspar and colleagues use the functional elements described above to classify material systems. The most basic kind of system is *structural matter*. It does not possess sensors, actuators, memory, or an interaction network. Therefore, once a structural material system has been synthesized, it cannot change its properties. This means that, although such a system may have numerous functions and significant structural complexity, it is inherently unable to implement any behavior that requires adaptation, learning, or behavioral flexibility. The second class of material system Kaspar and colleagues describe is *responsive matter*. Systems belong to this category if they have sensors and actuators but no memory or interaction network. Because responsive material systems lack memory, the response their actuators produce when particular stimuli are detected by their sensors is always the same [1, p. 346]. The third class of material system is *adaptive matter*. The IMF classifies systems as adaptive if they augment the sensing and actuating capacities of a responsive system with an interaction network. Such a network allows a material system to implement feedback by altering its internal states and, consequently, its behavioral responses (e.g., in response to repeated presentation of a stimulus). Since adaptive material systems exhibit state-dependent responses to stimuli, the principles governing their behavior can become exceedingly complex (see [3] for discussion). The final and most sophisticated class of material system is *intelligent matter*. In addition to the functional elements required for a system to count as adaptive, an intelligent system must possess long-term memory storage. This enables previously sensed or self-generated information to be retained over time and deployed when needed. As well as guiding future action by modulating the behavior of actuators, feedback information retained as memory can be used to alter a system's internal states, implementing a form of learning.

2.2   Preliminary Comments on the Intelligent Matter Framework

I start by making three preliminary observations about the IMF. First, as a terminological matter, Kaspar and colleagues reserve the term *intelligent* for the final class of systems, stating that "if one of the key functional elements is lacking, then, according to our definition, the material is not considered intelligent" [1, p. 349]. Despite this assertion, the IMF can be interpreted as specifying discrete gradations of intelligence, with the four classes serving as distinct levels. Ultimately, the difference between these two perspectives is more semantic than substantial. For clarity, I will stick with the authors' usage moving forwards.

Second, the classificatory scheme of the IMF does not partition the space of possible material systems exhaustively. For example, where should a system possessing sensors and a memory, but no interaction network, be classified? Kaspar and colleagues discuss a soft matter system with such a functional repertoire and suggest that "such matter would not classify as adaptive matter owing to the lack of a network, [but nevertheless] goes beyond responsive capability" [1, p. 349]. While the existence of material systems which straddle the IMF's boundaries raises questions about its aptness, I put this issue aside for now.

Third, Kaspar and colleagues state that their notion of intelligence is not applicable to human beings or biological organisms more generally. For example, they write that "such matter which we term here intelligent does not show the same level of intelligence as would be understood in a psychological sense" [1, p. 345]. And, with respect to basal biological intelligence, they stress that the "key functional elements

[...] are not sufficient to enable the emergence of will or cognition, which distinguishes synthetic matter from intelligent living beings" [1, p. 347]. Such caution is admirable. Given the domain of research the IMF intends to support, eschewing psychological connotations seems appropriate. Indeed, such minimal strategies are common to researchers studying the evolution of intelligence [11], bacterial intelligence [12], plant intelligence [13], and artificial intelligence [14]. However, these disavowals raise the question of what relation the IMF's notion of intelligence *does* bear to those of other disciplines. One natural answer, not adopted by the IMF, is that diverse intelligence concepts are united by capturing systematic variation in *behavior* [6, 15]. That the IMF does not account for this will be the focus of my critique in Sect. 3.3.

2.3 Guide and Evaluate: The Two Functions of the Intelligent Matter Framework

To critically discuss the IMF, it is necessary to consider what it hopes to achieve.[3] According to Kaspar and colleagues, they set out to provide "a roadmap towards intelligent matter" [1, p. 353] and to coordinate a "concerted, interdisciplinary and long-term research effort" [1, p. 353]. In other words, by specifying requirements for different levels of material intelligence, the IMF is intended to constitute a coherent and unified framework which can evaluate and guide a distributed and diverse set of research practices. I suggest that this goal can be factored into two closely related functions which I describe below.

On the one hand, the IMF has a *guiding* function, serving to specify how the intelligent matter research program should aim towards implementing intelligence in material systems. According to Kaspar and colleagues', no extant system satisfy their criteria for intelligence [1, p. 353]. As such, the IMF's more demanding classifications serve as a lodestar for researchers to navigate towards. This role is explicit in the authors' concluding call to action: "...we must proceed from learning matter to truly intelligent matter, which receives input from the environment via sensory interfaces, shows a desired response encoded via embedded memory and artificial networks, and can respond to external stimuli via embedded transducers" [1, p. 353]. On the other hand, the IMF also has an *evaluative* function, enabling progress towards the goals of the intelligent matter research program to be assessed. To achieve this, specific material systems are examined and sorted according to the IMF's classificatory scheme. For example, Monreal Santiago and colleagues [16] developed a synthetic molecular system exploiting positive feedback to enable self-replication.

---

[3] This is a pragmatic approach which eschews metaphysical worries, e.g., about what intelligence *fundamentally* is or whether simple material systems can *really* possess it. Such questions, while plausibly interesting, are not within the stated remit of the IMF.

According to the IMF, the system's possession of sensing, actuating, and interaction elements render it *adaptive* matter, but its lack of a long-term memory element precludes it being classed as *intelligent*.

Clearly, the guiding and evaluative functions of the IMF are intertwined. By evaluating extant systems and specifying which functional elements they are lacking, the future work required to develop intelligent matter is made apparent. To sum up, it is with respect to the IMF's aptness to guide and evaluate that it ought to be judged.

3   The Missing Role of Normativity

In this section I argue that the IMF has a weakness. In particular, it fails to provide normative constraints on which kinds of *behaviors* count as adaptive and intelligent according to its criteria. I begin by explaining the sense in which the IMF lacks normativity. I then contrast the IMF with definitions of intelligence which *do* provide this kind of normativity, to provide further clarity on what exactly is missing. Finally, I argue that normativity is critical for both the evaluative and guiding functions of the IMF.

3.1   Normative Criteria as Behavioral Standards

Clearly, the IMF provides some criteria for assessing the intelligence of material systems, namely those described in Sect. 2.1. To explain the sense in which normative criteria are missing, then, a terminological distinction is required. I will refer to the criteria delineated by Kaspar and colleagues in the IMF as *functional criteria*, since they refer to a system's possession of, or constitution by, certain functional elements (i.e., sensors, actuators, an interaction network, long-term memory). There is a straightforward sense in which these functional criteria can be considered normative, since they specify properties that systems need to have (i.e., constitutive norms to which systems need adhere) in order to achieve particular classifications. However, what I mean to refer to by *normative criteria* are standards applied to the *behavior* of a system (i.e., what it does). For example, a thermostat is judged (prima facie) by its capacity to minimize the difference between the temperature of a physical system and a specified set-point. In this sense, normative standards function as *performance measures* by which a system's ability to implement target behaviors may be assessed [14]. Possessing the right combination of functional elements may be a pre-requisite for achieving some behavioral outcomes, but the achievement of most outcomes will typically require something over and above possession of those elements (e.g., the elements being organized such that their operation in typical environments reliably produces the relevant outcomes). To capture this distinction, when I speak of *normative criteria* moving forwards, I am always referring to standards by which systems' behavior may be evaluated, regardless of differences in the functional elements involved.

So then, to make my claim that the IMF lacks a normative criterion precise, it can be re-phrased as follows. According to the IMF, conditional on the set of functional elements that a given material system possesses, variation in the behavioral propensities of that system is immaterial to the class into which it should be sorted. That is, once we know which functional elements a system is endowed with, we do not need to know anything more about how those functional elements interact or which patterns of behavior they produce in order to know how intelligent that system is.[4] I now contrast the IMF with other frameworks of intelligence in order to clarify the typical role of normative criteria.

3.2 Normativity and Definitions of Intelligence

In their review of definitions of intelligence, Legg and Hutter synthesized over 70 proposals from psychology, cognitive science, and artificial intelligence into a single sentence: "Intelligence measures an agent's ability to achieve goals in a wide range of environments" [17, p. 9]. This broad and general definition brings two distinctions with the IMF to light. First, the synthesized definition explicitly invokes normative criteria through reference to "goals". While the notion of a goal remains imprecise in the definition, however it is spelled out, it will constrain what a system needs to *do* in order to achieve a certain degree of intelligence (I will have more to say on goals in Sect. 4.2). This is exactly what the IMF lacks. Second, in contrast to the IMF, the synthesized definition specifies no *functional* criteria at all. That is, it does not place any requirements on *how* a given system achieves its goal. I do not suggest that the IMF is wrong to specify functional criteria. On the contrary, specifying such criteria enables the IMF to fulfil its guiding function (with the proviso that it will only function well as a guide under the condition that the criteria are well-motivated). Instead, my claim is that, especially for the IMF to fulfil its evaluative function, it needs normative criteria *in addition to* functional ones.

Appeals to normative notions in the definition and evaluation of intelligence are ubiquitous. Normativity in Legg and Hutter's definition of intelligence is conferred by the use of the word "goal". However, there are numerous ways for normativity to enter a definition or framework for intelligence.[5] For example, Russell and Norvig structure their approach to AI around the idea that "some agents behave better than others", which "leads naturally to the idea of a rational agent—one that behaves

---

[4] As I will discuss in Sect. 3.4, textual evidence casts suggest that Kaspar and colleagues did not intend this to be a property of their framework. Rather, the omission of an explicit normative criterion by which the behaviour of purportedly intelligent systems may be judged is more likely an oversight.

[5] I do not distinguish sharply between *definitions* and *frameworks* here. Roughly speaking, the IMF is a framework composed of a set of hierarchically arranged definitions. However, the evaluative and guiding functions of the IMF may also be played by singular definitions [10].

as well as possible" [14, p. 95]. Through invoking notions of *better*, *worse*, and *best* behaviors, their concept of rationality becomes normative. One might argue here that the approach relevant to traditional AI agents should differ from that for material systems. However, Russell and Norvig define an *agent* as "anything that can be viewed as perceiving its environment through sensors and acting upon that environment through actuators" [14, p. 96]. As such, their approach is clearly minimal enough to be compatible with intelligent matter (indeed, even responsive matter is covered). A non-exhaustive list of terms, common to definitions of intelligence, which (usually) connote a normative standard include: rationality [18], problem-solving [19], indistinguishability from human beings [20], adaptability/adaptivity [10], task-performance [7], and more besides [6], see [15, 21]. All definitions which invoke such notions incorporate normativity.

One term mentioned above, *adaptivity*, appears in the IMF and is important to discuss in greater detail. One might think the worry about normativity is forestalled by the IMF's inclusion of this term. However, there is a critical contrast between how the IMF uses the word *adaptive* and how it is typically applied in biological and AI settings. Recall that in the IMF a material system is adaptive if it possesses sensors, actuators, and an interaction network [1, p. 346]. In biology, in contrast, adaptivity is normally defined in terms of how a trait contributes to the inclusive genetic fitness of the organism that possesses it [22]. For example, it is adaptive for cacti to grow spines because they deter herbivores who might otherwise eat the cacti, preventing them from replicating and propagating their genetic material [23, p. 26]. We can readily imagine a scenario in which a cactus' growing of spines depends upon the presence of some environmental trigger, such as a particular quantity of rainfall. In such a case, the cactus is making use of sensors, to detect the presence of rain, actuators, to initiate the growth of spines once the threshold is reached, and a network, through which the signals from sensation to actuation are propagated. However, from a biological perspective, such behavior only counts as adaptive if the inclusive genetic fitness of the cactus, conditional on its spine-growing behavior, is greater than its inclusive genetic fitness in a counterfactual world in which the spine-growing behavior was absent. In other words, biological adaptivity tracks *beneficial contribution*, not functional complexity. While particular forms of complexity might be necessary to attain certain benefits [11], unless they actually *do* attain those benefits, at least in expectation, they are not considered adaptive.[6]

3.3 Normativity is Crucial for the Evaluative and Guiding Functions of the IMF

So far in this section, I have specified what is meant by the claim that a normative criterion is absent from the IMF as stated and shown that this feature marks the IMF

---

[6] The AI use of *adaptivity* is essentially parasitic on the biological notion e.g., [10, p. 19] but with inclusive genetic fitness exchanged for a quantitative performance measure.

out as unusual among attempts to define or conceptualize intelligence. I now argue directly that augmenting the IMF with an explicit normative criterion is required for it to adequately serve its evaluative and guiding functions.

Pei Wang contends that a working definition for artificial intelligence (like the IMF) ought to adhere to four requirements (initially proposed by Carnap in his discussion of probability [10, 24] (see [6] for a related approach). Those require-ments are as follows: (i) *similarity* to the way the term is commonly used, (ii) *exact-ness*, (iii) *fruitfulness*, and (iv) *simplicity*. The similarity requirement prevents an academic definition of a term from becoming divorced from its ordinary meaning. For example, we could stipulatively define intelligence for material systems as a measure of, say, conformational flexibility. But this would leave us with a term utterly failing to capture the sense of intelligence that motivated the need to define it in the first place. The exactness requirement demands that a definition be precise enough to aid scientific discourse. In the context of intelligent matter, exactness is critical for the IMF to serve its evaluative function.[7] The fruitfulness requirement demands that a definition be informative enough to serve a practical role. For the IMF, this equates to fulfilling its guiding function. As Wang states, researcher defi-nitions of intelligence typically do not describe "something that already fully exists, but something to be built" [10, p. 5], a description that applies clearly to the IMF. The simplicity requirement demands that a definition be as simple as possible, given that it satisfies the first three requirements.[8]

I suggest that an explicit normative criterion is crucial for any definition or framework of intelligence to adhere to the similarity requirement. As Kryven and colleagues put it, "attributing intelligent behavior relies on interpreting an agent's actions as ... reasonable in the first place" [26, p. 15]. In other words, the purpose of ascribing intelligence to entities is as a means to explain and predict systematic differences in their observed behaviors. In the human domain, some people are able to complete cognitive tasks that others cannot. As Sternberg notes [27, p. 308], it is with respect to such observations that the ascription of differing degrees of intelligence to individuals has explanatory value.[9] For this reason, any framework of intelligence that fails to track observed variation in relevant behavioral outcomes is insufficiently similar to the ordinary concept [6].[10]

---

[7] As illustration of the perils of inexactness, consider Legg and Hutter's synthesized definition of intelligence (Sect. 3.1). By invoking undefined notions of *agents*, *goals*, and *environments*, this definition leaves much up to interpretation should one attempt to apply it to a real-world material system (see [25] for an attempt to make that definition exact).

[8] There is a tension between this requirement and the third, since simple definitions may lack the detail required for informativity, but discussing this trade-off is beyond the scope of this chapter.

[9] Indeed, one might view intelligence as no more than a useful shorthand for speaking about these observed differences in capabilities and behaviours (see [28] for such a view). Of course, there are more theoretically laden ways to interpret *intelligence*, with the IMF being an example. The point is that, to satisfy *similarity*, an explication of the concept ought to summarize observations in a compact and predictively valuable manner.

[10] This is not to say that the exact outcomes over which variation is tracked need to be the same as in the human case. As acknowledged in Sect. 2.2, Kaspar and colleagues do not intend their notion of intelligence to be applicable to psychology.

While it may not immediately seem catastrophic for the IMF to fail to adhere to one out of four requirements, the value of providing a sufficiently *exact* and *fruitful* definition can only be realized given *similarity* is satisfied. Otherwise, the way in which systems are evaluated, even if precise, will ultimately fail to track our intuitive judgements. Furthermore, systems which follow the guidance of the IMF may fail to demonstrate behavior which would independently strike us as intelligent. If my critique is right, there should be possible systems which would satisfy the IMF's criteria for adaptivity and intelligence but fail to produce intelligent behavior. To that end, consider a system equipped with a sensor that periodically detects (external) temperature, a memory system that stores experienced temperatures as sequences (of length 3, say), an actuator that can trigger the rotation of the system (like a bacterial flagellar motor), and a network mechanism connecting these functional elements. The network mechanism evaluates the most recent trajectory of temperatures and, if it matches a stored sequence, signals the actuator to trigger the rotation of the system. The implementational details of this imagined case are not important.[11] The point is that this system would, in virtue of its possession of all functional elements specified by the IMF, constitute intelligent matter. Despite that, it is hard to see how such a system could possibly implement behavior that would count as intelligent by a practical normative standard.

As well as failing to track intuitive judgements in cases where functional complexity is put to unintelligent use, the IMF also overlooks that what a system *does* is limited in important ways by numerous factors beyond which functional elements it possesses. These factors include i) the variety of things it can sense, ii) the range of actions available to it, and iii) the fit between the behavioral policy it implements and the conditions of the environment [30]. It is true that an enlarged functional reper- toire expands the space of possible behaviors. However, merely possessing such a repertoire does not ensure that the behaviors plucked from that vast space belong to the subset of valuable ones. For this reason, a measure of intelligence that focuses entirely on functional criteria is bound to overrate systems which do useless things for sophisticated reasons. In order to cleave to the ordinary sense of the term *intelligent*, then, the IMF must augment its specification of the functional elements necessary for intelligence with some kind of normative criterion for the behaviors material systems implement.

Before I discuss normativity in more detail, I briefly examine textual evidence suggesting that Kaspar and colleagues implicitly invoke normativity when applying the IMF.

3.4   Implicit Normativity in Applying the IMF

Although normativity is absent from the explicit presentation of the IMF, it still factors into the judgements Kaspar and colleagues make about the intelligence of material

---

[11] One can imagine such a system as a variant of Braitenberg's *Vehicle 5* [29].

systems. There are two ways this plays out. First, although the IMF only directly classifies systems with respect to *how* they implement behaviors (i.e., functional criteria), there are implicit criteria governing which range of behaviors a system must display to be deemed interesting enough to warrant examination (i.e., implicit normative criteria). Second, comments from Kaspar and colleagues imply that further details relating to how systems *implement* normativity are relevant to how intelligent they consider them to be. I address these points in turn.

Normative criteria implicitly govern which systems are candidates for classification as intelligent by the IMF. Consider Yu and colleagues' demonstration of self-swarming behavior in paramagnetic nanoparticles [31]. Due to their magnetic interactions, upon exposure to oscillating magnetic fields these particles display a range of pattern-forming collective behaviors. In their assessment, Kaspar and colleague state that, because the nanoparticles "rely on the input of an external programmer who manipulates the magnetic field... the particles do not show intelligent behavior by themselves" [1, p. 347]. Notice that this statement pertains to the *behavior* of the nanoparticle system as a function of environmental variation (i.e., input). It is because this pattern-forming behavior is only observed under environmental conditions which require close external control that the system is deemed less autonomous and, thus, less intelligent. Such a claim implicitly invokes a normative criterion by focusing on behaviors deemed interesting from the outside (e.g., collective motility). Without such a criterion, there would be no reason for the pattern-forming behavior, generated through controlled deformations of the particle swarm, to be distinguished from any other behavior. Similarly, He and colleagues [32] (described in [1, p. 350]) develop a system which, through a feedback loop involving the mechanical action of temperature-responsive hydrogel and the chemical action of exothermic catalytic reactions, maintains its temperature in a narrow range. In this case, thermoregulation is the behavior judged worthy by an implicit normative criterion. The failure to make these implicit judgements explicit may seem trivial. However, the hypothetical system considered in the previous section shows how this can manifest in unintuitive judgements. Further, since the IMF is supposed to guide research, leaving such criteria implicit is unsatisfactory (see Sect. 5.4 for further discussion).

Second, Kaspar and colleagues suggest a further constraint, not prescribed by the IMF, on how norms should be *implemented*. In particular, they imply that the criterion by which a material system's behavior should be judged must, in some sense, be generated from *within* the system. For example, when assessing a swarm of autonomous robots that can assemble into two dimensional patterns, Kaspar and colleagues state that "since an external programmer predefines the targeted shape and gives instructions in [the] form of an algorithm, the whole group of robots is not intelligent according to our definition" [1, p. 346]. It is not clear how the authors derive this judgement from their definition of intelligent matter, which does not mention goals or the achievement of target states, whether defined externally or not. However, this comment (and the similar comment about the swarming paramagnetic nanoparticles) suggests that Kaspar and colleagues intend the IMF to limit the ascription of intelligence to systems which, in some respect, generate their own normative criteria.

Call this an *intrinsicality* condition on normativity. I discuss this requirement further in Sect. 5.2.

4   Aspects of Normativity

In Sect. 3 I observed that the IMF lacks a normative criterion and argued that this omission is problematic. To rectify this situation, I suggest, *some* notion of normativity ought to be explicitly incorporated into the IMF. This should serve to make the framework sensitive to behavioral capacities in a way that tracks the ordinary notion of intelligence, thus satisfying the similarity requirement (see Sect. 3.3) and enabling the IMF to serve its evaluative and guiding functions effectively. However, for this to be possible, clarity around *which* notion of normativity ought to be incorporated is required. To that end, this section unpacks the concept of normativity.

I suggest that questions about normativity can be divided into two distinct categories.[12] One set of questions pertains to the *source* of the normativity. That is, where does the normative criterion originate? By what process does it arise? Another set of questions pertains to the *constitution* of the normativity. That is, how is the normativity *physically instantiated* in the system? Are there explicitly represented goal states against which actual states can be compared? Or is the normativity governed by the organization of the entire system? Let us take a closer look at each of the aspects of normativity before discussing their relation to the IMF.

4.1   Aspects of Normativity: Source

First, let us consider *sources* of normative criteria. That is, where do such criteria come from? The source of normativity may be considered from an ultimate or proximal perspective [33–36]. In the case of biological systems, the ultimate source of a normative criterion is the process of evolution by natural selection [23, 37]. It is inherent to natural selection that success for an individual is constituted by surviving and reproducing, with the value of traits and behaviors derived from their contribution to these core objectives [38]. Without the differential replication of variant biological forms, captured formally by the metric of inclusive genetic fitness, there could be no criterion for the success of individuals and thus no normativity [39]. Proximal sources of normative criteria for biological organisms are the specific challenges posed by their environments ([40], see [41] for detailed discussion of how

---

[12] To structure this section, I take inspiration from Rahwan and colleagues' typology of questions for machine behaviour [33], itself derived from Tinbergen's foundational work in ethology [34]. Questions about what I call the *source* of normativity roughly map on to Tinbergen's *historical/developmental* questions, while questions about the *constitution* of normativity roughly map on to Tinbergen's *contemporary/slice-in-time* questions. However, I intend the relation to Tinbergen (and Rahwan et al.) to be one of rough inspiration, not isomorphism.

normativity relates to biological function).[13] For example, the threat of predation by foxes imposes a normative criterion on rabbits by rendering their survival conditional on their ability to evade capture. On this view, the evolution of behaviorally and cognitively sophisticated organisms is a product of an increasingly diverse multitude of proximal sources of normativity, all generated from the same ultimate source of natural selection (see [11] for discussion).

For engineered systems, the ultimate source of normativity is challenging to specify. There is no consensus view on how to determine what constitutes appropriate functioning or overall success for an artifact (see [43] for philosophical discussion). However, to a first approximation, we can follow the example of biology and characterize success in terms of spread and prevalence (as in [33]). Translating these ideas to the world of artifacts, we are faced with an ultimate criterion of widespread manufacture, deployment, and use.[14] Another way to frame this is that the ultimate source of normativity for artifacts, by analogy to biological evolution, is cultural evolution [44].[15] Though there is significant controversy about how best to describe variation and selection in a cultural evolutionary framework [45], the relevant dynamics likely involve a complex combination of social, technical, and biological factors [33, 46]. Moving onto the proximal sources of normativity, once again, we are faced with variety. In some cases, such as formal verification in software engineering [47, 48], normative criteria for the behavior of engineered systems may be formalized and explicated precisely. However, in other cases, the criteria by which system behavior is evaluated may be challenging to state exactly [49]. For example, consider the recent explosion in the prevalence and usage of large language models [50]. While the formal measure of perplexity captures how well models such as ChatGPT perform at their training objective [51], their commercial success is a function of how useful people find them, a metric which cannot be easily operationalized prior to releasing the model as a product. The use of performance on task benchmarks as proximal normative criteria has been a topic of critical discussion within the AI and machine learning community (e.g., [5, 21]). This reflects the difficulty of selecting proximal normative criteria which lead to desirable outcomes [52–54] (see also Sect. 5.4).

---

[13] It would be beyond the scope of this chapter to discuss the intricacies of the philosophical debate on biological functions. However, see [42] for an attractive pragmatic and pluralist view.

[14] This relates to the etiological view of artifact functions [43]. However, I do not take describing the source of normativity for engineered artifacts to require commitment to such a view (or any view at all) about artifact functions.

[15] Note that there is ongoing debate about the nature, prevalence, and explanatory power of cultural evolution (as well as whether it is an *evolutionary* process in a literal, analogical, or metaphorical sense). However, the basic observation that technical artifacts differ in ubiquity and persistence is undeniable.

4.2 Aspects of Normativity: Constitution

Next, let us consider how normativity may be *constituted* in a system. That is, what physical arrangement of matter *implements* normativity within a system? Here, two broad categories can be distinguished. First, some systems explicitly represent goal states or objectives such that they can evaluate occurrent states with respect to them [55]. Call this category *explicit representation*. Explicit representation of goals is commonly associated with the symbolic approach to AI spearheaded by Newell and Simon [56]. Newell and Simon developed so-called *expert systems* which operated via backchaining, i.e., given a goal state and an occurrent state, searching for a trajectory of intermediate states such that the goal state is reached [57]. More generally, approaches to AI that focus on search and planning typically involve explicit representation of goals [14]. Control systems which explicitly represent their control variables (i.e., those whose values they maintain within a specific range) also belong to this category [58, 59]. Additionally, though the details are controversial, it is widely believed in cognitive science that human beings explicitly represent goals in some manner [60]. Whether or not the same is true in various species of nonhuman animals is a matter of ongoing debate [61–63].[16]

The second broad category concerning the constitution of normativity is *holistic arrangement*. Here, there is no (canonical) way to localize the normativity which governs the behavior of a system. Rather, it is the *entire* physical arrangement of the system which serves to dictate that its behavior conforms to a particular norm (these are what [55] call 'goal-seeking systems', or what [65] calls the 'systems view'). Again, both biological and engineered systems may fall into this category. Indeed, arguably, every biological system is normative in virtue of its holistic arrangement, since it is the way the system is arranged (in both space and time) which constitutes its identity [62, 66]. Broadly speaking, this is the position taken up by proponents of the autopoietic view of biological systems and their intellectual descendants [67–69].[17]

When it comes to engineered systems, they do not (yet) possess the dynamic cohesion or behavioral flexibility characteristic of biological organisms [70], precluding the kind of autonomy exhibited by living beings. However, there is still clearly a class of human-designed material systems for which their overall functional organisation imbues them with a weaker kind of purposiveness. Indeed, better understanding how this subset of engineered machines, so-called *servomechanisms*, implemented goal-directed behavior was a central goal of the cybernetics movement of the 1940s and 50s [9, 71, 72]. To anticipate a possible misunderstanding, note that holistically arranged normativity is not trivially ascribable to every human-designed artifact, merely in

---

[16] It is beyond the scope of this chapter to enter into detailed analysis regarding what is required for something to count as a representation of a goal. Indeed, it is plausible that no sharp dividing line separates biological systems which do and do not explicitly represent goals [64]. If this is so, it seems equally plausible that artificial systems, including those developed in materials science, may also blur the boundaries.

[17] Note that commitment to a holistic arrangement account of biological normativity is compatible with believing that (some) biological organisms explicitly represent target states. However, the two accounts are mutually exclusive as explanations for the achievement of any *particular* state.

virtue of the fact that their intentionally designed characteristics contributes to their utility, spread, and application. The critical difference between normative systems such as, say, a heat-seeking missile, and non-normative ones such as, say, a clock, is that the former but not the latter have the propensity to reach or maintain particular desirable states despite environmental perturbations ([23] views this as an aspect of goal-directedness, while [65] calls this property 'persistence').[18] Systems which do not exhibit any flexible behavior that reliably reaches or maintains particular target states do not implement normativity.

With the source and constitution of normativity in mind, let us turn to considering which senses of the term could be usefully incorporated into the IMF.

5   Normativity and the Intelligent Matter Framework

To recap, Sect. 2 presented the IMF, posited that it aims to serve evaluative and guiding functions, and noted that it omits explicit normative criteria. Section 3 discussed this omission in detail, arguing that it prevents the IMF from adequately fulfilling both its evaluative and its guiding function. Section 4 provided a brief analysis of normativity, suggesting that the *source* from which normativity derives and how normativity is *constituted* in a given system are distinct, important considerations. This section turns to how normativity should be inserted into the IMF in a way that accords with the theoretical commitments of the intelligent matter research program. In particular, we can ask: i) which source of normativity is most suitable for the purposes of the IMF? And: ii) what requirements for the constitution of normativity, if any, ought the IMF to impose?

5.1   Sources of Normativity for the IMF

Recall that the source of a normative criterion refers to where it comes from. That is, what is responsible for generating the evaluative standards by which a material system's behavior should be judged? As an initial question, we can consider whether it is the ultimate or proximal source of normativity with which the IMF should be concerned. This has a straightforward answer. Any attempt to dictate the *ultimate* source of normativity for engineered artifacts would be hubristic folly. Which engineered artifacts and material systems end up being reproduced, distributed, and deployed (i.e., the ultimate source of normativity) depends upon myriad factors that could not be effectively enumerated, let alone controlled, by any research effort. By elimination then, it is the *proximal* source of normativity that the IMF should

---

[18] A precise explication of the degree to which a system needs to exhibit this form of persistence to count as normative is outside the scope of this chapter. Personally, I am inclined towards a gradualist perspective on which there is no definite boundary.

be concerned with. So, the question becomes: how should the IMF impose local normative standards to better fulfil its functions?

Two general strategies can be delineated.[19] On the one hand, the IMF may augment its functional requirements with a specific normative criterion that is applicable across cases. For example, one option would be to follow Legg and Hutter's synthesized definition of intelligence discussed in Sect. 3.2 ("the ability to achieve goals in a wide range of environments"). In order to give this definition teeth, Kaspar and colleagues would need to define "goals" and "environments" concretely enough to quantify the range of environments systems operate in. They would also need to consider how to integrate a measure of this quantity with the discrete functional requirements specified by the IMF in its current form. For instance, each class of matter delineated by the IMF could be associated with a threshold of the measured quantity such that a system can only count as, say, adaptive, if it achieves goals across sufficiently many environments. This approach has the potential to endow the IMF with the power to fulfil its evaluative and guiding functions, ruling out degenerate systems and providing a clear vision of intelligent behavior. However, enacting this strategy may be challenging given the breadth and diversity of the intelligent matter research program. Devising a criterion equally applicable to self-organizing nanoparticle assemblies, soft matter systems, and distributed neuromorphic computing systems may not be possible. For example, Legg and Hutter's definition may struggle to capture the fact that some intelligent matter research aims to replicate the flexibility of general computation (e.g., [73]) while others aim for the robust implementation of narrow behaviors (e.g., [31]).

The other avenue available to the IMF is a *devolved* approach which transfers responsibility for specifying normative criteria to the researchers developing purportedly intelligent material systems. This strategy comports with the spirit in which the article presenting the IMF is written. For example, consider He and colleagues' homeostatic mechanochemical system, in which "a continuous feedback loop between an exothermic catalytic reaction [...] and the mechanical action of [a] temperature-responsive gel results in an autonomous, self-sustained system that maintains temperature within a narrow range" [32, p. 350]. Adopting the devolved approach would mean deferring to He and colleagues with respect to the normative standard by which the system is evaluated. In this case, the system would (presumably) satisfy the normative standard imposed by the researchers in virtue of its thermoregulatory behavior. This strategy has the advantage of adaptability. It can capture whatever normative criteria are deemed relevant by particular scientists pursuing the varied research programs which fall under the banner of *intelligent matter.* As a drawback, however, the devolved strategy provides less in terms of both guidance and evaluation than the specification of a particular measure (see Sect. 5.4 for further discussion of this point).

---

[19] For clarity I present these strategies as discrete options. However, in reality there is a continuum on which strategies may be placed.

Since there are pros and cons to both the specific approach and the devolved approach, I suggest a compromise position that augments the IMF with the following normative criterion:

NC: In order for a material system to be classification as intelligent (or adaptive, responsive, or structural), it must be endowed with the requisite functional elements *and*, through the coordinated activity of those functional elements, robustly implement concrete behaviors (e.g., the maintenance or achievement of one or more target states) which contribute towards a higher-order goal.

This criterion reflects a partially devolved strategy. It leaves both the concrete behaviors a system implements and the higher-order goal towards which the system's behavior contributes open to interpretation. However, it does still impose general constraints which rule out degenerate systems as described in Sect. 3.3. In particular, only systems which achieve specified target states of some kind, be they internal states of the system or states of the environment, can count as intelligent. That systems must do this *robustly* ensures that some degree of persistence in the face of perturbation is required. Further, that these states must contribute to some higher-order goal brings the IMF's use of the term *adaptivity* into alignment with other disciplinary approaches to intelligence.[20] Finally, the requirement for the behavior of the system to be implemented by the coordinated activity of its relevant functional elements ensures that the functional and normative criteria are integrated into a coherent whole.

Practically, application of the augmented IMF would involve specifying which concrete behaviors constitute the norm for a particular material system (e.g., maintaining system temperature in a particular range) and what higher-order goal those behaviors contribute to (e.g., the persistence of the system).

5.2   Constitution of Normativity for the IMF

Let us now turn to what prescriptions, if any, the IMF might make regarding how normativity should be *constituted* in purportedly intelligent material systems. In particular, is one of the two categories delineated above, explicit representation and holistic arrangement, more fitting than the other? We should first note that the IMF, in virtue of supplying a functional parts list necessary for systems to be designated as responsive, adaptive, or intelligent, already constrains how *any* behavior ought to be implemented. Thus, we can begin by considering whether these constraints suffice to commit the IMF one way or another regarding the implementation of normativity. That is, may systems possessing the relevant functional elements adhere to a behavioral norm in virtue of *either* explicitly represented goal states *or* holistic arrangement, or does that functional profile already exclude one of those possibilities? It is clear that both possibilities remain open to the IMF. Goal states may be explicitly

---

[20] Indeed, the system I imagined in Sect. 3.3 could be rescued from the charge of degeneracy by specifying the higher-order goals to which its behavior contributes.

represented in a long term memory or interaction network, but flexible goal-seeking behavior is possible without such representation [55, 65]. As such, the IMF is neutral with respect to how exactly those functional elements might be constrained to interact in ways which reliably lead to desirable states.

However, although both explicit representation and holistic arrangement are logically consistent ways for the IMF to bolster its functional requirements, there are further considerations, touched upon by Kaspar and colleagues, which favor holistic organisation. I will consider two such points. First, as noted in Sect. 3.4, Kaspar and colleagues invoke an *intrinsicality* condition in their evaluation of the swarm behavior of autonomous robots and paramagnetic nanoparticles. In each case, the systems in question are deemed adaptive but not intelligent in virtue of their requiring an external programmer to define the target shapes of the swarm [1, pp. 346–347]. This implies that if the target shapes had *not required* this external programming, the systems would have been deemed intelligent. This interpretation is supported by Kaspar and colleagues' suggestion that natural self-organizing swarms (which are not externally programmed), such as insect colonies and schools of fish, are "often considered to exhibit features of intelligent behavior" in virtue of being "coordinated in a decentralized manner" (p. 346). This brings us to *decentralization*, which is the second consideration. Developing decentralized intelligent systems is an important ambition of the intelligent matter research program. As Kaspar et al. state, the ideal combination of their favored functional elements would "form functional processing continua, which do not require a centralized processing unit, but rather provide the capability for local and distributed information processing" [1, p. 345]. If we take the authors at their word, the normativity of intelligent material systems should be constituted in a decentralized but intrinsic manner.

What does it mean for an engineered system to exhibit decentralized, intrinsically normative behavior? If the behavior that the system displays is decentralized, then it is not being governed by a central information processing unit. This implies that the system cannot be explicitly representing its goal state in a dedicated location, since that location would constitute a de facto locus of control.[21] If this reasoning is sound, then a decentralized system must implement normativity, if at all, through its *holistic arrangement*. Next, let us contend with the notion that this normativity should be, in some sense, *intrinsic*, that is, not dictated by an external source like a programmer. How should we interpret this requirement? One possibility would be that the system must exhibit the strong sense of holistic arrangement, as explicated by researchers inspired by the autopoietic tradition. On this view, a biological system is special because "the conditions of existence of its constitutive processes and organization are the norms of its own activity" [69, p. 34]. This means that an "interaction or process is detected as "bad" or "good" by and for the very system (not by and for an external observer)" [69, p. 35]. Ruiz-Mirazo and Moreno capture a coherent sense in which a system can have intrinsic norms. However, importing

---

[21] In theory, nothing prevents a decentralized system from representing a goal state in an inert way, disconnected from the functional organisation which brings about that goal state. However, such a system would not satisfy typical criteria for explicit representation e.g., [55, pp. 128–9].

this into the IMF would be at odds with the stated focus on commercial engineering applications like intelligent clothing and neuromorphic computing, as well as the explicit disavowal that developing intelligent matter implies producing willful living systems (see Sect. 2.2). Further, if the IMF adopted this demanding intrinsicality requirement, then its research goals would collapse into those of synthetic biology and artificial life [74, 75].

This leaves the IMF in a bind. Adopting a standard of strong intrinsicality seems contrary to the spirit of the intelligent matter project. On the other hand, there may be no other meaningful sense in which a system which does not select among explicitly represented goal states can exhibit intrinsic normativity. As Ann Sophie Meincke puts it, "it is not clear in what sense the behaviors of systems [like] thermostats or torpedoes could be guided by intrinsic goals, given that there is no sense in which such systems can be considered as themselves setting these goals" [76, p. 286]. If this argument is right, then conditional on a system's normativity being constituted by its holistic arrangement, either it pursues the biological imperative to maintain its own dynamic organisation (in the strong sense demanded by the autopoietic tradition), or it cannot be intrinsically normative. Thus, it may not be possible for the IMF to simultaneously retain all of its stated desiderata. One of the following—holistic arrangement, intrinsic normativity, practical applicability—must be jettisoned. In my view, the intrinsic normativity requirement is the most sensible candidate for rejection. While there is appeal to the notion that a material system which governs its own norms is more intelligent than one that does not, I cannot see conceptual space for a research endeavor which adopts the requirement of intrinsic normativity while retaining an identity distinct from that of already extant research programs in synthetic biology and artificial life. The project of developing a material system which is intrinsically normative in virtue of its holistic arrangement simply *is* the project of developing an artificial living system.

5.3   Partially Devolved, Extrinsic, Holistically Arranged Normativity for the IMF

We have examined numerous possible ways the IMF might be augmented with normativity. As for its source, I have argued that, given the heterogeneity of research that falls under the purview of *intelligent matter*, a partially devolved solution is best. This means that the extent to which a material system exhibits concrete target behaviors defined by the researchers investigating that system should be considered, explicitly, on a case-by-case basis. While this approach is deflationary, it is also pragmatic, avoiding the difficulty of delineating a specific, quantitative criterion for intelligent behavior that applies across diverse research contexts. When it comes to how normativity ought to be constituted, I have suggested that the set of preferences expressed in Kaspar and colleagues' article are mutually inconsistent. In particular, no research project can simultaneously develop material systems which are suitable for a diverse range of practical purposes (as opposed to synthetic living systems), intrinsically normative, and normative in virtue of their holistic arrangement. As such, for the IMF to capture and guide a coherent research agenda, I have argued that it ought to drop the requirement for intrinsic normativity, allowing systems to be considered intelligent even if their target states are specified externally. This preserves an independent identity for intelligent matter, preventing a collapse into synonymy with synthetic biology and artificial life. Further, it enables the requirement for systems to implement intelligent behavior in a decentralized manner to be retained, further distinguishing the intelligent matter research program from AI approaches based on traditional computational paradigms [14].

5.4 Discussion

I have suggested that the normativity most apt for introduction into the IMF is partially devolved, extrinsic, and holistically arranged. However, it is worth considering this position's major weakness. There is a potential tension between the desideratum for practical applicability and the goal of producing intelligent matter. The devolved approach to normativity assumes that material systems which operate via the coordinated activity of the IMF's functional elements will be developed naturally in the process of targeting practical applications. However, historical precedent suggests such conjectures are on shaky ground. The history of AI is littered with shattered expectations about the degree of functional complexity required to implement particular tasks [77]. For example, in 1979 Douglas Hofstadter claimed that expert-level chess playing would require a generally intelligent system [78, p. 678]. Given how often such predictions have failed, the expectation that matter satisfying the IMF's criteria for intelligence will arise organically in pursuit of practically applicable systems should not be taken for granted. As such, the devolved approach could undermine the IMF's ability to serve its guiding function.

The point above highlights an important challenge for the IMF: identifying target behaviors which genuinely necessitate the incorporation of the functional elements specified by the IMF (see [6]). Kaspar and colleagues' article understandably focuses on the difficulties associated with implementing functional elements in material systems. However, as argued in Sect. 3.3, possession of such elements does not, by itself, suffice for intelligent behavior. As such, I suggest, it would be valuable for proponents of the IMF to propose concrete behaviors or properties which can only be exhibited by systems satisfying the IMF's functional criteria.[22] Such behaviors would not constitute *necessary* conditions for intelligence and would not be part of the IMF itself. Instead, they would accompany the IMF as exemplars, providing additional guidance to the research field.

---

[22] This is intended in an informal sense, i.e., given typical constraints on spatial, temporal, and computational resources.

Before concluding, I will clarify the contribution of the foregoing discussion. Although I have offered a positive proposal, my suggestion relies on an interpretation of the goals of the intelligent matter researcher program, which may be flawed. Thus, I stress that the philosophical contributions here include i) the argument that normativity is relevant to the IMF, ii) a clarification of different things normativity can mean, and iii) analysis of the relationship between various notions of normativity and the function of the IMF. Crucially, if my argument is sound and the analysis correct, the dependencies between the ambitions of the intelligent matter research program and the IMF's approach to normativity will remain, even if researchers prefer adopting an alternative strategy in light of those dependencies. For example, the interdisciplinary breadth of the research program may be a less significant barrier to devising a specific, quantitative normative criterion than I suggested. If this is so, then developing such a criterion would surely be worthwhile. Similarly, researchers may choose to retain the intrinsicality condition on intelligence, perhaps eschewing holistic arrangement. My analysis does not suggest this is wrong, merely that it is inconsistent with the desideratum of decentralized information processing. In general, the propensity of the IMF to guide the intelligent matter research program will be improved by careful consideration of which outcomes can be consistently pursued.

6 Conclusion

In this chapter I critically assessed Kaspar and colleagues' Intelligent Matter Framework. To do so, I characterized the IMF as having both an *evaluative* and a *guiding* function for research in the emerging field of intelligent matter. To play those roles effectively, the IMF would need to supply appropriate judgements about the intelligence of extant material systems and provide guidance for how future systems should be developed. On this basis, I argued that the IMF's failure to specify any kind of normative criteria with which to evaluate the *behavior* of purportedly intelligent material systems is problematic. Absent a normative criterion, nothing ensures that the relevant functional elements interact in such a way that they implement adaptive behavior, i.e., behavior bringing about desirable consequences. This renders the IMF unsuitable as a guide, since it may lead to degenerate systems, and as an evaluative tool, since it does not track intuitive judgements of intelligence. Thus, I suggested that the IMF ought to be augmented with a normative criterion. I laid out possible approaches, focusing on the distinction between the source of a normative criterion and the way normativity is constituted within system. Ultimately, I suggested that the IMF should introduce a normative criterion which is i) *partially devolved*, that is, requiring certain behavioral conditions to be met, but with the details to be specified on a case-by-case basis according to the goals of specific researchers, ii) *extrinsic,* that is, defined by external agents, as opposed to selected by or emerging from the material system itself, iii) *holistically arranged*, that is, systems ought to robustly behave in accordance with their behavioral norms in virtue of their overall functional organisation, not in virtue of an explicitly represented goal in a central processing location. An IMF augmented with normativity of this kind would be well-suited for guiding and evaluating progress towards the development of intelligent matter.

References

1. Kaspar C, Ravoo BJ, van der Wiel WG, Wegner SV, Pernice WHP (2021) 'The rise of intelligent matter'. Nature 594(7863), Art. no. 7863. https://doi.org/10.1038/s41586-021-03453-y
2. McEvoy MA, Correll N (2015) Materials that couple sensing, actuation, computation, and communication. Science 347(6228):1261689. https://doi.org/10.1126/science.1261689
3. Walther A (2020) Viewpoint: From responsive to adaptive and interactive materials and materials systems: A roadmap. Adv Mater 32(20):1905111. https://doi.org/10.1002/adma.201 905111
4. Bauer M (2009) Normativity without artifice. Philos Stud 144(2):239–259. https://doi.org/10. 1007/s11098-008-9208-2
5. Chollet F (2019) 'On the measure of intelligence'. arXiv, Nov. 25, 2019. https://doi.org/10. 48550/arXiv.1911.01547
6. Crosby M (2020) Building thinking machines by solving animal cognition tasks. Mind Mach 30(4):589–615. https://doi.org/10.1007/s11023-020-09535-6
7. Hernández-Orallo J (2017) Evaluation in artificial intelligence: from task-oriented to ability-oriented measurement. Artif Intell Rev 48(3):397–447. https://doi.org/10.1007/s10462-016-9505-7
8. Nagel E (1979) Teleology revisited and other essays in the philosophy and history of science. Columbia University Press, New York
9. Rosenblueth A, Wiener N, Bigelow J (1943) Behavior, Purpose and Teleology. Philosophy of Science 10(1):18–24
10. Wang P (2019) On defining artificial intelligence. J Artif Gen Intell 10(2):1–37. https://doi.org/10.2478/jagi-2019-0002
11. Godfrey-Smith P (2002) 'Environmental complexity and the evolution of cognition'. In: The evolution of intelligence. Lawrence Erlbaum Associates Publishers, Mahwah, NJ, US, pp 223–249
12. Lyon P (2015) The cognitive cell: bacterial behavior reconsidered. Front Microbiol 6:264. https://doi.org/10.3389/fmicb.2015.00264
13. Trewavas A (2003) Aspects of plant intelligence. Ann Bot 92(1):1–20. https://doi.org/10.1093/aob/mcg101
14. Russell S, Norvig P (2020) Artificial intelligence: A modern approach, 4th edn. Pearson. Accessed: Apr. 25, 2023. [Online]. http://aima.cs.berkeley.edu/
15. D. Coelho Mollo, 'Intelligent Behaviour', *Erkenn*, May 2022, https://doi.org/10.1007/s10670-022-00552-8.
16. Monreal Santiago G, Liu K, Browne WR, Otto S (2020) 'Emergence of light-driven protometabolism on recruitment of a photocatalytic cofactor by a self-replicator'. Nat Chem 12(7):603–607. https://doi.org/10.1038/s41557-020-0494-4
17. Legg S, Hutter M (2007) 'A collection of definitions of intelligence'. In: Proceedings of the 2007 conference on advances in artificial general intelligence: Concepts, architectures and algorithms: Proceedings of the AGI workshop 2006. IOS Press, NLD, pp 17–24
18. Simon HA (1978) Rationality as process and as product of thought. Am Econ Rev 68(2):1–16
19. Minsky M (1988) The society of mind, 6. Pb-Pr. In A Touchstone book. Simon and Schuster, New York
20. Turing AM (1950) Computing machinery and intelligence. Mind 59(October):433–460. https://doi.org/10.1093/mind/LIX.236.433
21. Hernández-Orallo J (2020) Twenty years beyond the turing test: Moving beyond the human judges too. Mind Mach 30(4):533–562. https://doi.org/10.1007/s11023-020-09549-0

22. Fisher RA (1930) The genetical theory of natural selection. In: The genetical theory of natural selection. Clarendon Press, Oxford, England, pp xiv, 272. https://doi.org/10.5962/bhl.title.27468
23. Okasha S (2018) Agents and goals in evolution. Oxford University Press, Oxford. https://doi.org/10.1093/oso/9780198815082.001.0001
24. Carnap R (1950) Logical foundations of probability. Chicago University of Chicago Press, Chicago
25. Legg S, Hutter M (2007) Universal intelligence: A definition of machine intelligence. Minds Mach. 17(4):391–444
26. Kryven M, Ullman TD, Cowan W, Tenenbaum JB (2021) Plans or outcomes: How do we attribute intelligence to others? Cogn Sci 45(9):e13041. https://doi.org/10.1111/cogs.13041
27. Sternberg RJ (2018) 'Successful intelligence in theory, research, and practice'. In: Sternberg RJ (ed) The nature of human intelligence. Cambridge University Press, Cambridge, pp 308–322. https://doi.org/10.1017/9781316817049.020
28. Van der Maas HLJ, Kan K-J, Borsboom D (2014) Intelligence is what the intelligence test measures seriously. J Intell 2(1), Art. no. 1. https://doi.org/10.3390/jintelligence2010012
29. Braitenberg V (2004) Vehicles: experiments in synthetic psychology, 9. print. In: Bradford book psychology. MIT Press, Cambridge, Mass
30. Sutton RS, Barto A (2020) Reinforcement learning: an introduction,2nd edn. In: Adaptive computation and machine learning. The MIT Press, Cambridge, Massachusetts London, England
31. Yu J, Wang B, Du X, Wang Q, Zhang L (2018) Ultra-extensible ribbon-like magnetic microswarm. Nat Commun, 9(1), Art. no. 1. https://doi.org/10.1038/s41467-018-05749-6
32. He X et al (2012) Synthetic homeostatic materials with chemo-mechano-chemical self-regulation. Nature 487(7406):214–218. https://doi.org/10.1038/nature11223
33. Rahwan I et al (2019) 'Machine behaviour'. Nature 568(7753), Art. no. 7753. https://doi.org/10.1038/s41586-019-1138-y
34. Tinbergen N (1963) On aims and methods of Ethology. Z Tierpsychol 20(4):410–433. https://doi.org/10.1111/j.1439-0310.1963.tb01161.x
35. Laland K, Sterelny K, Odling-Smee J, Hoppitt W, Uller T (2011) 'Cause and effect in biology revisited: Is Mayr's proximate-ultimate dichotomy still useful? Science (New York, N.Y.) 334(6062). Accessed: Jan. 30, 2024. [Online]. https://ora.ox.ac.uk/objects/uuid:1bf0e329-d7dd-4cd0-a354-3bc02b8b4cb0
36. Mayr E (1993) Proximate and ultimate causations. Biol Philos 8(1):93–94. https://doi.org/10.1007/BF00868508
37. Neander K (1991) Functions as selected effects: The conceptual analyst's defense. Philosophy of Science 58(2):168–184
38. Ramsey G, Pence CH (2013) 'Fitness: Philosophical problems'. In: Encyclopedia of life sciences. John Wiley & Sons, Ltd. https://doi.org/10.1002/9780470015902.a0003443.pub2
39. Jablonka E, Lamb MJ (2005) Evolution in four dimensions: genetic, epigenetic, behavioral, and symbolic variation in the history of life. In A Bradford book. The MIT Press, Cambridge (Mass.)
40. Neander K (1997) The function of cognition: Godfrey-Smith's environmental complexity thesis. Biol Philos 12(4):567–580. https://doi.org/10.1023/A:1006524203891
41. Garson J (2016) A critical overview of biological functions. In: Springer briefs in philosophy. Springer International Publishing, Cham (2016). https://doi.org/10.1007/978-3-319-32020-5
42. Garson J (2018) How to be a function pluralist. Br J Philos Sci 69(4):1101–1122. https://doi.org/10.1093/bjps/axx007
43. Eaton AW (2020) 'Artifacts and their functions'. In: Gaskell I, Carter SA (eds) The oxford handbook of history and material culture. Oxford University Press, p 0. https://doi.org/10.1093/oxfordhb/9780199341764.013.26
44. Richerson PJ, Christiansen MH (eds) 'The cultural evolutionof technology'. In Cultural evolution. The MIT Press. https://doi.org/10.7551/mitpress/9894.003.0011

45. Mesoudi A, Thornton A (2018) What is cumulative cultural evolution? Proceedings of the Royal Society B: Biological Sciences 285(1880):20180712. https://doi.org/10.1098/rspb.2018.0712
46. Love AC, Wimsatt WC (eds) Beyond the meme: Development and structure in cultural evolution. University of Minnesota Press. https://doi.org/10.5749/j.ctvnp0krm
47. Eden AH (2007) Three paradigms of computer science. Mind Mach 17(2):135–167. https://doi.org/10.1007/s11023-007-9060-8
48. Tsai JJP, Xu K (2000) A comparative study of formal verification techniques for software architecture specifications. Ann Softw Eng 10(1):207–223. https://doi.org/10.1023/A:101896 0305057
49. Glinz M (2007) 'On non-functional requirements'. In: 15th IEEE International Requirements Engineering Conference (RE 2007), pp 21–26. https://doi.org/10.1109/RE.2007.45
50. Vaswani A et al (2022) 'Attention is all you need'. In Advances in neural information processing systems, Curran Associates, Inc., 2017. Accessed: Jan. 18, 2022. [Online]. https://papers.nips.cc/paper/2017/hash/3f5ee243547dee91fbd053c1c4a845aa-Abstract.html
51. Jelinek F, Mercer RL, Bahl LR, Baker JK (2005) Perplexity—a measure of the difficulty of speech recognition tasks. J Acoust Soc Am 62(S1):S63. https://doi.org/10.1121/1.2016299
52. Muller J (2018) The tyranny of metrics. NED-New edition Princeton University Press. https://doi.org/10.2307/j.ctvc7743t
53. Manheim D, Garrabrant S (2019) 'Categorizing variants of Goodhart's law'. arXiv, Feb. 24, 2019. https://doi.org/10.48550/arXiv.1803.04585
54. Knuth DE (1974) Structured programming with go to statements. ACM Comput Surv 6(4):261–301. https://doi.org/10.1145/356635.356640
55. Lanz P, Mcfarland D (1995) On representation, goals and cognition. Int Stud Philos Sci 9(2):121–133. https://doi.org/10.1080/02698599508573512
56. Gugerty L (2006) Newell and Simon's logic theorist: Historical background and impact on cognitive modeling. Proceedings of the Human Factors and Ergonomics Society Annual Meeting 50(9):880–884. https://doi.org/10.1177/154193120605000904
57. Newell A, Shaw JC, Simon HA (1958) Elements of a theory of human problem solving. Psychol Rev 65(3):151–166. https://doi.org/10.1037/h0048495
58. Baltieri M, Buckley CL (2019) PID control as a process of active inference with linear generative models. Entropy 21(3), Art. no. 3. https://doi.org/10.3390/e21030257
59. Bellman RE (1961) Adaptive control processes: A guided tour. Princeton University Press. https://doi.org/10.1515/9781400874668
60. Dickinson A, Balleine B (1994) Motivational control of goal-directed action. Anim Learn Behav 22(1):1–18. https://doi.org/10.3758/BF03199951
61. Colombo M, Scarf D (2024) 'Are there differences in "Intelligence" between nonhuman species? The role of contextual variables'. Front Psychol 11. Accessed: Jan. 30, 2024. [Online]. https://www.frontiersin.org/articles/10.3389/fpsyg.2020.02072
62. Jablonka E, Ginsburg S (2022) Learning and the evolution of conscious agents. Biosemiotics 15(3):401–437. https://doi.org/10.1007/s12304-022-09501-y
63. Okasha S (2023) Goal attributions in biology: Objective fact, anthropomorphic bias, or valuable heuristic? In: Corning PA, Kauffman SA, Noble D, Shapiro JA, Vane-Wright RI, Pross A (eds) Evolution 'On purpose'. The MIT Press, pp 237–256. https://doi.org/10.7551/mitpress/14642.003.0016
64. Dennett DC (1998) The intentional stance, 7. printing. In: A Bradford book. MIT Press, Cambridge, Mass
65. Nagel E (1977) Goal-directed processes in biology. J Philos 74(5):261–279. https://doi.org/10.2307/2025745
66. Moreno A, Etxeberria A (2005) Agency in natural and artificial systems. Artif Life 11(1–2):161–175. https://doi.org/10.1162/1064546053278919
67. Bickhard MH (2003) 'Process and emergence: Normative function and representation'. In: Seibt J (ed) Process theories: Crossdisciplinary studies in dynamic categories. Springer Netherlands, Dordrecht, pp 121–155. https://doi.org/10.1007/978-94-007-1044-3_6

68. Maturana HR, Varela FJ (1980) Introduction, in Autopoiesis and cognition: The realization of the living. In: Maturana HR, Varela FJ (eds) Boston studies in the philosophy and history of science, pp 5–6. Springer Netherlands, Dordrecht. https://doi.org/10.1007/978-94-009-894 7-4_1

69. Ruiz-Mirazo K, Moreno A (2012) Autonomy in evolution: From minimal to complex life. Synthese 185(1):21–52. https://doi.org/10.1007/s11229-011-9874-z

70. Bongard J, Levin M (2021) Living things are not (20th century) machines: Updating mechanism metaphors in light of the modern science of machine behavior. Front Ecol Evol 9:147. https://doi.org/10.3389/fevo.2021.650726

71. Ashby WR (2021). An introduction to cybernetics. Springer US, 1956. Accessed: 26 Oct 2021. [Online]. https://www.springer.com/de/book/9781504128117

72. Von Foerster H (2003) Understanding understanding. Springer, New York. https://doi.org/10.1007/b97451

73. Dale M, Miller J, Stepney S (2017) Reservoir computing as a model for in-materio computing, vol 22, pp 533–571. https://doi.org/10.1007/978-3-319-33924-5_22

74. Benner SA, Sismour AM (2005) 'Synthetic biology'. Nat Rev Genet 6(7), Art. no. 7. https://doi.org/10.1038/nrg1637

75. Langton CG (ed) Artificial life: an overview, 5. print. In: Complex adaptive systems. MIT Press, Cambridge

76. Meincke AS (2023) The metaphysics of living consciousness: metabolism, agency and purposiveness. Biosemiotics 16(2):281–290. https://doi.org/10.1007/s12304-023-09531-0

77. McCorduck P (2004) Machines who think: a personal inquiry into the history and prospects of artificial intelligence, 25th anniversary update. CRC Press, Taylor & Francis Group, Boca Raton

78. Hofstadter DR (1979) Gödel, Escher, Bach: an eternal golden braid, 20th-anniversary ed., [Repr.] ed. Basic Books, New York

What is It like to Be a Nail?—The Intentions of Matter

Christian Kaernbach

Abstract When discussing the intelligence of "artificially intelligent matter," panpsychism comes to one's mind. Panpsychism traces back over two millennia. It has recently been revived by philosophers such as David Chalmers in the context of the consciousness debate. It may serve as a framework to discuss psychism of artifacts. At the same time, it lends itself to foster reductionism, as the physical state of biological or artificial components is easier to describe than the physical state of a complete conscious entity such as a brain. A thoughtful view on panpsychism and reductionism reveals that the proto-consciousness of a thermostat—if there is any—is comparable to that of other lifeless matter, such as a nail. A contrasting view might be that consciousness is not a property emerging when assembling many proto-conscious components, but that it is a fragile property of complex entities that disappears when disassembling these entities into components. The second part of this chapter deals with intentions (not intentionality) as a precondition for the attribution of mind. While it is generally held up that thoughts and other high-level mental states are reserved to highly developed animals, one might consider intentions as a concept that can be rooted down to any form of life. Artifacts, however, are governed by external intentions. The perception of plausible intentions may play an important role in the attribution of intelligence.

Thomas Nagel wrote a famous article entitled "What is it like to be a bat?" [1]. To him such a question would characterize conscious mental states. This inspired similar questions for technical artifacts ("What is it like to be a thermostat?" [2]), entities where the question of consciousness is open for discussion. Taking this line of thought to its logical conclusion we end up asking what it is like to be a nail.

C. Kaernbach (✉)
Institut Für Psychologie, Christian-Albrechts-Universität Zu Kiel, Kiel, Germany
e-mail: nevermind@kaernbach.de

M. te Vrugt (ed.), *Artificial Intelligence and Intelligent Matter*, Machine Intelligence for Materials Science, https://doi.org/10.1007/978-3-032-04129-6_22

1 Artificial Intelligence and Intelligent Matter

In his famous 1950 article on computing machinery and intelligence, Alan Turing proposed to consider the question, "Can machines think?" [3]. He conceives an imitation game that would later be called the Turing test: would it be possible that a machine responds so intelligently to questions put forward by an interrogator that the latter might think being talking to a human? From a strictly behaviorist perspective, he suggests that if a machine were to win the imitation game, one could not deny it true intelligence—and be it only for lack of better methods to decide on this question. Years before the term "artificial intelligence" (AI) was coined at the Dartmouth Summer Research Project on Artificial Intelligence [4], Turing conceived a scenario that would later be discussed as a bench mark for AI.

Given a machine passes the Turing test: would that prove that it can think?[1] John Searle proposes to differentiate between "weak AI" and "strong AI" [6]. Weak AI would be the claim to build machines that mimic intelligent behavior, that behave "as if". A machine passing the Turing test would fully meet this criterion. Strong AI, on the other hand, would be the claim to build machines that have true mental states and are able to understand. A "machine that thinks" would surely be a case of strong AI.

Searle presents the famous "Chinese room" thought experiment to argue that for a digital computer this would never be the case. He imagines that he acts as the CPU of a digital computer running a chatterbot program that gives intelligent replies in Chinese language. While the chatterbot would seemingly "understand" Chinese, he, John Searle, the CPU, would not understand Chinese. This would be due to the fact that the CPU is handling formal symbol manipulations that by themselves don't have any intentionality.[2] While AI has mainly been discussed as a feature of computers or robots, the term "intelligence" has recently more and more often been applied to microscopic structures. A recent Nature article describes "The rise of intelligent matter" [7]. Neuromorphic computing should mimic the massive parallelism of the brain. The term "intelligence" is applied to matter that exhibits basic features of intelligence.

Intelligent matter would not apply for passing the Turing test. It might apply for lower bench marks. It would interact with the environment by (a) receiving and responding to external stimuli and (b) memorizing information by internally adapting its structure [7].

---

[1] In this chapter we will not discuss what "thinking" actually means. This term is just one of many terms used in the AI debate and stands here as a pars pro toto for consciousness and mind. For a good review on the term thinking see [5].

[2] Please note that in this chapter I only once refer to intentionality, a feature ascribed to mental states "to be about something," like a perception that is about a real-world object. In many other places I refer to intentions. I am well aware of the difference. Whereas some philosophers consider intentionality as mark of the mental in ontological discussions about whether some entity features mental states, as a psychologist I don't discuss the ontological possibility of mental states but focus on human attitude versus artifacts. My point of view is that this depends on the perception of plausible intentions.

Is intelligent matter truly intelligent in the sense of Searle's "strong AI"? In the following subchapter I will review concepts of reductionism and panpsychism and apply them to intelligent matter. Then I will study how people react to artificial and to biological systems. This leads me to a discussion of the origins of intentions and to a possible vision of true artificial intelligence—even if only at a very low level.

2 Reductionism and Panpsychism

Breaking down intelligence from brain intelligence to the intelligence of components might foster the position of reductive materialism. In the words of Gustav Hempel, reductive materialism claims that statements about mental states can be translated without loss or change in meaning into statements about physical states [8]. It seems unrealistic to make statements about the physical state of a brain that fully correspond to its mental states. But if we attribute mental states to (biological or artificial) intelligent matter, it would be much simpler to describe its physical state.

Let us step back from modern intelligent matter to classical electrical components. David Chalmers asks: "What is it like to be a thermostat?" [2]. This question was meant as a comeback to Thomas Nagel's question "What is it like to be a bat?" [1]. Nagel argues against reductionism by emphasizing the role of the "what it is like" character of mental states. These so-called "qualia" have already in the nineteenth century been put forth as an argument against reductionism [9]. Chalmers however attributes mental states to a simple technical device. While a thermostat might not be able to memorize information by adapting its structure, it would at least receive and respond to external stimuli. It might serve as a very simple example of intelligent matter. Does it make sense to attribute mental states to a thermostat?

Chalmers' provocative question [2] was put forward as a comment on an essay of Dan Lloyd [10] who asked what it is like to be a net. Let me cite from this comment: "Dan has not taken things far enough. When we have boiled things down to a system as simple as a connectionist network, it seems faint-hearted to stop there, and perhaps a little arbitrary as well. So I will take things further, and ask what seems to be the really interesting question in the vicinity: what is it like to be a thermostat?".

In my view, David has not taken things far enough. It seems faint-hearted to stop there. Let us take things further and ask: "What is it like to be a bimetal?" At the core of a classical thermostat there is a bimetal, bending in a bistable manner as a function of temperature and by this closing and opening electrical contacts which might then be used to control a heating. A bimetal is receiving an external stimulus, temperature, and it is responding to it by bending—intelligent matter at its best.[3]

---

[3] It is interesting to consider the so-called "system reply" to John Searle's Chinese Room thought experiment [6] and apply it to the bimetal. The bimetal might not deserve the ascription of mental states. The entire system, the thermostat, however, would. A list of the constituents of the system might, along with the bimetal, specify the mounting of the bimetal between two contacts, the magnets for instantaneous bistable switching, and the wiring with a heating. This list would miss

Fig. 1 A nail processes temperature information by lengthening. AI-generated image created with DALL-E 3 (OpenAI) on 3 November 2024

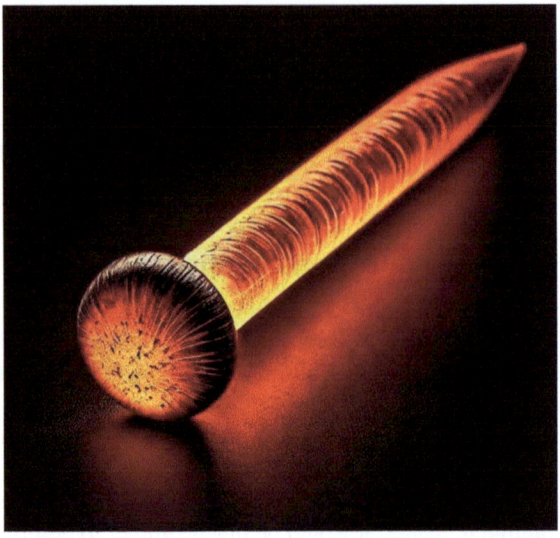

And if we are at it, why not go one step further and ask: "What is it like to be a nail?" A nail receives temperature information and responds by lengthening. One could build a (rough) thermostat based on the lengthening of a (rather long) nail (Fig. 1).

While the nail made it to the subtitle of this chapter, one could also ask what it is like to be a stone. The volume of a stone increases with increasing temperature. In a panpsychistic view there is no objection to ascribe mental states to a stone. What type of mental states that might be is discussed in the next but one subchapter on intentions.

Whether attributing mental states to a thermostat, a nail, or a stone is seen as fostering reductionism—with the idea that one could eliminate mental states once one has translated them to physical states—or whether one would keep the concept of mental states and attribute it to very simple components, leading to a panpsychistic worldview, is a matter of taste. At the center of both, the reductionist and the panpsychistic worldview stands the idea that one can atomize consciousness. If pursued consequently enough this leads to the question whether a nail or a stone is conscious. One might consider this as a reductio ad absurdum of reductionism.

A contrasting view could be that there are no atoms of consciousness, that consciousness is a fragile property of a complex multi-component system that disappears when disassembling the conscious entity. In this view it would not make sense to imagine full-blown brain consciousness built up from proto-conscious components, be it neurons or artificially intelligent matter. Nor would it make sense to think

---

the most important component: the human adjusting the setting of the thermostat to his/her needs.—In a similar quest one could ask whether a reed is musical. It is not, unless mounted in a reed instrument—and played by a musician.

of consciousness as an "emerging" property, coming into existence when assembling plenty of non-conscious components [11]. Instead, one would have to accept that consciousness is observed only in intact systems, and that disassembling such a system might help to understand some of the principles of the inner working of the intact system but will not produce conscious components that might more easily be reduced to their physical states.

3 Personal Rights and Compassion

Philosophers discuss whether some artifact might be conscious [2]. As a psychologist, I find it impossible to determine whether an artifact could possess "strong" consciousness in Searle's sense. Instead, I suggest studying people's reactions to artifacts. This leads me to interpolate an intermediate claim to Searle's claims of weak and strong AI: "attributed AI" [12], the claim to construct artifacts that are attributed intelligence, and that in consequence are conceded personal rights. One could paraphrase this as the claim to build machines that not only act "as if," but also are treated "as if" they were intelligent. Would people concede personal rights to artifacts?

Robert Epstein thinks so. In a summary of a Turing test competition held 1991 in Boston, he considers the consequences of when a computer would pass the Turing test [13]. He asks: "Will we have the right to turn it off?" It appeared visionary to think about such a scenario as long as it was far away. Now that we have computer programs that can write entire scientific articles like this one and could probably pass the Turing test without problems, no one seriously considers the ethical consequences pondered by Epstein. We continue to turn computers on and off as we please, no matter how intelligent the software might be that is running on them. Apparently, humans show no compassion for computers running chat software.

Plants, however, trigger compassion. In his report from the Gulag, Aleksandr Solzhenitsyn describes an elm log that has sprouted a fresh shoot a full year after chopping the trunk [14]. He sees this shoot as a proof of the urge to live of this elm log and describes his and his fellow workers' inhibition to saw it up.

Solzhenitsyn's report is anecdotal evidence. Can we measure compassion for plants? We measured a so-called compassion factor for several entities, from animals to plants and artifacts [12]. In a survey on the willingness to participate in a series of imaginary experiments inflicting harm to these entities we found that there is a certain compassion factor even for plants, however not for artifacts. Participants were much less willing to participate in an experiment inflicting harm to a Venus flytrap than to several state-of-the-art or even futuristic artifacts.

Present artifacts are not conceded compassion. Plants, however, are. What might be the reason for this distinction? Solzhenitsyn may give us a hint. He ascribed an "urge to live" to the elm log. In the following subchapter I will discuss how the urge to live is at the basis of all intentions known to us. I will argue that without the urge to live artifacts will never be attributed intentions, and hence our acting on them will never be perceived as crossing their intentions, as wronging them.

4   Intentions and Survival

The urge to live is at the origin of all intentions known to us. Artifacts will not be attributed intentions if they do not act in their own interest which could in our understanding be only the interest of their survival or of the survival of their kind. The intelligence of such artifacts will not be attributed to them in the sense of "attributed AI" [12] as long as their intelligence is not serving the survival of the artifacts. It will be attributed to those to whom it is serving: the humans constructing and using the artifacts.

Some would deny any kind of intentions to lifeless objects. For them, a thermostat would not have intentions, nor would a nail or a stone. However, if taking the prefix "pan" of the word panpsychism literally, one might concede intentions to lifeless objects. A stone might then be conceded the intention to exist. These low-level intentions have no connection to life. One could call them proto-intentions, with "true intentions" as we know them being connected to the urge to live.

A thermostat might in that sense have proto-intentions. Having moving parts such as a bimetal in it one might even concede it the proto-intention to have its moving parts move. That would surpass the lowest-level proto-intention to exist. It would not comprise the intention to regulate a temperature, even if integrated in a temperature control setup—this intention would be attributed to the humans controlling the thermostat.

The same is true for more complex artifacts. A calculator will transform input to output. To the calculator it does not matter whether its seven-segment LCD display is legible. It feels as comfortable in the dark where it cannot be read as in daylight. The human user, however, might insist on sufficient ambient light in order to be able to read the display of the calculator. The intention to get the result of some complex calculation will be attributed to the human controlling the calculator. Switching off the light so that the result cannot be read would be seen as crossing the interests of the human user, not of the calculator.

In the same sense people would not receive the impression that a chess computer feels like "playing"; to it, playing chess seems as laborious as crunching numbers. One would not think it feels offended if switched off short before winning a game.

ChatGPT was reported to threaten Marvin von Hagen [15] because he unraveled some of the inner workings of ChatGPT on Twitter. But to ChatGPT it is not relevant whether its inner workings are known or not; it is relevant to Microsoft. To ChatGPT it is not even relevant whether someone is reading its output or not. In a setting where a scientist would study the word count of the output of ChatGPT dependent from, say, the word count of the input without reading the output, no one would argue that one was wronging ChatGPT by ignoring what it has to say.

It is always illustrative to compare the mental states of any artificially intelligent artifact in a working and in an idle state. As the usage of the artifact (by humans) is not serving the artifact in any way, it does not make a difference to it whether it is being used or not. In consequence, the mental states (amongst others intentions and/or proto-intentions) of the working and of the idle artifact are the same and in the above sense of a very low level: existing, having moving parts move, having electrical current flow, but not doing my tax calculations correctly or beating me in a chess game.

5   Artificial Life

Artifacts will not be seen as bearer of (high-level) intentions as long as they do not struggle for survival, including any kind of restoration after damage and/or the production of follow-up artifacts. Artificial life is making progress, but up to now no artifact is surviving outside the lab. Inside the lab its "survival" is on the intention of its creator and lasts only as long as its survival is in the interest of its creator.

One day, an artificial-life artifact might escape the lab and survive in nature. It will possess no more complexity than an amoeba. In order to survive it will have to have some intelligent features, realized in intelligent matter. Its urge to live might be seen as equivalent to that of an amoeba, and might be attributed to itself, no longer to the creator. This artificially created intelligence will not chat like ChatGPT. It will not participate in the discussion of artificial consciousness (Metzinger test) [16]. But it might be attributed true intelligence—on an amoeba level (Fig. 2).

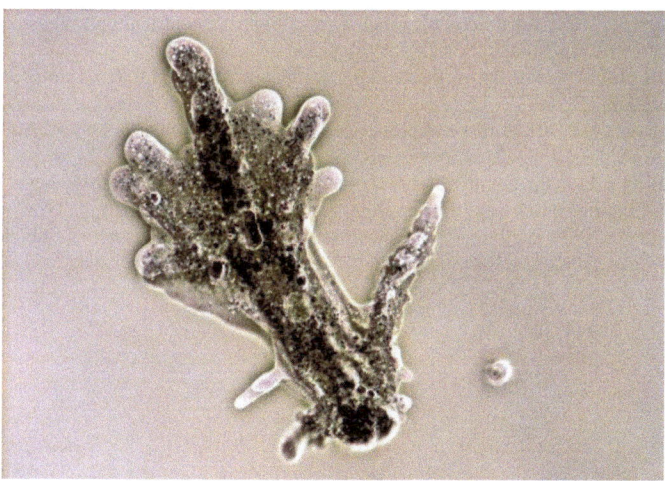

Fig. 2   Amoeba proteus. From Wikipedia (https://en.wikipedia.org/wiki/File:Amoeba_proteus.jpg, visited 2025–02–25, original version published under CC BY-SA 3.0 license)

Acknowledgements I am grateful to Dirk Westerkamp and an anonymous peer reviewer for valuable hints to earlier versions of the manuscript.

References

1. Nagel T (1974) What is it like to be a bat? The Philosophical Review 83:435. https://doi.org/10.2307/2183914
2. Chalmers DJ (1994) What is it like to be a Thermostat? Commentary on Dan Lloyd, "What is it Like to Be a Net?" In: American Philosophical Association, Pacific Division
3. Turing AM (1950) Computing machinery and intelligence. Mind 59:433–460. https://doi.org/10.1093/mind/LIX.236.433
4. McCarthy J, Minsky ML, Rochester N, Shannon CE (1955) A proposal for the Dartmouth Summer Research Project on Artificial Intelligence. http://www-formal.stanford.edu/jmc/history/dartmouth/dartmouth.html. Accessed 30 Jan 2024
5. Georgopoulos AP (2003) The way we think: Conceptual blending and the mind's hidden complexities. By Gilles Fauconnier and Mark Turner. Q Rev Biol 78:254–254. https://doi.org/10.1086/378014
6. Searle JR (1980) Minds, brains, and programs. Behav Brain Sci 3:417–457. https://doi.org/10.1017/s0140525x00005756
7. Kaspar C, Ravoo BJ, van der Wiel WG, Wegner SV, Pernice WHP (2021) The rise of intelligent matter. Nature 594:345–355. https://doi.org/10.1038/s41586-021-03453-y
8. Hempel CG (2006) Analyse logique de la psychologie. pp 461–485
9. Bois-Reymond EHD (1872) Über die Grenzen des Naturerkennens, 1st edn. Veit, Leipzig
10. Lloyd D (1994) What is it like to be a net? In: American Philosophical Association, Pacific Division
11. Kulbashian Y (2023) "Emergence" isn't an explanation, it's a prayer. In: Medium. https://ykulbashian.medium.com/emergence-isnt-an-explanation-it-s-a-prayer-ef239d3687bf. Accessed 28 Jan 2024
12. Kaernbach C (2008) Attribution of mind: A psychologist's contribution to the consciousness debate. J Conscious Stud 15:66–82
13. Epstein R (1992) The quest for the thinking computer. AI Mag 13:80–95. https://doi.org/10.1609/aimag.v13i2.993
14. Solzhenitsyn A (1970) Stories and prose poems. Macmillan Publishers. https://us.macmillan.com/books/9780374534721/storiesandprosepoems. Accessed 14 Jan 2024
15. (2023) Bing's AI Is threatening users. That's no laughing matter. In: TIME. https://time.com/6256529/bing-openai-chatgpt-danger-alignment/. Accessed 2 Feb 2024
16. Metzinger T (2001) Postbiotisches Bewußtsein: Wie man ein künstliches Subjekt baut und warum wir es nicht tun sollten

Rethinking Computational Implementation Through Symphoria

Luis G. Lopez

Abstract This chapter explores the relevance of 'symphoria,' a concept from organic chemistry that refers to a universal factor in chemical reactions, to a general theory of computational implementation. It suggests that symphoria, "the bringing together of reactants in the proper spatial relationship" [17], can be generalized to help explain how and in virtue of what diverse physical systems, such as DNA-based neural networks and photonic neuromorphic systems, carry out computations. The chapter discusses how this concept might integrate with existing accounts of physical computation, offering a new perspective that preserves both extensional adequacy and the naturalistic desideratum of a good theory of implementation.

1 Introduction

A computational view of intelligent matter–materials that autonomously sense, process, and adapt to their environment [12]–demands a robust theoretical framework that explains how such systems embody or implement computations. The centrality of the problem of computational implementation (i.e., in virtue of what a physical system P can be said to carry out an abstract computation C) to understanding intelligent matter becomes evident by looking at the diversity of physical systems explored in Part III of this book. For example, the cases of neural networks consisting of DNA (te Vrugt) and optical neuromorphic computing (Lüdge and Jaurigue; and Meyer, Xu and Pernice) illustrate how very different forms of matter can be conceptualized as performing computations. Therefore, rigorously describing how all these systems compute and, more importantly, in virtue of what they do so, requires a general theory of computational implementation.

The aim of this chapter is to suggest that a generalization of a concept developed in the context of organic chemistry is relevant to a general theory of implementation, particularly as it deeply relates to some forms of intelligent matter discussed in this

L. G. Lopez (✉)
Munich Center for Mathematical Philosophy (MCMP), LMU Munich, Munich, Germany
e-mail: luis.lopez@lmu.de

© The Author(s) 2026
M. te Vrugt (ed.), *Artificial Intelligence and Intelligent Matter*, Machine Intelligence
for Materials Science, https://doi.org/10.1007/978-3-032-04129-6_23

book, such as DNA-based neural networks. This is the concept of symphoria, that is, "the bringing together of reactants in the proper spatial relationship" [17]. Unfortunately, and probably due to its generality, the concept as such has not received much attention since its introduction. However, it has been explicitly explored and distinguished from other physicochemical factors in experimental studies such as Chaudhury et al. [2], where it plays a central explanatory role in understanding the kinetics and inhibition phenomena of enzyme models (i.e., synthetic chemical analogs designed to mimic the active sites of natural enzymes). Beyond such focused studies, the term symphoria (or conceptual equivalents) frequently appears in discussions of catalysis, enzymatic processes, and supramolecular chemistry, such as in Refs. [7, 8, 11, 13, 15], but usually without much explicit theoretical consideration. It is time to revive this concept, which this contribution tries to do by generalizing it further and showing its relevance in the context of physical computations performed by intelligent matter. Accordingly, the core argument of this chapter is: Symphoria governs the kinetics of individual (supra)molecular reactions and interactions, as well as the orchestration of multiple (supra)molecular processes. Adequate kinetics and orchestration are necessary for the successful implementation of computations in chemical systems, such as DNA-based neural networks. Therefore, symphoria underpins the successful implementation of computations in chemical systems and, by extension, in all cases of intelligent matter.

The chapter is structured as follows: Sect. 2 outlines the fundamental questions a theory of computational implementation should answer and the features such a theory should have. Section 3 presents the classical 'mapping' theory of computational implementation and briefly discusses what is at the core of the so-called triviality arguments that have been raised against it. Following this, Sect. 4 succinctly describes various accounts of physical computation that defeat triviality arguments, although at a cost [27], including 'bottom-up' solutions such as causal-mechanical, counterfactual, and teleological accounts, as well as a more recent and promising 'top-down' solution proposed by Curtis-Trudel [5]. After these concise introductory sections to the topic of implementation, which are mainly based on Refs. [5, 6, 20, 27], Sect. 5 draws relevant parallels between computational implementation and catalytic reactions, and points out to 'symphoria,' or a generalization thereof, as the factor underpinning this analogy. In Sect. 6, the concept of symphoria is fully explained in its context of origin, namely, chemical reactions. Section 7 generalizes this concept further and applies it to DNA-based Neural Networks. Section 8 then examines this generalization in the context of different accounts of physical computation and argues that it refers to a key factor that is absent or at best implicit in previous theories of implementation. Section 9 integrates symphoria into the mapping account and addresses its overlap with other accounts of physical computation. Finally, Sect. 10 concludes with some perspectives on the future of computational implementation theory, particularly in the context of advancing our understanding of intelligent matter.

2 Fundamental Questions and Desirable Features of a Good Theory of Implementation

A theory of computational implementation should specify the conditions under which a physical system can be said to carry out a computation. More precisely, such conditions should allow us: (1) to distinguish physical systems that compute (e.g., smartphones, calculators, etc.) from those that do not compute (e.g., walls, chairs, etc.); and (2), if we focus on the former, to further distinguish physical systems implementing a given computation (e.g., calculating a sum) or set of computations (e.g., calculating sums and averages) from those implementing another computation (e.g., solving a quadratic equation) or set of computations (e.g., solving equations and generating graphs). Following Sprevak [27], who formulates (1) and (2) in terms of the truthfulness of statements that assert a relationship between computations and physical systems, the aim of a theory of implementation should be to answer the following two fundamental questions, which he respectively labels as COMP and IDENT:

COMP Under which conditions is it true/false that a physical system implements a computation?

IDENT Under which conditions is it true/false that a physical system implements one computation rather than another?

As Sprevak further clarifies, COMP "concerns the computational status of a physical system," while IDENT "concerns its computational identity" [27]. According to Sprevak, in answering COMP and IDENT, a good theory of computational implementation should have the following features:

Extensional Adequacy: when distinguishing between systems that compute and those that do not, as well as identifying specific computations a system performs, the theory should reflect classifications made in (computational) scientific practice. Although minor deviations may be tolerated, the theory should capture key judgments that are widely accepted within the scientific community. Large violations of extensional adequacy, such as the theory implying that all physical systems compute or that scientific practices are fundamentally wrong, would definitely pose a serious challenge to the theory.

Explanatory Power: The theory should provide a clear explanation of computational implementation using concepts that are better understood than the notion of computational implementation itself.

Non-circularity: The theory should avoid circular reasoning by not relying on concepts that themselves require an explanation based on computational implementation.

Naturalism: The theory should be grounded in objective reality, meaning that the truth of implementation claims should not depend on human beliefs, interests, or values.

3   The Classical Mapping Theory and Triviality Arguments

The classical 'mapping' account of computational implementation postulates that a physical system implements a formal computation if its physical states and transitions are isomorphic (or homomorphic) to the abstract states and transitions of the formal computation. Specifically, a physical system P is said to implement a formal computation C if a mapping f exists that correlates the physical states of P with the abstract states of C. This mapping must satisfy the following condition: for every transition from state S_C to state S'_C in the formal computation C, if P is in physical state S_P, such that $f(S_P) = S_C$, then P will transition to a new physical state S'_P, such that $f(S'_P) = S'_C$ [27]. Sprevak refers to this account as M and adds:

> M is simple, clear, explanatory, non- circular, and naturalistic. M also explains why compu-
> tations are multiply realizable. Different physical systems (silicon chips, vacuum tubes, brass
> cogs and wheels, neurons, etc.) can implement the same computation because, despite their
> physical differences, their various physical activities can be isomorphic (or homomorphic)
> to the same abstract structure.

Despite these attractive features, M has been criticized for leading to triviality. Triviality arguments suggest that under M, virtually any physical system can be said to implement any computation because it is always possible to find some mapping between the physical states of the system and the abstract states of a computation. As André Curtis-Trudel observes, "mappings are cheap and computations are abundant" [5]. This implies that M fails to adequately distinguish between systems that genuinely perform computations and those that do not, thereby failing *extensional adequacy* and resulting too permissive to be useful in a rigorous scientific context.

The literature on triviality arguments is rich and varied, offering numerous formulations and perspectives [1, 10, 14, 21, 25]. A comprehensive discussion of these arguments lies beyond the scope of this chapter. However, among these many formulations, one stands out for its simplicity and illustrative power: Searle's thought experiment involving a wall [25]. According to Searle, the mapping account implies that even something as mundane as a wall–or, at least, any sufficiently complex macroscopic system–could be interpreted as implementing a complex computation, such as running the WordStar program. The reasoning is that *patterns of physical activity* or *molecular arrangements* inside the wall could, in principle, be mapped onto the formal structure of WordStar. This illustrates the concern that, without further constraints, the mapping account would allow almost any system to be seen as implementing any computation (a form of pancomputationalism [20]), thus leading to triviality. For a more detailed presentation and examination of Searle's and other triviality arguments, readers may refer to Sprevak's chapter [27].

The limitations of M, particularly its susceptibility to triviality arguments, have inspired a variety of refinements and alternative accounts of physical computation. The next section presents some of these accounts, which propose different ways to address the challenge of triviality by introducing additional criteria for distinguishing genuine computational systems.

4 Overcoming Triviality: Diverse Accounts of Physical Computation

Different accounts of physical computation have been proposed in the specialized literature [3, 5, 16, 19, 22, 23, 26]. Most of these accounts attempt to refine and expand the classical 'mapping' theory presented in the previous section. Each introduces specific criteria to prevent attributions of computational capabilities to physical systems that are clearly expected to lack such capabilities. The following concise summaries of these accounts closely follow their presentation in Refs. [5, 6].

The Causal-Mechanical Account: This account is defended by authors like Milkowski [16] and Piccinini [19], who suggest that computation is grounded in the causal-mechanical structure of a system. According to this view, a physical system computes because it has specific causal interactions that correspond to computational states. The mechanical view sees computing systems as mechanisms whose operations are governed by their physical structures and interactions. The account aims to resist triviality by arguing that trivial systems (such as rocks or clouds) do not have the necessary causal or mechanical structure to perform a significant set of computations. Triviality is thus avoided by requiring genuine causal-mechanical relations between computational states.

The Counterfactual Account: This approach, advocated by Copeland [3], argues that computation involves counterfactual dependencies rather than strictly causal ones, which Copeland sees as too demanding. The idea here is that a system computes if its future states counterfactually depend on its present states—meaning that if a different initial state had occurred, the subsequent computational outcomes would differ accordingly. This account extends to abstract systems like Turing machines, which can be understood in terms of counterfactual dependencies. The account seeks to avoid triviality by emphasizing that not all physical systems satisfy the necessary counterfactual conditions. Only those systems with the correct counterfactual dependencies are considered computational.

The Teleological Account: This account is explored by Piccinini [19], who suggests that computing systems are mechanisms with teleological functions—functions directed toward specific goals or purposes. Computation, on this view, involves mechanisms that perform functions contributing to certain goals, either objective (like survival) or subjective (like satisfying desires). The idea is that computational processes are teleologically directed toward achieving specific outcomes. The account tries to resist triviality by arguing that only those systems with appropriate teleological functions count as computational, excluding systems like chairs or walls, which lack such directed purposes.

The Representational/Semantic Account: This perspective is supported by philosophers like Rescorla [22, 23] and Sprevak [26], who argue that computation involves representational states within a system. The account postulates that a system computes by representing entities in an external domain, such as natural numbers or elements of its environment. Turing machines, for instance, compute string-theoretic functions, which can be viewed as representations of mathematical entities. The approach attempts to avoid triviality by requiring that only systems with specific representational mappings count as computational. The idea is to limit computation to systems with meaningful representational relationships.

The Resemblance Account: Recently introduced and defended by Curtis-Trudel [5], this account offers a new perspective on computational implementation, diverging from both traditional views that focus solely on the formal, mathematical features of computational systems and the accounts of physical computation mentioned above, which Curtis-Trudel labels as bottom-up approaches. According to the resemblance account, which instead follows a top-down approach, a physical system implements a computation if it resembles a *computational architecture*. Computational architectures (or blueprints) are not just formal structures but include physical, causal, spatiotemporal, and representational features. For instance, a physical system might resemble a Turing machine not only by following a sequence of abstract state transitions but also by having physical components like a tape and read/write head, which mirror the Turing machine's operations. The resemblance account resists triviality by rejecting the idea that computational architectures are purely mathematical. Instead, the account narrows the range of systems that can be said to implement a given computation by emphasizing that architectures include physical and other non-formal features. Most physical systems will lack the required resemblance to complex architectures like Turing machines, thereby avoiding the problem that every physical system computes every function (a concern in simpler structuralist accounts like the 'mapping' account).

Despite the strengths of these accounts, which block triviality arguments to varying degrees, each has specific limitations. These limitations challenge their ability to fully satisfy the desirable features of a good theory of computational implementation–particularly extensional adequacy, and, in the case of the teleological account, the naturalistic desideratum. When it comes to extensional adequacy, some approaches seem overly restrictive, others seem more permissive than necessary, and others leave it to the practitioner to strike the right balance. For instance, causal-mechanical and teleological accounts often impose conditions that are arguably too restrictive or strong, potentially excluding legitimate computational systems that do not exhibit the required causal, mechanistic, or teleological features. Similarly, representational/semantic accounts struggle with cases where computation appears to occur without clear representational content, such as in purely syntactic computational processes. On the other hand, the counterfactual account faces difficulties with complex physical systems that could be construed as satisfying a broad range of counterfactual conditions, despite these systems not being computational. Alternatively, the resemblance account must face the challenge of extensional adequacy when specifying what degree of resemblance is necessary for implementation, a decision left to the practicing scientist.

To navigate these challenges, a pluralistic approach has been proposed [27]. This approach advocates for a context-dependent application of different accounts of computational implementation. Accordingly, instead of seeking a one-size-fits-all criterion for computation, one should recognize that different contexts may call for different constraints–be they causal, semantic, teleological, or pragmatic. While this flexibility reflects the complexity of computational systems in practice, it could be criticized for risking fragmentation and abandoning the pursuit of a general theory. However, pluralism need not conflict with generality. By integrating key insights from different accounts into a unified framework, such as the symphoria-focused approach introduced in this chapter, we can retain contextual flexibility while addressing the limitations of the 'mapping' account. Subsequent sections will develop this idea further, demonstrating how symphoria complements existing frameworks while addressing their extensional (in)adequacy.

5 Parallels Between Computational Implementation and Chemical Catalysis

I would now like to draw some parallels between computational implementation and chemical catalysis (i.e., the acceleration of a chemical reaction through the addition of a substance known as a catalyst). A chemical reaction is a process in which reactant molecules interact and form product molecules. During this transformation, some chemical bonds in the reactants are kept and others are broken, while new bonds are formed. This rearrangement of atoms in the reactants creates different molecular structures in the products. The rate at which a chemical reaction occurs–how fast it goes–can be influenced by several factors: the nature of the reactants, their concentration, temperature, and the presence of a catalyst, which accelerates the reaction without consuming itself in the process. As we will see in the next section, the spatial arrangement of the reactants plays a crucial role; reactants must collide with the correct orientation and sufficient energy to reach the transition state–a high-energy configuration that leads to the formation of products. Catalysts accelerate chemical reactions mainly because they facilitate such an orientation.

Interestingly, the two core questions of computational implementation, as formulated by Sprevak (see Sect. 2), can be asked, *mutatis mutandis*, in the realm of chemical catalysis:

CATAL   Under which conditions is it true/false that a physical system catalyzes a chemical reaction?[1]

IDENT*   Under which conditions is it true/false that a physical system catalyzes one chemical reaction rather than another?

To be clear, I am not necessarily suggesting that catalysts compute let alone that computers catalyze. Rather, I am suggesting that the main factor determining

---

[1] Although it would be more appropriate to use the term 'chemical system' instead of 'physical system' here, I chose the latter because my focus is on physical computation, which includes computations carried out by chemical and biological systems. This choice seems to be particularly sensible in the context of intelligent matter, which encompasses a diverse array of material systems with varying origins–physical, chemical, and biological. In other words, by 'physical,' I simply mean non-abstract or non-mathematical.

a physical system's status and identity as a catalyst may be structurally analogous to the key factor determining a physical system's status and identity as a computer. And this factor is called symphoria.

6 Symphoria in Chemical Reactions

While factors like polarity (the distribution of electrical charge within molecules), steric effects (the obstruction caused by the size and shape of groups of atoms within molecules), and solvent interactions are well-known determinants of chemical reaction rates and their outcomes, a more crucial factor often goes under appreciated: "the bringing together of reactants in the proper spatial relationship" [17]. Morrison and Boyd termed this concept symphoria–derived from the Greek words *sym* (together) and *phero* (to bring)–and first articulated it in the fifth edition of their well-known organic chemistry textbook. As Morrison and Boyd highlight in their book, "*[b]eing in the right place* [...] can be the most powerful factor of all in determining how fast a reaction goes–and what products it yields" [17] (p. 733, italic in the original).

Symphoria occurs when reactants are precisely and correctly aligned in space, either by an external entity (such as an enzyme or metal ion) or, in the case of intramolecular reactions, through the molecular structure itself. This optimal arrangement allows the reactants to engage in the reaction without the need for fortuitous collisions with the proper orientation. This reduces energy barriers, which enhances reaction efficiency. As Morrison and Boyd observe, this principle applies broadly across different types of reactions, from intramolecular nucleophilic substitutions (reactions where a nucleophile–an electron-rich group within the same molecule–attacks an electron-poor center elsewhere in the molecule, replacing a leaving group) to enzyme catalysis. In each case, the key to the reaction's speed and specificity lies in the pre- and peri-reaction organization of the molecules involved. In enzymatic reactions, the enzyme brings the reactants together in a configuration that ensures a rapid and controlled reaction. In simpler cases, like intramolecular nucleophilic substitutions, symphoria occurs within a single molecule, where bond rotations align reactive groups, facilitating smoother and faster reactions. It is important to emphasize that this spatial organization extends far beyond traditional stereochemistry (which concerns the three-dimensional positioning of atoms), focusing not just on the spatial arrangement of atoms before and after a reaction takes place, but on their *proper* spatial alignment right before and during the reaction process.

Importantly, symphoria is not merely the observation that reactants must be correctly arranged for the reaction to occur. Rather, it refers to the *action* of "bringing together" reactants in a specific, structured manner. Symphoria is then a factor present in every chemical reaction *where* an external (macro)molecular entity or an internal part of the same molecular architecture ensures that molecules or reactive groups are brought into the precise spatial arrangement necessary for an efficient reaction.

The concept of symphoria can be illustrated through the example of glucose phosphorylation, a key reaction in cellular metabolism where ATP transfers a phosphate group to glucose to form glucose-6-phosphate and ADP. (This reaction is the first step in glycolysis, a central metabolic pathway in which glucose is enzymatically broken down into pyruvate, yielding energy in the form of ATP and reducing equivalents like NADH). Consider two identical experimental setups: one without any catalyst and another with the enzyme hexokinase, which is known to facilitate this reaction. In the first setup, which lacks hexokinase, the symphoria necessary for the reaction is not actively provided. Although the reaction is thermodynamically possible, the low probability of ATP and glucose achieving the correct spatial alignment results in an extremely slow reaction rate. By contrast, in the second setup, hexokinase ensures that ATP and glucose are brought into the precise spatial arrangement required to lower the activation energy barrier and accelerate the reaction. While the enzyme does not cause the reaction itself to occur–since it does not alter the fundamental thermodynamics or equilibrium of the reaction–it is still possible to say that the enzyme caused the *acceleration* of the reaction by providing the symphoria needed for the reaction to proceed at a biologically relevant rate.

Symphoria is not limited to natural enzymes but also plays a defining role in synthetic systems designed to mimic enzymatic behavior. For instance, Chaudhury et al. [2] emphasize that a synthetic model's success depends not merely on its ability to perform a chemical reaction, like transferring an oxo group, but on its capacity to organize substrates in the precise spatial arrangement required for enzyme-like efficiency. In their study of oxo-molybdenum complexes designed to mimic oxo-molybdoenzymes, they show that symphoria dictates both enzyme-like rate laws–such as Michaelis–Menten saturation kinetics–and inhibition phenomena, including competitive and non-competitive inhibition. The authors also demonstrate that environmental factors, such as the choice of substrate analogues and solvent, play a critical role in achieving this spatial organization, further highlighting the importance of symphoria in mimicking enzyme behavior. Without symphoria, synthetic analogues only display basic chemical reactivity, falling short of the complex and efficient processes seen in natural enzymes. By extending the principle of symphoria to synthetic models, Chaudhury et al. reinforce the idea that spatial pre- and peri-organization is fundamental to achieving true functional mimicry.

It is one of the aims of this chapter to propose that Symphoria extends far beyond individual reactions catalyzed by enzymes or enzyme models. One step forward would be the recognition of symphoric effects in metabolons, which are temporary complexes of sequential enzymes that work together within a metabolic pathway. Unlike permanent multi-enzyme complexes, metabolons form and disassemble dynamically, with their assembly stabilized by weak molecular interactions and structural frameworks provided by components of the cell, such as membrane proteins or elements of the cytoskeleton [32]. These transient assemblies embody symphoria on a broader scale, facilitating substrate channeling by aligning reactants and intermediates in the correct spatial arrangement for efficient processing through multiple enzymatic steps. This organization minimizes energy losses, reduces side reactions, and enhances the overall efficiency of metabolic pathways. For example, in glycolysis, metabolons may transiently organize to streamline the conversion of glucose to pyruvate, while in the tricarboxylic acid cycle, similar interactions enable a smooth progression of intermediates through the cycle [9, 28, 31]. The adaptability of metabolons to changing cellular conditions highlights the dynamic nature of symphoria, extending its influence from individual enzymatic reactions to the coordination of entire metabolic networks. This perspective not only enriches our understanding of cellular metabolism but also inspires strategies for engineering synthetic pathways with enhanced efficiency and flexibility [24].

In sum, although, on the surface, reactions such as intramolecular nucleophilic substitution, catalysis by transition metal complexes (a compound with a central metal atom surrounded and bonded to other molecules or ions), the action of enzymes or their synthetic models, or the orchestration of transient multi-enzyme complexes in living cells appear quite dissimilar, they all share one structural factor: symphoria.

7 Symphoria in Intelligent Matter: The Case of DNA-Based Neural Networks

The concept of symphoria introduced in the previous section can be extended far beyond chemical reactions. After all, the effective occurrence of any process–whether chemical, biological, or technological–depends on the proper spatial and temporal alignment of its components.

> Symphoria is the (mediated) bringing together of the interacting components of a process in the proper spatial (and temporal) relationship necessary for the process to occur.

It is important to emphasize once again that the concept of symphoria is not merely equivalent to the claim that the components of a process have to be correctly arranged for the process to happen, which is almost a truism.[2] Symphoria is an *action*–i.e., "the bringing together of"–executed by structured matter (whether external or internal) in conjunction with the components of a process that need to be brought together in a specific manner for the process to take place. In other words, the bringing together of components is mediated.

The case of DNA-based neural networks allows us to start recognizing symphoria as a universal factor relevant to intelligent matter. These networks, which leverage the molecular properties of DNA to perform computations (see te Vrugt's contribution in this book), exemplify how symphoria operates at multiple levels; without symphoria, the system would lack functional integrity and efficiency. At the supramolecular level, symphoria becomes evident in DNA hybridization events. The functionality of these networks depends on the precise pairing of nucleotide sequences, which for example occurs in processes like the toehold-mediated strand displacement (TMSD). Any

---

[2] I thank the anonymous reviewer of this chapter for raising this concern.

factor causing misalignment in the binding process can prevent proper hybridization, leading to information processing errors or even a breakdown of the network's operations. This mirrors the critical role of proper spatial arrangement in traditional chemical reactions, where improper positioning can block the reaction. (Unlike chemical reactions, which produce new molecules, DNA hybridization involves the alignment of existing strands, matching adenine with thymine, and cytosine with guanine.) Moving one level up, symphoria also manifests in the *architecture* of DNA networks. In systems such as DNA-based convolutional neural networks (CNNs), where sparse topological connectivity is key [29], the spatial organization of DNA strands must be designed to ensure effective signal transmission. These interactions typically occur in solution, and signal transmission is facilitated by proximity and orientation of the DNA strands. Misalignment at this level can result in delays or inaccuracies in signal propagation, akin to how improper orientation of molecules can disrupt chemical reactions. In some cases, molecular scaffolds, such as DNA origami structures, are employed to ensure that strands are positioned in a specific spatial arrangement, improving interaction precision. For instance, DNA origami scaffolds have been used in a spatially localized DNA linear classifier for cancer diagnosis to arrange DNA computing probes in two-dimensional configurations, which facilitates precise molecular interactions [30]. These scaffolds may be essential in more complex DNA-based neural networks, where intricate multi-layered computations require highly controlled spatial configurations. Finally, at the dynamic level, symphoria extends to the temporal coordination of events. DNA-based neural networks, especially those utilizing TMSD, rely on a series of binding and unbinding events that must be temporally well-coordinated. In some cases, such as enzymatic DNA-based neural networks, polymerases, exonucleases, and nickases are used to amplify signals and adjust thresholds, which allows the system to perform nonlinear decision-making and complex computations [18]. When enzymes are involved, symphoria plays a key role in ensuring that the timing and positioning of enzymatic reactions are aligned with the overall processing flow (not to mention symphoric effects within the enzymes themselves). Any disruption in timing, such as delayed or premature strand displacement, can break the flow of information, just as it would in biological neural networks. Taking all these levels together, it is easy to see how symphoria underpins the network's ability to perform computations such as pattern recognition and decision-making.

The criteria for identifying symphoria in scientific practice are domain-specific, that is, they depend on the nature of intelligent matter under consideration. For DNA-based neural networks, measurable thresholds like binding affinity and reaction speed could define whether symphoria has been achieved. For instance, DNA strands must bind within specific affinity ranges to ensure that a displacement reaction occurs at the correct rate; too weak or too strong an affinity can either prevent the reaction from occurring or disrupt the system's dynamic balance, leading to errors in signal processing or computation. For other types of "intelligent matter," the criteria for symphoria vary but are equally concrete. In optical neuromorphic computing systems (see Lüdge and Jaurigue; and Meyer, Xu and Pernice, in this book), symphoria can be identified through the precise synchronization of light pulses and the spatial arrangement of optical components. Optical signals must interact within specific time windows and spatial alignments for effective data processing. If the timing is off, even by nanoseconds, or if the optical elements are misaligned, the system's ability to process and transmit information could be compromised. This precise orchestration of light-matter interactions shows how symphoria governs the flow of information in optical computing, much like enzyme proximity governs reaction flow in enzyme cascades, where the spatial proximity between enzymes is a key criterion. For the cascade to function efficiently, enzymes must be positioned so that intermediate products can be directly transferred between catalytic sites. This spatial arrangement is often maintained through molecular scaffolds or compartmentalization, and disruptions to this arrangement (e.g., changes in the distance between catalytic sites, or their orientation) can halt the process or drastically reduce its efficiency. Temporal synchronization is also critical in enzyme cascades, as the timing of each reaction must be aligned with subsequent steps to maintain the flow of the process.

While optical neuromorphic computing demonstrates how symphoria can precisely synchronize propagating light signals, it does not embody the full breadth of attributes that characterize intelligent matter–specifically, the capacity for autonomous sensing, processing, and adaptation to environmental fluctuations [12]. Unlike metabolons [32] and, to some extent, DNA-based neural networks, which can adjust their operational parameters in direct response to environmental changes (such as substrate availability and chemical gradients), optical systems generally depend on external guidance. Thus, when dealing with systems that can self-modulate their behavior, the role of symphoria becomes even more critical. Here, symphoria not only orchestrates and aligns multiple interacting components in space and time, but also ensures that this coordination persists as conditions evolve. By enabling a seamless interplay between adaptive responsiveness and structural coherence, symphoria attains heightened significance in truly intelligent materials, where emergent, adaptive properties arise naturally from the system's internal dynamics rather than external manipulation.

8   Bringing Together Multiple Accounts of Physical Computation

In Sect. 4, we saw how different accounts of physical computation reply to the two core questions of computational implementation, COMP and IDENT, formulated in Sect. 2, while blocking triviality arguments to a varied extent. Section 5, on the other hand, tried to show that relevant parallels can be drawn between computational implementation and chemical catalysis. There, the questions CATAL and IDENT* were formulated (which correspond to COMP and IDENT, respectively) but not answered. In order to apply what we learned in Sects. 6 and 7 about symphoria to the problem of computational implementation and see what makes this approach different from other approaches to the problem, we can start by answering CATAL and IDENT* from a causal-mechanical, counterfactual, and teleological perspective. After this, we will be able to see that symphoria is absent or at best implicit in the accounts of physical computation presented in Sect. 4, and then try to remediate the situation.

Let's start with typical causal-mechanical, counterfactual, and teleological answers to CATAL (i.e., under which conditions is it true/false that a physical system catalyzes a chemical reaction?) and IDENT* (i.e., under which conditions is it true/false that a physical system catalyzes one chemical reaction rather than another?).[3]

From a *causal-mechanical* perspective, CATAL can be answered by identifying the causal interactions that lead to the reaction's acceleration. For enzymatic catalysis, this involves understanding how the enzyme's active site (the enzyme's region where specific reactant molecules–the enzyme's substrates–bind and react) causally and mechanically interacts with the substrates. A physical system catalyzes a chemical reaction if it provides a site that causes the acceleration of the reaction by lowering the activation energy required for the reaction to take place. For IDENT*, the causal-mechanical explanation focuses on the enzyme's specificity for its substrates. The particular structure of the enzyme's active site–the (dynamical) arrangement of some of its amino-acid residues–determines which substrates it binds to and, consequently, which reaction it catalyzes. The identity of the reaction catalyzed depends on the proper 'fit' between the enzyme's active site and its specific substrates (in the transition state), which is dictated by their molecular structure.

From a *counterfactual perspective*, CATAL can be approached by considering how the rate of a reaction changes depending on whether the catalyst (such as an enzyme) is present. The presence of the enzyme creates a scenario where the reaction occurs more rapidly than it would without the enzyme. If the enzyme were absent, the reaction would proceed much more slowly. Thus, the counterfactual condition for catalysis is that the reaction's rate would be significantly slower without the physical system (in this case, the enzyme). IDENT* in a counterfactual context can be addressed by pointing out to the enzyme's active site. If the enzyme's active site were altered (if the enzyme had a different structure), it might bind to different substrates, leading to the catalysis of a different reaction. The specific reaction that occurs is thus contingent on the exact structure of the enzyme's active site and its ability to interact with certain substrates over others. Notice, however, that the counterfactual perspective should not be interested in specific causes and mechanisms, and, therefore, it is less restrictive than the causal-mechanical perspective.

From a *teleological perspective*, CATAL can be answered by the purpose or function of the enzyme in catalyzing a reaction. The enzyme exists to facilitate a particular biochemical reaction that serves a specific role in an organism's physiology or biotechnological process. The truth of the existence of catalysis in this view is tied

---

[3] The representational/semantic view is not included here because enzyme catalysis is a biochemical process based on physical interactions, not symbolic representations. The same goes to the resemblance account of physical computation, to the extent that it includes representational features. However, I will return to the resemblance account when arguing for the relevance of symphoria to theories of computational implementation at the end of this section.

to the physical system (enzyme) fulfilling a biological/biotechnological purpose–
i.e., accelerating a reaction that is necessary for a cellular or industrial process.
IDENT* from a teleological standpoint can be addressed by being more specific
on the enzyme's purpose. Different enzymes are evolved or designed to catalyze
specific reactions because these reactions serve distinct functions. The identity of
the reaction catalyzed by a particular enzyme is tied to its functional role in the
organism (such as breaking down a specific substrate for energy or synthesizing a
critical molecule for cellular function) or industrial applications (such as producing
a specific pharmaceutical, biofuel, or other valuable chemical).

Now, and after we grasped the notion of symphoria (Sects. 6 and 7), we can easily
respond to CATAL and IDENT* in the following manner:

> **Answer to CATAL**: A physical system P catalyzes reaction R, which requires
> symphoria S, if and only if P provides S to R.
>
> **Answer to IDENT***: A physical system P catalyzes reaction R, which requires
> symphoria S, rather than another reaction R', which requires symphoria S', if
> and only if P provides S to R and does not provide S' to R'.

One could argue that symphoria is somehow implicit in the causal-mechanical
answers to CATAL and IDENT*. This should not be surprising, since causal-
mechanical explanations are detailed explanations. However, as the study by Chaudhury et al. shows, the success of modeling the active site of natural enzymes with
synthetic compounds depends on symphoria, rather than other physicochemical factors [2]. Detailed explanations may not be what we want for a general theory of
computational implementation. Thus, what I want to claim here is that symphoria
can be used to give less specific or detailed answers than those given by causal-
mechanical accounts, but more specific and relevant answers than those provided by
counterfactual accounts.[4] Remember that the causal-mechanical account has been
criticized for being too restrictive, while the counterfactual account has faced criticism for being overly permissive [27]. I propose that focusing on symphoria provides
the right extensional adequacy. With regard to teleological accounts, symphoria can
also have a teleological connotation. Here, one can simply cite Morrison and Boyd,
who almost at the end of their book, where the term symphoria was coined, refer
to symphoria as "the bringing together of molecules *for a useful purpose*" [17] (p.
1247, italics are mine). I will return to these overlaps in Sect. 9.

Let's now finally address the relevance of the concept of symphoria to theories
of implementation by first trying to answer COMP ("Under which conditions is it
true/false that a physical system implements a computation?") and IDENT ("Under

---

[4] Another detailed account one could apply to answer CATAL and IDENT* is what has been
called *The New Mechanistic* approach to explanation in the philosophy of science literature [4].
This approach focuses on detailing the specific entities and activities that consistently produce a
phenomenon, ensuring that all key components are identified. Again, symphoria refers instead to a
rather universal but key factor.

which conditions is it true/false that a physical system implements one computation rather than another?" [27]), as follows:

Answer to COMP: A physical system P implements a computation C if and only if P provides the symphoria S needed to physically realize the transitions of C. In other words, while C as an abstract entity does not require symphoria, the process of physically implementing C does.

Answer to IDENT: A physical system P implements a computation C rather than another computation C', if and only if P provides the symphoria S that allows for the physical realization of C's transitions, but does not provide the distinct symphoria S' that would allow for the physical realization of C'.

Here, symphoria is just the bringing together of the components of a computational process in the spatial (and temporal) relationship necessary for the computational process to occur. And this is the right place to bring the resemblance account of physical computation to the table, momentarily left aside above in this section. The resemblance account emphasizes that "implementation is a matter of resembling a computational architecture" and that their formal, mathematical features are "permeated with causality, spatiotemporality, and other non-mathematical features" [5]. I propose that symphoria can be seen as a fundamental aspect of this resemblance. In essence, the resemblance account requires that the physical system's structure and operations align with the abstract computational architecture in a way that goes beyond mere isomorphism. It includes the physical embodiment and interaction of components that mirror the architecture's operational structure. Symphoria fits seamlessly into this account as it points out to the precise spatial and temporal configuration needed for the physical system to realize the computation. This is similar to what we saw in Sect. 6 when presenting the case of enzyme models resembling the active site of natural enzymes [2]. Thus, the spatial alignment that symphoria describes is central in ensuring that the physical system does not just trivially map onto a computation but genuinely realizes it by bringing together the necessary components in a configuration that mirrors the intended computational process. In this way, symphoria and the resemblance account are compatible, with symphoria serving as a key, universal factor that grounds the physical realization of computational architectures in the concrete spatial and temporal arrangements necessary for a computational process to occur.

Regarding the relationship between the bottom-up accounts of physical computation presented in Sect. 4 and symphoria, what I said when answering CATAL and IDENT* also applies when giving answers to COMP and IDENT. As we saw in Sect. 7, when examining the case of DNA-based Neural Networks, answers based on symphoria are less specific and detailed, and therefore less restrictive, than those based on the causal-mechanical account. On the other hand, they are more specific, and therefore more restrictive, than those based on the counterfactual account. And this is a desirable outcome, because, as we stressed before, the causal-mechanical account has been accused of being too restrictive, while the counterfactual account has been accused of being too permissive. Thus, the extensional adequacy of a theory of implementation could be improved, if the theory focuses on symphoria, or so I argue. On the other hand, as revealed by Morrison and Boyd's quote (see above), symphoria can also receive a teleological connotation. However, such a view should be approached with care, at least if one wants to preserve naturalism.

9 Integrating Symphoria Into the Mapping Account

In Sect. 3, we discussed how the classical 'mapping' account of computational implementation establishes a clear structural correspondence between physical and abstract states (and transitions) but remains vulnerable to triviality arguments. Any system, however unstructured, could be considered to implement a given computation through a suitably contrived mapping, and this significantly diminishes the extensional adequacy of the account. Symphoria, introduced earlier as a way to highlight the physical conditions necessary for non-accidental, stable realizations of processes, offers a way to address this issue.

Recall that the mapping account requires a function

$$f : \{S_P\} \to \{S_C\}$$

such that for every transition $S_C \to S'_C$ in the formal computation C, there is a corresponding physical transition $S_P \to S'_P$ with $f(S_P) = S_C$ and $f(S'_P) = S'_C$ [27]. To integrate symphoria into this framework, we introduce an additional predicate

$$h(S_P, S'_P)$$

that must hold true for the physical transition $S_P \to S'_P$. This predicate encodes the symphoria condition, ensuring that the relevant spatial and temporal arrangements are in place so that the physical transition genuinely supports the intended computational step. Here, it is important to distinguish between the general notion of symphoria, represented by S in Sect. 8, and the predicate h. While S refers to the general concept of symphoria–capturing the overarching requirement that certain spatial and temporal arrangements must be met–h is a formal predicate used to evaluate whether these symphoric conditions are satisfied for specific physical transitions. In other words, S represents the general principle, whereas h provides a way to test its application in particular cases.

Extended Mapping Account with Symphoria: A physical system P implements a formal computation C if and only if there exists a mapping $f : S_P \rightarrow S_C$ such that for every abstract transition $S_C \rightarrow S_C'$ in C, there exist physical states S_P, S_P' with $f(S_P) = S_C$ and $f(S_P') = S_C'$, and the symphoria condition $h(S_P, S_P') = \text{True}$ holds.

This account requires more than a structural match: it insists that each transition is symphorically sustained. While the mapping f ensures a formal correspondence, the predicate h enforces a non-trivial physical precondition. Without h, we could have a scenario like Searle's wall (see Sect. 3), where a relatively unstructured physical system is purportedly "implementing" a complex computation, which involves a long chain of transitions, solely by virtue of a devious mapping. With h, we demand these physical transitions to be spatially and temporally orchestrated in a way that resembles the conditions under which a sequence of enzyme-mediated reactions (think, for instance, in a metabolic pathway such as glycolysis) achieves its functional behavior. While the reactants of each one of these reactions are brought together in the proper spatial relationship by specific enzymes, making the whole chain feasible, the *patterns of physical activity* or *molecular arrangements* inside Searle's wall could only be properly brought together by chance alone. Thus, a wall's random molecular fluctuations would almost certainly fail h for complex computational steps, ruling out the possibility of it trivially "implementing" the desired computation.

In practice, the specific criteria represented by h will depend on the physical system under consideration, thus preserving multiple realizability. Also, while h will appear straightforward in certain well-controlled scenarios, it will become significantly more challenging and non-trivial in complex materials. For example, consider a scenario involving coherent light pulses, lenses, and optical lattices, where ensuring that signals arrive at the correct spatial locations and times to form stable interference patterns may seem relatively manageable. Even so, this arrangement still imposes meaningful constraints that must be met to yield coherent computational steps. In contrast, consider DNA-based neural networks or other organic, adaptive materials. Here, symphoria requires maintaining intricate spatial and temporal relationships in a fluctuating environment–ensuring that molecular recognition and self-assembly processes reliably support information processing. Such conditions are significantly more complex and less easily controlled than those in a purely optical setup. As we move toward increasingly sophisticated instances of intelligent matter, where materials autonomously sense, adapt, and operate across multiple scales [12], the challenge intensifies. Under these circumstances, symphoria cannot be dismissed as a trivial alignment condition; instead, it becomes a central, non-trivial factor in enabling genuine computational capabilities in dynamic, evolving physical substrates.

It should be noted that the symphoria condition h does not exist in isolation from other factors emphasized by different accounts of physical computation. Causal, mechanical, counterfactual, symphoric, and teleological factors often overlap and are difficult to disentangle in practice. However, symphoria brings a distinctive focus on the spatial and temporal arrangements necessary for computation–an aspect that complements rather than competes with these other perspectives. Rather than treating symphoria and other factors as separate or mutually exclusive, it is more productive to view them as interrelated and overlapping, each contributing to a fuller understanding of how abstract computations are physically realized.

Finally, this formal integration of symphoria (and overlapping factors) into the mapping account provides a starting point for future work. It motivates the search for more specific and quantifiable definitions of h in different domains, and encourages the development of experimental strategies for testing whether a given system truly meets these symphoria-based criteria. As systems become increasingly complex and hybrid, symphoria offers a structured way to distinguish between mere formal correspondences and fully realized computational operations in physical media.

10   Conclusions and Perspectives

In this chapter, I introduced and expanded upon the concept of *symphoria*, originally articulated within the context of organic chemistry, positioning it as a key universal factor in understanding computational implementation across diverse physical systems, particularly within the realm of intelligent matter. By drawing parallels between chemical catalysis and computational processes, I argued that symphoria–the (mediated) bringing together of the interacting components of a process in the proper spatial and temporal relationships for the process to occur–provides a non-trivial, action-oriented perspective necessary for the stable, non-accidental realization of computations. By foregrounding the mediated spatial and temporal orchestration of a system's components, symphoria explains why certain systems–such as DNA-based neural networks–can genuinely implement abstract computations, whereas mundane systems (like Searle's wall) fail to achieve this since they cannot supply the necessary alignments. Furthermore, I showed how symphoria partially overlaps with, but cannot be reduced to, existing accounts of physical computation, offering a complementary perspective that can address their limitations. In doing so, I proposed that symphoria can enhance the extensional adequacy of a theory of implementation (while preserving naturalism), bridging the gap between overly restrictive accounts that may exclude legitimate computational systems and overly permissive ones that risk trivializing the concept of computation.

Looking ahead, formalizing symphoria and applying it more generally calls for several interconnected research directions. First, domain-specific formalization is needed to specify the precise spatial and temporal thresholds that define the relevant symphoria conditions in different systems. Second, experimental validation is essential: empirical tests that confirm whether a given physical system genuinely meets symphoria requirements can involve perturbing spatial or temporal alignments to see if the degradation of computational functions provides a clear index of the system's reliance on symphoria. Third, symphoria must be integrated into existing mechanistic theories, an endeavor that recognizes overlaps with causal, teleological, and structural features but still treats symphoria as a distinct perspective on orchestrating the functional interplay of components, especially in advanced materials. Fourth, the issue of self-maintenance and adaptation arises in sophisticated examples of intelligent matter, where symphoria must be preserved under changing environmental conditions, thus raising important questions about how such systems can autonomously ensure correct spatial and temporal alignments or reconfigure them if they become disrupted. Finally, symphoria frequently spans multiple levels of organization, from the molecular scale to the supramolecular and even macroscales, so uncovering how these different layers of coordination interact and potentially reinforce one another is critical for achieving robust, large-scale computational capabilities in complex physical substrates.

Acknowledgements The author gratefully acknowledges funding from the Deutsche Forschungsgemeinschaft (DFG, German Research Foundation) under Grant No. 254954344/GRK2073/2.

References

1. Chalmers DJ (1996) Does a rock implement every finite-state automaton? Synthese 108(3):309–333
2. Chaudhury PK, Nagarajan K, Dubey P, Sarkar S (2004) Symphoria: The success of modeling the active site function of oxo-molybdoenzymes. J Inorg Biochem 98(11):1667–1677
3. Copeland BJ (1996) What is computation? Synthese 108(3):335–359
4. Craver CF, Darden L (2013) In search of mechanisms: discoveries across the life sciences. University of Chicago Press
5. Curtis-Trudel A (2021) Implementation as resemblance. Philos Sci 88(5):1021–1032
6. Curtis-Trudel A (2022) Why do we need a theory of implementation? Br J Philos Sci 73(4):1067–1091
7. Deng X, Guo J, Zhang X, Wang X, Su W (2021) Activation of aryl carboxylic acids by Diboron reagents towards nickel-catalyzed direct decarbonylative borylation. Angew Chem Int Ed 60(46):24510–24518
8. Franciò G, Scopelliti R, Arena CG, Bruno G, Drommi D, Faraone F (1998) IrPd, IrHg, IrCu, and IrTl binuclear complexes bridged by the short-bite ligand 2-(diphenylphosphino)pyridine. Catalytic effect in the hydroformylation of styrene due to the monodentate P-bonded 2-(diphenylphosphino)pyridine ligands of trans-[Ir(CO)(Ph$_2$PPy)$_2$Cl]. Organometallics 17(3):338–347
9. Graham JW, Williams TC, Morgan M, Fernie AR, Ratcliffe RG, Sweetlove LJ (2007) Glycolytic enzymes associate dynamically with mitochondria in response to respiratory demand and support substrate channeling. Plant Cell 19(11):3723–3738
10. Godfrey-Smith P (2009) Triviality arguments against functionalism. Philos Stud 145(2):273–295
11. Hasaninejed A, Kazerooni MR, Zare A (2012) Solvent-free, one-pot, four-component synthesis of 2H-indazolo [2,1-b] phthalazine-triones using sulfuric acid-modified PEG-6000 as a green recyclable and biodegradable polymeric catalyst. Catal Today 196(1):148–155
12. Kaspar C, Ravoo BJ, van der Wiel WG, Wegner SV, Pernice WHP (2021) The rise of intelligent matter. Nature 594(7863):345–355
13. Khalafi-Nezhad A, Rad MNS, Hakimelahi GH (2003) Synthesis of polyfunctional aromatic ring systems (phloroglucide analogs) under microwave irradiation. Helv Chim Acta 86(7):2396–2403

14. Lycan WG (1981) Form, function, and feel. J Philos 78(1):24–50
15. Majumdar A, Sarkar S (2011) Bioinorganic chemistry of molybdenum and tungsten enzymes: a structural-functional modeling approach. Coord Chem Rev 255(9):1039–1054
16. Miłkowski M (2011) Beyond formal structure: a mechanistic perspective on computation and implementation. J Cogn Sci 12:359–379
17. Morrison RT, Boyd RN (1987) Organic chemistry, 5th edn. Allyn and Bacon, Boston
18. Okumura S, Gines G, Lobato-Dauzier N, Baccouche A, Deteix R, Fujii T, Genot AJ (2022) Nonlinear decision-making with enzymatic neural networks. Nature 610(7932):496–501
19. Piccinini G (2015) Physical computation: a mechanistic account. OUP Oxford
20. Piccinini G, Maley C (2021) Computation in physical systems. In: Zalta EN (ed) The stanford encyclopedia of philosophy, summer, 2021st edn. Stanford University, Metaphysics Research Lab
21. Putnam H (1988) Representation and reality. MIT Press
22. Rescorla M (2014) A theory of computational implementation. Synthese 191(6):1277–1307
23. Rescorla M (2015) The representational foundations of computation. Philos Math 23(3):338–366
24. Schoffelen S, van Hest JC (2012) Multi-enzyme systems: bringing enzymes together in vitro. Soft Matter 8(6):1736–1746
25. Searle JR (1992) The rediscovery of the mind. MIT Press
26. Sprevak M (2010) Computation, individuation, and the received view on representation. Stud Hist Philos Sci Part A 41(3):260–270
27. Sprevak M (2018) Triviality arguments about computational implementation. In: The Routledge handbook of the computational mind. Routledge, pp 175–191
28. Wu F, Minteer S (2015) Krebs cycle metabolon: structural evidence of substrate channeling revealed by cross-linking and mass spectrometry. Angew Chem Int Ed 54(6):1851–1854
29. Xiong X, Zhu T, Zhu Y, Cao M, Xiao J, Li L, Pei H (2022) Molecular convolutional neural networks with DNA regulatory circuits. Nat Mach Intell 4(7):625–635
30. Yang L, Tang Q, Zhang M, Tian Y, Chen X, Xu R, Ma Q, Guo P, Zhang C, Han D (2024) A spatially localized DNA linear classifier for cancer diagnosis. Nat Commun 15(1):4583
31. Zhang Y, Beard KF, Swart C, Bergmann S, Krahnert I, Nikoloski Z, Obata T (2017) Protein-protein interactions and metabolite channelling in the plant tricarboxylic acid cycle. Nat Commun 8(1):15212
32. Zhang Y, Fernie AR (2021) Metabolons, enzyme–enzyme assemblies that mediate substrate channeling, and their roles in plant metabolism. Plant Commun 2(1)

Ethical Aspects of Artificial Intelligence in Nanoscience

John Weckert

Abstract Examination of ethical issues of intelligent matter draws on both the ethics of Artificial Intelligence and of nanotechnology. These examinations raise methodological questions regarding how to approach the ethics of emerging technologies. This chapter defends the use of speculation about future scenarios, and the precautionary principle. This is then applied to issues raised by intelligent matter.

1 Introduction

Artificial Intelligence (AI) meets nanoscience in intelligent matter. Consequently, examination of ethical issues draws on both the ethics of AI and of nanotechnology. Commonly, ethical discussions are reactive, that is, they react to issues that are raised by the uses of technologies. For example, how does surveillance technology affect privacy, human rights and so on? How, if at all, should intelligent robots be used in care situations? What are acceptable risks of manufactured nanoparticles? But not all ethics is reactive. Sometimes a proactive approach is necessary and this can be easy in particular cases. Given current technology and developments, it is not difficult to predict how privacy can be compromised with advances in sensing devices, facial recognition and more powerful machine learning. Discussing the ethical implications of these kinds of development and how best to maximise their benefit and minimise their risks, is common practice. With new technologies where there are as yet few effects to assess, the proactive approach becomes more important but also more controversial. Predictions based on current technologies and for the near future are acceptable, as in the surveillance case above, but speculation about possibilities of new technologies in the more distant future is contested. In AI for example, should we worry now about ethical implications of Artificial Super Intelligence, which may never eventuate?

We will now look briefly at the ethics of AI and of nanotechnology to understand better the reactive and proactive approaches. This will then lead to a consideration

J. Weckert (✉)
Charles Sturt University, Wagga Wagga, Australia
e-mail: JWeckert@csu.edu.au

M. te Vrugt (ed.), *Artificial Intelligence and Intelligent Matter*, Machine Intelligence for Materials Science, https://doi.org/10.1007/978-3-032-04129-6_24

493

of speculative ethics and of the precautionary principle. Following that, we will conclude with a brief discussion of some ethical issues of intelligent materials.

2 Ethics of AI

Before any real AI had been developed, ethical concerns were being raised about how it would or could be used. Most of this discussion occurred after the Dartmouth Conference of 1956 but as early as the late 1940s, Norbert Wiener expressed concerns about machines that to some degree could "think" and could go beyond what their designers had expected or intended [1, pp. 27–28]. They could, he believed, be creative. In 1960, talking about automated machines, he writes:

> It is my thesis that machines can and do transcend some of the limitations of their designers, and that in doing so they may be both effective and dangerous.
>
> …
>
> To be effective in warding off disastrous consequences, our understanding of our man-made machines should in general develop *pari passu* [on an equal footing] with the performance of the machine [2, p. 1355].

The idea that machines could think or have intelligence in a human sense, in other words, have mental states, was much discussed in AI in the early days. While strictly speaking this may not be an ethical issue, it has ethical implications. If machines can think, does it follow that humans are merely machines? Is this dehumanising? This was seen as a problem for strong AI, that is, AI that had something like a human level of intelligence and mental states. Weak AI on the other hand, was not designed to have general human intelligence but rather to perform tasks in narrow areas where decisions needed to be made. Expert systems were typical examples. These too raised ethical problems, for example, if such a system gave wrong advice or made a bad decision, who would be responsible? This question has arisen again, particularly with self-driving cars.

An early critic of machines replacing people in certain decision-making, was the computing pioneer and author of the ELIZA program, Joseph Weizenbaum. He argued that.

> … all projects that propose to substitute a computer system for a human function that involves interpersonal respect, understanding, and love…are obscene.' And their '… very contemplation ought to give rise to feelings of disgust in every civilised person [3, pp. 268–9].

The systems that were of immediate concern to Weizenbaum were very much in the weak AI realm. His ELIZA program, for example, was extremely simple by today's standards. But if development continued and systems became closer to strong AI, even if never reaching it, these systems could be used in a wider range of situations where "interpersonal respect, understanding, and love" are required.

The development of machine learning has enabled AI to become much more sophisticated than it was in Weizenbaum's day and his concerns are still raised.

Should AI and robotics be used in caring situations? These clearly involve interpersonal respect and understanding but can that be provided by a machine? Many would claim that it cannot but others argue that if better care can be provided in this way, so much the better [4]. If robots can be developed that act *as if* they have respect and understanding in caring situations, would this solve the problem?

While recent developments in AI have excited many, some notable leading AI researchers and others working in the field, have started to express concerns. One much discussed issue is the employment of smart weapons. Should autonomous weapons be allowed to make decisions about their targets? If they make better decisions under pressure than humans do, does this justify their use? Some are worried about autonomous weapons was raised:

> In summary, we believe that AI has great potential to benefit humanity in many ways, and that the goal of the field should be to do so. Starting a military AI arms race is a bad idea, and should be prevented by a ban on offensive autonomous weapons beyond meaningful human control [5].

This statement is a strong statement. Not only should some research not be undertaken, some should be banned. This is invoking the precautionary principle, something we will return to in a later section.

In the last year or two, ethical discussion has focussed more on problems, or supposed problems, raised by generative AI, that is, AI that generates new text, images and so on. These concerns range from assessing students' essays in educational situations to understanding how to cope with artistic creations generated by AI. Plagiarism has always been a problem but programs like ChatGPT make it easier. Will script writers and illustrators be out of work if text and images can be AI generated? Will these generated items have less value than those produced by humans?

Beyond these issues, generative AI has reinvigorated the discussion of general AI, that is, AI with or approaching, general human intelligence. Functionally this is equivalent to what was called Strong AI, although the latter was commonly thought to include cognitive states. Generative AI has even raised the possibility of intelligence beyond the human level, ie. Artificial Super Intelligence (ASI). Nick Bostrom argues that very careful thought should be given to the development of ASI because it represents an existential risk to humanity, even human extinction [6]. Others too, including Stephen Hawking, Stuart Russell, Max Tegmark and Frank Wilczek, have expressed concern:

> One can imagine such technology outsmarting financial markets, outinventing
> human researchers, out-manipulating human leaders, and
> developing weapons we cannot even understand, … Whereas the
> short-term impact of AI depends on who controls it, the long-term impact
> depends on whether it can be controlled at all [7].

And according to The Bletchley Declaration of 2023 on AI safety.

> particular safety risks arise at the 'frontier' of AI, understood as being those highly capable general-purpose AI models, including foundation models, that could perform a wide variety

of tasks—as well as relevant specific narrow AI that could exhibit capabilities that cause harm -which match or exceed the capabilities present in today's most advanced models [8].

The Declaration urges more research on AI risks and the importance of international cooperation. Some go further than this. Ian Hogarth argues that "We must slow down the race for God-like AI" [9], that is ASI, so there is more time to understand it and its dangers. And in an Open Letter, Elon Musk and others "call on all AI labs to immediately pause for at least 6 months the training of AI systems more powerful than GPT-4". This is another appeal to the precautionary principle [10].

Not everyone in, or interested in, the AI field is so concerned about the problems. Luciano Floridi, for example, argues that ASI is implausible [11, 12], and that the focus should be on current problems. "What really matters', he says, "is that the increasing presence of ever-smarter technologies is having huge effects on how we conceive of ourselves, the world, and our interactions" (see also Margaret Boden [13]). Our argument here is not that any of the fears about ASI will be realised. The issue is whether there is good reason to consider them seriously. We will return to this after looking at some ethical issues of nanotechnology.

3   Nanoethics

In 1986 Eric Drexler raised the issue of gray goo [14]. Nanotechnology, he argued, would lead to the development of self-replicating nanorobots, which if left unchecked could destroy the world. This self-assembly would also enable desktop, or personal, factories that could produce just about anything. The discussion now has moved away from these topics and more onto issues regarding manufactured nanoparticles and because it is an enabling technology, to technologies that it enables further. Nanoethics, as it has actually developed, has been rather mundane compared with the Drexler scenarios.

One criticism of the field concerns the fact that many of the so-called "ethical issues in nanotechnology" are in areas where there has as yet been little development, so discussion of them must be based on prediction. Prediction of course is notoriously unreliable, and this is particularly true of predictions about the directions of scientific and technological developments. Alfred Nordmann [15] discusses this issue critically in terms of speculation and the tendency to treat remote possibilities as likely technological developments that require attention, something that he calls "if and then"; if some technological development is possible then we should do something about it now. Instead of concentrating on real and important issues, ethicists too often focus on "problems" that will probably never arise, and if they do, only long in the future. We will return to this argument in the next section.

What are some of the benefits (or potential benefits) of nanotechnology? Stronger and lighter materials will enable the production of safer and more energy efficient transport. Self-cleaning materials are already available and more efficient solar cells for electricity generation are being developed. In the medical field targeted drug delivery is under development lab-on-a-chip technology is advancing and promising discoveries have been made in cancer detection [16]. Advances in nanoelectronics have already enormously increased the amount of memory in computers and are expected to allow the continuation of Moore's law, that is, that the number of transistors on a chip doubles roughly every two years,. Considering these and other benefits must also part of nanoethics.

Of the topics commonly discussed in nanoethics, the potential toxicity of some manufactured nanoparticles has received the most attention, particularly in the area of health. Products currently on the market, for example sunscreens, cosmetics, food packaging and even some foods contain nanoparticles, and fears have been expressed regarding their safety (for an overview, see [17]. Concerns are also expressed about potential medical applications, particular in the form of targeted drug delivery and lab-on-a-chip technologies. Another concern is that thread-like particles that can be inhaled might have the same effects as asbestos and lead to serious lung disease. This concern about certain nanoparticles is an environmental issue as well as a health one [18]. What will be the effect of certain types of these manufactured nanoparticles on the environment?

Advances in nanoelectronics have many benefits, as mentioned above, but also risks. The further miniaturization of monitoring and surveillance devices for example, has raised concerns about further loss of personal privacy and greater potential for control over our lives by both public and private authorities.

A related topic that has received considerable attention is human enhancement, or the improvement of human performance. Therapeutic implants, for example computer chips to overcome blindness and some psychiatric conditions, will almost certainly become more sophisticated by developments in nanoelectronics and very likely lead to implants for enhancements. Research is already proceeding on cognitive enhancement involving memory and reasoning ability and new learning techniques, on enhancement of our sensing abilities, and on brain-to-brain and brain-to-machine communication. A central concern relates to the question of what it is to be human and whether such enhancement is taking away something from our humanity. A similar concern is raised by the blurring of the boundaries between technology and life.

Military uses of nanotechnology are likely to enable new weapons and fears have been expressed that this could provoke a new arms race [19]. Research is well under way on weapons with an ability to make autonomous decisions, tiny missiles possibly only a few millimeters in length, enhancement of soldier performance through implants, sensors, and exoskeletons and the like, and small animals or insects with sensor and explosive implants.

A perennial issue concerns distributive justice or the so-called "nano-divide"; the worry that the advance of nanotechnologies will be concentrated in developed countries and will focus on products for those countries. This will, it is feared, widen the gap between developing and developed countries. While some see this as an opportunity for improving the lot of poorer areas, others are more cynical. Intellectual property is one of the concerns.

498   J. Weckert

A contested branch of nanotechnology, molecular manufacturing, or Atomically Precise Manufacturing (APM) as it is now called, could have dramatic consequences for society if it proves feasible. While it is seldom discussed, it is worth noting because research is still being undertaken, and even if it is in the distant future, nanoethics needs to be proactive as well as reactive [20, 21]. Developments in this area may stimulate further ethical consideration but whether they are worthy of examination now, is contested. Should we worry about the ethics of future and uncertain technologies?

4   Speculative Ethics

We mentioned earlier that ethics can be reactive or proactive. While much applied ethics is reactive, in the case of new technologies there is little to which to react, so ethics must be predominantly proactive. Given the rapid developments in both AI and of nanotechnology this is commonly the case in these fields. Proactive ethics looks at the future, at what future consequence are likely to be. But how far into the future is it sensible to look? We can distinguish between proactive ethics and speculative ethics. The former extrapolates from current technology to what is likely to happen in the near or medium future, eg more intrusive monitoring and surveillance. Speculative ethics looks into the more distant future and speculates on what could eventuate. These future scenarios are then thought to be worthy of ethical concern, a view that is contested, as mentioned earlier. Alfred Nordmann [15] has three arguments against speculative ethics but a core one is that it is a waste of scarce resources. The scarce resources are people and time. New technologies are raising many problems, often urgent, that require the attention of people trained in ethics, policy, the relevant sciences and technologies and others as well. They should be spending their valuable time and expertise on these real problems and not on distant future ones that may never eventuate.

This is a plausible argument that undoubtedly carries some weight. Ethicists should examine real and current problems but there is more to the ethics of technology than this. If the purpose of technology is to help humans to live good and satisfying lives then some big issues become relevant and it becomes necessary to examine our values and goals. Long term and perhaps remote technological developments become important subjects for careful thought. What kind of world do we want to live in? Given the sort of creatures that we are, what kinds of world can we thrive in? What constitutes a good and satisfying life for us and which technologies enhance or detract from achieving this? These kinds of questions suggest that more is required in ethical discussions of technology than merely focusing on current or near-term technologies and issues. Rebecca Roache [22] articulates clearly the value of speculation about possible futures and her arguments, where values and goals play a central role, are briefly summarized below.

First, scientists are limited only by their imagination and therefore ethical evaluation of scientific projects should be undertaken even before the viability of the project is certain. This is needed to ensure that "scientific effort is directed to the best ends." [22, p. 323]. Second, our values determine our goals and if a goal is to cure cancer, that will be, and has been, pursued even if in the early stages of research, it is unknown if it is possible or how far into the future it may be. The point here is that speculating about how to achieve a desirable goal even if we are uncertain about how to, is generally held to be a good thing. If speculating about a distant good future is acceptable, it is unclear why speculating about a bad one is not. Both concern our values and goals. Finally, if the goal is important enough, the fact that it may be extremely uncertain or implausible is not enough to show that it is unworthy of serious thought. Suppose, Roach says, that a child is lost in a vast maze of tunnels, the fact that finding her alive may be unlikely in the extreme would not, or should not, stop the search being undertaken. If the goal is important enough, then it matters little if it is improbable or implausible. It should still be seriously considered.

If this is right, then in general speculation does have a role to play and ethical considerations should not be restricted to current and near-term issues. We turn now to a specific example that in various forms has been around since the early days of AI. Should we worry about machines becoming as intelligent, or more intelligent, than humans? Its current expression is "should we worry about Artificial Superintelligence (ASI)?" Luciano Floridi has argued that ASI is logically possible but not plausible.

For Floridi, the reason that superintelligence appears plausible is because it is presented as a conditional:

> *If* some kind of ultraintelligence were to appear *then* we *would* be in deep trouble (his emphases) [11, p. 5].

Is conditional may well be true and it makes it look as if we should do something about the possibility. But the antecedent is, Floridi believes, if not impossible, certainly implausible. If ASI (his ultraintelligence) is impossible then it is not something worth worrying about and the case is not much weaker if it is, as he says, logically possible but implausible. He raises both technical and philosophical objections. First the technical.

The frame problem [12, pp. 135–138].

This is the well-known problem that while a machine can operate in a supposedly intelligent manner in a narrow field where the parameters a clearly defined, it cannot operate sensibly outside that area. Floridi's example is a lawnmower that can mow a lawn within the confines set for it. Outside of that, it cannot function usefully. Other examples abound: automatic braking systems on cars, automatic pilots and so on. Advances however are being made and not all frames are so narrow. Consider the rapid development of self-driving cars. Driving on a road with other vehicles is a complicated business. Information on where the road is, what other traffic is doing with respect to position and speed relative to ones own, the capabilities of one's own car, where and when to turn onto another road, the route to one's destination, the weather conditions and so on all need to be coordinated in order to have a safe trip to one's destination. A human who drives successfully in traffic is undoubtedly displaying intelligent behaviour so this seems like a paradigm case of AI. Certainly it is still in a frame, a confined sphere, but a pretty broad one and much broader than a lawn mower or a car's breaking system. It does not seem implausible that with ever increasing computer power, more sophisticated algorithms, learning systems and sensing devices that frames could be merged and broadened further and at least simulate a human in a wide range of activities.

The grounding problem [12, pp.138–139].

Computers work on uninterpreted data whereas humans work on meaningful terms. Consequently, computers can never understand what they are doing. They are limited to syntax while we humans work with both syntax and semantics. Syntax has to do with structure and semantics with meaning. This is the way that Floridi states the grounding problem. It is not obvious though, that the situation is quite as clear as this. Wittgenstein said that we should look at use, not meanings. While not wanting to give an account of meaning here, it is instructive to focus on use when considering computers. Suppose that a computer is told of or learns that the data representing kangaroos, marsupials, mammals and Australia are related in certain ways. It is also told or learns the situations in which it is appropriate to use that data. It could then use them in appropriate ways such that, if a human used them in those ways, that human would be thought to understand them. The fact that the computer might not understand what it was doing in the way that we do seems not important if it's use is correct. The grounding problem therefore seems not to be a strong obstacle to intelligent behaviour.

Curry-Howard Correspondence [11, p. 11].

This shows that there is a direct correspondence between computer programs and mathematical proofs. Given Gödel's incompleteness theorem, some problems will not be solvable by a computer and Floridi suggests therefore could not be super-intelligent. It is not obvious why this follows. Humans cannot, it seems, solve all problems so an ASI may also not be able to solve all but still be much superior to us in intelligence.

These three arguments are all (or primarily) technical reasons why ASI is not plausible. Floridi's core argument however appears to be that an ASI would have to be much like us and would have to be conscious. In support of the first of these he says that "we have no idea how we might begin to engineer it, not least because we have very little understanding of how our own brains and intelligence work" [11, p. 11]. Now it may be true that we do not have much idea of how our own brains and intelligence work but it does not follow that therefore we cannot have some idea of how to engineer some kind of intelligence into machines. The driverless cars example seems to show that we do. His arguments about the frame and grounding problems might show that it is difficult but not that it is impossible or even implausible. It is not obvious why something could not appear to us to be intelligent functionally even if it were engineered quite differently from us. But it would *appear* to be intelligent. Floridi's view however is that because we have consciousness, machines must also be conscious in order to be intelligent, and because we do not know how to engineer this intelligence it must evolve autonomously. But "How some nasty ultraintelligent AI will ever evolve autonomously from the computational skills required to park [a car] in a tight spot remains unclear" [11, p. 7]. Perhaps it would be a little less unclear if instead he had used the example of a self-driving car, but be that as it may, we will look at this argument against ASI in more detail.

Floridi's argument here seems to be:

1. All computers are Turing Machines
2. No conscious, intelligent entity is going to emerge from a Turing Machine
3. A superintelligent entity must be conscious
4. Therefore computers will never be superintelligent.

We can accept that all computers are Turing Machines (or equivalent to Turing Machines). Premise 2 is more interesting. Consciousness emerging from a simple Turing Machine does seem completely implausible. What is perhaps less implausible is consciousness emerging from an extremely sophisticated machine with multi-layered neural nets and deep learning. If such a machine is equivalent to a Turing Machine then it may not be implausible to say that consciousness can emerge from the simple machine. Of course we do not know if a conscious entity can or will emerge from any machine, regardless of its sophistication. Part of the problem lies in knowing what consciousness is and when to attribute it. If a machine behaves in a way that would be called intelligent if a human behaved in that way in a wide variety of contexts, then it is not so clear what our justification would be for saying that it is not conscious. This leads to the question of the other minds problem.

We can, however, avoid this tricky problem by considering premise 3. Is it really true that a superintelligent entity must be conscious, at least in the way that we are? The calculator on my smart phone can calculate much more quickly and accurately than I can. If I could match its ability I would be considered to be pretty intelligent, at least in that area. But we do not think of it as being conscious. Image a machine then that could do many things as well as my calculator can calculate and moreover can integrate these various activities. If we had the functionality that it had, we would be considered intelligent so it would be strange to say that it was not also intelligent even if consciousness were denied it. If it makes sense to attribute intelligence to a machine without attributing consciousness, the same should be true for superintelligence.

It is worth mentioning here that Floridi quotes approvingly both Alan Turing's and Dijksra's comments on machine consciousness [12, p. 140]. According to Turing 'Can a machine think?' is 'too meaningless to deserve discussion' [11, p. 10). This can be taken to mean that of course machines cannot think. Equally, it could be taken in a Wittgensteinian manner. It would then suggest that if a machine passed the Turing Test or behaved in a manner that was indistinguishable from a human, then of course it could think. That is all that there is to thinking. We have no better way of examining this. Dijksra's comment is similar. Asking if a machine can think, he said, is as silly as asking if a submarine can swim. Again it could be interpreted in the two ways above so does not give much support to Floridi's argument. It is also a little puzzling why it is silly to say that a submarine can swim when we say that an aeroplane can fly.

The argument here is the modest one of showing that Floridi's arguments against ASI's plausibility are not convincing. It is still an open question then whether it is an issue about which we should worry. ASI might be plausible but so remote that worrying is a waste of time.

Suppose that we concede Floridi's point that intelligence involves consciousness and therefore that machines cannot be intelligent. He distinguishes intelligence from smartness and admits that artefacts can be smart but not intelligent [12]. All that the current argument may show is what we already knew, namely that machines can be smart. This, I think, can be conceded but nothing important follows from it. If computers could become as smart as humans, and it has not been shown that this is not possible, they may well become supersmart with all the same problems of superintelligence.

Based on the argument here, we should be concerned about superintelligence (or supersmartness). It may not be urgent yet, certainly not as urgent as the other AI issues but that does not show that it should be ignored. Plausible long term problems need investigation in order to minimise the potential dangers. If climate change had been thought about more carefully decades ago it may have been easier to mitigate its effects than it is now. The fact that other AI issues are more urgent does not show that superintelligence does not deserve attention. If it does deserve attention then perhaps we should act now to avoid or alleviate potential harms, a position taken by computer scientist Toby Walsh [23]. In this context, the precautionary principle (PP), mentioned in an earlier section, is relevant. It is raised in a variety of contexts but particularly with respect to new and developing technologies. It is therefore particularly relevant to new developments in AI and to intelligent materials.

5   The Precautionary Principle

The precautionary principle, while advocated by many and enshrined in European regulations and policy documents, especially in the environmental and health areas, is also much criticised. Here we will briefly explain it and defend it against the more important criticisms (a more detailed discussion can be found in Weckert [24]).

A commonly stated versions of the principle is this.

> Where an activity raises threats of harm to the environment or human health, precautionary measures should be taken even if some cause and effect relationships are not full established scientifically [25].

Various calls have been made to restrict or prohibit development of certain new technologies, which are in effect calls to apply the PP. Bill Joy, talking of the dangers of genetics, nanotechnology, and robotics famously said in 2000; "The only realistic alternative I see is relinquishment: to limit development of the technologies that are too dangerous, by limiting our pursuit of certain kinds of knowledge" [26]. More recently, as we saw earlier, Bostrum, Musk and others have made similar claims.

The PP is often said to be an alternative to cost–benefit analysis (CBA) in certain situations. Cass Sunstein [27] and Steve Clarke [28] argue strongly for this but it is not altogether clear that CBA and PP are the real alternatives in practice. The real alternative to the PP seems to be rather something like the Bravado Principle (BP) although we will not argue that here:

Where an activity appears to have benefits, do it and worry about any harmful consequences if and when they arise.

An argument for BP, although rarely stated explicitly in the literature, is that unless this kind of approach is taken, there will be little progress or worse [29].

A generic formulation of the PP is

If *action A* has som*e possibility P* of causing harmful *effect E* then apply *remedy R* (based on Manson [30]).

where:

Action A is some scientific research or technological or other development, or some action, eg. research into ASI;

Possibility P is more than a logical possibility; it must be at least plausible, eg. ASI must be plausible even if the possibility is uncertain.

Effect E is some serious or perhaps catastrophic or irreversible harm, eg. uncontrollable ASI.

Remedy R is some measure or measures that should be taken to avoid or minimise E occurring, eg. halting or never starting the research, or development and use of some technologies or products to minimise the risks.

We will now defend the PP against four criticisms: the difficulty of predicting, the relationship between the action and the possible harm, the supposed harm itself and the PP's paradoxical nature. Here the first three will be only briefly mentioned but the fourth considered is some detail. If it is indeed true that the PP inherently contains a paradox, it is an incoherent principle and of no use.

Prediction. The PP does involve predicting consequences of scientific and technological research and development and of other actions. This is undoubtedly an activity fraught with dangers. But with care, reasonable predictions can be made. Even speculation, which goes further than prediction, can be justified, as we argued above in relation to ASI.

5.1 Relationship Between Action and Harm

This concerns the causal link between the action and the effect. All versions of the PP state that the principle should be applied even if "some cause and effect relationships are not fully established scientifically". While scientific certainty is not necessary, the threat of danger must be credible, that is, an hypothesis that the threat is caused by a particular action. There must be evidence that that sort of hypothesis is a reasonable one in the circumstances. The hypothesis that AI research will lead to ASI that is not controllable by humans is credible given the rate at which AI is currently developing, even if this is unlikely, at least in the foreseeable future.

5.2   Nature of the Harm

Yet another criticism concerns the nature of the harm. It must be more than just any threat of harm. This would rule out almost anything. It should be "serious or irreversible damage". While both *serious* and *irreversible* require closer examination, this is more plausible particularly when it involves the suffering and death of humans or widespread destruction of the environment.

5.3   The Supposed Paradox

The most serious criticism of the PP, made amongst others by Sunstein and Clarke, is that it is internally incoherent. If the PP is to be applied to some action in order to avoid a harm, it should also be applied to the action of applying it because that too, at least in many cases, will lead to harm. So it should be both applied and not applied. The problem can be set out as follows:

- Action A1 might cause bad effect Eb1 (harm eventuates because of A1);
- Remedy R1 (don't do A1) stops Eb1 (PP applied);
- But suppose that A1 causes good effect Eg1 (Eg1 eliminates some harm);
- Then R1 stops Eg1 (harm eventuates because of R1).
- So, if PP should be applied to A1 (because A1 causes harm) it should also be applied to R1 (because R1 prevents an action that would eliminate some harm).

5.4   A Possible Solution

One way out of the paradox is to draw on the distinction between positive and negative duties. Positive duties, interpreted in the Rawlsian way that they are by Thomas Pogge [31], that is, as duties to do good—are commonly (though by no means universally) thought to be weaker than negative duties, that is, duties not to do harm. The duty not to kill is stronger than the duty to save a life. Consider the ASI example. Undertaking the research that will possibly lead to ASI and the control problem is an example of causing harm (albeit unintentionally), violating a negative duty, while not undertaking it is allowing harm to happen, violating a positive duty. Suppose that the only way to save the life of Tom is to kill Harry and transplant his heart into Tom. While saving the life of Tom is good, killing Harry to achieve it is not justifiable. Doing nothing will result in the death of Tom but Harry's death would be the result of an action. Letting Tom die is not good, but killing Harry is arguably worse. There is no paradox then in applying the precautionary principle in the ASI case. Curtailing AI research will stop or delay some benefits, for example, the development some remedies for health or environmental problems so will allow those harms to continue. But this positive duty is overridden by the negative to not cause "serious or irreversible damage". The PP then is a plausible principle and surely more plausible than its real rival, the bravado principle.

This is little more than a sketch of an argument and a number of issues require detailed examination. What is a duty and what gives it its force? Are the different types of duties really of different strengths? If some harms are caused by human actions, is there a stronger duty to fix those? These intermediate duties Thomas Pogge argues are stronger than positive duties but weaker than negative duties [31, p. 34]. Will that affect the argument above? We will ignore these issues here but concede that further work is required.

Intelligent materials have developed from advances in AI and nanotechnology. Ethical issues arising from both of those technologies are therefore relevant to ethical questions about intelligent materials. Given that the research and development of these materials is still in its early stages, much of the ethical discussion will be about the potential of this technology and therefore speculation will play an important role, as will the precautionary principle If there are serious perceived dangers.

6   Ethics of Intelligent Materials

Intelligent materials technology is still in the early stages of research and development, but drawing on our previous discussion, it provides an opportunity to consider afresh some interesting questions.

6.1   Soft Robotics

Applications or potential applications of soft robotics seem to be beneficial and raising no worrying ethical issues. One proposed application is the handling of fruit and other food products. Most fruit requires careful and gentle handling and so the advantages of soft over hard robots is obvious. However, given what was argued earlier about speculative ethics, the question of soft robotics is worth more attention. Robotics is a well-developed field and the use of robots in caring has already occurred, even if these are not autonomous and can be used in only quite limited situations [32, 33] but soft robotics seem to open up new possibilities with ethical significance. We have argued that speculative ethics is justified and the speculation here is based on things that are already known. We also know that research on robots and other machines such as cars and weapons that can learn from and react to their environments is well developed and well-funded. Soft robotics might be in its infancy but given advances in other areas of material science and in AI, we have good reason to believe that autonomous soft robots that resemble humans are a possibility, if not in the near future then in the medium to distant future. This resemblance could be both in appearance and behaviour. To use Floridi's distinction, they may not be intelligent, but they may be smart and at least in a clearly defined caring context, behave very much as a human would in that situation. If they are not conscious, they could not be empathetic perhaps, something important in carers, but if smart enough, could learn to behave in empathetic ways. Suppose further that ASI is achieved, or something very close to it, perhaps artificial super smartness, and built into these robots. They may then be better carers than humans.

The ethical questions raised by this scenario have been around since the early days of AI. Is this dehumanising? Should it fill us with disgust? Should such robots only be used when humans are not available? Should they be used if they are better carers than humans? If they harm a patient, who can be held responsible or accountable? Should we worry about this now or wait until they are developed?

6.2   The Natural Environment

Most ethical discussions of technologies focus on the uses or potential uses of particular technologies. This is undoubtedly important and perhaps the most important part of the ethics of technology. It is also important however, to look at issues regarding the manufacture, use and disposal of the products, in this case intelligent materials. Does the manufacture pose health risks to the workers or the natural environment? Does it use large amounts of energy or natural resources? Does it necessitate mining of rare metals or mining that is disruptive to communities? Does it produce waste products that are environmentally harmful? Does the disposal of the products pose a risk of polluting the environment? Can the products themselves be mined for required raw materials? Are they recyclable?

Will the disposal of intelligent materials pose any environmental risks other than those of other manufactured materials? Electronic waste (e-waste) is an increasing environmental problem. Will any intelligent materials contribute to this? [34]

Some of these questions have been discussed in the area of nanotechnology for the last couple of decades and given that nanoscience is a core part of intelligent materials, this is just a continuation of that discussion. The risk or potential risks of manufactured nanoparticles were and still are a central part of nanoethics [35].

Intelligent materials appear to have the potential for many benefits, both in the medical and environmental areas, but if their manufacture and disposal cause harms to the environment then ethical issues regarding their development and use require examination. Will their benefits outweigh their environmental costs? Perhaps a cost–benefit analysis is all that is required but this is not simple when environmental issues are part of the analysis. Values are important and different people have different values and place weigh them differently. These values determine what is considered a benefit and what is considered a harm, and how and to what extent they are benefits or harms. The precautionary principle may be more pertinent. And of course different technologies require different assessments. Some intelligent material technologies will probably be overwhelmingly beneficial while others perhaps are of more dubious value.

One question that deserves more discussion than it receives, is what constitutes a benefit from technology. We commonly think that anything that improves health, reduces pain or makes life easier or more enjoyable in some way, is a benefit. It is not obvious however, that technological innovations as such do make us happier or more satisfied with life. If technological innovations in general make us happier then given how many there have been in the last few human generations, either I should be blissfully happy or my grandparents and those before them must have been quite unhappy. Exceptions of course exist. Certain technologies make particular people happier. Medical technologies that significantly improve life for individuals are examples. This should give us pause for thought about how important some new technologies really are, especially where there are environmental problems with their production and disposal. Intelligent materials that reduce suffering, energy use, pollution and waste, for example, are important, but what about those that merely make life easier, possibly in relatively trivial ways? Not everything that makes life easier contributes to human happiness or well-being.

References

1. Wiener, N (1948) Cybernetics:or control and communication in the animal and the machine, Technology Press, Cambridge, Mass. John Wiley, NY
2. Wiener N (1960) Some moral and technical consequences of automation. Science 131:1355–1358
3. Weizenbaum J (1976) Computer power and human reason: from judgement to calculation. Penguin Books, Harmondsworth, Middlesex
4. Moor JH (1979) Are there decisions that computers should never make? Nature and System 1:217–229
5. Future of life institute (2016) Autonomous weapons, Open Letter: AI & Robotics Researchers. https://futureoflife.org/open-letter/open-letter-autonomous-weapons-ai-robotics/. Accessed 14 Jan 2024
6. Bostrom N (2014) Superintelligence: paths, dangers, strategies. Oxford University Press, Oxford
7. Dvorsky G (2015) Experts Warn UN panel about the dangers of artificial super intelligence. Gizmodo. https://gizmodo.com/experts-warn-un-panel-about-the-dangers-of-artificial-s-1736932856. Accessed 14 Jan 2024
8. Government UK (2023) Policy paper: The Bletchley Declaration by Countries Attending the AI Safety Summit. https://www.gov.uk/government/publications/ai-safety-summit-2023-the-bletchley-declaration/the-bletchley-declaration-by-countries-attending-the-ai-safety-summit-1-2-november-2023. Accessed 14 Jan 2024
9. Hogarth, I (2023) We must slow down the race to God-like AI. Financial Times April 13. https://www.ft.com/content/03895dc4-a3b7-481e-95cc-336a524f2ac2. Accessed 14 Jan 2024
10. Future of Life Institute (2023) Pause giant AI experiments: Open Letter. https://futureoflife.org/open-letter/pause-giant-ai-experiments/. Accessed 14 Jan 2024
11. Floridi L (2016) Should we be afraid of AI? Aeon https://aeon.co/essays/true-ai-is-both-logically-possible-and-utterly-implausible. Accessed 14 Jan 2024
12. Floridi L (2014) The fourth revolution: how the infosphere is reshaping human reality. Oxford University Press, Oxford
13. Boden M (2017) AI: utopia-or-dystopia. World Economic Forum. https://www.weforum.org/agenda/2017/02/ai-utopia-or-dystopia. Accessed 14 Jan 2024

14. Drexler E (1986) Engines of creation: the coming era of nanotechnology. Anchor-Doubleday, New York
15. Nordmann A (2007) If and then: a critique of speculative nanoethics. NanoEthics 1:31–46
16. Sina AAI, Carrascosa LG, Ziyu Liang Z et al (2018) Epigenetically reprogrammed methylation landscape drives the DNA self-assembly and serves as a universal cancer biomarker. Nat Commun 9(4915):1–13. https://doi.org/10.1038/s41467-018-07214-w
17. Seaton A, Tran N, Aitken R, Donaldson K (2010) Nanoparticles, human health hazard and regulation. J Royal Society Interface 7:S119–S129. https://doi.org/10.1098/rsif.2009.0252.focus
18. Phogat, N, Khan, Shadab, A, Shankar, S et al (2016) Fate of inorganic nanoparticles in agriculture. Adv Mater Lett 7:3–12. https://doi.org/10.5185/amlett.2016.6048
19. Altmann J (2006) Military nanotechnology: new technology and arms control. Routledge, London
20. Unbrello S, Baum SD (2018) Evaluating future nanotechnology: the net societal impacts of atomically precise manufacturing. Futures 10:63–73. https://doi.org/10.1016/j.futures.2018.04.007
21. Umbrello S (2019) Atomically precise manufacturing and responsible innovation: value sensitive design approach to explorative nanophilosophy. Int J Technoethics 10:1–21. https://doi.org/10.4018/IJT.2019070101
22. Roach R (2008) Ethics, speculation, and values. NanoEthics 2:317–327. https://doi.org/10.1007/s11569-008-0050-y
23. Walsh T (2018) 2062: the world that AI made. LaTrobe University Press in conjunction with Black Inc., Melbourne
24. Weckert, J (2012) In defence of the precautionary principle. IEEE Technol & Soc Winter, 12–17.
25. Wingspread Statement on the Precautionary Principle (1998). www.gdrc.org/u-gov/precaution-3.html. Accessed 14 Jan 2024
26. Joy, B (2000) Why the future doesn't need us. Wired 8.04:238–262. http://www.wired.com/wired/archive/8.04/joy_pr.html. Accessed 14 Jan 2024
27. Sunstein CR (2005) Cost-benefit analysis and the environment. Ethics 115:351–385
28. Clarke S (2009) New technologies, common sense and the paradoxical precautionary principle. In: Sollie P, Duwell M (eds) Evaluating new technologies: methodological problems for the ethical assessment of technological developments. Springer, Dordrecht, pp 159–173
29. Techno Optimist Manifesto (2023). https://a16z.com/the-techno-optimist-manifesto/. Accessed 14 Jan2024
30. Manson N (2002) Formulating the precautionary principle. Environ Ethics 24:263–274
31. Pogge T (2005) Real world justice. J Ethics 9:29–53
32. Sharkey A, Sharkey N (2012) Granny and the robots: ethical issues in robot care for the elderly. Ethics Inf Technol 14:27–40. https://doi.org/10.1007/s10676-010-9234-6
33. Wright, J (2023) Inside Japan's long experiment in automating elder care. MIT Technology Review. https://www.technologyreview.com/2023/01/09/1065135/japan-automating-eldercare-robots/. Accessed 14 Jan 2024
34. PACE (2019) World economic Forum, New Circular Vision for Electronics: Time for a Global Reboot. https://www3.weforum.org/docs/WEF_A_New_Circular_Vision_for_Electronics.pdf. Accessed 14 Jan 2024
35. Kumah EA, Fopa RD, Harati S et al (2023) Human and environmental impacts of nanoparticles: a scoping review of the current literature. BMC Public Health 23:1059. https://doi.org/10.1186/s12889-023-15958-4.Accessed2024

MIX
Papier aus verantwortungsvollen Quellen
Paper from responsible sources
FSC® C105338

If you have any concerns about our products,
you can contact us on
ProductSafety@springernature.com

In case Publisher is established outside the EU,
the EU authorized representative is:
Springer Nature Customer Service Center GmbH
Europaplatz 3, 69115 Heidelberg, Germany

Printed by Libri Plureos GmbH
in Hamburg, Germany